Principles of
RELATIVITY PHYSICS

Principles of *RELATIVITY PHYSICS*

JAMES L. ANDERSON

DEPARTMENT OF PHYSICS
STEVENS INSTITUTE OF TECHNOLOGY
HOBOKEN, NEW JERSEY

ACADEMIC PRESS *New York · London*

ACADEMIC PRESS INC.
111 Fifth Avenue, New York, New York 10003

United Kingdom Edition published by
ACADEMIC PRESS INC. (LONDON) LTD.
Berkeley Square House, London W.1

LIBRARY OF CONGRESS CATALOG CARD NUMBER: 66-30140

PRINTED IN THE UNITED STATES OF AMERICA

To Helene

Preface

The physical world-pictures known today as the special theory and the general theory of relativity were, perhaps more than any other successful world-picture, the products of a single man's imagination. As a consequence, his development of these theories has served as the basis for many of the standard presentations of these subjects. Today both are accepted as valid theories, capable of describing phenomena that cannot be encompassed by the older world-picture of Newtonian mechanics. One is, therefore, no longer called upon to justify them on the same grounds employed for the purpose by their creator. Furthermore, both theories were developed before the advent of modern quantum theory and their early applications were mainly to classical systems. However, the current usages of the special theory in particular have been more and more directed to the subatomic domain where it plays a dominant role. Finally, considerable effort in recent years has gone into developing the consequences of the general theory and in obtaining a better understanding of its foundations.

Because of the above considerations it seems desirable to give an account of relativity physics that differs from the traditional presentations in a number of ways. In particular a deductive approach seems to be more appropriate at this time than the more usual inductive approach. We have therefore tried to set forth what we consider to be the essential elements of both special and general relativity (and also Newtonian mechanics) without trying to justify them on *a priori* grounds and only later to supply the justification by relating the consequences of these theories to the physical world. In doing so we have placed considerable emphasis on the notion of the symmetry group of a physical theory. We will see that Newtonian mechanics can be characterized as the collection of all physical theories that have the inhomogeneous

Galilean group as a symmetry group. Likewise special relativistic theories have the inhomogeneous Lorentz or, as it is now called, the Poincaré group as a symmetry group. Finally, general relativistic theories all have the group of arbitrary mappings of the space-time manifold onto itself as a symmetry group.

In developing the consequences of the special and general theories we have been guided by the current usages of the theories. While quantum phenomena is beyond the scope of this work, we have placed considerable emphasis on those aspects of relativity physics that are needed in a description of the subatomic world. Thus we discuss the structure of the Dirac equation as it occurs both in special and in general relativity and discuss at length the relation between the direct particle or action-at-a-distance and the field description of particle interactions. We also discuss some astrophysical applications of general relativity and include a chapter on cosmological applications.

The material for this book has been developed over a number of years during which the author taught introductory graduate courses in relativity physics at Stevens Institute of Technology and at New York University. It can be covered in a one-year course but also can be used, with some judicious choice of topics, as a one-semester course in either special relativity or general relativity. We have assumed that the student has some familiarity with the content of the usual first- and second-year graduate physics courses and that at some point has been exposed to a few of the ideas of special relativity. From time to time we have made reference to notions that might not be familiar to all readers. If these notions are not explained at the time, the reader can usually disregard them if he is not familiar with them. We have also included problems in the text proper at places where they seem most appropriate. A number of these problems are "finger exercises" and can be skipped by the reader who already has some familiarity with the techniques involved. We have also not hesitated to state results whose meanings should be clear to the reader but which require for their proofs techniques that would not be familiar to the average reader. In most cases we have supplied a reference where the detailed proofs can be found. Finally, we have not tried to give exhaustive treatments of a number of the more current developments in relativity physics but rather we have set forth what we believe to be the underlying ideas involved and have referred the reader to the literature or to one or another of the excellent survey articles now available on the subject.

The material proper in this work has been divided into three parts. The first part presents the underlying ideas needed for the last two parts on the special and general theories, respectively. In Part I, the first three chapters cover the mathematical tools needed for what follows and is in effect a short introduction to differential geometry. In this presentation we have used the standard tensor notation and techniques. While the more modern approaches are perhaps more elegant they lack familiarity and, we feel, are less intuitively appealing than the older presentations.

The fourth chapter is in many ways the foundation chapter for the whole work and should be read whether one is interested in the special or the general theory. In it we define and discuss in detail the notions of covariance and symmetry groups. While many readers may not agree with the uses of these terms made there, we have tried to give a rigorous (at least to a physicist) definition for them and to adhere to these definitions throughout the rest of the work. In this chapter we also discuss the relation between symmetry and conservation laws that derives from Nöther's theorem.

The second part of the book deals with the special theory of relativity. It opens with a brief review of the elements of Newtonian physics in a form that is somewhat different from that given by Newton. In particular we have tried to emphasize those elements that get revised or eliminated in the special and general theories. The next chapter deals with the properties of Minkowski geometry and the properties of the Lorentz and Poincaré groups. It also contains a discussion of the measurement process in special relativity based on the use of light clocks and light signals. Then follows a chapter on relativistic particle mechanics which contains a discussion of the Wheeler-Feynman action-at-a-distance description of electrodynamics. Part II ends with two chapters on relativistic continuum mechanics.

The general theory of relativity is the subject matter of Part III. The first chapter of this part deals with the foundations of the theory and in particular discusses the role of the principle of general invariance. In it we argue that this principle is not devoid of physical content as some authors have claimed but, properly interpreted, can serve as the sole basis for the theory. The remaining chapters deal in turn with exact solutions of the Einstein field equations, experimental tests of the theory, the problems of radiation and conservation laws, and finally cosmological considerations.

A number of people have contributed indirectly to this text. We would first like to thank those authors who have contributed unknowingly. We have tried to give credit where credit is due but after working in a field for so long one sometimes tends to forget the sources of all one's discussions and proofs. For those we have not acknowledged we hope that imitation will be considered a sincere form of flattery. We would also like to thank the numerous colleagues with whom we have had discussions concerning the ideas that went into this work and especially David Findelstein and our teacher, Peter Bergmann. Finally, we would like to thank Banesh Hoffmann who kindly and patiently read the manuscript of this work and made many helpful and useful comments.

Teaneck, J.L.A.
New Jersey

Notation

For the most part we have tried to adhere to current usage in our notation. Lower-case Greek suffixes take the values 0, 1, 2, 3 while lower-case Latin suffixes take the values 1, 2, 3. If a suffix is repeated in an expression, it is to be summed over its range of values (Einstein summation convention.) If an unsummed suffix appears in an equation, the equation is understood to hold for each value of the suffix.

We will use the symbol x^μ to designate both a point of the space-time manifold and the coordinates of this point. Partial derivatives will be denoted by a comma notation so that $f_{,\mu} = \partial f/\partial x^\mu$. Four-dimensional covariant derivatives will be denoted by a semicolon, $f_{\mu;\nu}$ and three-dimensional covariant derivatives will be denoted by a bar, $f_{r|s}$. A function of the four coordinates x^μ will be denoted by $f(x)$ while a function of the three coordinates x^r will be denoted by $f(\mathbf{x})$. The components of a four-component object A^μ will sometimes be given in the form $(A^0, \mathbf{A})$ so that $x^\mu = (x^0, \mathbf{x})$. We will also employ standard three-dimensional vector notation when appropriate. The cross-product of two vector $\mathbf{A}$ and $\mathbf{B}$ will be denoted by $\mathbf{A} \times \mathbf{B}$, their dot- or scalar-product by $\mathbf{A} \cdot \mathbf{B}$, the gradient of a scalar φ by $\nabla\varphi$, the divergence of a vector A by $\nabla \cdot \mathbf{A}$ and its curl by $\nabla \times \mathbf{A}$. Unless otherwise stated, our equations will hold at each point of the space-time manifold.

The notation $\det f_{\mu\nu}$ will denote the determinant of $f_{\mu\nu}$ considered as a matrix. The symbol $\eta_{\mu\nu}$ will denote the Minkowski matrix with nonzero components $\eta_{00} = -\eta_{11} = -\eta_{22} = -\eta_{33} = 1$. We will designate the components of a diagonal matrix $a_{\mu\nu}$ by $\mathrm{diag}(a_{00}, a_{11}, a_{22}, a_{33})$. Thus $\eta_{\mu\nu} = \mathrm{diag}(1, -1, -1, -1)$.

The Kronecker delta will be denoted by $\delta_B{}^A$ for any suffix set and the one-dimensional Dirac delta function of the variable y by $\delta(y)$. Finally, we will make use of the abbreviations MMG for the manifold mapping group, kpt for a kinematically possible trajectory, and dpt for a dynamically possible trajectory.

Contents

Part III. DYNAMICAL SPACE-TIME THEORIES

PART
I

GEOMETRICAL FOUNDATIONS FOR SPACE-TIME THEORIES

1

Geometrical Structures

One of the tasks of the physicist is to construct mathematical models or theories that can be made to correspond to elements of the physical world. The criteria he uses for judging the success of a given theory vary from time to time and from field to field. In general the more varied the class of phenomena the theory is able to describe, the more successful or fundamental it is considered to be. Simplicity is another criterion of success and some physicists would include the criterion of beauty. Of all models used to describe the physical world, the most successful to date are those that make use of the space-time concept. All the great world-pictures—for example, Newtonian mechanics, special relativity, quantum mechanics—make use of this concept and, indeed, most physicists today consider a theory to be fundamental only if it does make explicit use of this concept. Thus thermodynamics is not considered to be a fundamental theory by itself but rather a consequence of a more fundamental theory, statistical mechanics, which does make use of the space-time concept. It is the purpose of this book to set forth the basic principles that underlie all space-time theories—that is, all theories that make explicit use of the space-time concept—and to describe the general features of the classical world-pictures that comprise these theories.

In this first chapter we describe the basic geometrical structures that are common to all space-time theories. We begin with the primitive notion of a space-time point and the assumption that the totality of all such points together constitute a manifold, the space-time manifold. This basic assumed property of the collection of all space-time points allows us to introduce a coordinate system on the manifold and with its help to define the notion of a mapping of the manifold onto itself. We next introduce the concept of a geometrical object in terms of its transformation properties under such a mapping. The concept of a geometrical object will play a key role in all future discussions, since it is the various geometrical objects that are ultimately to be associated with the physical elements of a space-time system. Of special interest to us will be the geometrical objects known familiarly as scalars, vectors, and tensors, that have linear, homogeneous transformation laws under a space-time mapping.

1-1. The Space-Time Manifold

The common feature of all space-time theories is the use of the concept of the *space-time point*. It corresponds to where and when a physical event, for example, the collision of two mass points, takes place. The collection of all such points, that is, the collection of the wheres and whens of all possible physical events, constitutes *space-time*. The distinguishing feature of a particular point of this space-time is that it has no distinguishing features; all points of space-time are assumed to be equivalent.

The question of the independent existence of space-time has a long and involved history with its beginnings going back to the times of the early Greeks and Hebrews. It was revived during the Renaissance, reached a climax with the famous Leibniz-Clarke controversy, and has continued down to the present day.[1] For us, however, the question is not important. What is important is that a large class of successful physical theories make use of the space-time concept as a fundamental building block. In what follows we shall refer to such theories as *space-time theories*.

As we have described it so far, space-time is merely a collection of points. As it is used in physical theories, however, it is endowed with a number of properties. Of prime importance are its assumed *topological* properties. Roughly speaking, the topological properties of a space consist of those properties that are unaffected by arbitrary deformations of the space.[2] A space that is topologically equivalent to a torus is quite different from one that is equivalent to a sphere, since it is not possible to deform a torus into a sphere. For the most part, the topological properties of space-time have been taken to be those of the *Euclidean four-plane* consisting of all possible quadruplets

[1] For a short history of this question the interested reader is referred to: Max Jammer, *Concepts of Space* (Harvard University Press, Cambridge, 1957).

[2] For the reader who is unfamiliar with the topological concepts employed here, no finer introduction can be suggested than P. Alexandroff's *Elementary Concepts of Topology* (Dover, New York, 1961). In style, it is the antithesis of the abstract school of writing in mathematics.

of real numbers. We shall see later, however, that the global topology of space-time is not entirely at our disposal but is restricted to a certain degree by the type of geometry we wish to impose on it. What is common to all space-time theories is the assumption that locally the topology of space-time is that of the Euclidean four-plane. As a consequence it is possible to map the points of any small but finite region of space-time onto the points of a corresponding region of the Euclidean four-plane in a one-to-one bicontinuous manner. This property of space-time is equivalent to assuming that the points of space-time constitute a manifold, the *space-time manifold.* One should, however, keep in mind that this is only a local property of the manifold. In particular, it may be impossible to map the whole manifold onto a single Euclidean four-plane.

1-2. Coordinates and Coordinate Coverings

The assumption that the points of space-time constitute a manifold makes possible a procedure that will greatly simplify our discussion of the various geometrical structures that we will deal with. Because of this assumption, we are able to *coordinatize* space-time. We do this by assigning to each point of a space-time region the coordinates of its image point in the Euclidean four-plane. In what follows we will designate the coordinates of an arbitrary point of space-time by x^μ, where the superscript μ takes on the values 0, 1, 2, 3. We also use the symbol x^μ to designate the point whose coordinates are x^μ. In assigning coordinates to the points of space-time we must, of course, make sure that each point of a space-time region has a unique set of coordinates. The totality of coordinates for any given region constitutes a *coordinate patch.*

In assigning coordinates to the points of space-time a difficulty will arise if the topology of the manifold in the large is not equivalent to a single Euclidean four-plane. If this is the case, it will not be possible to use a single coordinate patch to cover the entire manifold. One is familiar with this situation in geography, where the longitude and latitude system becomes singular at the poles. In such a situation it is necessary to use a number of overlapping coordinate patches, which together constitute a *coordinate covering* of the manifold. For the earth, two coordinate patches would suffice. One patch could cover the entire surface, with the exception of a small region around one pole, while another patch could be used to cover the polar region.

Since the points of the space-time manifold can be associated with those of the Euclidean four-plane in many different ways, one sees that the coordinatization of the manifold is not unique. This lack of uniqueness is, of course, a consequence of our basic assumption of the indistinguishability of the points of space-time. As a consequence we must require that any results obtained with the help of a particular coordinatization should be independent of this coordinatization. This requirement is known as the requirement of *coordinate covariance*.

Whenever we introduce a coordinate covering onto our manifold, certain subsets of points will be covered by two overlapping coordinate patches. A point of such a subset will therefore have associated with it two sets of coordinates, x^μ and x'^μ. We now make the further assumption that the x^μ are continuous and differentiable functions of the x'^μ, and vice versa. Such being the case, we say that the points of space-time constitute a *differentiable manifold*.

1-3. Space-Time Mappings

Once we have coordinatized the space-time manifold, it is possible to characterize a *mapping* of this manifold onto itself in a simple manner. Under such a mapping each point of the manifold is associated with some other point of the manifold. Since we want any such mapping to preserve the topological properties of the manifold, we must limit ourselves to mappings that are one-to-one and bicontinuous. It is this notion of a mapping that will be used later when we come to define a geometrical object. The physical interpretation of a space-time mapping will be discussed at length in Section 6-16, where we will show that it corresponds to the relation between the space and time measurements made with different measuring devices.

Let us assume that we have coordinatized our manifold and that under a particular mapping the point x^μ is associated with the point x'^μ. In general the x'^μ will be some functions of the x^μ. By giving these functions for all points of the manifold, we can thereby characterize a particular mapping and write

$$x^\mu \to x'^\mu = x'^\mu(x), \tag{1-3.1}$$

where the x^μ are the coordinates of the point that is mapped onto the point with coordinates x'^μ. In order that the mapping be one-to-one and bicontinuous we must require that the mapping functions $x'^\mu(x)$ be single-valued functions of their arguments and possess a nonvanishing Jacobian. If this is the case we can invert Eq. (1-3.1) and write

$$x'^\mu \to x^\mu = x^\mu(x'). \tag{1-3.2}$$

The four functions $x^\mu(x')$ can be interpreted as defining the inverse mapping to that defined by the functions $x'^\mu(x)$. In what follows, mappings that are one-to-one and bicontinuous will be called *allowed mappings* of the manifold onto itself.

If the topology of the space-time manifold is such as to require only a single coordinate patch for its coordinatization—that is, if it has the global topology of the Euclidean four-plane—the mapping functions $x^{\mu}(x')$ appearing in Eq. (1-3.1) are restricted only by the conditions that they be single-valued and have a nonvanishing Jacobian. However, if the topology is such as to require more than one coordinate patch for the purposes of coordinatization, we must take care to insure that the mapping functions do not map a point lying in the overlap region of two or more patches onto more than one other point of the manifold. In such cases the mapping functions will be further restricted, the form of the restrictions depending on the topology of the manifold and the coordinatization employed.

As an example of the kind of restrictions that the topological structure of the manifold can impose, let us consider the simple case of a two-dimensional manifold with the topology of a cylinder. We can coordinatize this manifold in such a way that one of the coordinates, say x_1, runs along a generator of the cylinder while the second coordinate, x_2, runs around the circumference of the cylinder and takes on values in the interval $(0, 2\pi)$. With this coordinatization we see that the points lying along the "seam" of the coordinate system are characterized by two different values of x_2, namely, 0 and 2π. Consequently, in order that the allowed mappings be one-to-one, all mapping functions must satisfy certain periodicity conditions in x_2. More specifically, they must have the form

$$x'_1 = x'_1(x),$$

$$x'_2 = x_2 + x'_2(x),$$

where $x'_1(x)$ and $x'_2(x)$ are periodic functions of x_2 with period 2π. If we had used a different coordinatization, the conditions on the mapping functions would be different but equivalent to those stated above.

1-4. Groups of Mappings

An important property of the set of all allowed mappings of the space-time manifold onto itself is that they form a group.[3] In order that the elements of a set form a group, there must be defined a rule of multiplication that associates with any two members of the set another member. Furthermore, this multiplication must be associative (but not necessarily commutative). There

[3] For an eminently readable review of elements of the group theory that will play a role in our considerations here, the reader is referred to F. Gürsey, "Introduction to Group Theory" in *Relativity, Groups, and Topology*, Les Houches, 1963 (Gordon and Breach, New York, 1964).

must also exist an identity element with the property that its product with any element of the set is equal to that element. Finally, the inverse of any element, defined so that its product with the element in question is equal to the identity element, must also be a member of the set.

These requirements are satisfied in the case of the manifold mappings discussed above if we take the product of two mappings to be the mapping that results from performing first the one mapping and then the other. Since the result of performing two mappings in succession depends in general on the order in which they are performed, the group of mappings is non-commutative. In addition, since the topology of the manifold restricts the allowed mappings, this group will also depend on the topology and in fact can be used in part to characterize this topology. In what follows we shall refer to this group as the *manifold mapping group* (hereafter abbreviated as MMG).

Because of the assumed continuity of the mapping functions it is possible to build up any finite mapping with positive Jacobian by an integration process involving a series of infinitesimal mappings. (Since mappings with singular Jacobians are not allowed, one could never construct a mapping with a negative Jacobian by such an integration process.) This is an important result, since it allows us to discuss many of the properties of geometrical objects in terms of infinitesimal mappings rather than finite ones, thereby simplifying these discussions considerably. For an infinitesimal mapping, Eq. (1-3.1) takes the form

$$x^\mu \to x'^\mu = x^\mu + \xi^\mu(x), \tag{1-4.1}$$

where the $\xi^\mu(x)$ are small but otherwise arbitrary functions of the x^μ, called the *descriptors* of the mapping. In particular, the dependence of the $\xi^\mu(x)$ on the coordinates is not restricted by the topology of the manifold in the large, since in all cases this topology is assumed to be locally Euclidean. The inverse mapping is given by

$$x'^\mu \to x^\mu = x'^\mu - \xi^\mu(x'), \tag{1-4.2}$$

correct to first order in the descriptors.

Before we continue our discussion of infinitesimal mappings it is desirable to introduce a notational device, the *Einstein summation convention*, that will greatly simplify the writing of our equations. The convention is simply this: Whenever an index appears twice in an expression it is to be summed over its range of values. Thus if the index A takes integer values from 1 to N, the expression $y_A w_A$ is shorthand for $\sum_{A=1}^{N} y_A w_A$. The exceptional case where such a summation is not to be performed will be indicated in the text.

Let us now consider the net effect of performing two infinitesimal mappings, first in one order and then in the reversed order. Under the first mapping

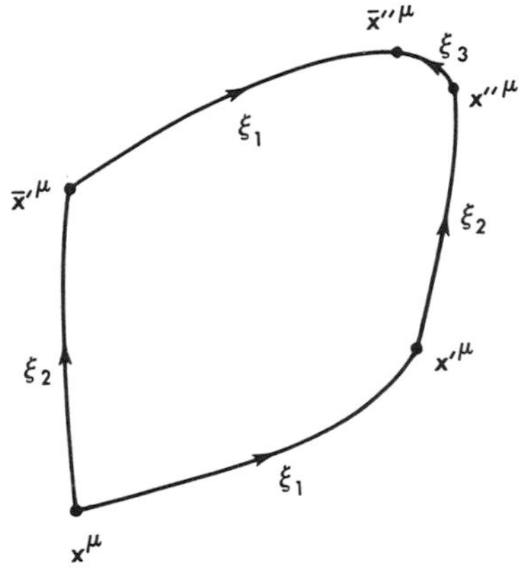

Fig. 1.1.

with descriptors $\xi_1{}^\mu$ the point x^μ gets mapped onto the point x'^μ (see Fig. 1.1). The second mapping, with descriptors $\xi_2{}^\mu$, then sends the point x'^μ over onto the point x''^μ so that the combined mapping sends x^μ over onto x''^μ. Thus we have

$$x'^\mu = x^\mu + \xi_1{}^\mu(x')$$

and

$$\begin{aligned} x''^\mu &= x'^\mu + \xi_2{}^\mu(x') \\ &= x^\mu + \xi_1{}^\mu(x) + \xi_2{}^\mu(x) + \xi^\mu_{2,\nu}(x)\xi_1{}^\nu(x), \end{aligned}$$

correct to second order in the descriptors.[4] Likewise, if the second mapping maps x^μ onto $\bar{x}'^\mu$ while the first maps $\bar{x}'^\mu$ onto $\bar{x}''^\mu$, we have

$$\bar{x}'^\mu = x^\mu + \xi_2{}^\mu(x)$$

and

$$\begin{aligned} \bar{x}''^\mu &= \bar{x}'^\mu + \xi_1{}^\mu(\bar{x}') \\ &= x^\mu + \xi_2{}^\mu(x) + \xi_1{}^\mu(x) + \xi^\mu_{1,\nu}(x)\xi_2{}^\nu(x). \end{aligned}$$

The difference, $\bar{x}''^\mu - x''^\mu$, is thus given by

$$\begin{aligned} \bar{x}''^\mu - x''^\mu &= \xi^\mu_{1,\nu}\xi_2{}^\nu - \xi^\mu_{2,\nu}\xi_1{}^\nu \\ &= \xi_3{}^\mu, \end{aligned} \tag{1-4.3}$$

where $\xi_3{}^\mu$ are the descriptors of the mapping that maps x''^μ directly onto $\bar{x}''^\mu$.

[4] In this equation and all that follows we make use of the comma notation to denote ordinary differentiation: $,\mu \equiv \partial/\partial x^\mu$. Likewise $,\mu\nu \equiv \partial^2/\partial x^\mu\,\partial x^\nu$, etc.

The mapping characterized by the descriptors $\xi_3{}^\mu$ is called the *commutator* of the two mappings characterized by the descriptors $\xi_1{}^\mu$ and $\xi_2{}^\mu$. Its importance lies in the fact that it determines the structure of the group of mappings in the vicinity of the identity element. In particular, a subset of mappings with all descriptors of a particular type will form a group only if the descriptors of all commutators are of this same type. This property of the commutator will prove especially important when we come to consider subgroups of mappings that constitute a Lie group.

A subgroup of mappings of the MMG constitutes a Lie group when the mappings of this subgroup are characterized by one or more parameters that take on a continuous range of values. For many purposes it is convenient to consider these parameters as constituting the coordinates of a point in a manifold, called the *group space*. Once the commutator structure of a Lie group is specified in the neighborhood of the identity, a knowledge of the topology of the group space in the large is sufficient to determine the overall structure of the group. The requirement that the commutator of two mappings of a given type be again of this type in order for these mappings to form a group becomes an integrability condition in the case of Lie groups. Suppose that we wish to determine the mapping to be associated with some point of the group space that is not in the neighborhood of the identity. In general there are a great many ways in which this mapping can be built up out of a series of infinitesimal mappings by an integration process, each of which can be represented by a curve in the group space connecting the origin (which corresponds to the identity element) to the point in question. The above-stated requirement on the commutator of two infinitesimal mappings can then be shown to be a necessary condition that the mapping to be associated with a given point in the group space be independent of the path of integration used to arrive at this point. It is therefore somewhat analogous to the condition that a given force field be derivable from a potential.

As an example of a Lie group, consider the *rotation group* in three dimensions. One form of the infinitesimal mappings corresponding to this group is given by

$$x^r \to x'^r = x^r + \varepsilon^{rs} x^s \qquad r, s = 1, 2, 3, \tag{1-4.4}$$

where $\varepsilon^{rs} = -e^{sr}$. Since an arbitrary antisymmetric 3×3 matrix has three arbitrary components, we see that we are dealing with a three-parameter Lie group. The commutator of two such mappings is obtained with the help of Eq. (1-4.3), suitably modified so as to apply in three dimensions, and has the form

$$\xi_3 = \varepsilon_3^{rt} x^t = (\varepsilon_1^{rs}\varepsilon_2^{st} - \varepsilon_2^{rs}\varepsilon_1^{st})x^t.$$

It is an easy matter to show that $\varepsilon_3^{rt} = \varepsilon_1^{rs}\varepsilon_2^{st} - \varepsilon_2^{rs}\varepsilon_1^{st}$, the matrix obtained in this manner, is indeed antisymmetric. The commutator of two infinitesimal mappings of the form (1-4.4) is thus seen to be also of this form.

If the infinitesimal parameters ε^{rs} are unrestricted by any symmetry requirements, we again have a group, the *full linear group* in three dimensions. It is a nine-parameter Lie group and contains the rotation group as a subgroup. In four dimensions, the full linear group is a sixteen parameter group.

Problem 1.1. Do mappings of the form (1-4.4), but with $\varepsilon^{rs} = \varepsilon^{sr}$, generate a group?

Problem 1.2. Show that the infinitesimal mappings

$$x'^{\mu} = x^{\mu} + \varepsilon_{\nu}{}^{\mu} x^{\nu} + \varepsilon^{\mu} \tag{1-4.5}$$

generate a group when the ε^{μ} are arbitrary parameters and the $\varepsilon_{\nu}{}^{\mu}$ are such that

$$\eta_{\mu\rho}\varepsilon_{\nu}{}^{\rho} + \eta_{\nu\rho}\varepsilon_{\mu}{}^{\rho} = 0,$$

where $\eta_{\mu\nu}$ is a diagonal matrix with $\eta_{\mu\nu} = \mathrm{diag}(1, -1, -1, -1)$ and will be referred to as the *Minkowski matrix*. (The group generated by these infinitesimal mappings is the *Poincaré* or *inhomogeneous Lorentz group* and is seen to be a ten-parameter Lie group.)

Problem 1.3. Show that the infinitesimal mappings

$$\begin{aligned} x'^{0} &= x^{0} + \varepsilon, \\ x'^{r} &= \varepsilon^{rs} x^{s} + \zeta^{r} x^{0} + \varepsilon^{r}, \end{aligned} \tag{1-4.6}$$

generate a group when ε, ε^{r} and ζ^{r} are arbitrary infinitesimal parameters and $\varepsilon^{rs} = -\varepsilon^{sr}$. (The group generated by these infinitesimal mappings is the *inhomogeneous Galilean group*.)

The expressions for $\xi^{\mu}(x)$ given in the above problems for the Galilean and Poincaré groups are the *Cartesian* forms for these quantities. However, they are not the only forms that yield the commutator structure of these two groups. In fact, given any Lie group and one particular form of the $\xi^{\mu}(x)$ that reproduces the commutator structure of this group, we can construct an infinity of other ξ^{μ} that have this property. In general such a ξ^{μ} will have the form

$$\xi^{\mu}(x) = \varepsilon^{i} f_{i}^{\mu}(x) \qquad i = 1, 2, \cdots, N, \tag{1-4.7}$$

where the ε^{i} are the group parameters. Our basic requirement is that the commutator of two such infinitesimal mappings must again be of this form. By making use of Eq. (1-4.3) we find for this commutator:

$$\xi_{3}{}^{\mu} = (\varepsilon_{2}{}^{i}\varepsilon_{1}{}^{j} - \varepsilon_{1}{}^{i}\varepsilon_{2}{}^{j}) f^{\mu}_{i,\nu} f_{j}{}^{\nu}\,. \tag{1-4.8}$$

In order that it be of the form (1-4.7) the functions f_i^{μ} must be related to each other by an equation of the form

$$f^{\mu}_{i,\nu} f_j^{\mu} = c^k_{ij} f_k^{\mu}, \tag{1-4.9}$$

where the c^k_{ij} are constants independent of the ε^i and the x^{μ}. They are called the *structure constants* of the group and serve to characterize it in a manner that is independent of the particular form taken by the f_i^{μ}. Consequently, two groups that have the same set of structure constants are equal (isomorphic) to each other, at least in the neighborhood of the identity element. If we substitute Eq. (1-4.9) back into Eq. (1-4.8), we obtain

$$\xi_3^{\mu} = (\varepsilon_2{}^i\varepsilon_1{}^j - \varepsilon_1{}^i\varepsilon_2{}^j)c^k_{ij}f_k^{\mu}$$

so that the infinitesimal parameters $\varepsilon_3{}^k$ of the commutator are given by

$$\varepsilon_3{}^k = c^k_{ij}(\varepsilon_2{}^i\varepsilon_1{}^j - \varepsilon_1{}^i\varepsilon_2{}^j). \tag{1-4.10}$$

From the manner of their construction we see that the structure constants are antisymmetric in the two lower indices, that is,

$$c^k_{ij} = -c^k_{ji}. \tag{1-4.11}$$

They can also be shown to satisfy the *Jacobi identity*:

$$c^k_{ij}c^m_{kl} + c^k_{li}c^m_{kj} + c^k_{jl}c^m_{ki} = 0. \tag{1-4.12}$$

One can, of course, reparameterize a given group by introducing new parameters that are functions of the original ones. If one does so, then the new infinitesimal parameters must be linearly related to the original infinitesimal parameters:

$$\varepsilon'^i = \alpha_j{}^i\varepsilon^j,$$

where $\alpha_j{}^i$ is a matrix with nonvanishing determinant. One can then express the ξ^{μ} in terms of the ε'^i as

$$\xi^{\mu}(x) = \varepsilon'^i f'^{\mu}_i(x),$$

where

$$f'^{\mu}_i(x) = \bar{\alpha}^j_i f_j^{\mu}(x),$$

with $\bar{\alpha}_j{}^i$ the inverse of $\alpha_j{}^i$; that is,

$$\alpha_k{}^i\bar{\alpha}_j{}^k = \delta_j{}^i \qquad \left(\delta_j{}^i = \begin{cases}1, & i = j\\ 0, & i \neq j\end{cases}\right)$$

The commutator of two mappings again leads to an expression of the form (1-4.10) for the ε' but with a different set of structure constants appearing therein. A short calculation shows that the new structure constants are related to the original ones by

$$c'^{k}_{ij} = \alpha_n{}^k \bar{\alpha}_i{}^l \bar{\alpha}_j{}^m c^n_{lm}. \tag{1-4.13}$$

Since a change of parameter does not change the basic structure of a group, it follows that two groups with structure constants related by Eq.(1-4.13) must be considered to be the same group.

Problem 1.4. Reparameterize the rotation group by taking, as new infinitesimal parameters, $\varepsilon^1 = \varepsilon^{23}$, $\varepsilon^2 = \varepsilon^{31}$, and $\varepsilon^3 = \varepsilon^{12}$ and calculate the structure constants for these parameters.

Suppose now that we have a set of $f_i{}^\mu$ that satisfy Eq. (1-4.9) and that we introduce a new set of variables $x'^\mu = x'^\mu(x)$. Then the descriptors ξ^μ, when expressed in terms of these new variables, have the form

$$\begin{aligned}\xi'^\mu(x') &= \frac{\partial x'^\mu}{\partial x^\rho}\,\xi^\rho(x)\\ &= \varepsilon^i g_i{}^\mu(x'),\end{aligned}$$

where

$$g_i{}^\mu \equiv x'^\mu{}_{,\rho} f_i^\rho.$$

We will now show that the transformed ξ' have the same commutator structure as the original ξ. To this end we calculate the commutator as before to obtain

$$\xi'^\mu_3 = (\varepsilon_2{}^i\varepsilon_1{}^j - \varepsilon_1{}^i\varepsilon_2{}^j) g^\mu_{i,\nu} g_j{}^\nu,$$

which must be equal to $\varepsilon_3{}^k g_k{}^\mu$ if the group structure is to be preserved by the transformation. Now it follows after a short calculation that

$$g^\mu_{i,\nu} g_j{}^\nu = x'^\mu{}_{,\rho\sigma} f_i^\rho f_j^\sigma + x'^\mu{}_{,\rho} f^\rho_{i,\sigma} f_j{}^\sigma,$$

from which we obtain

$$\xi'^\mu_3 = (\varepsilon_2{}^i\varepsilon_1{}^j - \varepsilon_1{}^i\varepsilon_2{}^j)(x'^\mu{}_{,\rho\sigma} f_i^\rho f_j^\sigma + x'^\mu{}_{,\rho} f^\rho_{i,\sigma} f_j{}^\sigma).$$

Since the first bracketed term on the right side of this equation is antisymmetric in i and j, and the first term in the second bracketed expression is symmetric in i and j, the product of these two terms is zero; so, we have, with the help of Eq. (1-4.9),

$$\begin{aligned}\xi'^\mu_3 &= (\varepsilon_2{}^i\varepsilon_1{}^j - \varepsilon_1{}^i\varepsilon_2{}^j) x'^\mu{}_{,\rho} c^k_{ij} f_k^\rho\\ &= \varepsilon_3{}^k g_k{}^\mu,\end{aligned}$$

which was to be shown. Since the relation between the x^μ and the x'^μ is just that between the coordinates of two points of the manifold which are related to each other by the mapping $x^\mu \to x'^\mu = x'^\mu(x)$, it follows that the intrinsic group structure of a subgroup of the MMG is preserved under all mappings of this latter group.

1-5. Geometrical Objects

By itself the bare space-time manifold is not very useful for the description of physical systems. One could discuss in a limited way the trajectories of point particles and the topological relations between them (for example, how many times they cross), but that is about all. In order to make an adequate correspondence with the physical world it is necessary to introduce the notion of a *geometrical object* defined on the manifold. In many ways the bare space-time manifold is like the blank canvas with which the artist starts when he paints a picture, and the geometrical objects are analogous to the paint he applies to this canvas. And like him, the physicist attempts to create a representation of the physical world as he perceives it. The chief difference between the endeavors of the two lies in the rules each follows and the criteria for success. (One is tempted to stretch the analogy a bit further by noting the inability of the one to comprehend much of the recent work of the other.)

It is possible to characterize all the geometrical objects that we will encounter in a purely axiomatic way. In fact, this was just the procedure employed by Euclid in the construction of his geometry. Such a procedure has the advantage of emphasizing the geometrical nature of the objects so defined. However, it does not lend itself readily to the use of standard analytic techniques. We shall therefore follow the procedure of defining such objects in terms of their transformation properties under the allowed mappings of the space-time manifold onto itself.

The central requirement that we shall impose on all geometrical objects is that they constitute the basis of a *realization* of the MMG. Under a mapping of this group a geometrical object, which we denote by the symbol y, will undergo a transformation. The transformed object will be denoted by y'. It will then constitute the basis of a realization of the MMG, provided the following conditions are satisfied:

(1) If y is transformed into y' by a mapping characterized by the functions $x'^\mu(x)$ and y' is transformed into y'' by the mappings $x''^\mu(x)$, then y will be transformed into y'' by the mapping $x''^\mu(x'(x))$, that is, the product mapping.

(2) y is transformed into itself by the identity mapping.

(3) If y is transformed into y' by a given mapping, then y' will be transformed into y by the inverse of this mapping.

The condition that all geometrical objects constitute a realization of the mapping group will ensure that the intrinsic properties of these objects and all relations between them are preserved under the mappings of the group.

The simplest example of a geometrical object is just a point of the manifold. A more general class of objects, which includes the point as a special case, comprises the *local* geometrical objects. These objects consist of a set of numbers, called its *components* and designated by $y_A(x)$, that are associated with a point x^μ of the manifold. The index A serves to distinguish the various components and assumes usually a discrete set of values $1, 2, \cdots, N$, where N is some finite integer.

Suppose now that we have a geometrical object defined at the point x^μ and a mapping that sends x^μ into x'^μ. We want to associate with the new point a set of components that are related to those at x^μ in such a way that the components constitute the basis of a realization of the mapping group. The relation between the components $y_A(x)$ and $y'_A(x')$ is called the *transformation law* for the components. This transformation law will then characterize part of the geometrical properties of the object in question. In general the transformed components will depend on the original components and on the particular mapping under consideration, the latter being characterized by the functions $x'^\mu(x)$ appearing in Eq. (1-3.1). Consequently the law of transformation will have the general form

$$y_A(x) \to y'_A(x') = Y_A(y(x), x'(x)). \tag{1-5.1}$$

The transformation functions Y_A appearing here are not completely arbitrary, of course, but must be of such a form that the object constitutes the basis of a realization of the group. Such will be the case, provided that for the identity mapping $Y_A(y(x), x) = y_A(x)$ and

$$Y_A(y(x), x''(x'(x))) = Y_A(Y(y(x), x'(x)), x''(x')). \tag{1-5.2}$$

Eq. (1-5.2) is seen to be the mathematical expression of requirement (1) above for the object to constitute the basis of a realization of the MMG. For an infinitesimal mapping characterized by the descriptors $\xi^\mu(x)$, the transformation law for the components $y_A(x)$ takes the form

$$y_A(x) \to y'_A(x') = y_A(x) + \delta y_A(x), \tag{1-5.3}$$

where the infinitesimal quantities $\delta y_A(x)$ are assumed to depend on the $y_A(x)$, the $\xi^\mu(x)$, and their derivatives.

Even with the restriction on the transformation laws of the y_A as embodied in Eq. (1-5.2), there are still an enormous number of different kinds of geometrical objects. However, there is a special subclass of these objects that have proved to be especially useful in representing physical quantities. For this subclass the Y_A are linear functions of the components y_A of the geometrical objects. Even more important is the case where they are also homogeneous functions of these components. In this case the object is said to constitute the basis of a *representation* of the MMG.

We now turn our attention to a study of geometrical objects with these transformation properties.

1-6. Tensors

The simplest geometrical object possessing a linear homogeneous transformation law is the *scalar*. It is a single component object, $y(x)$, with a transformation law of the form

$$y'(x') = y(x). \tag{1-6.1}$$

Note that it is not enough to define a scalar simply as a number assigned to a point of the manifold. One must also give the law of transformation (1-6.1), since there are other one-component objects that have different transformation laws (for example, pseudoscalars). It is clear that a scalar constitutes the basis of a representation (albeit a rather trivial one) of the MMG, since the transformation law (1-6.1) satisfies Eq. (1-5.2).

After the scalars, the next most simple local geometrical objects are the *covectors* and *contravectors*. (In most texts these objects are called covariant and contravariant vectors, respectively.) As an example of a contravector consider two neighboring points of the manifold. (Since we have not as yet introduced the notion of distance on our manifold, nearness of points is to be understood in the topological sense of the word.) Being near to each other, we can assume that if the coordinates of one of the points are x^μ, those of the other will be $x^\mu + dx^\mu$, where the dx^μ are four coordinate differentials. We can then show that the dx^μ can be considered to be the components of a geometrical object at the point x^μ. To do so let us consider a mapping that sends the point x^μ into the point $x'^\mu(x)$. Because of the assumed continuity of the mapping functions, this mapping will send the point $x^\mu + dx^\mu$ into a point $x'^\mu + dx'^\mu$, where

$$\begin{aligned} x'^\mu + dx'^\mu &= x'^\mu(x + dx) \\ &= x'^\mu(x) + \frac{\partial x'^\mu}{\partial x^\nu}\, dx^\nu. \end{aligned}$$

In arriving at this last result we have expanded $x'^\mu(x + dx)$ in a Taylor series about the point x^μ and dropped all terms of order 2 or more in the dx^μ. We see thereby that the components dx'^μ at the point x'^μ are related to the components dx^μ at the point x^μ by

$$dx'^\mu = \frac{\partial x'^\mu}{\partial x^\nu}\, dx^\nu. \tag{1-6.2}$$

One can now show that the transformation law (1-6.2) for the differentials dx^μ satisfies Eq. (1-5.1), and hence these quantities constitute the components of a local geometrical object. By extension we now define a general contravector as a four-component object, $A^\mu(x)$ at the point x^μ, whose components transform according to the law

$$A'^\mu(x') = \frac{\partial x'^\mu}{\partial x^\nu} A^\nu(x). \tag{1-6.3}$$

Problem 1.5. Show that a contravector is a geometrical object.

By making use of Eq. (1-6.3) we can calculate the change $\delta A^\mu(x)$ in the components $A^\mu(x)$ when we perform an infinitesimal mapping. Using the general expression (1-4.2) for such a mapping we find that

$$A'^\mu(x') = \frac{\partial}{\partial x^\nu}(x^\mu + \xi^\mu(x))A^\nu(x)$$

$$= A^\mu(x) + \xi^\mu{}_{,\mu} A^\mu(x)\,,$$

since $\partial x^\mu/\partial x^\nu = \delta_\nu{}^\mu$, where $\delta_\nu{}^\mu$ is the *Kronecker delta* with values given by

$$\delta_\nu{}^\mu = \begin{cases} +1, & \mu = \nu \\ 0, & \mu \neq \nu \end{cases}. \tag{1-6.4}$$

Consequently

$$\delta A^\mu(x) = \xi^\mu{}_{,\nu}(x) A^\nu(x). \tag{1-6.5}$$

In addition to contravectors, there are other four-component, linear, homogeneous objects, among which are the covectors. As an example of such an object consider the derivatives $\phi_{,\mu}$ of a scalar field $\phi(x)$ defined on our manifold; that is, a collection of scalars defined at each point of some region of the manifold. By making use of the chain rule of differentiation we can derive the law of transformation for these derivatives. We have simply

$$\phi'_{,\mu}(x') = \frac{\partial x^\nu}{\partial x'^\mu} \phi_{,\nu}(x), \tag{1-6.6}$$

where we have made explicit use of the fact that $\phi(x)$ is a scalar and so has a law of association given by Eq. (1-6.1). Again we can generalize and define an arbitrary covector as a four-component object $B_\mu(x)$ localized at the point x^μ, with a law of association given by

$$B'_\mu(x') = \frac{\partial x^\nu}{\partial x'^\mu} B_\nu(x). \tag{1-6.7}$$

Following a procedure similar to that used in the case of a contravector, we see that

$$\delta B_\mu = -\xi^\nu{}_{,\mu} B_\nu . \tag{1-6.8}$$

Once having obtained the transformation law (1-6.7) for an arbitrary covector field as a generalization of the transformation law for the gradient of a scalar field, we can reverse our argument and say that by considering the numbers $\phi_{,\mu}(x)$ to be the components of a covector at the point x^μ, we preserve their intrinsic property of being the gradients of a scalar field under a mapping. We could have just as well taken the $\phi_{,\mu}$ as the components of a contravector at x^μ. However, if we did so, the transformed components would no longer constitute the gradients of a scalar field. Likewise, we could have taken the displacements dx^μ to be the components of a covector, but then the transformed components would not represent a displacement. We see from this discussion that the decision to consider a given set of numbers as the components of a contra- or covector depends on how these numbers were obtained, that is, on their intrinsic or geometrical properties.

Scalars, contravectors, and covectors are special cases of a general class of geometrical objects with linear, homogeneous transformation laws, called *tensors*. A tensor of *rank* r is a 4^r component object, $T^{\mu\nu\rho\cdots}_{\varepsilon\kappa\cdots}$, with p contra-indices $\mu\nu\rho\cdots$ and q co-indices $\varepsilon\kappa\cdots$, where $p + q = r$. In the law of association for this object each index contributes a factor according to its type. Thus a third rank tensor $T^{\mu\nu}_{\lambda}$ transforms as

$$T'^{\mu\nu}_{\lambda}(x') = \frac{\partial x'^\mu}{\partial x^\rho}\frac{\partial x'^\nu}{\partial x^\sigma}\frac{\partial x^\kappa}{\partial x'^\lambda}\, T^{\rho\sigma}_{\kappa}(x).$$

For such an object we have

$$\delta T^{\mu\nu}_{\lambda} = \xi^\mu{}_{,\rho} T^{\rho\nu}_{\lambda} + \xi^\nu{}_{,\rho} T^{\mu\rho}_{\lambda} - \xi^\rho{}_{,\lambda} T^{\mu\nu}_{\rho},$$

and similarly for higher-rank tensors. We note in passing that a geometrical object may have nontensor indices in addition to any tensor indices it may possess.

One can perform a number of operations with and on tensors that result in new tensors. We will now list the more important of these and assert without proof that the result of each operation is again a tensor, since it is fairly obvious in all of the cases.

(a) Addition of Tensors

One can add or subtract tensors of the same type defined at the same point of the manifold to produce a new tensor of the same type at that point.

(b) Multiplication of Tensors

The collection of numbers obtained by multiplying in all combinations the components of two tensors of rank r and s at a point of the manifold give a new tensor of rank $r + s$ at that point. In particular, multiplication of a tensor by a scalar yields a new tensor of the same rank.

(c) Contraction

From a tensor of rank r we can construct a new tensor of rank $r - 2$ by summing over one contra-index and one co-index. Thus $D^\rho \equiv D^{\mu\rho}_\mu$ is a tensor (in this case a contratensor) formed from the mixed tensor $D^{\mu\rho}_\nu$ by summing over the indices μ and ν. By summing over ρ and ν one could construct another contravector. In general these two vectors would not be equal to each other unless $D^{\mu\rho}_\nu = D^{\rho\mu}_\nu$, that is, unless $D^{\mu\rho}_\nu$ is symmetric in the index pair $\mu\rho$. Note that contraction always involves summation over a contra- and a co-index. Summation over two contra- or two co-indices would not result in a tensor.

Problem 1.6. Show that $\sum_\mu D_{\mu\mu}$ is not a geometrical object if $D_{\mu\nu}$ is a cotensor.

An invariant property of all tensors is their *symmetry* with respect to the interchange of indices. If upon the interchange of two contra- or two co-indices (but not a contra- with a co-index, or vice versa) the tensor remains unchanged in value, we say that it is *symmetric* in these two indices, while if it changes sign only, it is *antisymmetric*. Thus, if

$$T^{\cdots\mu\cdots\nu\cdots}_{\cdots} = T^{\cdots\nu\cdots\mu\cdots}_{\cdots},$$

the tensor is symmetric in μ and ν. Sometimes the symmetries can be multiple. Thus we will have to deal later with a tensor $B_{\mu\nu\rho\sigma}$, which is antisymmetric in μ and ν, and in ρ and σ, and symmetric in the double set $\mu\nu$ and $\rho\sigma$, so that

$$B_{\mu\nu\rho\sigma} = B_{\rho\sigma\mu\nu}.$$

The important thing about these symmetries is that they are preserved under a mapping of the manifold onto itself. Thus, if $C_{\mu\nu} = -C_{\nu\mu}$ and $C'_{\mu\nu}$ is the transform of $C_{\mu\nu}$, then $C'_{\mu\nu} = -C'_{\nu\mu}$.

It sometimes happens that certain components or linear combinations of components of a tensor (and also other geometrical objects) transform among themselves. Thus, from the components of a tensor $T^{\mu\nu}$, we can construct its symmetric part $T^{(\mu\nu)}$ and its antisymmetric part $T^{[\mu\nu]}$ according to

$$\begin{aligned} T^{(\mu\nu)} &= \tfrac{1}{2}(T^{\mu\nu} + T^{\nu\mu}) \\ T^{[\mu\nu]} &= \tfrac{1}{2}(T^{\mu\nu} - T^{\nu\mu}). \end{aligned} \tag{1-6.9}$$

Similarly we can construct the transformed symmetric part $T'^{(\mu\nu)}$ and antisymmetric part $T'^{[\mu\nu]}$ from the transformed $T'^{\mu\nu}$. Then, one can show that $T'^{(\mu\nu)}$ is a function of $T^{(\mu\nu)}$ and the mapping function only, and similarly for $T^{[\mu\nu]}$. Whenever a geometrical object can be broken up into parts that transform among themselves, we say that we have a *reducible* object. If no such decomposition is possible, we have an *irreducible* object.

The construction of $T^{(\mu\nu)}$ and $T^{[\mu\nu]}$ from $T^{\mu\nu}$ given above can be extended to higher rank tensors. Thus, from a tensor $T_{\mu\nu\rho}$ we can construct a completely symmetric tensor (symmetric under interchange of any two of its indices) given by

$$T_{(\mu\nu\rho)} = \frac{1}{3!}(T_{\mu\nu\rho} + T_{\rho\mu\nu} + T_{\nu\rho\mu} + T_{\nu\mu\rho} + T_{\mu\rho\nu} + T_{\rho\nu\mu})$$

and a completely antisymmetric tensor (antisymmetric under interchange of any two of its indices) given by

$$T_{[\mu\nu\rho]} = \frac{1}{3!}(T_{\mu\nu\rho} + T_{\rho\mu\nu} + T_{\nu\rho\mu} - T_{\nu\mu\rho} - T_{\mu\rho\nu} - T_{\rho\nu\mu}).$$

Clearly, if $T_{\mu\nu\rho}$ is itself completely symmetric, then $T_{(\mu\nu\rho)} = T_{\mu\nu\rho}$ while if it is completely antisymmetric, $T_{[\mu\nu\rho]} = T_{\mu\nu\rho}$. The operation of constructing a completely symmetric or completely antisymmetric tensor from a given tensor can be extended to tensors of higher rank in an obvious manner. In the first case the operation is called *symmetrization* of a tensor while in the second case it is called *antisymmetrization* of a tensor. Of course, for fifth- or higher-rank tensors antisymmetrization leads to an identically vanishing tensor. We can also apply these operations to the product of two or more tensors since such a product is again a tensor. Thus

$$A_{[\mu}B_{\sigma]}{}^{\rho} = \tfrac{1}{2}(A_{\mu}B_{\sigma}{}^{\rho} - A_{\sigma}B_{\mu}{}^{\rho}),$$

etc.

Problem 1.7. Show that a mixed second-rank tensor $T_{\nu}{}^{\mu}$ is reducible and that its irreducible parts consist of its trace $T_{\mu}{}^{\mu}$ and a traceless tensor with components $T_{\nu}{}^{\mu} - \frac{1}{4}\delta_{\nu}{}^{\mu}T_{\rho}{}^{\rho}$, where $\delta_{\nu}{}^{\mu}$ is the Kronecker delta.

Finally we describe a method that is occasionally needed for deciding the tensor character of a given object. If we know that when it is multiplied by an arbitrary tensor of a particular type, the result is another tensor, then in general we can conclude that the original object is a tensor. As an example, consider the four numbers B^{μ} defined at a point. If for *any* covector C_{μ} defined at this point we are told that $B^{\mu}C_{\mu}$ is a scalar, then it follows that B^{μ} is a contravector. If $B^{\mu}C_{\mu}$ is a scalar, then under a mapping

$$(B^{\mu}C_{\mu})' = B'^{\mu}C'_{\mu} = B^{\mu}C_{\mu}.$$

Since, under this mapping,

$$C'_\mu = \frac{\partial x^\rho}{\partial x'^\mu} C_\rho ,$$

it follows that

$$B'^\mu \frac{\partial x^\rho}{\partial x'^\mu} C_\rho = B^\rho C_\rho .$$

Since C_ρ is arbitrary, we can conclude, therefore, that

$$B'^\mu \frac{\partial x^\rho}{\partial x'^\mu} = B^\rho .$$

Then, since

$$\frac{\partial x^\rho}{\partial x'^\mu} \frac{\partial x'^\mu}{\partial x^\sigma} = \delta_\sigma{}^\rho ,$$

by the chain rule for differentiation, we have

$$B'^\mu = \frac{\partial x'^\mu}{\partial x^\rho} B^\rho ;$$

hence B^μ is a contravector.

Problem 1.8. Show that if $B^{\mu\nu} C_\nu$ transforms like a contravector, where C_ν is an arbitrary covector, then $B^{\mu\nu}$ is a contratensor.

A particularly useful tensor can be constructed with the help of the Kronecker delta, $\delta_\nu{}^\mu$, with values given by Eq. (1-6.4). If we take the numbers $\delta_\nu{}^\mu$ to be the values of the components of a mixed second-rank tensor, as the indices indicate, it has the important property that under a mapping, $\delta'^\mu_\nu = \delta_\nu{}^\mu$, that is, it has the same component values at all points of the manifold. It is therefore unnecessary to specify where on the manifold this tensor resides. There are, of course, co- and contratensors that have these component values at a point, for example, $h^{\mu\nu} = 1$ for $\mu = \nu$ and zero otherwise. It would not be sufficient, however, to specify only these component values to define the tensor as we can in the case of the mixed tensor; we would have to specify as well the point of the manifold where it has these component values.

In addition to the Kronecker delta itself, we can also construct other useful generalizations. One such generalization is the tensor $\delta^{\mu\nu}_{\rho\sigma}$ with component values given by

$$\delta^{\mu\nu}_{\rho\sigma} = \begin{cases} +1 & \mu \neq \nu,\ \rho = \mu,\ \sigma = \nu \\ -1 & \mu \neq \nu,\ \rho = \nu,\ \sigma = \mu \\ \ \ 0 & \text{otherwise} \end{cases} \qquad (1\text{-}6.10)$$

One can show that

$$\delta^{\mu\nu}_{\rho\sigma} = \begin{vmatrix} \delta_\rho{}^\mu & \delta_\rho{}^\nu \\ \delta_\sigma{}^\mu & \delta_\sigma{}^\nu \end{vmatrix} = \delta_\rho{}^\mu \delta_\sigma{}^\nu - \delta_\sigma{}^\mu \delta_\rho{}^\nu,$$

which immediately demonstrates the tensor character of $\delta^{\mu\nu}_{\rho\sigma}$. Likewise, one has

$$\delta^{\mu\nu\lambda}_{\rho\sigma\tau} = \begin{vmatrix} \delta_\rho{}^\mu & \delta_\rho{}^\nu & \delta_\rho{}^\lambda \\ \delta_\sigma{}^\mu & \delta_\sigma{}^\nu & \delta_\sigma{}^\lambda \\ \delta_\tau{}^\mu & \delta_\tau{}^\nu & \delta_\tau{}^\lambda \end{vmatrix}, \tag{1-6.11}$$

etc. In four dimensions the generalized Kronecker delta with five indices vanishes identically. Clearly one has

$$A_{[\mu\nu]} = \tfrac{1}{2}\delta^{\rho\sigma}_{\mu\nu} A_{\rho\sigma},$$

$$A_{[\mu\nu\lambda]} = \frac{1}{3!}\,\delta^{\rho\sigma\tau}_{\mu\nu\lambda} A_{\rho\sigma\tau},$$

etc.

Before we consider other linear homogeneous objects, we should point out that not all apparently simple operations on tensors lead to new tensors or even to new geometrical objects. As an example, consider the ordinary derivatives $A_{\mu,\nu}$ of a covector field A_μ. By direct calculation we find that

$$A'_{\mu,\nu} = \frac{\partial x^\rho}{\partial x'^\mu}\frac{\partial x^\sigma}{\partial x'^\nu} A_{\rho,\sigma} + \frac{\partial^2 x^\rho}{\partial x'^\mu\, \partial x'^\nu} A_\rho .$$

Thus, in order to calculate $A'_{\mu,\nu}$, we must know not only the $A_{\rho,\sigma}$ but also the A_ρ at each point. Since, at a point, these quantities are independent of each other, the association law for $A_{\mu,\nu}$ depends upon more than just the $A_{\mu,\nu}$ and the functions describing the mapping. Consequently $A_{\mu,\nu}$ is not even a geometrical object, let alone a tensor. In spite of this, there are a number of differential operations that we can perform on tensors to produce new tensors. We again state the results without proof.

(1) If A_μ is a covector field, then $A_{\mu,\nu} - A_{\nu,\mu} = \delta^{\rho\sigma}_{\mu\nu} A_{\rho,\sigma}$ is an antisymmetric cotensor field of rank 2. This cotensor is sometimes called the *curl* of the covector.

(2) If $A_{\mu\nu}$ is an antisymmetric cotensor of the second rank, the cyclic divergence

$$A_{\mu\nu,\rho} + A_{\rho\mu,\nu} + A_{\nu\rho,\mu} = \delta^{\iota\kappa\lambda}_{\mu\nu\rho} A_{\iota\kappa,\lambda}$$

is a completely antisymmetric cotensor of the third rank.

(3) If $A_{\mu\nu\rho}$ is a completely antisymmetric cotensor of the third rank, then

$$A_{\mu\nu\rho,\sigma} - A_{\sigma\mu\nu,\rho} + A_{\rho\sigma\mu,\nu} - A_{\nu\rho\sigma,\mu} = \delta^{\iota\kappa\lambda\tau}{}_{\nu\rho\sigma} A_{\iota\kappa\lambda,\tau}$$

is a completely antisymmetric cotensor of the fourth rank. This last construction is as far as one can go with this process in four dimensions.

1-7. Tensor Densities

The totality of all types of tensors does not exhaust the collection of geometrical objects with linear, homogeneous transformation laws. In fact they are but a subclass of a more general type of object called a *tensor density*. These are objects that transform like tensors but with the additional feature that the Jacobian of the mapping function raised to some power appears as a factor in the law of transformation. Thus a tensor density $\mathfrak{T}^{\mu\cdot}_{\nu} \cdots$ transforms according to the law

$$\mathfrak{T}'^{\mu\cdots}_{\nu\cdots} = \left|\frac{\partial x}{\partial x'}\right|^W \frac{\partial x'^\mu}{\partial x^\rho} \cdots \frac{\partial x^\sigma}{\partial x'^\nu} \cdots \mathfrak{T}^{\rho\cdots}_{\sigma\cdots} \tag{1-7.1}$$

where $|\partial x/\partial x'|$ is the Jacobian of the mapping and W is a constant. It is called the *weight* of the density. Thus tensors are tensor densities of weight zero. The expression for $\delta\mathfrak{T}^{\mu\cdots}_{\nu\cdots}$ differs from that of an ordinary tensor by the addition of a term due to the appearance of the determinant in the transformation law (1-7.1). To find this additional term we note first that, to first order in the descriptors,

$$\left|\frac{\partial x}{\partial x'}\right| = |\delta_\beta{}^\alpha - \xi^\alpha{}_{,\beta}|$$

$$\simeq 1 - \xi^\alpha{}_{,\alpha},$$

and hence

$$\left|\frac{\partial x}{\partial x'}\right|^W \simeq 1 - W\xi^\alpha{}_{,\alpha}.$$

Therefore

$$\delta\mathfrak{T}^{\mu\cdots}_{\nu\cdots} = -W\xi^\alpha{}_{,\alpha}\mathfrak{T}^{\mu\cdots}_{\nu\cdots} + \xi^\mu{}_{,\rho}\mathfrak{T}^{\rho\cdots}_{\nu\cdots} + \cdots - \xi^\rho{}_{,\nu}\mathfrak{T}^{\mu\cdots}_{\rho\cdots} - \cdots .$$

Problem 1.9. Show that a scalar density of weight W is a geometrical object.

All the operations that can be performed on tensors can also be performed on densities, with the following difference. When we multiply two densities of weights W_1 and W_2 we again get a new density with a weight $W_1 + W_2$. We can also form new densities by differentiation, with somewhat different results than in the case of tensors. The derivative of a scalar density is no longer a density or even a geometrical object. However, we have the following results:

(1) If $\mathfrak{A}^{\mu}$ is a contradensity of weight $+1$, then its divergence $\mathfrak{A}^{\mu}{}_{,\mu}$ is a scalar density of weight $+1$.

(2) If $\mathfrak{A}^{\mu\nu}$ is an antisymmetric contradensity of weight $+1$, then its divergence $\mathfrak{A}^{\mu\nu}{}_{,\nu}$ is a contradensity of weight $+1$.

(3) If $\mathfrak{A}^{\mu\nu\rho}$ is a completely antisymmetric contradensity of weight $+1$, then its divergence $\mathfrak{A}^{\mu\nu\rho}{}_{,\rho}$ is also a contradensity of weight $+1$.

(4) If $\mathfrak{A}^{\mu\nu\rho\sigma}$ is a completely antisymmetric contradensity of weight $+1$, then its divergence $\mathfrak{A}^{\mu\nu\rho\sigma}{}_{,\sigma}$ is also a contradensity of weight $+1$.

Having discussed the general properties of tensor densities, we now consider a very useful example of such. Consider the set of numbers $\varepsilon^{\mu\nu\rho\sigma}$ with values $+1$ and -1, depending upon whether $\mu\nu\rho\sigma$ is an even or an odd permutation of 0123 and zero otherwise. If we take these numbers to be the components of a completely antisymmetric tensor density of weight $+1$, this density has the useful property that (like the Kronecker tensor) the values of its components are unchanged under a mapping. Likewise, one can also form the density $\varepsilon_{\mu\nu\rho\sigma}$ with components determined in the same way as those of $\varepsilon^{\mu\nu\rho\sigma}$, but with weight -1. Then it also has the property that its components are unchanged under a mapping. These densities are usually referred to in the literature as the *Levi-Civita densities.*

Problem 1.10. Show that the components of the Levi-Civita densities are unchanged under a mapping.

The Levi-Civita densities are useful for a number of purposes. In particular, one can use them to form the determinant of a second-rank density. Thus the determinant of the contradensity $\mathfrak{A}^{\mu\nu}$, det $\mathfrak{A}^{\mu\nu}$, is given by

$$\det \mathfrak{A}^{\mu\nu} = \frac{1}{4!}\, \varepsilon_{\gamma\iota\mu\rho}\varepsilon_{\delta\kappa\nu\sigma}\mathfrak{A}^{\gamma\delta}\mathfrak{A}^{\iota\kappa}\mathfrak{A}^{\mu\nu}\mathfrak{A}^{\rho\sigma}. \tag{1-7.2}$$

For the mixed density $\mathfrak{A}_{\nu}{}^{\mu}$, we have

$$\det \mathfrak{A}_{\nu}{}^{\mu} = \frac{1}{4!}\, \varepsilon_{\gamma\iota\mu\rho}\varepsilon^{\delta\kappa\nu\sigma}\mathfrak{A}_{\delta}{}^{\gamma}\mathfrak{A}_{\kappa}{}^{\iota}\mathfrak{A}_{\nu}{}^{\mu}\mathfrak{A}_{\sigma}{}^{\rho}, \tag{1-7.3}$$

while for the codensity $\mathfrak{A}_{\nu}{}^{\mu}$,

$$\det \mathfrak{A}_{\mu\nu} = \frac{1}{4!}\, \varepsilon^{\gamma\iota\mu\rho}\varepsilon^{\delta\kappa\nu\sigma}\mathfrak{A}_{\gamma\delta}\mathfrak{A}_{\iota\kappa}\mathfrak{A}_{\mu\nu}\mathfrak{A}_{\rho\sigma}. \tag{1-7.4}$$

We see from the above equations that, since the right-hand sides are obviously scalar densities of weight $4W - 2$, $4W$, and $4W + 2$, respectively, where W is the weight of the original density, the determinants are also densities with these weights. If $\mathfrak{A}^{\mu\nu}$ and $\mathfrak{A}_{\mu\nu}$ are antisymmetric, we have the further results that

$$\sqrt{\det \mathfrak{A}^{\mu\nu}} = \tfrac{1}{8}\varepsilon_{\mu\nu\rho\sigma}\mathfrak{A}^{\mu\nu}\mathfrak{A}^{\rho\sigma} \tag{1-7.5}$$

and

$$\sqrt{\det \mathfrak{A}_{\mu\nu}} = \tfrac{1}{8}\varepsilon^{\mu\nu\rho\sigma}\mathfrak{A}_{\mu\nu}\mathfrak{A}_{\rho\sigma}. \tag{1-7.6}$$

With the help of the Levi-Civita densities we can also construct the inverse of a given second-rank tensor. We illustrate the process in the case of a cotensor $g_{\mu\nu}$. First we construct the minor $M^{\mu\nu}$ of $g_{\mu\nu}$ in the determinant $g = \det g_{\mu\nu}$ according to

$$M^{\mu\nu} = \frac{\partial g}{\partial g_{\mu\nu}} = \frac{1}{6}\,\varepsilon^{\mu\gamma\iota\rho}\varepsilon^{\nu\delta\kappa\sigma}g_{\gamma\delta}g_{\iota\kappa}g_{\rho\sigma},$$

and from this the inverse $g^{\mu\nu}$ is given by

$$g^{\mu\nu} = \frac{1}{g}\,M^{\mu\nu}. \tag{1-7.7}$$

From the theory of determinants, or by direct calculation, it follows that

$$g^{\mu\nu}g_{\lambda\nu} = \delta_\lambda{}^\mu.$$

It also follows, either from its definition or from Eq. (1-7.7) that $g^{\mu\nu}$ is a contratensor.

We conclude this discussion of the Levi-Civita densities by stating certain relations between them, again without proof. They are

$$\frac{1}{0!}\,\varepsilon^{\mu\nu\lambda\tau}\varepsilon_{\iota\kappa\rho\sigma} = \delta^{\mu\nu\lambda\tau}_{\iota\kappa\rho\sigma}, \tag{1-7.8}$$

$$\frac{1}{1!}\,\varepsilon^{\mu\nu\lambda\tau}\varepsilon_{\iota\kappa\rho\tau} = \delta^{\mu\nu\lambda}_{\iota\kappa\rho}, \tag{1-7.9}$$

$$\frac{1}{2!}\,\varepsilon^{\mu\nu\rho\sigma}\varepsilon_{\iota\kappa\rho\sigma} = \delta^{\mu\nu}_{\iota\kappa}, \tag{1-7.10}$$

$$\frac{1}{3!}\,\varepsilon^{\mu\nu\rho\sigma}\varepsilon_{\iota\nu\rho\sigma} = \delta_\iota{}^\mu, \tag{1-7.11}$$

$$\frac{1}{4!}\,\varepsilon^{\mu\nu\rho\sigma}\varepsilon_{\mu\nu\rho\sigma} = 1. \tag{1-7.12}$$

The question arises naturally now: Are there linear homogeneous local geometrical objects other than tensors and tensor densities? The answer is "no"; on a bare manifold these are in fact the only finite dimensional local objects that are realizations of the MMG.

1-8. Tensor Fields

While the geometrical objects that we have considered up to now constitute, by construction, the bases of realization of the MMG (manifold mapping group), they do not constitute the bases of faithful realizations. We say that a realization is *faithful* if there is a one-to-one relation between the transformed components of an object and the elements of the MMG. Since there are many different mappings for which $\partial x^\mu/\partial x'^\nu$ have a given set of values at a single point of the manifold, we see that a single local object by itself can never constitute the basis of a faithful realization of the MMG. Rather, in the case of the tensors and tensor densities, they constitute the basis of a faithful representation of the full linear group in four dimensions, since the quantities $\partial x^\mu/\partial x'^\nu$ are all quite arbitrary at any one point of the manifold. In order to obtain a faithful realization of the MMG, we need a *field* of objects. As the name implies, a field is a collection of objects, all of the same type, defined at each point of the manifold. In general a field as a whole will constitute a faithful realization of the MMG. Thus, faithful realizations of this group are all infinite dimensional.

When working with fields it is convenient to re-express the group property embodied in Eq. (1-5.2) in terms of infinitesimal mappings. To this end we define the quantities $\bar{\delta} y_A(x)$ by

$$\bar{\delta} y_A(x) \equiv y'_A(x) - y_A(x). \tag{1-8.1}$$

Unlike the quantities $\delta y_A(x)$ defined by Eq. (1-5.3), these latter quantities have the important property that

$$\begin{aligned} \bar{\delta}(y_{A,\mu}(x)) &\equiv y'_{A,\mu}(x) - y_{A,\mu}(x) \\ &= (\bar{\delta} y_A(x))_{,\mu}, \end{aligned}$$

that is, $\bar{\delta}$ (considered as an operator) and differentiation are commutative. Consequently, given any function $\mathfrak{E}(x)$ of the $y_A(x)$ and their derivatives that maintains its functional form under a mapping, it is an easy matter to compute $\bar{\delta}\mathfrak{E}(x)$. One has

$$\begin{aligned} \bar{\delta}\,\mathfrak{E}(x) &\equiv \mathfrak{E}'(x) - \mathfrak{E}(x) \\ &= \frac{\partial\,\mathfrak{E}(x)}{\partial y_A(x)}\,\bar{\delta} y_A(x) + \frac{\partial\,\mathfrak{E}(x)}{\partial y_{A,\mu}(x)}\,(\bar{\delta} y_A(x))_{,\mu} \\ &\quad + \cdots. \end{aligned} \tag{1-8.2}$$

We see from its definition that $\bar{\delta} y_A(x)$ is the difference between the original components $y_A(x)$ at the point x^μ and the transformed components at a point that is mapped onto x^μ by the mapping.

In order to obtain a relationship between the $\bar{\delta} y_A(x)$ and the $\delta y_A(x)$, we substitute for x^μ its value $x'^\mu - \xi^\mu(x')$ in the first term on the right-hand side of Eq. (1-8.1) and expand in a Taylor series about the value x'. Thus

$$\begin{aligned}\bar{\delta} y_A(x) &= y'_A(x' - \xi(x')) - y_A(x) \\ &= y'_A(x') - y'_{A,\mu}(x')\xi^\mu(x') - y_A(x).\end{aligned}$$

We can now drop all primes in the second term of the above equation, since it is already a first-order term and we are working only to first order in the infinitesimal quantities $\xi^\mu(x)$. It then follows from the definition of $\delta y_A(x)$ that

$$\bar{\delta} y_A(x) = \delta y_A(x) - y_{A,\mu}(x)\xi^\mu(x). \tag{1-8.3}$$

For a scalar field $\phi(x)$ we have

$$\bar{\delta}\phi(x) = -\phi_{,\mu}(x)\xi^\mu(x);$$

for a covector field $A_\mu(x)$ we have, making use of Eq. (1-6.8),

$$\bar{\delta} A_\mu(x) = -A_\nu(x)\xi^\nu{}_{,\mu}(x) - A_{\mu,\nu}(x)\xi^\nu(x),$$

etc.

With the aid of the $\bar{\delta} y_A(x)$ it is now possible to restate the condition that the $y_A(x)$ constitute a realization of the MMG. Such will be the case, provided

$$\bar{\delta}_3 y_A(x) = \bar{\delta}_2(\bar{\delta}_1 y_A(x)) - \bar{\delta}_1(\bar{\delta}_2 y_A(x)), \tag{1-8.4}$$

where $\bar{\delta}_1 y_A$ is the infinitesimal change induced by the mapping generated by $\xi_1{}^\mu$, and similarly for $\bar{\delta}_2 y_A$ and $\bar{\delta}_3 y_A$, and where $\xi_1{}^\mu$, $\xi_2{}^\mu$ and $\xi_3{}^\mu$ are related by Eq. (1-4.3).

Problem 1.11. By performing two infinitesimal mappings first in one order and then the other, show that Eq. (1-8.4) is a necessary condition that there exist a third infinitesimal mapping that transforms the resultant components of one such sequence of mappings into those of the other.

As an application of Eq. (1-8.4) let us check that it is satisfied for a scalar field $\phi(x)$. For such a field, $\bar{\delta}\phi = -\phi_{,\mu}\xi^\mu$. Consequently,

$$\begin{aligned}\bar{\delta}_2(\bar{\delta}_1\phi) &= -\bar{\delta}_2(\phi_{,\mu}\xi_1{}^\mu) \\ &= -(\bar{\delta}_2\phi)_{,\mu}\xi_1{}^\mu \\ &= (\phi_{,\nu}\xi_2{}^\nu)_{,\mu}\xi_1{}^\mu.\end{aligned}$$

Likewise,

$$\bar{\delta}_1(\bar{\delta}_2\phi) = (\phi_{,\nu}\xi_1{}^\nu)_{,\mu}\xi_2{}^\mu,$$

so that

$$\bar{\delta}_2(\bar{\delta}_1\phi) - \bar{\delta}_1(\bar{\delta}_2\phi) = -\phi_{,\nu}(\xi^\nu_{1,\mu}\xi_2{}^\mu - \xi^\nu_{2,\mu}\xi_1{}^\mu)$$
$$+ \phi_{,\mu\nu}(\xi_2{}^\nu\xi_1{}^\mu - \xi_1{}^\nu\xi_2{}^\mu).$$

Since differentiation is commutative, $\phi_{,\mu\nu} = \phi_{,\nu\mu}$; hence the last term vanishes and we have the result

$$\bar{\delta}_3\phi = -\phi_{,\nu}\xi_3{}^\nu$$

where $\xi_3{}^\nu$ is given by Eq. (1-4.3). Since $\bar{\delta}_3\phi$ has the correct form for a scalar, we see that Eq. (1-8.4) is indeed satisfied for a scalar field.

Problem 1.12. Show that Eq. (1-8.4) is satisfied for a covector field.

1-9. Integrals and Stokes' Theorem

In addition to the purely local geometrical objects discussed above, we will also need to consider objects that associate numbers with two or more points of the manifold and even whole submanifolds, for example, curves and surfaces. As a simple example, consider the sum of two scalars $\psi(x_1)$ and $\psi(x_2)$ associated with the points $x_1{}^\mu$ and $x_2{}^\mu$, respectively. In effect, this sum, $\psi(x_1) + \psi(x_2)$, can be thought of as a single number associated with these two points. Because of the transformation law for scalars we have that under a mapping

$$\psi'(x'_1) + \psi'(x'_2) = \psi(x_1) + \psi(x_2),$$

and hence this two-point scalar is a geometrical object. Likewise, the sum of any number of scalars defined at different points of the manifold is a geometrical object. However, only scalars can be added in such a manner; it is clear, for instance, that the sum of two vectors associated with different points of the manifold is not a geometrical object, since a knowledge of the sum of the components is not sufficient to determine the components of the sum under a mapping.

From the above discussion one might conclude that the integral of a scalar field over a region R of the manifold is a geometrical object. However, such is not the case. The reason for this is that d^4x, the element of volume, is not itself a scalar but rather, as we will see, a scalar density of weight -1. Consequently, the quantities $\psi(x)d^4x$ that are added together to form an integral are not scalars. By the same token, however, we can expect that the integral of a scalar density of weight $+1$ over a region of the manifold is a geometrical object. One can also form other scalars from products of tensors and coordinate differentials that can be integrated over submanifolds by forming line, surface, and hypersurface integrals. Furthermore, several of these integrals can be related to each other by generalizations of Stokes' theorem. In much of the following discussion we will have occasion to make use of these integrals and the relations between them, so we discuss them briefly in the remainder of this section.

In describing a submanifold or even the full manifold it is convenient to make use of a parametric representation. If the dimensionality of this submanifold is $M(\leq 4)$, then one can label the points contained therein by M parameters λ_i, $i = 1, \cdots, M$. The coordinates of the points in this submanifold are then given parametrically by the equations

$$x^\mu = \phi^\mu(\lambda_1, \cdots, \lambda_M). \tag{1-9.1}$$

Let us now form the element of "area"

$$d\tau^{\mu_1\mu_2\cdots\mu_M} \equiv D^{\mu_1\mu_2\cdots\mu_M}\, d\lambda_1\, d\lambda_2 \cdots d\lambda_M, \tag{1-9.2}$$

where

$$D^{\mu_1\mu_2\cdots\mu_M} = \delta^{\mu_1\mu_2\cdots\mu_M}_{\nu_1\nu_2\cdots\nu_M} \phi^{\nu_1}{}_{,1} \phi^{\nu_2}{}_{,2} \cdots \phi^{\nu_M}{}_{,M}. \tag{1-9.3}$$

Here we again use the comma notation to denote differentiation, that is, $\phi^\mu{}_{,i} = \partial\phi^\mu/\partial\lambda_i$. As defined, the element $d\tau^{\mu_1\mu_2\cdots\mu_M}$ is a scalar under arbitrary parameter changes and an Mth rank contratensor under coordinate mappings. Hence, if $f_{\mu_1\mu_2\cdots\mu_M}$ is an Mth rank cotensor, $f_{\mu_1\mu_2\cdots\mu_M}\, d\tau^{\mu_1\mu_2\cdots\mu_M}$ is a scalar under both parameter changes and coordinate mappings. Consequently we can form the integral

$$I = \int_{\Omega_M} f_{\mu_1\mu_2\cdots\mu_M}\, d\tau^{\mu_1\mu_2\cdots\mu_M}. \tag{1-9.4}$$

For $M = 1, 2, 3$ and 4, I is respectively a line, surface, hypersurface, and volume integral.

For some purposes it is convenient to re-express the element of area $d\tau$ in terms of coordinate differentials. Since the components $d_i x^\mu$ of the displacement $d\lambda_i$ are given by

$$d_i x^\mu = \phi^\mu{}_{,i}\, d\lambda_i, \qquad \text{(no sum on } i\text{)}$$

we have

$$d\tau^{\mu_1\mu_2\cdots\mu_M} = \delta^{\mu_1\mu_2\cdots\mu_M}_{\nu_1\nu_2\cdots\nu_M}\, d_1x^{\nu_1}\, d_2x^{\nu_2} \cdots d_Mx^{\nu_M}. \tag{1-9.5}$$

If we are dealing with a two-dimensional submanifold (for example, a surface), the element of area is given by

$$\begin{aligned} d\tau^{\mu\nu} &= \delta^{\mu\nu}_{\rho\sigma}\, d_1x^\rho\, d_2x^\sigma \\ &= d_1x^\mu\, d_2x^\nu - d_2x^\mu\, d_1x^\nu. \end{aligned}$$

If, in particular, the surface is the surface $x^0 = x^3 = 0$ and the parameters λ_1 and λ_2 are taken to coincide with the coordinates x^1 and x^2, respectively, then the only nonvanishing components of $d\tau^{\mu\nu}$ are $d\tau^{12} = -d\tau^{21} = dx^1\, dx^2$.

We now state without proof the generalization of Stokes' theorem. If Ω_M is a region of a submanifold of dimension M, and Ω_{M-1} is the $M-1$ dimensional submanifold that forms the boundary of Ω_M, then

$$\int_{\Omega_{M-1}} f_{\mu_1\mu_2\cdots\mu_{M-1}}\, d\tau^{\mu_1\mu_2\cdots\mu_{M-1}} = \int_{\Omega_M} f_{\mu_1\mu_2\cdots\mu_{M-1},\mu_M}\, d\tau^{\mu_M\mu_1\cdots\mu_M}. \tag{1-9.6}$$

The proof is somewhat involved, but follows along the lines of the usual proof of Stokes' theorem in three dimensions. As in the three-dimensional case, Ω_M must be simply connected in order for the theorem to be valid. This means that it must be possible to shrink the boundary Ω_{M-1} to a point without passing outside of Ω_M.

Problem 1.13. Show that $f_{\mu_1\cdots\mu_{M-1},\mu_M}\, d\tau^{\mu_M\mu_1\cdots\mu_M}$ is a scalar.

Of particular importance is the case when $M = 4$. Then Eq. (1-9.6) can be put into a particularly convenient and familiar form. To do so, let us introduce the dual $\mathfrak{F}^\mu$ to $f_{\mu\nu\rho}$ by

$$\mathfrak{F}^\mu = \frac{1}{3!}\,\varepsilon^{\mu\nu\rho\sigma} f_{\nu\rho\sigma}$$

and also the dual elements

$$dS_\mu = \frac{1}{3!}\,\varepsilon_{\mu\nu\rho\sigma}\, d\tau^{\nu\rho\sigma}$$

and

$$dS = \frac{1}{4!}\,\varepsilon_{\mu\nu\rho\sigma}\, d\tau^{\mu\nu\rho\sigma}.$$

Then Eq. (1-9.6) for $M = 4$ can be rewritten in the form

$$\int_{\Omega_3} \mathfrak{F}^\mu\, dS_\mu = \int_{\Omega_4} \mathfrak{F}^\mu{}_{,\mu}\, dS. \tag{1-9.7}$$

If, further, we take the parameters associated with the points in Ω_4 to be the coordinates of these points, then

$$dS = dx^0\, dx^1\, dx^2\, dx^3 = d^4x,$$

and dS_μ has components given by

$$dS_\mu = (dx^1\, dx^2\, dx^3,\, dx^0\, dx^2\, dx^3,\, dx^0\, dx^1\, dx^3,\, dx^0\, dx^1\, dx^2). \quad (1\text{-}9.8)$$

In this form we have the four-dimensional analogue of Gauss' theorem.

1-10. Geometrical Interpretation of Tensors

In this section we give a geometrical interpretation of tensors. The material is not crucial for what follows and may be omitted at a first reading. We begin by defining the tangent plane at a point. Consider the totality of curves passing through a point $x_0{}^\mu$ described parametrically by the equations

$$x^\mu = \phi^\mu(\lambda)$$

with

$$x_0{}^\mu = \phi^\mu(\lambda_0).$$

The tangent to a given curve is a geometrical object called a *contravector* with components $\dot{\phi}^\mu(\lambda) \equiv d\phi^\mu(\lambda)/d\lambda$. The totality of all such vectors form a linear space $S(x_0)$ at $x_0{}^\mu$, the *tangent plane* at this point. We can introduce a basis in this space. It is a collection of four linearly independent vectors $\mathbf{e}_{(\alpha)}$. Any other vector $\mathbf{A}^{\bullet}$ in this space can be represented by a linear combination of the basis vectors as

$$\mathbf{A}^{\bullet} = A^\mu \mathbf{e}_{(\mu)}.$$

The A^μ are the components of $\mathbf{A}^{\bullet}$ with respect to this basis. A particular basis is formed by taking the $\mathbf{e}_{(\mu)}$ to be tangent to the four coordinate lines through $x_0{}^\mu$, that is, the lines

$$x^0 = \phi^0(\lambda), \qquad x^1 = x_0{}^1, \qquad x^2 = x_0{}^2, \qquad x^3 = x_0{}^3,$$

with $\phi^0(\lambda_0) = x^0$, and similarly for the other lines. The base vector $\mathbf{e}_{(0)}$ thus has the contravariant components in the particular coordinate system employed, given by

$$\mathbf{e}_{(0)} = (\dot{\phi}^0(\lambda_0), 0, 0, 0),$$

and similarly for the other tangent vectors. We can always adjust the parameters along these lines so that $\dot{\phi}^0(\lambda_0) = 1$, etc., so that the base vectors have components

$$\mathbf{e}_{(0)} = (1, 0, 0, 0),$$

$$\mathbf{e}_{(1)} = (0, 1, 0, 0),$$

$$\mathbf{e}_{(2)} = (0, 0, 1, 0),$$

$$\mathbf{e}_{(3)} = (0, 0, 0, 1).$$

Then the components of $\mathbf{A}^{\boldsymbol{\cdot}}$ with respect to this basis are just the components of a contravector, and we call any vector in $S(x_0)$ a contravector.

To see that the A^μ are in fact the components of a contravector, consider a mapping of the manifold onto itself. Under the mapping

$$x'^\mu = x'^\mu(x),$$

the basis vectors at $x_0{}^\mu$ get carried over to the point $x'^\mu_0 = x'^\mu(x_0)$. In terms of the basis vectors $\mathbf{e}'_{(\mu)}$ at this associated point, we have

$$e_{(\mu)} = a_\mu{}^\nu e'_{(\nu)} .$$

Now the components of $\mathbf{e}_{(\mu)}$ with respect to the basis $\mathbf{e}'_{(\mu)}$ at x'^μ_0 are[5]

$$\frac{\partial x'^\nu}{\partial x^\rho} e^\rho_{(\mu)} = \frac{\partial x'^\nu}{\partial x^\mu},$$

so that

$$a_\mu{}^\nu = \frac{\partial x'^\nu}{\partial x^\mu}$$

since the $\mathbf{e}'_{(\mu)}$ have the same components at x'^μ_0 as do the $e_{(\mu)}$ at the point $x_0{}^\mu$. Since

$$A = A^\mu e_{(\mu)} = A^\mu a_\mu{}^\nu e'_{(\nu)} = A'^\nu e'_{(\nu)},$$

the components A'^ν at x'^μ_0 referred to the basis at this point are related to the components A^μ by

$$A'^\nu = a_\mu{}^\nu A^\mu = \frac{\partial x'^\nu}{\partial x^\mu} A^\mu$$

which is just the transformation properties of the components of a contravector.

Given the tangent space $S(x_0)$ we can associate with it a dual vector space. Any linear function $\underset{\cdot}{B}(A^{\boldsymbol{\cdot}})$ of the vectors in $S(x_0)$ possesses the linear superposition property characteristic of vectors and therefore defines a new type

[5] The reason that the components of $\mathbf{e}_\mu$ at the point x'^μ_0 are as given is that under the mapping, all curves through the point $x_0{}^\mu$ get mapped onto curves through x'^μ_0. In particular, the line $x^0 = \phi^0(\lambda)$, $x^1 = x_0{}^1$, $x^2 = x_0{}^2$, $x^3 = x_0{}^3$ gets mapped onto the line

$$x'^\mu = x'^\mu(\phi^0(\lambda), x_0{}^1, x_0{}^2, x_0{}^3),$$

so that the tangent to this line at the point x'^μ_0 has the components

$$\dot\phi'^\mu = \frac{\partial x'^\mu}{\partial x^0} \dot\phi^0 = \frac{\partial x'^\mu}{\partial x^0}$$

and similarly for the other coordinate lines.

of vector, which we call a *covector* and which we designate by the symbol $B_{\bullet}$. Since $B(A^{\bullet})$ is a linear function of the vectors of $S(x_0)$, we have that

$$B(\alpha_1 A_1^{\bullet} + \alpha_2 A_2^{\bullet}) = \alpha_1 B(A_1^{\bullet}) + \alpha_2 B(A_2^{\bullet}).$$

It follows that any linear combination $\beta_1 B_1 + \beta_1 B_2$ of the functions $B_1(\mathbf{A}^{\bullet})$ and $B_2(\mathbf{A}^{\bullet})$ also has this property. The totality of all such vectors $\mathbf{B}_{\bullet}$ spans the space $S^*(x_0)$ dual to $S(x_0)$. Any vector in $S^*(x_0)$ is called a *covector*. We can introduce a basis in $S^*(x_0)$ consisting of four linearly independent covectors $\mathbf{e}^{*(\mu)}$. By definition the vector $\mathbf{B}_{\bullet}$ vanishes if the function $B(\mathbf{A}_{\bullet})$ vanishes for any $\mathbf{A}^{\bullet}$, and two covectors are equal if $B_1(\mathbf{A}^{\bullet}) = B_2(\mathbf{A}^{\bullet})$ for any $\mathbf{A}^{\bullet}$.

We now define a scalar product: The value of the function $B(\mathbf{A}^{\bullet})$ for a particular contravector $\mathbf{A}^{\bullet}$ is the scalar product of the vector $\mathbf{B}_{\bullet}$ with the vector $\mathbf{A}^{\bullet}$ and is written

$$(\mathbf{B}_{\bullet}, \mathbf{A}^{\bullet}) \equiv B(A^{\bullet}).$$

Since, for a given $A^{\bullet}$, $B(\mathbf{A}^{\bullet})$ must be of the form

$$B(\mathbf{A}^{\bullet}) = B_\mu A^\mu,$$

the B_μ will constitute the components of B in the space S^* and hence

$$(\mathbf{B}_{\bullet}, \mathbf{A}^{\bullet}) = B_\mu A^\mu.$$

Consider in particular the case where $\mathbf{A}$ is $\mathbf{e}_0$. We choose a basis in S^* so that

$$(\mathbf{e}^*_{(\mu)}, e_{(\nu)}) = \delta_{\mu\nu} = \begin{cases} 1 & \mu = \nu \\ 0 & \mu \neq \nu \end{cases}.$$

Consequently the $\mathbf{e}^*_{(\mu)}$ have components

$$e^*_{(0)} = (1, 0, 0, 0),$$

$$e^*_{(1)} = (0, 1, 0, 0),$$

$$e^*_{(2)} = (0, 0, 1, 0),$$

$$e^*_{(3)} = (0, 0, 0, 1).$$

Under a mapping the components of the $e^{*(\mu)}$ transform like the components of covectors, and hence the components B_μ of any vector in S^* referred in this basis transform like the components of a covector; hence the designation.

We have seen that the special basis chosen in $S(x_0)$ corresponds to choosing as basis vectors the tangents to the coordinate curves through $x_0{}^\mu$. These curves are nothing more than the intersections at the various triplets of coordinate surfaces through $x_0{}^\mu$, that is, the surfaces

$$x^\mu = x_0{}^\mu.$$

Likewise, the special basis chosen for $S^*(x_0)$ can be interpreted as being formed by using the normals to the coordinate surfaces. (Given any surface

$$\phi(x) = 0$$

in the manifold, the normal to $\phi(x)$ is by definition the covector with components $\phi_{,\mu}$.) We justify this designation by noting that any infinitesimal displacement dx^μ lying wholely in the surface must satisfy

$$\phi_{,\mu}\, dx^\mu = 0.$$

Since a displacement $\mathbf{dx}$ is a vector in S with components dx^μ, we see that the scalar product of the normal to a surface with a displacement lying in the surface is zero. With this definition of the vector normal to a surface, we see that the components of the $\mathbf{e}^*_{(\mu)}$ are just the components of the coordinate surface normals.

In order to give a geometric interpretation to tensors, we introduce the notion of the tensor (or Kronecker) product of two vector spaces. Let $\mathfrak{B}_1$ and $\mathfrak{B}_2$ be any two vector spaces. Take a vector $\mathbf{A}_1$ of one and a vector $\mathbf{A}_2$ of the other. From $\mathbf{A}_1$ and $\mathbf{A}_2$ are formed the *tensor product* $\mathbf{A}_1\mathbf{A}_2$ (as distinguished from the scalar product, which might not be defined for the two spaces considered) with the properties that it is commutative,

$$\mathbf{A}_1\mathbf{A}_2 = \mathbf{A}_2\mathbf{A}_1,$$

and distributive with respect to the sum. If, in addition, we require that

$$\mathbf{A}_1 = \beta\mathbf{B}_1 + \gamma\mathbf{C}_1, \qquad \mathbf{B}_1 \text{ and } \mathbf{C}_1 \in \mathfrak{B}_1,$$

then

$$\mathbf{A}_1\mathbf{A}_2 = \beta\mathbf{B}_1\mathbf{A}_2 + \gamma\mathbf{C}_1\mathbf{A}_2,$$

and if

$$\mathbf{A}_2 = \beta\mathbf{B}_2 + \gamma\mathbf{C}_2, \qquad \mathbf{B}_2 \text{ and } \mathbf{C}_2 \in \mathfrak{B}_2,$$

then

$$\mathbf{A}_1\mathbf{A}_2 = \beta\mathbf{A}_1\mathbf{B}_2 + \gamma\mathbf{A}_1\mathbf{C}_2.$$

The totality of all vector products $\mathbf{A}_1\mathbf{A}_2$ formed by taking the tensor product of all vectors in $\mathfrak{B}_1$ and $\mathfrak{B}_2$ span a new vector space, the space $\mathfrak{B}_1 \otimes \mathfrak{B}_2$ which one calls the *tensor product* of the vector spaces $\mathfrak{B}_1$ and $\mathfrak{B}_2$. One can form a basis in $\mathfrak{B}_1 \otimes \mathfrak{B}_2$ by taking the tensor products of a set of basis vectors from $\mathfrak{B}_1$ with a set from $\mathfrak{B}_2$, that is, the totality of vector products $\mathbf{e}_{(\alpha)}\mathbf{e}_{(\beta)}$, where α runs from 1 to N_1, the dimensionality of $\mathfrak{B}_1$, and β runs from 1 to N_2, the dimensionality of $\mathfrak{B}_2$. Thus the dimensionality of $\mathfrak{B}_1 \otimes \mathfrak{B}_2$ is N_1N_2.

If both $\mathfrak{B}_1$ and $\mathfrak{B}_2$ correspond to $S(x_0)$, the tensor product $S(x_0) \otimes S(x_0)$ is then the space of second-rank contravectors at the point $x_0{}^\mu$. The vectors

$\mathbf{e}_{(\mu)}\mathbf{e}_{(\nu)}$ form a basis for this space and an arbitrary vector (contratensor) in this space can be written as

$$\mathbf{A}^{\cdot\cdot} = A^{\mu\nu}\mathbf{e}_{(\mu)}\mathbf{e}_{(\nu)}.$$

The $A^{\mu\nu}$ are the components of $\mathbf{A}^{\cdot\cdot}$ and transform as the components of a second-rank contratensor. Likewise, the vector product $S^*(x_0) \otimes S^*(x_0)$ is the vector space of second-rank cotensors $\mathbf{A}_{\cdot\cdot}$ with a basis formed from the vector products $\mathbf{e}^*{}_{(\mu)}\mathbf{e}^*{}_{(\nu)}$. Thus a vector $\mathbf{A}_{\cdot\cdot}$ in this space can be written as

$$\mathbf{A}_{\cdot\cdot} = A_{\mu\nu}e^*{}_{(\mu)}e^*{}_{(\nu)},$$

where the $A_{\mu\nu}$ are the components of $\mathbf{A}_{\cdot\cdot}$ with respect to this basis and transform as the components of a second-rank cotensor. Finally, one can form the vector product $S^*(x_0) \otimes S(x_0)$, the vector space of second-rank mixed tensors $\mathbf{A}_{\cdot}^{\cdot}$ with a basis formed from the vector products $e^*{}_{(\mu)}e_{(\nu)}$. Thus

$$\mathbf{A}_{\cdot}^{\cdot} = A_{\mu}{}^{\nu}e^*{}_{(\mu)}e_{(\nu)},$$

with the $A_{\mu}{}^{\nu}$ transforming as do the components of a mixed tensor. One can proceed in this way to build up vector spaces corresponding to tensors of arbitrary rank and type. One could also form more general tensors by forming the tensor product of vector spaces at different space-time points; for example, $S(x_0) \otimes S^*(x_1)$. However, such objects will not be needed in our considerations.

It is also possible to give a geometric interpretation to the tensor densities of weight $+1$. The function $\mathfrak{A}(x)$ defined over the manifold will be a scalar density of weight $+1$ if the integral $\int \mathfrak{A}(x)\, d^4x$ taken over an infinitesimally small region is a scalar. Likewise, $\mathfrak{A}$ is a co- or contravector density of weight $+1$ if $\int \mathfrak{A}\, d^4x$ over an infinitesimally small region about the point $x_0{}^{\mu}$ is a co- or contravector at $x_0{}^{\mu}$, that is, lies in $S^*(x_0)$ or $S(x_0)$. The co- and contradensities at $x_0{}^{\mu}$ each span a vector space, $\mathfrak{S}^*(x_0)$ and $\mathfrak{S}(x_0)$. By forming tensor products of these spaces with themselves and each other, one can proceed to build up the vector spaces of tensor densities of arbitrary rank and type of weight $+1$, just as in the case of the tensors of weight zero. There does not appear to be any simple geometrical interpretation of tensors of weight other than 0 and $+1$.

We conclude this discussion by pointing out that in general there is no way of making an association between vectors in $S(x_0)$ and $S^*(x_0)$ until one introduces a metric onto the manifold.

1-11. Internal Transformations

In addition to the transformations of geometrical objects associated with space-time mappings, we sometimes have to deal with other transformation groups. Such groups will be termed *internal groups*; the transformations, *internal transformations*. An example of such a group is the group of gauge transformations of electrodynamics. If A_μ represents the four-potential from which one obtains the electromagnetic field, then in addition to its space-time transformation properties, it can also undergo a gauge transformation in which

$$A_x \to A'_\mu = A_\mu + \lambda_{,\mu}. \tag{1-11.1}$$

It is clear that the gauge transformations form a group.

In many important applications the internal group is a Lie group. Of special interest is the case where this group is SL_2, the group of linear transformations in a complex two-dimensional space of determinant unity. It is a six-parameter group and the covering group of the proper Lorentz group. The importance of this group lies in the fact that it possesses double-valued representations and so allows us to introduce spinors on the space-time manifold. Of course, at this point, there is no relation between the internal group SL_2 and the group of coordinate mappings. In addition to SL_2, one also has occasion to use the groups SU_2 and SU_3, the groups of unitary linear transformations of determinate unity in complex two- and three-dimensional spaces. These latter groups play a role in the classification schemes of elementary particles developed in recent years.

If ψ_α forms the basis of a representation of some Lie group, then under an infinitesimal transformation of the group,

$$\psi_\alpha(x) \to \psi'_\alpha(x) = \psi_\alpha(x) + i\varepsilon^A(x)L^\beta_{A\alpha}\psi_\beta(x) \tag{1-11.2}$$

where the $\varepsilon^A(x)$ are the group parameters at the point x^μ and the $L^\beta_{A\alpha}$ are fixed Hermitian matrices, the *generators* of the group. The discussion of Section 1-8 applies with equal force to the case of the internal transformations; Eq. (1-8.4) is again the condition that the $\psi_\alpha(x)$ constitute the basis of a representation of the internal group. In the present case,

$$\begin{aligned} \bar{\delta}\psi_\alpha(x) &\equiv \psi'_\alpha(x) - \psi_\alpha(x) \\ &= i\varepsilon^A(x)L^\beta_{A\alpha}\psi_\beta(x). \end{aligned} \tag{1-11.3}$$

With $\bar{\delta}\psi_\alpha$ given by Eq. (1-11.3), the group condition (1-8.4) will be satisfied, provided the generators satisfy

$$\begin{aligned} [L_A, L_B]_\alpha{}^\gamma &\equiv L^\beta_{A\alpha}L^\gamma_{B\beta} - L^\beta_{B\alpha}L^\gamma_{A\beta} \\ &= C^C_{AB}L^\gamma_{C\alpha} \end{aligned} \tag{1-11.4}$$

where the C are the structure constants of the group. While the form of the generators will in general vary from representation to representation, the structure constants are the same for all representations.

As in the case of the Lie groups of manifold mappings, the structure constants satisfy the antisymmetry condition

$$C^C_{AB} = -C^C_{BA} \tag{1-11.5}$$

and the Jacobi identity

$$C^C_{AB}C^E_{CD} + C^C_{DA}C^E_{CB} + C^C_{BD}C^E_{CA} = 0. \tag{1-11.6}$$

The infinitesimal parameters $\varepsilon_3{}^A$ associated with the commutator of two infinitesimal internal transformations characterized by parameters $\varepsilon_1{}^A$ and $\varepsilon_2{}^A$ are related to these parameters by an equation analogous to Eq.(1-4.10):

$$\varepsilon_3{}^C = C^C_{AB}(\varepsilon_2{}^A\,\varepsilon_1{}^B - \varepsilon_1{}^A\varepsilon_2{}^B). \tag{1-11.7}$$

Problem 1.14. Consider the case where the internal group in question is N_3, the real three-dimensional rotation group. The change in ψ_i under an infinitesimal element of this group when i is an internal vector index is usually given by

$$\bar{\delta}\psi_i = \varepsilon_{ij}\psi_j, \qquad i, j = 1, 2, 3,$$

with

$$\varepsilon_{ij} = -\varepsilon_{ji}, \qquad |\varepsilon_{ij}| \ll 1.$$

Introduce new parameters ε_A with $\varepsilon_1 = \varepsilon_{23}$, $\varepsilon_2 = \varepsilon_{31}$, and $\varepsilon_3 = \varepsilon_{12}$; construct the matrices L_{Ai} and find the structure constants C^C_{AB}.

Problem 1.15. Show that the two-component objects ψ_α, $\alpha = 1, 2$, constitute the basis of a representation of the rotation group when $\bar{\delta}\psi_\alpha$ has the form

$$\bar{\delta}\psi_\alpha = i\varepsilon^A\tau^\beta_{A\alpha}\psi_\beta, \qquad A = 1, 2, 3,$$

and the generators $\tau^\beta_{A\alpha}$ are the Pauli matrices given by

$$\tau_1 = \begin{pmatrix} 0 & 1 \\ 1 & 0 \end{pmatrix}, \qquad \tau_2 = \begin{pmatrix} 0 & -i \\ i & 0 \end{pmatrix}, \qquad \tau_3 = \begin{pmatrix} 1 & 0 \\ 0 & -1 \end{pmatrix}.$$

Show that the structure constants calculated with these generators, using Eq. (1-11.4), are the same as the structure constants of the preceding problem.

Since the transformation groups we have been discussing here are not generated by coordinate mappings (there is no "transport" term $-\psi_{\alpha,\mu}\xi^\mu$ in the expression for $\bar{\delta}\psi_\alpha$), we must also specify the transformation properties of objects with internal transformation properties under a coordinate mapping, in order to specify completely its transformation properties. In general, then, a geometrical object can have both space-time indices, which indicate its transformation properties under mappings, and "internal" indices, which indicate how it transforms under an internal group.

2

Affine Geometry

We saw that, except for a few special combinations, the ordinary derivatives of tensors and densities did not form the components of new geometrical objects. The reason for this difficulty arises in the first instance from the fact that one cannot add or subtract tensors at different points of the manifold, an operation necessary to the construction of a derivative. This is awkward, since one would like to have some geometrical means for deciding how a tensor field changes as it moves over the manifold. In order to remedy this situation in part, one introduces the notion of the local parallel transport of a tensor, which in term leads to the construction of an affine geometry on the manifold.

2.1. Parallel Transport and the Coderivative

Consider a contravector field $A^\mu(x)$. The components of A^μ at the point $x^\mu + dx^\mu$ are related to the components at the point x^μ by the equation

$$\begin{aligned} A^\mu(x + dx) &= A^\mu(x) + dA^\mu(x) \\ &= A^\mu(x) + \frac{\partial A^\mu}{\partial x^\nu}\, dx^\nu. \end{aligned}$$

The $dA^\mu(x)$, being the difference of two vectors, $A^\mu(x + dx) - A^\mu(x)$, located at different points of the manifold, do not constitute the components of any geometrical object. What is needed for our purposes is another vector at the point $x^\mu + dx^\mu$ which, by definition, we take to be parallel to $A^\mu(x)$. For infinitesimally close points this vector will have components $A^\mu(x) + \delta A^\mu(x)$.

Again, the $\delta A^\mu(x)$, being the difference of two vectors at different points, will not be a vector. However, we shall construct δA^μ in such a manner that the difference

$$A^\mu(x) + dA^\mu(x) - [A^\mu(x) + \delta A^\mu(x)] = dA^\mu(x) - \delta A^\mu(x)$$

is a vector. Figure 2.1 illustrates this procedure.

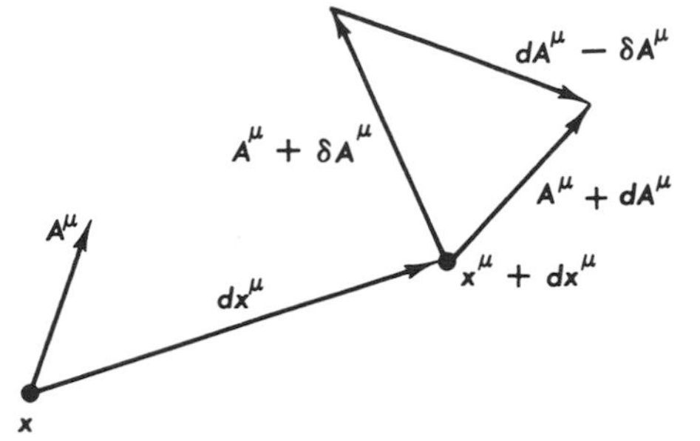

Fig. 2.1.

In constructing $\delta A^\mu(x)$ we will require that it vanish whenever either $A^\mu(x)$ or dx^μ vanishes, so as to conform with our usual notions of parallel vectors in Euclidean geometry. This can be accomplished most easily by demanding that $\delta A^\mu(x)$ be bilinear in $A^\mu(x)$ and dx^μ. Thus we are led to form $\delta A^\mu(x)$ according to

$$\delta A^\mu(x) \equiv -\Gamma^\mu_{\rho\sigma}(x) A^\rho(x)\, dx^\sigma, \tag{2-1.1}$$

the negative sign being introduced here so as to accord with convention.

The quantities $\Gamma^\mu_{\rho\sigma}(x)$ constitute the components of a new geometrical object defined on the manifold, called the *affine connection* or the *affinity* for short. Within certain bounds, the 64 components of $\Gamma^\mu_{\rho\sigma}(x)$ are arbitrary. Once they are given we say that we have constructed an *affine geometry* on the manifold. With their help we can then proceed to construct a new kind of derivative, the *coderivative* $A^\mu{}_{;\nu}$ of a contravector $A^\mu(x)$, according to

$$A^\mu{}_{;\nu} \equiv \frac{A^\mu + dA^\mu - A^\mu - \delta A^\mu}{dx^\nu}$$

$$= A^\mu{}_{,\nu} + \Gamma^\mu_{\rho\nu} A^\rho. \tag{2-1.2}$$

Having defined the components of the affinity on the manifold, we must now decide how these components transform under a mapping. We could demand that under a mapping the components transform like those of a tensor according to the position of the indices. However, if we did so, the $A^\mu{}_{;\nu}$ would not transform like the components of any geometrical object. To determine the transformation properties of the affinity, we shall demand instead that $dA^\mu - \delta A^\mu$ transform like the components of a contravector. We have that

$$dA^\mu = \frac{\partial A^\mu}{\partial x^\nu} dx^\nu$$

$$= \frac{\partial}{\partial x'^\kappa}\left(\frac{\partial x^\mu}{\partial x'^\iota} A'^\iota\right) \frac{\partial x'^\kappa}{\partial x^\nu}\frac{\partial x^\nu}{\partial x'^\lambda} dx'^\lambda$$

$$= \frac{\partial x^\mu}{\partial x'^\iota} A'^\iota{}_{,\kappa}\, dx'^\kappa + \frac{\partial^2 x^\mu}{\partial x'^\iota\, \partial x'^\kappa} A'^\iota\, dx'^\kappa. \tag{2-1.3}$$

In deriving this result we have made use of the fact that

$$\frac{\partial x'^\kappa}{\partial x^\nu}\frac{\partial x^\nu}{\partial x'^\lambda} = \delta_\lambda{}^\kappa$$

and have changed the dummy index (the index to be summed over) from ι to ν. We also have that

$$\delta A^\mu = -\Gamma^\mu_{\rho\sigma} A^\rho\, dx^\sigma$$

$$= -\Gamma^\mu_{\rho\sigma}\frac{\partial x^\rho}{\partial x'^\iota} A'^\iota \frac{\partial x^\sigma}{\partial x'^\kappa} dx'^\kappa. \tag{2-1.4}$$

Combining the results expressed by Eqs. (2-1.3, 2-1.4) gives

$$dA^\mu - \delta A^\mu = \frac{\partial x^\mu}{\partial x'^\nu} dA'^\nu + \left(\frac{\partial^2 x^\mu}{\partial x'^\iota\, \partial x'^\kappa} + \Gamma^\mu_{\rho\sigma}\frac{\partial x^\rho}{\partial x'^\iota}\frac{\partial x^\sigma}{\partial x'^\kappa}\right) A'^\iota\, dx'^\kappa. \tag{2-1.5}$$

We can then rewrite Eq. (2-1.5) in the form

$$dA^\mu - \delta A^\mu = \frac{\partial x^\mu}{\partial x'^\nu}\{dA'^\nu + \Gamma'^\nu_{\iota\kappa} A'^\iota\, dx'^\kappa\} \tag{2-1.6}$$

provided the transformed components $\Gamma'^\nu_{\iota\kappa}$ satisfy

$$\frac{\partial x^\mu}{\partial x'^\nu}\Gamma'^\nu_{\iota\kappa} = \Gamma^\mu_{\rho\sigma}\frac{\partial x^\rho}{\partial x'^\iota}\frac{\partial x^\sigma}{\partial x'^\kappa} + \frac{\partial^2 x^\mu}{\partial x'^\iota\, \partial x'^\kappa}.$$

If we multiply both sides of this equation by $\partial x'^\lambda/\partial x^\mu$ we obtain

$$\Gamma'^\lambda_{\iota\kappa} = \frac{\partial x'^\lambda}{\partial x^\mu}\frac{\partial x^\rho}{\partial x'^\iota}\frac{\partial x^\sigma}{\partial x'^\kappa}\Gamma^\mu_{\rho\sigma} + \frac{\partial x'^\lambda}{\partial x^\mu}\frac{\partial^2 x^\mu}{\partial x'^\iota\, \partial x'^\kappa}. \tag{2-1.7}$$

Returning now to Eq. (2-1.6) we see that if $\Gamma'^\nu_{\iota\kappa}$ are taken to be the components of the affinity after the mapping, we have that

$$dA^\mu - \delta A^\mu = \frac{\partial x^\mu}{\partial x'^\nu}(dA'^\nu - \delta A'^\nu),$$

which is just the transformation law for a contravector. Thus, if the components of the affinity transform as in Eq. (2-1.7), $dA^\mu - \delta A^\mu$ will transform as a contravector, as we originally demanded. We see that, because of the second term on the right-hand side of Eq. (2-1.7), the affinity does not transform as a tensor. Nevertheless, the transformed components depend only upon the components before the mapping and the transformation functions. It is a linear, inhomogeneous object.

Problem 2.1. Show that, with the law of transformation (2-1.7), the $\Gamma^\mu_{\rho\sigma}$ constitute the components of a geometrical object.

Because of their transformation law, the calculus of affinities is different from that of tensors. Thus we cannot add two affinities to get a new affinity, although the sum of an affinity and a tensor with the same index structure is again an affinity. Since we will have little occasion to consider more than one affinity on the manifold at a time, this result is not of great interest to us. One property of an affinity that is of importance is its symmetry, or lack thereof, under an interchange of the two lower indices. (We do not call these "co-indices," since we have reserved that term for indices that have tensor character.) If $\Gamma^\mu_{\rho\sigma}$ is symmetric in ρ and σ, then this property is preserved under a mapping. This is not true if it is antisymmetric in these two indices. However, the antisymmetric part, $\frac{1}{2}(\Gamma^\mu_{\rho\sigma} - \Gamma^\mu_{\sigma\rho})$, transforms by itself as a tensor, as follows immediately from the transformation law given by Eq. (2-1.7).

With the help of the affinity we have defined the coderivative $A^\mu{}_{;\nu}$, of a contravector in Eq. (2-1.2). With the transformation law for the affinity given in Eq. (2-1.7), one can show that the $A^\mu{}_{;\nu}$ form the components of a mixed, second-rank tensor. We would now like to extend our definition of covariant differentiation so as to apply to arbitrary tensors and densities. One could, of course, introduce a second affinity $\Delta^\rho_{\mu\nu}$ for covectors and define the change δB_μ in such an object to be given by

$$\delta B_\mu = \Delta^\rho_{\mu\nu} B_\rho \, dx^\nu.$$

However, it is neither necessary nor convenient to do so. Rather we shall impose additional conditions on the coderivative that will lead to expressions for the coderivative of arbitrary tensors and densities. We shall require first that the coderivative of a scalar is equal to the ordinary derivative, so that

$$A_{;\mu} = A_{,\mu}$$

for any scalar A. In addition we shall require that the product rule of differentiation hold for covariant differentiation. With the help of these additional rules we can then proceed to construct a coderivative for a covector A_μ. To this end we form, with the help of an arbitrary contravector B^μ, the scalar

$A_\mu B^\mu$. Then we have that

$$A_{\mu,\nu}B^\mu + B^\mu{}_{,\nu}A_\mu = A_{\mu;\nu}B^\mu + B^\mu{}_{;\nu}A_\mu$$
$$= A_{\mu;\nu}B^\mu + B^\mu{}_{,\nu}A_\mu + A_\rho \Gamma^\rho_{\mu\nu} B^\mu,$$

so that

$$(A_{\mu;\nu} - A_{\mu,\nu} + \Gamma^\rho_{\mu\nu}A_\rho)B^\mu = 0.$$

Since B^μ is completely arbitrary, we can conclude that our conditions are satisfied if

$$A_{\mu;\nu} = A_{\mu,\nu} - \Gamma^\rho_{\mu\nu}A_\rho, \tag{2-1.8}$$

which gives us a rule for forming the coderivative of a covector. Again one can check that the $A_{\mu;\nu}$ form the components of a cotensor of the second rank.

One can proceed in this way to construct the coderivative of an arbitrary tensor $T^{\mu\cdots}_{\nu\cdots}$ by first multiplying by an appropriate number of arbitrary co- and contravectors so as to form a scalar, and then applxing the product rule of differentiation as we did before. In addition to the ordinary derivative term there is an additional term for each tensor index in the expression for $T^{\mu\cdots}_{\nu\cdots;\rho}$ and thus

$$T^{\mu\cdots}_{\nu\cdots;\rho} = T^{\mu\cdots}_{\nu\cdots,\rho} + \Gamma^\mu_{\sigma\rho}T^{\sigma\cdots}_{\nu\cdots} - \Gamma^\sigma_{\nu\rho}T^{\mu\cdots}_{\sigma\cdots} + \cdots. \tag{2-1.9}$$

The way to remember this procedure is to assume the co-index as "below" and minus, as in $\pm$, and the opposite for the contra-indices. Also, the index referring to the differentiation (ρ in the above) always comes last in affine symbols. This latter rule is not too important, since we will in general work with symmetric affinities.

Problem 2.2. Show that $\delta^\mu_{\nu;\rho} = 0$.

In order to extend the notion of covariant differentiation to densities, we must impose still more conditions on this type of derivative. To this end we require that the coderivative $\mathfrak{A}_{;\mu}$, of a scalar density of weight W, transform like a codensity of rank 1 and weight W. Such a requirement is in keeping with the fact that, in the case of tensors, covariant differentiation in effect added another co-index to the tensor, as far as its transformation properties are concerned. In addition we must keep in mind the requirement that, for a scalar, ordinary and covariant differentiation give the same results. Taking these two requirements into account, one can show that they are satisfied if the coderivative of $\mathfrak{A}$ is given by

$$\mathfrak{A}_{;\mu} = \mathfrak{A}_{,\mu} - W\Gamma^\rho_{\rho\mu}\mathfrak{A} \tag{2-1.10}$$

where W is the weight of $\mathfrak{A}$. Proceeding now as in the case of tensors, we multiply a density, for whose coderivative we wish to determine the formula, by an appropriate number of arbitrary co- and contravectors and a scalar density of appropriate weight so as to form a scalar, and then again apply the product rule for covariant differentiation. By doing so, one finds that, if $\mathfrak{A}^{\mu\cdots}_{\nu\cdots}$ is a tensor density of weight W, then

$$\mathfrak{A}^{\mu\cdots}_{\nu\cdots;\rho} = \mathfrak{A}^{\mu\cdots}_{\nu\cdots,\rho} + \Gamma^{\mu}_{\sigma\rho}\mathfrak{A}^{\sigma\cdots}_{\nu\cdots} - \Gamma^{\sigma}_{\nu\rho}\mathfrak{A}^{\mu\cdots}_{\sigma\cdots} + \cdots - W\Gamma^{\sigma}_{\sigma\rho}\mathfrak{A}^{\mu\cdots}_{\nu\cdots}. \tag{2-1.11}$$

In Section 1-6 we saw that it is possible, in some instances, to construct new tensors from old ones by ordinary differentiation. In all instances these tensors are equal to those formed in the same way, but now using the coderivative instead of the ordinary derivative, provided only that the affinity is symmetric. Thus, for instance,

$$A_{\mu;\nu} - A_{\nu;\mu} = A_{\mu,\nu} - A_{\nu,\mu},$$

and similarly for the rest. These results do not hold, however, if the connection is not symmetric.

While the coderivative shares many of the properties of the ordinary derivative, it differs from it in one fundamental respect. Unlike ordinary differentiation, codifferentiation is not commutative. Thus, for an arbitrary contravector k^{μ}, we have that

$$\begin{aligned} k^{\rho}{}_{;\mu\nu} &\equiv (k^{\rho}{}_{;\mu})_{;\nu} \\ &= k^{\rho}{}_{,\mu\nu} + \Gamma^{\rho}_{\sigma\mu,\nu}k^{\sigma} + \Gamma^{\rho}_{\sigma\mu}k^{\sigma}{}_{,\nu} + \Gamma^{\rho}_{\lambda\nu}k^{\lambda}{}_{,\mu} + \Gamma^{\rho}_{\lambda\nu}\Gamma^{\lambda}_{\sigma\mu}k^{\sigma} - \Gamma^{\lambda}_{\mu\nu}k^{\rho}{}_{;\lambda}. \end{aligned}$$

Consequently one finds, for the coderivative commutator of a contravector k^{μ},

$$k^{\rho}{}_{;\mu\nu} - k^{\rho}{}_{;\nu\mu} = (\Gamma^{\rho}_{\sigma\mu,\nu} - \Gamma^{\rho}_{\sigma\nu,\mu} - \Gamma^{\rho}_{\lambda\mu}\Gamma^{\lambda}_{\sigma\nu} + \Gamma^{\rho}_{\lambda\nu}\Gamma^{\lambda}_{\sigma\mu})k^{\sigma} - (\Gamma^{\lambda}_{\mu\nu} - \Gamma^{\lambda}_{\nu\mu})k^{\rho}{}_{;\lambda}, \tag{2-1.12}$$

which, even in the case of a symmetric affinity, does not vanish in general. The bracketed expression in the first term on the right-hand side of the equation will play a very important role in many of our considerations. We represent it by the symbol $B^{\rho}_{\sigma\mu\nu}$; thus[1]

$$B^{\rho}_{\sigma\mu\nu} \equiv \Gamma^{\rho}_{\sigma\mu,\nu} - \Gamma^{\rho}_{\sigma\nu,\mu} - \Gamma^{\rho}_{\lambda\mu}\Gamma^{\lambda}_{\sigma\nu} + \Gamma_{\lambda\nu}\Gamma^{\lambda}_{\sigma\mu}. \tag{2-1.13}$$

We have said previously that the antisymmetric part of $\Gamma^{\sigma}_{\rho\nu}$ transforms like a tensor. Consequently the second term on the right-hand side of Eq. (2-1.12) transforms like a tensor; the left-hand side is constructed to transform like a tensor. It therefore follows that the first term on the right-hand side also transforms like a tensor. Then, since k^{σ} is an arbitrary contravector, it follows

[1] Note that some authors define $B^{\rho}_{\sigma\mu\nu}$ to be negative of our $B^{\rho}_{\sigma\mu\nu}$.

from our discussion in Section 1-3 that $B^{\rho}_{\sigma\mu\nu}$ is itself a tensor. The tensorial properties of $B^{\rho}_{\sigma\mu\nu}$ can also be checked by a lengthy but direct computation.

The tensor $B^{\rho}_{\sigma\mu\nu}$ is known as the *Riemann tensor*. Its vanishing, in the case of a symmetric affinity, is the necessary and sufficient condition for codifferentiation to be commutative. We have proved this, of course, only in the case of a contravector. However, it can be shown that the coderivative commutator of any tensor or tensor density can be written as a linear combination of $B^{\rho}_{\sigma\mu\nu}$ and its contraction $B^{\rho}_{\sigma\rho\nu}$, for a symmetric affinity. Hence this commutator will vanish in all cases when $B^{\rho}_{\sigma\mu\nu}$ vanishes.

Problem 2.3. Compute $\phi_{;\mu\nu} - \phi_{;\nu\mu}$ for the case that ϕ is a density of weight W and show that this quantity vanishes when $B^{\rho}_{\sigma\mu\nu}$ vanishes.

Problem 2.4. Show that the coderivative commutator of a tensor $T_{\mu\nu}$ is given by

$$T_{\mu\nu;\rho\sigma} - T_{\mu\nu;\sigma\rho} = -B^{\lambda}_{\mu\rho\sigma}T_{\lambda\nu} - B^{\lambda}_{\nu\rho\sigma}T_{\mu\lambda}$$

for a symmetric affinity.

2-2. The Coderivative for Internal Groups

In Section 1-11 we saw that in addition to the transformation properties of geometrical objects under space-time mappings, we also had to consider their transformation properties under the transformations of internal groups. When these internal degrees of freedom exist, we must extend our definition of coderivative to these objects. If ψ_α is such an object, we shall require that the coderivative $\psi_{\alpha;\mu}$ transform like ψ_α under internal transformations and like a tensor (covector if ψ_α is a scalar for space-time mapping) under space-time mappings. However, unlike the extension of the coderivative to arbitrary tensor densities, we must introduce additional elements in constructing $\psi_{\alpha;\mu}$.

For simplicity let us take ψ_α to be a space-time scalar and to transform, under an infinitesimal mapping of the internal group, according to

$$\bar{\delta}\psi_\alpha = i\varepsilon^A L^{\beta}_{A\alpha}\psi_\beta. \tag{2-2.1}$$

Since the group parameters ε^A are in general arbitrary space-time functions, $\psi_{\alpha,\mu}$ will not be a geometrical object. Again the reason is that the sum of two objects with the same index structure is not a geometrical object unless both objects reside at the same space-time point of the manifold. As before, we must transport ψ_α from the point x^μ to the point $x^\mu + dx^\mu$. The change $\delta\psi_\alpha$ in ψ_α under this transport will be assumed to be of the form

$$\delta\psi_\alpha = g\Gamma^{\beta}_{\mu\alpha}\psi_\beta\, dx^\mu, \tag{2-2.2}$$

where g is some constant. In terms of this transport we define the coderivative to be

$$\psi_{\alpha;\mu} \equiv \frac{\psi_\alpha + d\psi_\alpha - \psi_\alpha - \delta\psi_\alpha}{dx^\mu}$$

$$= \psi_{\alpha,\mu} - g\Gamma^\beta_{\mu\alpha}\psi_\beta . \qquad \text{(2-2.3)}$$

In order to determine the transformation of $\Gamma^\beta_{\mu\alpha}$ we require that

$$\bar{\delta}\psi_{\alpha;\mu} = i\varepsilon^A L^\beta_{A\alpha}\psi_{\beta;\mu}$$

for infinitesimal transformations of the internal group. At the same time it follows from Eq. (2-2.3) that

$$\bar{\delta}\psi_{\alpha;\mu} = (\bar{\delta}\psi_\alpha)_{,\mu} - g\bar{\delta}\Gamma^\beta_{\mu\alpha}\psi_\beta - g\Gamma^\beta_{\mu\alpha}\bar{\delta}\psi_\beta .$$

By comparing these two expressions for $\bar{\delta}\psi_{\alpha;\mu}$ one finds that

$$\bar{\delta}\Gamma^\beta_{\mu\nu} = \frac{1}{g}\,\varepsilon^A{}_{,\mu}L^\beta_{A\alpha} + i\varepsilon^A[L_A , \Gamma_\mu]^\beta_\alpha , \qquad \text{(2-2.4)}$$

where $[L_A, \Gamma_\mu]^\beta_\alpha$ is defined by

$$[L_A , \Gamma_\mu]^\beta_\alpha \equiv L^\gamma_{A\alpha}\Gamma^\beta_{\mu\gamma} - \Gamma^\gamma_{\mu\alpha}L^\beta_{A\gamma} .$$

A particular class of internal affinities is obtained when one takes

$$\Gamma^\beta_{\mu\alpha} = iL^\beta_{A\alpha}B^A_\mu . \qquad \text{(2-2.5)}$$

By making use of the commutation relations between the $L^\beta_{A\alpha}$ given by

$$[L_A , L_B]^\beta_\alpha = C^C_{AB}L^\beta_{C\alpha}$$

and Eq. (2-2.4), one finds that

$$\bar{\delta}B_\mu{}^A = \frac{1}{g}\,\varepsilon^A{}_{,\mu} + iC^A_{BC}\varepsilon^B B_\mu{}^C . \qquad \text{(2-2.6)}$$

Since the structure constants satisfy the Jacobi identity (1-11.6), their negatives satisfy the condition (1-11.4) that they be generators of the group; that is,

$$[L_A , L_B]^D_C = C^E_{AB}L^D_{EC} ,$$

where $L^C_{AB} = -C^C_{AB}$. As a consequence we see that $B_\mu{}^A$ transforms like an internal vector with the additional term $(1/g)\varepsilon^A{}_{,\mu}$ in its transformation law. It is also clear that in order for $\psi_{\alpha;\mu}$ to transform like a covector, it is necessary that $B_\mu{}^A$ transform like a covector under space-time mappings.

The extension of the coderivative to objects with both space-time and internal indices is a straightforward generalization of the methods used to form the coderivative in the special cases discussed above. Thus, if μ is a co-index and α an internal index, the coderivative of $\mathfrak{Y}_{\mu\alpha}$ would be given by

$$\mathfrak{Y}_{\mu\alpha;\nu} = \mathfrak{Y}_{\mu\alpha,\nu} - \Gamma^{\rho}_{\mu\nu}\mathfrak{Y}_{\rho\alpha} - igL^{\beta}_{A\alpha}B_{\nu}{}^{A}\mathfrak{Y}_{\mu\beta}\,.$$

If the internal group is SU_2, $B_\mu{}^A$ is just the B-field of Yang and Mills.[2] If the internal group is a one-parameter group, and therefore an Abelian Lie group, the structure constants C^A_{BC} are zero, and we have in this case

$$\bar{\delta}B_\mu = \frac{1}{g}\,\varepsilon_{,\mu}\,,$$

which is just the transformation law for the vector potential of electrodynamics under a gauge transformation. This quantity can therefore be viewed as an internal affinity for the group of rotations in a two-dimensional internal vector space. By taking the two components of a vector in this space to be the real and imaginary parts of a complex scalar, the rotations can be made to correspond to multiplying this scalar by an arbitrary phase factor $e^{i\varepsilon}$. But this is the usual transformation law for complex fields, scalar or otherwise, corresponding to a gauge transformation of the vector potential.

As in the case of the coderivative of a space-time object, the coderivative of an object with internal degrees of freedom is not commutative. A straightforward computation yields the result that

$$\psi_{\alpha;\mu\nu} - \psi_{\alpha;\nu\mu} = -igL^{\beta}_{A\alpha}G^{A}_{\mu\nu}\psi_\beta \tag{2-2.7}$$

with

$$G^A_{\mu\nu} = B^A_{\mu,\nu} - B^A_{\nu,\mu} + igC^A_{BC}B_\mu{}^B B_\nu{}^C. \tag{2-2.8}$$

From the way in which it is constructed we see that $G^A_{\mu\nu}$ transforms as an antisymmetric cotensor under space-time mappings, and under an infinitesimal transformation of the internal group the change in $G^A_{\mu\nu}$ is given by

$$\bar{\delta}G^A_{\mu\nu} = -i\varepsilon^B L^A_{BC}G^C_{\mu\nu}\,, \tag{2-2.9}$$

where $L^A_{BC} = -C^A_{BC}$. $G^A_{\mu\nu}$ is the analogue of the Riemann tensor; its vanishing is the necessary and sufficient condition that internal codifferentiation is commutative. For a one-parameter Lie group, $G^A_{\mu\nu}$ reduces to the form

$$G_{\mu\nu} = B_{\mu,\nu} - B_{\nu,\mu}\,,$$

which is just the usual relation, except for sign, between the vector potential A_μ and the electromagnetic field $F_{\mu\nu}$. Thus the vector potential is the affinity of the gauge group and the electromagnetic field is the corresponding Riemann tensor.

[2] R. L. Mills and C. N. Yang, *Phys. Rev.*, **96**, 191 (1954).

2-3. Affine Geodesics

Once we are given an affinity over a manifold, and hence an affine geometry on this manifold, we can construct an important family of curves, the *affine geodesics*. Such curves are defined by the following condition: Given a point P lying on the curve, take any vector proportional to the tangent to the curve at P and transport this vector parallel to itself along the curve to another point P' also lying on the curve. Then, if the parallel-transported vector is proportional to the tangent to the curve at P' for every point P' and every starting point P, the curve is an affine geodesic of the affinity. One can therefore describe the affine geodesic between two points as being the "straightest" line between these two points.

In order to translate this definition into algebraic terms and at the same time derive a set of equations the affine geodesics must satisfy, let the coordinates of the points lying on the curve be given parametrically as

$$x^\mu = \xi^\mu(\tau),$$

where τ is any continuous parameter defined along the curve, the only restriction on τ being that it must increase monotonically as one proceeds along the curve in a fixed direction. The tangent to the curve at a point with parameter value τ is then given as $d\xi^\mu/d\tau$. Since the differentials $d\xi^\mu$ transform like the components of a contravector and the parameter is assumed to be a scalar, the tangent is thus a contravector.

The contravector that results from the parallel transport of $d\xi^\mu/d\tau$ from the point ξ^μ to the point $\xi^\mu + d\xi^\mu$ is given by

$$\frac{d\xi^\mu}{d\tau} - \Gamma^\mu_{\rho\sigma}\frac{d\xi^\rho}{d\tau}\,d\xi^\sigma = \frac{d\xi^\mu}{d\tau} - \Gamma^\mu_{\rho\sigma}\frac{d\xi^\rho}{d\tau}\frac{d\xi^\sigma}{d\tau}\,d\tau,$$

while the tangent to the curve at the point $\xi^\mu + d\xi^\mu$ has the components

$$\frac{d\xi^\mu}{d\tau} + \frac{d^2\xi^\mu}{d\tau^2}\,d\tau.$$

From the definition of the geodesic it follows that these vectors must be proportional to each other. Therefore

$$\frac{d\xi^\mu}{d\tau} - \Gamma^\mu_{\rho\sigma}\frac{d\xi^\rho}{d\tau}\frac{d\xi^\sigma}{d\tau}\,d\tau = K(\tau, d\tau)\left(\frac{d\xi^\mu}{d\tau} + \frac{d^2\xi^\mu}{d\tau^2}\,d\tau\right),$$

where $K(\tau, d\tau)$ is the proportionality constant. When $d\tau = 0$ we must have that $K(\tau, 0) = 1$. Hence $K(\tau, d\tau)$ must be of the form

$$K(\tau, d\tau) = 1 + \alpha(\tau)\,d\tau.$$

Consequently we can conclude that the $\xi^\mu(\tau)$ must satisfy

$$\frac{d^2\xi^\mu}{d\tau^2} + \Gamma^\mu_{\rho\sigma}\frac{d\xi^\rho}{d\tau}\frac{d\xi^\sigma}{d\tau} = \alpha(\tau)\frac{d\xi^\mu}{d\tau}. \tag{2-3.1}$$

To a large extent the parameterization of the curve is arbitrary. Given one parameterization in terms of τ, we can introduce a new parameter $s(\tau)$ without affecting the geodesic nature of the curve. If we introduce this new parameter, s, Eq. (2-3.1) becomes

$$\frac{d^2\xi^\mu}{ds^2} + \Gamma^\mu_{\rho\sigma}\frac{d\xi^\rho}{ds}\frac{d\xi^\sigma}{ds} = \frac{\alpha s' - s''}{s'^2}\frac{d\xi^\mu}{ds}; \tag{2-3.2}$$

here s' and s'' are the first and second derivatives of s with respect to τ. We see that it is always possible to find a parameter $s(\tau)$ such that the right-hand side of Eq. (2-3.2) vanishes; it is one for which $\alpha s' - s'' = 0$. It is well known that this equation always possesses a solution for arbitrary $\alpha(\tau)$. Thus there exists a parameter s, for which the equation of a geodesic reads

$$\frac{d^2\xi^\mu}{ds^2} + \Gamma^\mu_{\rho\sigma}\frac{d\xi}{ds}\frac{d\xi^\sigma}{ds} = 0. \tag{2-3.3}$$

We see that, with the above choice of the parameter s along the geodesic, the tangent is parallel-transported into itself along the curve and is not merely proportionately transferred, as in the case of an arbitrary parameter. As a consequence the parameter s plays a preferred role among all possible parameters and is referred to as an *affine parameter*. (Actually a small set of parameters has this property; if s does, then so does $as + b$, where a and b are arbitrary constants.) We can therefore use the affine parameter s (or $as + b$) to define the length of a segment of the geodesic between the points P and P' as $\int_P^{P'} ds$. We thus have a means of comparing lengths along a particular geodesic. We cannot, however, compare lengths of segments belonging to different geodesics because of the arbitrariness in the parameter mentioned above. In order to perform such a comparison, it is necessary to introduce a metrical relation onto the manifold.

From the above definition we see that through a given point, there passes only one geodesic in a given direction. In fact we can construct this geodesic by taking an infinitesimal vector dx^μ pointing in this direction and parallel-transport it from the original point x^μ to the point $x^\mu + dx^\mu$, giving a new vector $d'x^\mu$. Then parallel-transport this vector from the point $x^\mu + dx^\mu$ to the point $x^\mu + dx^\mu + d'x^\mu$, and so on until the entire curve is laid out. Thus, corresponding to every direction at a point, there passes a geodesic. Furthermore, for points lying in the neighborhood of this point, there passes one and only one geodesic through this point and one of these neighboring points.

However, this statement no longer holds in the large where a number of geodesics might pass through two given points, as in the case of a sphere when the geodesics are taken to be great circles. Through two antipodal points one can draw an infinity of geodesics on such a surface.

Given an affine connection defined on the manifold, we have seen that one can determine uniquely the set of geodesics associated with this affinity. The converse, however, is not true. Since only the symmetric part of the affinity enters into the equations for a geodesic, two affinities with the same symmetric part, but different antisymmetric parts, will lead to the same system of geodesics. However, if we limit ourselves to symmetric affinities and further require that the path parameters along the various geodesics be given, then the affinity is uniquely determined.

2-4. Distant Parallelism and Affine Flatness

Two affinities $\Gamma'^{\mu}_{\rho\sigma}$ and $\Gamma^{\mu}_{\rho\sigma}$, related to each other through an equation of the form (2-1.7), describe the same affine geometry set down on the manifold in two different ways. Because of the inhomogeneous term in the transformation law for $\Gamma^{\mu}_{\rho\sigma}$, it is therefore possible to place a given affinity on the manifold in such a way that all 64 of its components can be made to vanish at a given point of the manifold. To see this, let x'^{μ}_0 be the point in question. In order to arrange matters so that, after a mapping, the affinity at this point vanishes, we must find another point $x_0{}^{\mu}$ with affinity $\Gamma^{\mu}_{0\rho\sigma}$ such that, under a mapping that associated $x_0{}^{\mu}$ with x'^{μ}_0, $\Gamma'^{\mu}_{0\rho\sigma}$ given by Eq. (2-1.7) vanishes. To show that this is always possible for a symmetric affinity, let us expand the inverse mapping functions $x^{\mu}(x')$ in a Taylor series about the point x'^{μ}_0 according to

$$x^{\mu}(x') - x^{\mu}(x'_0) = (x'^{\mu} - x'^{\mu}_0) + \tfrac{1}{2}a^{\mu}_{\rho\sigma}(x'^{\rho} - x'^{\rho}_0)(x'^{\sigma} - x'^{\sigma}_0) + \cdots.$$

Then, at the point x'^{μ}_0, we have that

$$\Gamma'^{\mu}_{0\rho\sigma} = \Gamma^{\mu}_{0\rho\sigma} + a^{\mu}_{\rho\sigma}.$$

We can therefore accomplish our objective of having $\Gamma'^{\mu}_{0\rho\sigma} = 0$ by choosing the $a^{\mu}_{\rho\sigma}$ so that

$$a^{\mu}_{\rho\sigma} = -\Gamma^{\mu}_{0\rho\sigma}.$$

Not only is it possible to arrange the affinity on the manifold so that it vanishes at a single point, but one can arrange it so that it vanishes along any one geodesic curve. But this is as far as one can go in general. Except for very special affinities it is not possible to arrange the affinity on the manifold so that it vanishes at every point. When it is possible to do so, we are dealing with a very special type of affine geometry, a *flat affine geometry*. Since this particular type of affine geometry will be important for some of our later considerations, we develop here some of its properties and a criterion for recognizing such a geometry.

Given an affinity on the manifold, one way of deciding whether or not it is the affinity of a flat geometry would be to try to find a mapping that would transform the affinity to zero everywhere. This, however, is hardly a practical way of proceeding. Actually it is possible to develop a very simple criterion for flatness. It is based on the notion of distant parallelism of vectors, to which we now turn our attention.

A given affinity allows us to construct a vector parallel to a given vector at a point infinitesimally separated from the point where the given vector is defined. An affine geometry therefore supplies us with the notion of *local parallelism*. However, we cannot in general decide whether two vectors defined at finitely separated points are or are not parallel. To decide such a question one would have to parallel-transport one of the vectors along some curve connecting the two points from its point of definition to the point of definition of the other vector. Unfortunately the resultant transported vector would in general depend upon the path chosen in going from the one point to the other. In other words, the affinity would not be integrable. If, however, the resultant vector were independent of the path of transport, we would have an integrable affinity and we could speak of the notion of *distant parallelism*.

Suppose that we do have an integrable affinity. Then, given a vector h^μ, defined at some point P of the manifold, we can construct a unique vector field $h^\mu(x)$ by transporting h^μ parallel to itself to each point of the manifold. (If the manifold is multiply-connected, we may not be able to define such a field uniquely everywhere, but only over some finite subregion of the manifold.) The change dh^μ in this field in going from the point x^μ to the point $x^\mu + dx^\mu$ is, by construction, equal to δh^μ, so that we can write

$$\frac{\partial h^\mu}{\partial x^\nu}\, dx^\nu = -\Gamma^\mu_{\rho\nu} h^\rho \, dx^\nu.$$

Since dx^μ is quite arbitrary, we can conclude that

$$h^\mu{}_{;\nu} = h^\mu{}_{,\nu} + \Gamma^\mu_{\rho\nu} h^\rho = 0; \tag{2-4.1}$$

that is, h^μ is covariantly constant throughout the region of the manifold over which it is defined. These, then, are the equations for a parallel vector field. Alternatively, the conditions that they possess solutions are the necessary and sufficient conditions that the affinity be integrable. A necessary condition for the existence of a solution is that $h^\rho{}_{,\mu\nu} = h^\rho{}_{,\nu\mu}$. Thus we must have

$$-(\Gamma^\rho_{\sigma\mu} h^\sigma)_{,\nu} + (\Gamma^\rho_{\sigma\nu} h^\sigma)_{,\mu} = (-\Gamma^\rho_{\sigma\mu,\nu} + \Gamma^\rho_{\sigma\nu,\mu}) h^\sigma - \Gamma^\rho_{\sigma\mu} h^\sigma{}_{,\nu} + \Gamma^\rho_{\sigma\nu} h^\sigma{}_{,\mu} = 0.$$

By making use of Eq. (2-4.1) we can rewrite these conditions in the form

$$(\Gamma^{\rho}_{\sigma\mu,\nu} - \Gamma^{\rho}_{\sigma\nu,\mu} - \Gamma^{\rho}_{\lambda\mu}\Gamma^{\rho}_{\sigma\nu} + \Gamma^{\rho}_{\lambda\nu}\Gamma^{\lambda}_{\sigma\mu})h^{\sigma} \equiv B^{\rho}_{\sigma\mu\nu}h^{\sigma} = 0.$$

Since the initial values one assigns to h^{μ} at the point P are quite arbitrary, the above conditions must hold for arbitrary values of h^{μ} at every point of the manifold. Consequently we can conclude that a necessary condition for the integrability of an affinity is that

$$B^{\rho}_{\sigma\mu\nu} = 0 \tag{2-4.2}$$

at every point of the manifold.

The vanishing of the Riemann tensor is also a sufficient condition for the integrability of an affinity, although to show this is a little harder. The proof follows along the same lines as that for showing that the vanishing of the curl of a vector is a sufficient condition for the vector to be the gradient of a scalar. We show first that the vanishing of the Riemann tensor is a sufficient condition that the affinity is integrable around a small closed loop. Once having shown this, it then follows that the integration of the affinity along two neighboring curves connecting the points P and P' yields the same results, since these two curves can be formed from the addition of a number of infinitesimal loops. Finally we can conclude that integration along any two curves connecting points P and P' that can be continuously deformed into one another will yield the same result.

To prove that the affinity is integrable around a small closed loop when $B^{\rho}_{\sigma\mu\nu} = 0$, consider the small polygon with vertices at the points x^{μ}, $x^{\mu} + dx^{\mu}$, $x^{\mu} + \delta x^{\mu}$, $x^{\mu} + dx^{\mu} + \delta x^{\mu}$, as in Fig. 2.2. Parallel transport of an arbitrary

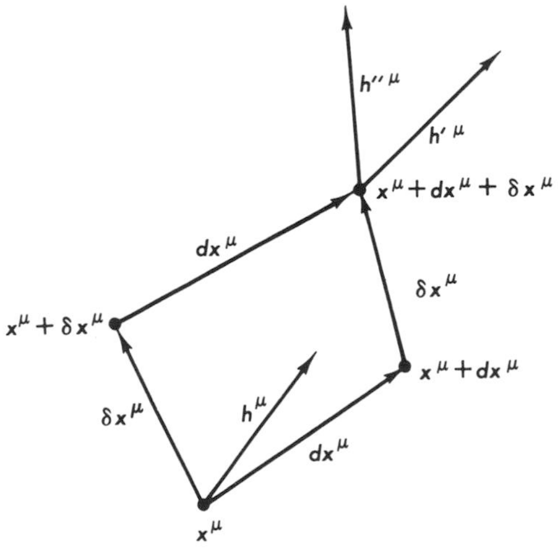

Fig. 2.2.

vector h^μ from x^μ to $x^\mu + dx^\mu$ results in a vector with components $h^\rho - \Gamma^\rho_{\sigma\mu} h^\sigma\, dx^\mu$. In transporting this vector to the point $x^\mu + dx^\mu + \delta x^\mu$, we must use the value of the affinity at $x^\mu + dx^\mu$, namely $\Gamma^\rho_{\sigma\mu} + \Gamma^\rho_{\sigma\mu,\nu}\, dx^\nu$. In this way we can construct a vector h'^ρ at the point $x^\mu + dx^\mu + \delta x^\mu$ with components given by

$$h'^\rho = h^\rho - \Gamma^\rho_{\sigma\mu} h^\sigma\, dx^\mu - (\Gamma^\rho_{\sigma\mu} + \Gamma^\rho_{\sigma\mu,\tau}\, dx^\tau)(h^\sigma - \Gamma^\sigma_{\lambda\nu} h^\lambda\, dx^\nu)\, \delta x^\mu.$$

We could continue to parallel-transport this vector back to x^ν via the point $x^\mu + \delta x^\mu$ and compare its components with the original components at this point. However, we can test equally well the integrability of the affinity by transporting h^μ from x^μ to $x^\mu + dx^\mu + \delta x^\mu$ via the point $x^\mu + \delta x^\mu$. When we do this we obtain a vector h''^μ with components similar to those of h'^μ except that the roles of dx^μ and δx^μ are reversed. If these two vectors are equal to one another, then the affinity is integrable; otherwise, it is not. We have

$$h''^\rho - h'^\rho - B^\rho_{\sigma\mu\nu} h^\sigma\, \delta x^\mu\, dx'^\nu = 0.$$

Therefore we see that the vanishing of the Riemann tensor is also a sufficient condition for the integrability of the affinity. The important thing about this criterion is that it is independent of how we place the affinity on the manifold. Since $B^\rho_{\sigma\mu\nu}$ is a tensor, if it vanishes for one setting of the affinity, it will vanish for all affinities that are related to this one by the transformation law (2-1.7). Thus integrability is an invariant property of an affinity.

Let us now return to our original question of when it is possible to find a mapping such that the transformed components of a given affinity all vanish. Assume that this affinity is integrable, so that Eq. (2-4.1) possesses a solution for every vector given at P. Let us construct at P four linearly independent vectors $h_\alpha{}^\mu$. Here the subscript α is not a co-index but merely a label to distinguish the four different vectors. Starting with these four vectors at P, we proceed to construct four vector fields, $h_\alpha{}^\mu(x)$, of parallel vectors. Since linear independence is a linear relation between the vectors at P, it will be preserved by parallel transport. We thus have 64 independent equations of the form

$$h^\mu_{\alpha,\sigma} = -\Gamma^\mu_{\rho\sigma} h_\alpha{}^\rho, \tag{2-4.3}$$

which we can solve for the 64 components of $\Gamma^\mu_{\rho\sigma}$ at each point of the manifold. Since the $h_\alpha{}^\mu$ are linearly independent, it follows from the theory of linear equations that the determinant formed from them is nonzero. Hence we can find an inverse to the array $h_\alpha{}^\mu$, $\bar{h}_\mu{}^\alpha$, which satisfies

$$\bar{h}_\mu{}^\alpha h_\alpha{}^\nu = \delta_\mu{}^\nu. \tag{2.4-4}$$

(We use the summation convention here as well to indicate a summation over the index α.) Now multiply Eq. (2-4.3) by $\bar{h}_\nu{}^\alpha$ and sum over α. Making use of Eq. (2-4.4), we obtain thereby that

$$\Gamma^{\mu}_{\nu\sigma} = -\bar{h}_{\nu}{}^{\alpha} h^{\mu}_{\alpha,\sigma} \,. \tag{2-4.5}$$

Thus we see that an integrable affinity can be written in terms of the 16 quantities $h_{\alpha}{}^{\mu}$ and their inverse at each point of the manifold. Since a general affinity is specified by 64 numbers at each point of the manifold, we see again that an integrable affinity is a very special type of affinity.

If the affinity is symmetric we can effect a further reduction in the number of independent fields needed for its specification. In this case we have

$$\Gamma^{\mu}_{\sigma\nu} = -\bar{h}_{\sigma}{}^{\alpha} h^{\mu}_{\alpha,\nu} = \Gamma^{\mu}_{\nu\sigma} \,.$$

Then, since

$$\bar{h}^{\alpha}_{\sigma,\nu} h_{\alpha}{}^{\mu} = -\bar{h}_{\sigma}{}^{\alpha} h^{\mu}_{\alpha,\nu}$$

as follows from Eq. (2-4.4), we can conclude that

$$h_{\alpha}{}^{\mu}(h^{\alpha}_{\sigma,\nu} - \bar{h}^{\alpha}_{\nu,\sigma}) = 0.$$

Since the determinant of $h_{\alpha}{}^{\mu}$ is nonzero, it then follows that

$$\bar{h}^{\alpha}_{\sigma,\nu} = \bar{h}^{\alpha}_{\nu,\sigma} \,.$$

But this is just the necessary and sufficient condition that, within the same region where the affinity is integrable, the $\bar{h}_{\sigma}{}^{\alpha}$ can be expressed as the gradient of four scalars f^{α}. Thus

$$\bar{h}^{\alpha}_{\sigma}, = f^{\alpha}{}_{,\sigma} \,.$$

Let us now use these four scalars to construct a mapping of the manifold onto itself so that the coordinates of point x'^{μ} to be associated with the point x^{μ} are given by

$$x'^{\mu} = f^{\mu}(x).$$

Then

$$\frac{\partial x'^{\mu}}{\partial x^{\nu}} = f^{\mu}{}_{,\nu} = \bar{h}_{\nu}{}^{\mu},$$

and consequently

$$\frac{\partial x^{\mu}}{\partial x'^{\nu}} = h_{\nu}{}^{\mu}.$$

If we substitute these expressions together with the expression for $\Gamma^{\mu}_{\rho\sigma}$ into Eq. (2-1.7) we find that, under this particular mapping, $\Gamma'^{\mu}_{\rho\sigma} = 0$.

Thus we see that we can always find a mapping that will result in the transformed components of a symmetric integrable affinity being all zero. The corresponding affine geometry is thus a flat affine geometry, and vice versa, and the condition for all this to be true is that the Riemann tensor constructed from the affinity in question should vanish. Furthermore we see that in a flat affine geometry, codifferentiation is commutative and in general independent of the tensor or tensor density being differentiated, as asserted before. Finally, we conclude that every flat affinity can be obtained from the affinity with all its components zero by some mapping of the manifold onto itself.

Results analogous to those we have stated above for the space-time affinity also hold for the internal affinity. In particular, one can show that the vanishing of the internal Riemann tensor $G^A_{\mu\nu}$, given by Eq. (2-2.8), is the necessary and sufficient condition for the integrability of the internal affinity. Likewise its vanishing is the necessary and sufficient condition that one can always find a set of transformation functions ε^A such that $B_\mu{}^A = 0$.

3

Metrical Geometry

The ideas of affine geometry developed in the previous few sections are sufficient for a geometrical description of Newtonian mechanics. However, in addition to the notion of parallelism there are other equally important geometrical concepts such as distance and angle, which have no meaning within an affine geometry but which are necessary for a discussion of the two relativity theories of Einstein. They find their proper expression in a metrical geometry.

3-1. The Metric Tensor

One can give a metrical structure to a manifold by assigning to each pair of neighboring points x^μ and $x^\mu + dx^\mu$ a distance ds. In general this distance will depend upon both x^μ and dx^μ. We should expect that when $dx^\mu = 0$, the distance should also vanish. Since this is a local distance we are defining, just as we were able to define local parallelism only in general, this suggests that ds should be homogeneous of degree 1 in dx^μ. While the requirement of homogeneity limits the possibilities for ds considerably, there is still much arbitrariness. The simplest choice, and the one made by Riemann, is that ds^2 is given by the expression

$$ds^2 = g_{\mu\nu}(x)\, dx^\mu\, dx^\nu \tag{3-1.1}$$

in analogy to the Pythagorean expression for distance in Euclidean geometry. In the literature, this expression for ds is often referred to as the *line element*.

A metrical geometry with ds^2 given by Eq. (3-1.1) is called a *Riemannian geometry*. In such a geometry it is possible to define the *norm* of a contravector A^μ as

$$A^2 = g_{\mu\nu} A^\mu A^\nu, \tag{3-1.2}$$

where all quantities are taken at the same point of the manifold. If $A^2 > 0\,(<0)$ for arbitrary nonzero contravectors, the metric is said to be *positive* (*negative*) *definite*. Otherwise the metric is said to be indefinite. For the most part, the metrics used in our physical models will be indefinite. Some care must be employed when dealing with such metrics, since a number of familiar properties of definite metrics do not hold in the indefinite case. One can also define the angle between two contravectors A^μ and B^μ as

$$\cos(A, B) = \frac{g_{\mu\nu} A^\mu B^\nu}{\sqrt{g_{\alpha\beta} A^\alpha A^\beta}\sqrt{g_{\gamma\delta} B^\gamma B^\delta}}. \tag{3-1.3}$$

With this definition of angle it is possible to say when two vectors are perpendicular to each other in some cases. If A^2 and B^2 are nonzero, then A^μ is said to be perpendicular to B^μ if $\cos(A, B) = 0$.

Problem 3.1. Show that, in the case of a positive definite metric $\cos(A, B) < 1$. If we are dealing with an indefinite metric this angle may turn out to be imaginary in some cases. Nevertheless we continue to speak of two vectors A^μ and B^μ as being perpendicular whenever $g_{\mu\nu} A^\mu B^\nu = 0$.

In order to decide what kind of geometrical object $g_{\mu\nu}(x)$ is, we must impose additional requirements on ds. The most natural requirement is that the distance between two neighboring points should remain unchanged under a mapping. This will be the case if ds^2 is a scalar, and hence if $g_{\mu\nu}$ is a cotensor of the second rank. Furthermore, we assume that it is symmetric. Although Einstein considered asymmetrical metrics in his later theories, a full discussion of such a geometry would lead us too far afield at the present time. Furthermore, the addition of an antisymmetric part to the metric would not change ds^2.

Given $g_{\mu\nu}$ we can construct several other useful quantities. First, we can construct the determinant g of $g_{\mu\nu}$:

$$g \equiv \det g_{\alpha\beta}.$$

With its help we can then form the inverse to $g_{\mu\nu}$ according to

$$g^{\mu\nu} = \frac{1}{g}\frac{\partial g}{\partial g_{\mu\nu}}. \tag{3-1.4}$$

Since

$$g_{\mu\rho} g^{\nu\rho} = \delta_\mu{}^\nu, \tag{3-1.5}$$

it follows that $g^{\mu\nu}$ is a contratensor of rank 2 and is also symmetric if $g_{\mu\nu}$ is symmetric. We see also that g is a scalar density of weight $+2$.

Problem 3.2. Given that $g_{\mu\nu} = 0$ for $\mu \neq \nu$, express $g^{\mu\nu}$ directly in terms of $g_{\mu\nu}$.

Given the metric tensor $g_{\mu\nu}$ and its inverse, we can introduce the operations of *raising* and *lowering* indices. For a tensor $T^{\cdots\mu\cdots}_{\cdots\ \cdots}$ we can construct another tensor $T^{\cdots\ \cdots}_{\cdots\mu\cdots}$ according to

$$T^{\cdots\ \cdots}_{\cdots\mu\cdots} = g_{\mu\nu} T^{\cdots\nu\cdots}_{\cdots\ \cdots},$$

while from $U^{\cdots\ \cdots}_{\cdots\mu\cdots}$ one can construct $U^{\cdots\mu\cdots}_{\cdots\ \cdots}$ according to

$$U^{\cdots\mu\cdots}_{\cdots\ \cdots} = g^{\mu\nu} U^{\cdots\ \cdots}_{\cdots\nu\cdots}.$$

Similarly we can raise and lower indices on densities. In a sense, two tensors that differ from each other only through the raising and lowering of all or some of their indices can be considered as the same tensor. Once one introduces the operations of raising and lowering indices, it is important to keep track of their relative positions. Thus $T_\nu{}^\mu = g_{\nu\sigma} T^{\mu\sigma}$ and $T_\nu{}^\mu = g_{\nu\sigma} T^{\sigma\mu}$ are not equal unless $T^{\mu\nu}$ is symmetric.

3-2. Metric Geodesics and the Metric Affinity

The definition of a geodesic associated with a given affinity was a generalization of one of the properties of a straight line in Euclidean geometry. We now propose to define a *metric geodesic* by generalizing another property of a straight line, namely: It is the shortest distance between two points. We accordingly define a metric geodesic between two points P and P' as that curve joining the two points whose distance is stationary under small variations that vanish at the end points. We do not demand that the length of the curve be a minimum, since for a very important class of metrics this length is a maximum. We now proceed to derive a set of equations that will determine the metric geodesics of the space.

Let $\xi^\mu(\lambda)$ be the coordinates of points lying on a curve joining P and P', given as functions of some parameter λ along the curve. The length of the curve is then given by

$$s_{PP'} = \int_P^{P'} ds = \int_P^{P'} \frac{ds}{d\lambda}\, d\lambda,$$

where

$$\frac{ds}{d\lambda} = \sqrt{g_{\mu\nu}\dot{\xi}^\mu\dot{\xi}^\nu}$$

and where we use the abbreviation $\dot{\xi}^\mu = d\xi^\mu/d\lambda$. We wish to find that curve which makes $s_{PP'}$ stationary with respect to small variations that vanish at the end points. We are thus dealing with a problem in the calculus of variation. The change in the value of $ds_{PP'}$ due to a small variation $\delta\xi^\mu$ is given by

$$\delta s_{PP'} = \int_P^{P'} \left\{\frac{1}{2}\frac{d\lambda}{ds} g_{\mu,\nu\sigma}\dot{\xi}^\mu\dot{\xi}^\nu - \frac{d}{d\lambda}\left(\frac{d\lambda}{ds} g_{\mu\sigma}\dot{\xi}^\mu\right)\right\}\delta\xi^\sigma\, d\lambda + \frac{d\lambda}{ds} g_{\mu\sigma}\dot{\xi}^\mu\delta\xi^\sigma\Bigg|_P^{P'}.$$

Since the variation vanishes at the end points, the end-point term does not contribute to $\delta s_{PP'}$. Then, since $\delta\xi^\sigma$ is otherwise arbitrary, it follows after some rearrangement of terms that, if $\delta s_{pp'} = 0$ for all such variations, then

$$-\{\mu\nu, \sigma\}\dot{\xi}^\mu\dot{\xi}^\nu - g_{\mu\sigma}\ddot{\xi}^\mu - \frac{d^2\lambda/ds^2}{d\lambda/ds} g_{\mu\sigma}\dot{\xi}^\mu = 0 \qquad (3\text{-}2.1)$$

are the Euler-Lagrange equations for the metric geodesics of the space. The quantities $\{\mu\gamma, \sigma\}$ introduced here are called *Christoffel symbols of the first kind* and are given by the equation

$$\{\mu\nu, \sigma\} \equiv \tfrac{1}{2}(g_{\mu\sigma,\nu} + g_{\sigma\nu,\mu} - g_{\mu\nu,\sigma}). \qquad (3\text{-}2.2)$$

Equation (3-2.1) can be simplified if we choose the parameter λ to coincide with the distance s along the curve. In that case the last term vanishes in this equation. Finally, it is customary to solve Eq. (3-2.1) for the second derivatives of the ξ^μ, which can be accomplished by multiplying Eq. (3-2.1) by $g^{\sigma\rho}$ and summing over σ. The net result is the geodesic equation

$$\frac{d^2\xi^\rho}{ds^2} + \left\{\begin{matrix}\rho\\ \mu\nu\end{matrix}\right\}\frac{d\xi^\mu}{ds}\frac{d\xi^\nu}{ds} = 0. \qquad (3\text{-}2.3)$$

The quantities $\{{}^{\rho}_{\mu\nu}\}$ appearing here are called *Christoffel symbols of the second kind* and are related to those of the first kind by

$$\left\{\begin{matrix}\rho\\ \mu\nu\end{matrix}\right\} = g^{\rho\sigma}\{\mu\nu, \sigma\}. \qquad (3\text{-}2.4)$$

Since the parameter s appearing in the geodesic equation is the metrical distance along the geodesic, we must restrict the solutions to Eq. (3-2.3) by the supplementary condition that

$$g_{\mu\nu}\frac{d\xi^\mu}{ds}\frac{d\xi^\nu}{ds} = 1. \qquad (3\text{-}2.5)$$

In arriving at the above result we have made a tacit assumption that $ds^2 \neq 0$. Actually we will have to deal with metrics for which there is a whole class of geodesics, the *null* geodesics for which the distance between any two points on the curve is zero. If this is the case, then we can no longer use s as the parameter along the curve. However, it is still possible to find a parameter u such that the equation of a null geodesic is given by Eq. (3-2.3) with s replaced by u and with Eq. (3-2.5) replaced by

$$g_{\mu\nu} \frac{d\xi^\mu}{du} \frac{d\xi^\nu}{du} = 0. \tag{3-2.6}$$

In either case the geodesics have the important property that under a mapping, geodesics are mapped onto geodesics and null geodesics are mapped onto null geodesics of the transformed metric. The property that a curve is a geodesic is thus an intrinsic geometric property of that curve.

Problem 3.3. Show that Eq. (3-2.5) or Eq. (3-2.6) is a first integral of Eq. (3-2.3).

Problem 3.4. Show that if u is a possible parameter along a null geodesic such that the geodesic equation has the form (3-2.3), then so is $u' = au + b$, where a and b are arbitrary constants.

Problem 3.5. Find the general solution of the geodesic equations for null geodesics in a two-dimensional space with $ds^2 = (dx^1)^2 - (dx^2)^2$.

Problem 3.6. On the surface of a two-dimensional sphere with angle variables θ and ϕ, and ds given by

$$ds^2 = d\theta^2 + \sin^2\theta \, d\phi^2,$$

show by example that the geodesics are great circles.

In addition to imposing a metric on the manifold, we can also introduce an affine connection by giving $\Gamma^\rho_{\mu\nu}(x)$. Logically, the two structures are unrelated and can be imposed independently of each other. There are several ways to make a connection between these two constructs. We shall do so by requiring that the affinity be so chosen that the affine geodesics, determined by Eq. (2-3.3), coincide with the metric geodesics determined by Eqs. (3-2.3) and (3-2.5). We mentioned in Section 2-3 that if we restrict ourselves to symmetric affinities, the affinity is uniquely determined if we give the family of affine geodesics together with the parameters along the various curves. Comparing Eq. (2-3.3) with Eq. (3-2.3) we see that the two geodesics will coincide if we set

$$\Gamma^\rho_{\mu\nu} = \begin{Bmatrix} \rho \\ \mu\nu \end{Bmatrix}. \tag{3-2.7}$$

With the parameterization determined by Eq. (3-2.5) or (3-2.6), this choice becomes unique.

With the choice of affinity as indicated in Eq. (3-2.7) one can show immediately that

$$g_{\mu\nu;\rho} \equiv 0, \tag{3-2.8}$$

so that the metric is affinely constant over the manifold. This has as a consequence that if $A_\mu = g_{\mu\nu}A^\nu$, then

$$A_{\mu;\nu} = g_{\mu\rho}A^\rho{}_{;\nu}. \tag{3-2.9}$$

This equation has the following interpretations: If we parallel-transport both A_μ and A^μ from x^μ to $x^\mu + dx^\mu$, we obtain the vectors $A_\mu + \delta A_\mu$ and $A^\mu + \delta A^\mu$. From Eq. (3-2.9) it then follows that

$$A_\mu + \delta A_\mu = g_{\mu\nu}(x + dx)(A^\nu + \delta A^\nu);$$

that is, the result of parallel transport is independent of whether we transport the co-components of a vector or the contra-components. In fact, by demanding that this be true we could also have the relation expressed by Eq. (3-2.7).

Problem 3.7. Prove the above statement by showing that it implies the validity of Eq. (3-2.8) and that this equation in turn implies the relation (3-2.7).

As a consequence of Eq. (3-2.8) one can also derive the two additional results that

$$g^{\mu\nu}{}_{;\rho} \equiv 0$$

and

$$g_{;\rho} \equiv 0.$$

Both results are of use in calculations involving codifferentiation of expressions involving the metric.

3-3. Metric Flatness

Since $g_{\mu\nu}$ is symmetric by assumption, it is always possible to find a mapping such that, at a given point of the manifold, the metric can be reduced to a diagonal form with the values of the diagonal terms being either $+1$ or -1. The excess of plus over minus elements is called the *signature* of the metric. Since the metric is assumed to be a continuous function of the coordinates imposed on the manifold and must also be nonsingular (that is, $g \neq 0$), it follows that the signature of the metric at any one point serves to determine the signature everywhere and, furthermore, that the signature is an invariant of the metric.

We now ask a question similar to one we asked concerning an affinity. What are the conditions that assure that there exists a mapping of the manifold such that, under the mapping, the transformed metric is everywhere diagonal, with the diagonal elements taking on the values ± 1? By analogy with the affine case we call any metric that can be so transformed a *flat metric* and call the corresponding geometry a *flat metric geometry*. The necessary and sufficient condition for affine flatness is the vanishing of the Riemann tensor $B^{\rho}_{\sigma\mu\nu}$ everywhere on the manifold. If this condition is satisfied we can always find a mapping such that $\Gamma^{\rho}_{\mu\nu} = 0$ everywhere. These same conclusions obviously apply if the affinity is a metric affinity. From the fact that $g_{\mu\nu;\rho} = 0$ for a metric affinity, it follows immediately that

$$g_{\mu\nu,\rho} = g_{\sigma\nu}\Gamma^{\sigma}_{\mu\rho} + g_{\mu\sigma}\Gamma^{\sigma}_{\rho\nu}, \tag{3-3.1}$$

so that when $\Gamma^{\rho}_{\mu\nu} = 0$ everywhere, the metric is everywhere constant. Since we can always find a mapping such that the metric is diagonal with ± 1 values for the diagonal elements at some one point of the manifold, it follows that the vanishing of the Riemann tensor everywhere is a sufficient condition for the metric to take on this form everywhere and hence is a sufficient condition for flatness. As we will see, it is also a necessary condition.

Problem 3.8. Let a two-dimensional surface of a sphere be characterized by the angle variables θ and ϕ. If ds is given by

$$ds^2 = d\theta^2 + \sin^2\theta \, d\phi^2,$$

show that the space is not metrically flat.

3-4. The Riemann-Christoffel Tensor

Because of its transcendent importance for all future discussions, we now discuss some of the properties of the Riemann tensor. We note first that for an arbitrary affinity,

$$B^{\rho}_{\sigma\mu\nu} = -B^{\rho}_{\sigma\nu\mu}. \tag{3-4.1}$$

In addition, if the affinity is symmetric,

$$B^{\rho}_{\sigma\mu\nu} + B^{\rho}_{\nu\sigma\mu} + B^{\rho}_{\mu\nu\sigma} = 0. \tag{3-4.2}$$

As a consequence of these two conditions, it follows that $B^{\rho}_{\sigma\mu\nu}$ has 80 independent components.

Let us now specialize to the case of a metric affinity. We shall use the symbol $R^{\rho}_{\sigma\mu\nu}$ to designate the components of the Riemann tensor formed from such an affinity. This special Riemann tensor is often called the *Riemann-Christoffel tensor*. From it we can construct the totally covariant tensor $R_{\rho\sigma\mu\nu}$ by lowering the contra-index of $R^{\rho}_{\sigma\mu\nu}$:

$$\begin{aligned} R_{\rho\sigma\mu\nu} &= g_{\rho\lambda}R^{\lambda}_{\sigma\mu\nu} \\ &= \tfrac{1}{2}(g_{\rho\mu,\sigma\nu} + g_{\sigma\nu,\rho\mu} - g_{\rho\nu,\sigma\mu} - g_{\sigma\mu,\rho\nu}) \\ &\quad + g^{\alpha\beta}(\{\rho\mu, \alpha\}\{\sigma\nu, \beta\} - \{\rho\nu, \alpha\}\{\sigma\mu, \beta\}). \end{aligned} \tag{3-4.3}$$

We see, incidentally, that if $g_{\mu\nu,\rho} = 0$, then both $\{\mu\nu, \rho\}$ and $R_{\rho\sigma\mu\nu}$ are also zero, thereby proving the necessity of the flatness condition.

The tensor $R_{\mu\nu\rho\sigma}$ has additional symmetries over and above those of $B^{\rho}_{\sigma\mu\nu}$. The simplest way to find these symmetries is to carry out a mapping such that $\{\mu\nu, \sigma\} = 0$ at some point of the manifold, and examine the properties of $R_{\mu\nu\rho\sigma}$ at that point. Since the point is arbitrary and since symmetry properties of tensors are preserved under mappings, we are able to conclude that any symmetry properties found in this manner will be generally true everywhere and will be independent of the values of $\{\mu\nu, \rho\}$. Such a procedure is very useful in learning about the geometrical properties of geometrical objects in general.

One looks for a mapping that simplifies the components of the object under investigation and then studies its geometrical properties in this simplified form. Since such properties are preserved under a mapping, one can conclude that any geometrical properties found in the simplified form must also apply in the general case. When one carries out such a procedure in the case of $R_{\rho\sigma\mu\nu}$, one readily sees that there are the additional symmetries:

$$R_{\rho\sigma\mu\nu} = -R_{\rho\sigma\nu\mu} = -R_{\sigma\rho\mu\nu} = R_{\mu\nu\rho\sigma}. \tag{3-4.4}$$

As a consequence of these additional symmetries it follows that the number of independent components of $R_{\rho\sigma\mu\nu}$ is reduced to 20.

In addition to satisfying the above algebraic identities, $R_{\mu\nu\rho\sigma}$ also satisfies the differential *Bianchi identities*:

$$R_{\mu\nu\rho\sigma;\tau} + R_{\mu\nu\tau\rho;\sigma} + R_{\mu\nu\sigma\tau;\rho} \equiv 0. \tag{3-4.5}$$

Again one can prove these relations by the method outlined above for the proof of the algebraic identities. As identities, these equations impose no special restrictions on the metric tensor, being satisfied by any such tensor.

Problem 3.9. Prove the Bianchi identity (3-4.5) by the method outlined above.

From $R_{\rho\sigma\mu\nu}$ one can form additional tensors by the process of contraction. The *Ricci tensor* is given by

$$\begin{aligned} R_{\mu\nu} &= R^{\rho}_{\mu\rho\nu} = g^{\rho\sigma} R_{\sigma\mu\rho\nu} \\ &= \begin{Bmatrix} \rho \\ \mu\rho \end{Bmatrix}_{,\nu} - \begin{Bmatrix} \rho \\ \mu\nu \end{Bmatrix}_{,\rho} - \begin{Bmatrix} \sigma \\ \mu\nu \end{Bmatrix}\begin{Bmatrix} \rho \\ \sigma\rho \end{Bmatrix} + \begin{Bmatrix} \rho \\ \sigma\nu \end{Bmatrix}\begin{Bmatrix} \sigma \\ \mu\rho \end{Bmatrix} \end{aligned} \tag{3-4.6}$$

and is symmetric. Note that any other single contractions either give zero, such as $g^{\rho\sigma}R_{\rho\sigma\mu\nu}$, or are equal to $R_{\mu\nu}$ or its negative. A further contraction yields the *curvature scalar* R, given by

$$R = g^{\mu\nu} R_{\mu\nu}. \tag{3-4.7}$$

The scalar R has the important property that, aside from a trivial constant, it is the only scalar that depends on the $g_{\mu\nu}$ and their first second derivatives and is linear in the second derivative. In all there are 14 independent scalars that one can construct from the metric and its first and second derivatives. When $R_{\mu\nu} = 0$, all but four of them vanish. One can also show that the only symmetric second-rank cotensors that contain no higher than second derivatives and which are linear in these quantities, are $R_{\mu\nu}$ and $Rg_{\mu\nu}$. Of particular importance is the *Einstein tensor*

$$G_{\mu\nu} \equiv R_{\mu\nu} - \tfrac{1}{2} g_{\mu\nu} R. \tag{3-4.8}$$

It is obviously symmetric and satisfies the contracted Bianchi identity

$$(g^{\mu\nu} G_{\nu\sigma})_{;\mu} \equiv 0 \tag{3-4.9}$$

as follows directly from the Bianchi identity (3-4.5).

The last important tensor constructed from the Riemann-Christoffel tensor is the *Weyl conformal tensor*:

$$\begin{aligned} C_{\rho\sigma\mu\nu} &= R_{\rho\sigma\mu\nu} - \tfrac{1}{2} g_{\rho\mu} R_{\nu\sigma} + \tfrac{1}{2} g_{\rho\nu} R_{\mu\sigma} + \tfrac{1}{2} g_{\sigma\mu} R_{\nu\rho} \\ &\quad - \tfrac{1}{2} g_{\sigma\nu} R_{\mu\rho} - \tfrac{1}{6} g_{\rho\nu} g_{\mu\sigma} R + \tfrac{1}{6} g_{\rho\mu} g_{\nu\sigma} R. \end{aligned} \tag{3-4.10}$$

The Weyl tensor has the special property that

$$g^{\rho\mu} C_{\rho\sigma\mu\nu} \equiv 0. \tag{3-4.11}$$

Furthermore, if $C_{\rho\sigma\mu\nu} = 0$ everywhere, then the metric is *conformally flat*; that is, there exists a mapping such that $g_{\mu\nu}$ can be diagonalized, with $\pm\gamma(x)$ appearing in the diagonal positions and where $\gamma(x)$ is some space-time function. This follows from the fact that $C_{\rho\sigma\mu\nu}$ can be expressed entirely in terms of the density $\tilde{g}_{\mu\nu} = g_{\mu\nu}/g^{1/4}$ and its inverse, and is equal to the Riemann-Christoffel tensor formed by replacing $g_{\mu\nu}$ by $\tilde{g}_{\mu\nu}$. Consequently the vanishing of the Weyl tensor implies the vanishing of this latter quantity, which in turn implies that there exists a mapping such that $\tilde{g}_{\mu\nu}$ is everywhere diagonal, with

± 1 appearing along the diagonal. Only g is arbitrary and $\pm g^{1/4}(x)$ appears along the diagonal of $g_{\mu\nu}$. As in the case of the special properties of the Riemann-Christoffel tensor, the above properties of the Weyl tensor can be most easily proved by mapping to a point where $g_{\mu\nu,\rho} = 0$.

One can give a relatively simple geometrical interpretation of the curvature tensor. At a point P let ζ^μ and η^μ be two unit vectors. These two vectors span a subspace of the tangent space at P, namely, the space of all vectors of the form $\alpha\zeta^\mu + \beta\eta^\mu$. Now, through P, consider the totality of all geodesics whose tangents lie in the subspace. They form a surface called a *geodesic surface*. One can then show that the Gaussian curvature[1] at P of this surface is equal to

$$-K = \frac{R_{\mu\nu\rho\sigma}\zeta^\mu\eta^\nu\zeta^\rho\eta^\sigma}{(g_{\mu\rho}g_{\nu\sigma} - g_{\mu\sigma}g_{\nu\rho})\zeta^\mu\eta^\nu\zeta^\rho\eta^\sigma}. \tag{3-4.12}$$

Riemann defined the Riemannian curvature at P in the directions ζ^μ, η^μ, to be this Gaussian curvature. If the curvature K at P is independent of these directions, then necessarily

$$R_{\mu\nu\rho\sigma} = -K(g_{\mu\rho}g_{\nu\sigma} - g_{\mu\sigma}g_{\nu\rho}).$$

A Riemannian space for which the curvature tensor can be written in this form for all points of the space is called a *space of constant curvature*, and provided the dimensionality of this space is >2, K is everywhere constant in that space. If, in particular, $K = 0$, we have a flat space, as follows from our previous definition of a flat space.

One can give a similar interpretation to R and $R_{\mu\nu}$. At a point let four orthogonal directions be given from which we can form six geodesic surfaces as before, using these vectors for ζ^μ and η^μ. If $K_{(\alpha\beta)}$ is the curvature of the geodesic surface formed from the directions $\zeta^\mu_{(\alpha)}$ and $\zeta^\mu_{(\beta)}$, then the curvature scalar is given by the double sum

$$R = 2\sum_{\alpha<\beta} K_{(\alpha\beta)}.$$

The sum is independent of the particular directions chosen. Thus R can be thought of as being the mean curvature of the space at the point in question.

[1] The Gaussian curvature at a point P of a two-dimensional surface embedded in a flat three-dimensional space is given by $G = 1/R_1R_2$, where R_1 and R_2 are two principal radii of curvature of this surface. Gauss was able to show that $G = \lim_{S\to 0} E/S$, where S is the area of a small geodesic rectangle containing P, and E is the excess of the sum of its interior angles over 2π. The importance of this result rests in the fact that while the values of R_1 and R_2 depend on how the surface is embedded in the flat three-dimensional space, G does not and is determined solely by the geometry of the two-dimensional surface, that is, by its metric.

Now let an additional direction be specified by the vector ξ^μ. Then one can show that

$$\sum_\alpha K_{(\alpha)} \sin^2 \theta_\alpha = \frac{R_{\mu\nu}\xi^\mu\xi^\nu}{g_{\mu\nu}\xi^\mu\xi^\nu},$$

where $K_{(\alpha)}$ is the curvature formed from the directions ξ^μ and $\zeta^\mu_{(\alpha)}$ and $\theta_{(\alpha)}$ is the angle between these two directions.

3-5. Algebraic Classification of the Weyl Tensor

Because of its symmetry properties and the fact that $g^{\mu\nu}C_{\mu\rho\nu\sigma} \equiv 0$, it is possible to classify the Weyl tensor at a point by algebraic means. The same classification applies to the Riemann-Christoffel tensor itself when $R_{\mu\nu} = 0$, since in this case the two tensors are equal. The classification was first carried out by Petrov[2] and, together with similar classifications, has been used by many authors in discussing the problem of gravitational radiation. To effect the classification at a point, one constructs from the Weyl tensor three 3×3 matrices M_{AB}, N_{AB}, and P_{AB} according to the following scheme: An index A of M corresponds to an index pair $\mu\nu$ of $C_{\mu\nu\rho\sigma}$ according to

$$(1, 2, 3) \Rightarrow (23, 31, 12) \tag{3-5.1}$$

so that

$$M_{11} = C_{2323}, \quad M_{12} = C_{2331}, \quad M_{13} = C_{2312}, \quad M_{21} = C_{3123},$$

etc.; an index A of P corresponds to an index pair $\mu\nu$ of $C_{\mu\nu\rho\sigma}$ according to

$$(1, 2, 3) \Rightarrow (10, 20, 30), \tag{3-5.2}$$

so that

$$P_{11} = C_{1010}, \quad P_{12} = C_{1020}, \quad P_{13} = C_{1030}, \quad P_{21} = C_{2010},$$

etc.; the first index of N corresponds to an index pair $\mu\nu$ of $C_{\mu\nu\rho\sigma}$ as in Eq. (3-5.1), while the second index corresponds to an index pair $\mu\nu$ as in Eq. (3-5.2), so that

$$N_{11} = C_{2310}, \quad N_{12} = C_{2320}, \quad N_{13} = C_{2330}, \quad N_{21} = C_{3110},$$

etc. A knowledge of the elements of these three matrices is completely equivalent to a knowledge of $C_{\mu\nu\rho\sigma}$.

Because of the symmetry properties of $C_{\mu\nu\rho\sigma}$, it follows that M and P are symmetric while N is asymmetric and trace-free, that is,

$$\operatorname{tr} N = \sum_A N_{AA} = 0.$$

[2] A. Z. Petrov, *Sci. Nat. Kazan State University*, **114**, 55–69 (1954).

Furthermore, since $g^{\mu\nu}C_{\mu\rho\nu\sigma} = 0$, it follows that

$$N_{AB} = N_{BA},$$

$$P_{AB} = -M_{AB},$$

and

$$\operatorname{tr} P \equiv \sum_A P_{AA} = 0,$$

so that both P and M are symmetric and trace-free

One proceeds with the classification by forming the traceless symmetric complex matrix

$$Q_{AB} = M_{AB} + iN_{AB},$$

and looks for the eigenvalues λ and eigenvectors n_A of this matrix, that is, solutions of

$$Q_{AB}n_B = \lambda n_A. \tag{3-5.3}$$

In general both λ and n_A will be complex. Since Q is traceless it follows that the sum of the three roots of Eq. (3-5.3) is zero. If Q has respectively 3, 2, or 1 independent eigenvectors, the Petrov type is respectively I, II, or III. For Weyl tensors of type I there are two independent complex eigenvalues. If two of the eigenvalues are equal, then the Petrov type is D or type I degenerate. For Petrov type II there is one independent eigenvalue. If this eigenvalue vanishes, then the Petrov type is N or type II null. For Petrov type III all the eigenvalues are equal and hence vanish. By adding the case when $C_{\mu\nu\rho\lambda} = 0$ (type 0), one can form a Penrose[3] diagram:

$$\begin{array}{ccccc} & & \mathrm{I} & & \\ & \swarrow & & \searrow & \\ & \mathrm{II} & \rightarrow & D & \\ \swarrow & & \searrow \;\; \swarrow & & \searrow \\ \mathrm{III} & \rightarrow & N & \rightarrow & 0 \end{array}$$

with arrows pointing in the direction of increasing specialization. In the literature, type I is called *algebraically general*; the other types together are called *algebraically special*, for obvious reasons. By a suitable mapping, the various types can be put into the following canonical forms with nonvanishing elements indicated:

[3] R. Penrose, *Ann. Phys. Princeton*, **10**, 171–201 (1960).

Type I

$$M = \begin{pmatrix} m_1 & \cdot & \cdot \\ \cdot & m_2 & \cdot \\ \cdot & \cdot & m_3 \end{pmatrix}, \qquad N = \begin{pmatrix} n_1 & \cdot & \cdot \\ \cdot & n_2 & \cdot \\ \cdot & \cdot & n_3 \end{pmatrix},$$

with $\sum m_A = \sum n_A = 0$. For type D, $m_2 = m_3$, $n_2 = n_3$.

Type II

$$M = \begin{pmatrix} -2m & \cdot & \cdot \\ \cdot & m+\sigma & \cdot \\ \cdot & \cdot & m-\sigma \end{pmatrix}, \qquad N = \begin{pmatrix} -2n & \cdot & \cdot \\ \cdot & n & \sigma \\ \cdot & \sigma & n \end{pmatrix}.$$

For type N, $m = n = 0$.

Type III

$$M = \begin{pmatrix} \cdot & -\sigma & \cdot \\ -\sigma & \cdot & \cdot \\ \cdot & \cdot & \cdot \end{pmatrix}, \qquad N = \begin{pmatrix} \cdot & \cdot & \sigma \\ \cdot & \cdot & \cdot \\ \sigma & \cdot & \cdot \end{pmatrix}.$$

We emphasize that the Petrov classification is of a local nature, and unlike the signature of $g_{\mu\nu}$, can change as one ranges over the manifold.

3-6. Geometry of Subspaces

In Section 1-9 we considered submanifolds of the space-time manifold. A number of important applications involve a three-dimensional submanifold, sometimes called a *hypersurface* and defined by the equation

$$\phi(x) = 0. \tag{3-6.1}$$

The normal n_μ to this manifold at the point x^μ is defined to be

$$n_\mu = \phi_{,\mu}. \tag{3-6.2}$$

A displacement lying wholly in the hypersurface satisfies

$$\phi_{,\mu}\, dx^\mu = 0,$$

and hence it follows from Eq. (3-6.2) that such a displacement is orthogonal to n_μ:

$$n_\mu\, dx^\mu = 0.$$

Alternatively, we can characterize the hypersurface by specifying the points that lie in it. These points are given by

$$x^\mu = x^\mu(\lambda^i) \tag{3-6.3}$$

where λ^i, $i = 1, 2, 3$, are a set of parameters. A displacement lying in the hypersurface is then given by

$$dx^\mu = x^\mu{}_{,i}\, d\lambda^i, \tag{3-6.4}$$

where we have made use of the comma notation to denote differentiation with respect to the parameters. If we now use Eq. (3-6.3) to substitute for x^μ in Eq. (3-6.1) we must obtain an identity in the λ^i. Consequently we must have

$$\phi_{,i} = \phi_{,\mu} x^\mu{}_{,i} = 0, \tag{3-6.5}$$

thereby demonstrating that the two definitions of a displacement in the hypersurface are equivalent.

For all practical purposes we can consider the parameters λ^i as forming a coordinate system on our hypersurface. Consider now two neighboring points on the hypersurface, λ^i and $\lambda^i + d\lambda^i$. The space-time coordinates of these two points are then $x^\mu(\lambda)$ and $x^\mu(\lambda + d\lambda) = x^\mu(\lambda) + x^\mu{}_{,i}(\lambda) d\lambda^i$. The distance between these two points is

$$\begin{aligned} ds^2 &= g_{\mu\nu}(x(\lambda))\, dx^\mu\, dx^\nu \\ &= g_{\mu\nu}(x(\lambda)) x^\mu{}_{,i} x^\nu{}_{,j}\, d\lambda^i\, d\lambda^j, \end{aligned}$$

which we can write as

$$ds^2 = \gamma_{ij}(\lambda)\, d\lambda^i\, d\lambda^j,$$

where

$$\gamma_{ij}(\lambda) \equiv g_{\mu\nu}(x(\lambda)) x^\mu{}_{,i} x^\nu{}_{,j} \tag{3-6.6}$$

is the metric *induced* on the hypersurface by $g_{\mu\nu}$.

Problem 3.10. Show that γ_{ij} transforms like a three-dimensional tensor under arbitrary mappings of the hypersurface onto itself.

Problem 3.11. Find the metric induced on a two-dimensional surface $(x^1)^2 + (x^2)^2 + (x^3)^2 = a^2$ embedded in a flat three-dimensional space.

The metric γ_{ij} defined on a hypersurface is sometimes referred to as the *first fundamental form* of the hypersurface. It describes its intrinsic geometry, but not how it is embedded in the space-time manifold, that is, its extrinsic geometry. There are several ways that one can characterize this extrinsic geometry, all of which involve a quantity Ω_{ij} called the *second fundamental form*. Like γ_{ij}, Ω_{ij} transforms like a cotensor under arbitrary mappings of the hypersurface onto itself.

To see how the second fundamental form arises, let us consider a geodesic curve on the hypersurface emanating from a given point P in some direction that is tangent to the subspace at P. Being a geodesic of the subspace, it depends only on γ_{ij} and the Christoffel symbols $\{{}^{\,i}_{jk}\}_\gamma$ associated with this three-dimensional metric. Thus, if

$$\lambda^i = \zeta^i(s) \tag{3-6.7}$$

defines the points lying on this geodesic, with the metric length s computed using γ_{ij} along it as measured from P, then

$$\frac{d^2\zeta^i}{ds^2} + \begin{Bmatrix} i \\ jk \end{Bmatrix}_\gamma \frac{d\zeta^j}{ds}\frac{d\zeta^k}{ds} = 0. \tag{3-6.8}$$

Except in special cases, a geodesic of a hypersurface will not be a geodesic of the manifold in which it is embedded. If

$$x^\mu = \xi^\mu(s) \equiv x^\mu(\zeta(s))$$

are the points of this manifold that lie on the geodesic, then it can be shown that

$$\frac{d^2\xi^\mu}{ds^2} + \begin{Bmatrix} \mu \\ \rho\sigma \end{Bmatrix}_\mu \frac{d\xi^\rho}{ds}\frac{d\xi^\sigma}{ds} = e\Omega_{ij}\frac{d\zeta^i}{ds}\frac{d\zeta^j}{ds}\hat{n}^\mu, \tag{3-6.9}$$

where $\hat{n}^\mu$ is a unit normal to the submanifold at P and $e = \pm 1$, depending on whether the norm of $\hat{n}^\mu$ is ± 1 (we assume that the norm of $n^\mu \neq 0$). The quantity Ω_{ij} defined by this equation is thus seen to be a measure of how much a geodesic of the submanifold deviates from being a geodesic of the manifold in which it is embedded. It follows from Eq. (3-6.9) that Ω_{ij} transforms like a three-dimensional cotensor under mappings of the submanifold onto itself.

Problem. 3.12. Consider a three-dimensional submanifold with induced metric γ_{ij}, given by Eq. (3-6.6), and a second fundamental form Ω_{ij}, a point P of this submanifold, and a displacement dx^μ at P lying in the submanifold. Let two geodesics, one defined on the submanifold and satisfying Eq. (3-6.8) and the other defined on the full manifold, pass through P in the direction defined by dx^μ. If the parameters along the two manifolds are taken to be the respective arc lengths as measured from P, show that distance between two points on the two geodesics at the same distance s from P is equal to $\frac{1}{2}(\Omega_{ij}\dot{\zeta}^i\dot{\zeta}^j s^2)$ for small s. (The $\zeta^i(s)$ are the points of the submanifold lying on the geodesic in it, and the dot indicates differentiation with respect to s.)

3-7. Characteristic Surfaces

If the metric of the space-time manifold is indefinite, there exists a special class of hypersurfaces $\omega(x) = \text{const}$ with the property that a displacement lying in the hypersurface at a point is parallel to the normal to the hypersurface at that point. Thus if dx^μ is such a displacement, it will be proportional to $g^{\mu\nu}\omega_{,\nu}$ and hence we must have

$$g^{\mu\nu}\omega_{,\mu}\omega_{,\nu} = 0 \tag{3-7.1}$$

as a consequence of the equation $\omega = \text{const}$. Surfaces satisfying Eq. (3-7.5) either identically or as a consequence of their defining equation are called *characteristic surfaces* and play an important role in the propagation of electromagnetic radiation and other similar wave-type phenomena. Clearly the metric must be indefinite for such surfaces to exist. In what follows we will assume that the signature of the metric is minus three to conform to the signature of the metric we will employ in our space-time theories.

Given a characteristic surface $\omega = \text{const}$ we can construct a family of curves called *rays* or *bicharacteristics* with the property that the tangent to a ray through a point is proportional to $g^{\mu\nu}\omega_{,\nu}$ at that point. If the points on a ray are given parametrically by

$$x^\mu = z^\mu(\lambda), \tag{3-7.2}$$

then, by a suitable choice of the path parameter λ, we have

$$\frac{dz^\mu}{d\lambda} = g^{\mu\nu}\omega_{,\nu}. \tag{3-7.3}$$

It then follows from Eq. (3-7.1) that

$$\frac{dz^\mu}{d\lambda}\omega_{,\nu} = 0, \tag{3-7.4}$$

showing that a displacement along a ray lies in the characteristic surface used in its construction. We can easily derive the equation of a ray. From Eq. (3-7.3) it follows, with the help of Eq. (3-7.1), that

$$\frac{d^2z^\mu}{d\lambda^2} + \begin{Bmatrix} \mu \\ \rho\sigma \end{Bmatrix} \frac{dz^\rho}{d\lambda}\frac{dz^\sigma}{d\lambda} = 0 \tag{3-7.5}$$

and since it also follows from these equations that

$$g_{\mu\nu}\frac{dz^\mu}{d\lambda}\frac{dz^\nu}{d\lambda} = 0, \tag{3-7.6}$$

we see that a ray is a null geodesic.

If in addition to being given a characteristic surface $\omega = \text{const}$, we are also given another surface $\phi = \text{const}$ with the property that $g^{\mu\nu}\phi_{,\mu}\phi_{,\nu} > 0$ (since the signature of $g_{\mu\nu}$ is -3 such surfaces can in general be constructed), we can define a two-dimensional surface called a *wave front*. Such a surface is defined to be the surface formed by the intersection of the characteristic hypersurface and the $\phi = \text{const}$ hypersurface. Given a wave front lying in a $\phi = \text{const}$ hypersurface it is possible to construct two unique characteristic hypersurfaces that contain it. To do so we construct first at each point of the wave front the unit normal to the $\phi = \text{const}$ hypersurface—that is, the vector

$$n^\mu = \frac{g^{\mu\nu}\phi_{,\nu}}{\sqrt{g^{\alpha\beta}\phi_{,\alpha}\phi_{,\beta}}}. \tag{3-7.7}$$

In addition we construct a unit vector l^μ normal to the wave front and tangent to the $\phi = \text{const}$ hypersurface. For the purposes of this construction let the points on this hypersurface be given parametrically by

$$x^\mu = x^\mu(\lambda^r) \qquad r = 1, 2, 3 \tag{3-7.8}$$

and the equation of the wave front by

$$\psi(\lambda^r) = \text{const}. \tag{3-7.9}$$

Further, let $\gamma_{rs}(\lambda^r)$ given by Eq. (3-6.6) be the metric induced on the $\phi = \text{const}$ hypersurface. Since $g^{\mu\nu}\phi_{,\mu}\phi_{,\nu} > 0$ it follows that $\gamma^{rs}\psi_{,r}\psi_{,s} < 0$. The vector l^μ is then given by

$$l^\mu = x^\mu{}_{,r}q^r \tag{3-7.10}$$

where

$$q^r = \frac{\gamma^{rs}\psi_{,s}}{\sqrt{-\gamma^{ab}\psi_{,a}\psi_{,b}}}.$$

Since $n^2 = 1$, $l^2 = -1$, and $g_{\mu\nu}n^\mu l^\nu = 0$, it follows that the vector

$$k^\mu = n^\mu + l^\mu \tag{3-7.11}$$

satisfies $k^2 = 0$. (Note that k^μ is not unique since $-l^\mu$ is also a unit normal to the wave front. This accounts for there being two characteristic surfaces associated with a given wave front.) One now takes k^μ at a point on the wave front to be the tangent to a ray passing through that point. Since the ray through a point with a given tangent is unique, the totality of all such rays constitutes the desired characteristic surface.

As an example of the above construction let us consider a flat metric and map so that $g_{\mu\nu} = \text{diag}(1, -1, -1, -1)$. We take the $x^0 = \text{const}$ hypersurface to be the $\phi = \text{const}$ hypersurface and the equation of the wave front in this hypersurface to be

$$(x_0{}^1)^2 + (x_0{}^2)^2 + (x_0{}^3)^2 = a^2. \qquad (3\text{-}7.12)$$

The unit normal to the x^0 = const hypersurface is $n^\mu = (1, 0, 0, 0)$ while the normal to the wave front perpendicular to n^μ is $l^\mu = (0, x^1/a, x^2/a, x^3/a)$ where for definiteness we have taken the outward pointing normal. Since, with the assumed form for $g_{\mu\nu}$, $\{{}^{\mu}_{\rho\sigma}\} = 0$, the solution of Eq. (3-7.5) is of the form

$$x^\mu - x_0{}^\mu = k^\mu \lambda,$$

the ray passing through the point $x^\mu = (0, x_0{}^1, x_0{}^2, x_0{}^3)$ with $k^\mu = n^\mu + l^\mu$ is given by

$$x^r - x_0{}^r = \frac{x_0{}^r}{a}\lambda$$

$$x^0 = \lambda.$$

Therefore, eliminating λ and making use of Eq. (3-7.12) we obtain, as the equation of the desired characteristic surface,

$$(x^1)^2 + (x^2)^2 + (x^3)^2 = (a + x^0)^2. \qquad (3\text{-}7.13)$$

Thus the wave fronts corresponding to the intersection of other x^0 = const hypersurfaces with this characteristic hypersurface are spheres of radius $a + x^0$.

Problem 3.13. Construct the characteristic hypersurface that contains the wave front $\alpha_r x_0{}^r = 0$ in the hypersurface $x^0 = 0$ when $g_{\mu\nu} = \text{diag}(1, -1, -1, -1)$.

4

Structure of Space-Time Theories

In this chapter we shall discuss some of the general properties of physical theories that make use of the geometrical concepts developed in the previous three chapters. Of particular importance will be the covariance properties of these theories, leading to the notion of the covariance group of a theory. We shall see that the requirement that a theory admit a given group as a covariance group imposes severe limitations on the possible forms this theory can assume. In particular it limits the possible types of quantities that appear in the theory and also the dynamical laws of the theory. We shall also argue that the manifold mapping group must be a covariance group of all space-time theories.

In addition to the notion of a covariance group we shall introduce the notion of a symmetry group of a physical theory. This will be done by noting that in a physical theory there are in general two types of quantities, which we call absolute and dynamical, that describe the trajectories of the theory. As the name implies, the dynamical quantities depend on the absolute elements but not vice versa. The symmetry group of a theory is then defined as the largest subgroup of the covariance group of the theory that leaves invariant the absolute elements of the theory.

As we shall see, the requirement of symmetry with respect to a given group is even more restrictive than the requirement of covariance under this group. Thus in later chapters we shall show that the requirement of symmetry with respect to the Poincaré group characterizes all of the physical theories of special relativity. In addition to the restrictions it imposes on the form of a physical theory, the requirement of symmetry under a group has additional important consequences. If the dynamical laws that restrict the possible values of the dynamical quantities of a theory can be derived from a variational principle, then for each symmetry of the system there is a corresponding equation of continuity—that is, a conservation law.

Special problems arise if the elements of a symmetry group are characterized by one or more arbitrary space-time functions as in the case of the MMG and the gauge group of electrodynamics. One finds that the equations of motion for the dynamical objects are not all independent but rather satisfy a number of Bianchi-type identities. The effect of the existence of these identities on the initial value problem for the theory are discussed in the last section of this chapter.

4-1. The Elements of a Physical Theory

Every physical theory attempts to associate mathematical quantities of some kind with the elements of the physical system the theory is supposed to describe. How one makes this association is one of the most difficult parts of physics, since a theory cannot, by itself, tell us how it is to be made. Nevertheless, unless such an association can be made the theory is vacuous as far as the physicist is concerned. Furthermore, it must be possible, at least in principle, to associate each of the mathematical quantities that enter into the description of a physical system with some element of that system. Otherwise some of these quantities are superfluous to the description and can be eliminated from the theory. For the class of theories we have called "space-time theories" the mathematical quantities employed will belong to the class of geometrical objects discussed in Part I of this work. For instance, we will see later that the electromagnetic field can be associated with an antisymmetric contratensor $F^{\mu\nu}(x)$ defined on the space-time manifold.

In general, the mathematical quantities that we associate with the elements of a physical system can take on a range of values. The specification of one possible set of values of these quantities constitutes what we shall call a *kinematically possible trajectory* (hereafter abbreviated kpt) of the physical system. Of course, not every kpt will correspond to an actual trajectory of the physical system. A kpt that could, at least in principle, be realized by the system will be termed a *dynamically possible trajectory* (hereafter abbreviated dpt).

Since only a subset of kpt correspond to the dpt of a physical system, a specification of this subset is necessary to complete a physical theory. This specification will be referred to in what follows as the *dynamical laws* or *equations of motion* of the theory. These dynamical laws then, together with a specification of the kpt, constitute what we have called a "physical theory."

In the case of space-time theories such laws are usually, although not always, given in the form of local relations between the geometrical objects that describe the kpt of the theory. In addition, these laws may also include boundary conditions that are satisfied by all of the dpt—for example, the outgoing radiation condition of electrodynamics. Of course, whatever form the dynamical laws take, they must admit at least one kpt as a dpt. At the same time, they should be maximal, in the sense that any additional relations between the quantities that describe the kpt would lead to an overdetermined system admitting no kpt as dpt.

4-2. Covariance Properties of Physical Theories

One of the most important concepts that we shall make use of in our discussion of space-time theories is that of a *covariance group*. We should warn the reader at the outset that various authors have attached different meanings than the one we will employ to the term *covariance*. Likewise, what we have called a "covariance group" has been called by other names in the literature. In what follows we shall try to give a precise definition of this term and then adhere to it throughout the remainder of the text.

A physical theory will be said to admit a group G as a covariance group or to be G-covariant provided two conditions are satisfied. First, the kpt of the theory must constitute the basis of a faithful realization of the group G. The restriction to a faithful realization is, of course, necessary since, for example, any group can be represented unfaithfully by associating a given kpt with itself for each element of the group. The second condition is that the realization associate dpt with dpt. Thus, if the kpt of the theory are described by a set of quantities y_A and y'_A is associated with y_A by an element of the covariance group, then if y_A is a dpt, y'_A is also a dpt. It is, of course, possible that the kpt of a theory constitute the realization of some group that does not associate dpt with dpt. Unless this second condition is also satisfied, however, this group will not be a covariance group of the theory.

As we shall see throughout this work, the requirement that the theory which is to describe a particular physical system admit a given group as a covariance group imposes severe restriction on the possible forms of this theory. Suppose, for example, we have decided that the kpt of a theory are to be described by a field $F^{\mu\nu}(x)$ that is antisymmetric in the indices μ and ν. Such a field can serve as the basis of a representation of the Poincaré group by considering it to transform as a tensor under this group. (Since an antisymmetric tensor field can serve as the basis of a representation of the MMG it can, *a fortiori*, serve as the basis of a representation of any subgroup of the MMG.) If then the Poincaré group is required to be a covariance group of the theory in question, there is only one possible set of dynamical laws that are linear in $F^{\mu\nu}$ and of first differential order. These laws are of the form

$$F^{\mu\nu}{}_{,\nu} = 0$$

and

$$F_{[\mu\nu,\rho]} = 0 \tag{4-2.1}$$

where $F_{\mu\nu}$ is related to $F^{\mu\nu}$ by

$$F_{\mu\nu} = \eta_{\mu\rho}\eta_{\nu\sigma}F^{\rho\sigma}$$

with $\eta_{\mu\nu}$ the Minkowski matrix. Later we will see that these equations of motion are equivalent to the empty-space Maxwell equations.

To demonstrate the Poincaré-covariance of the equations of motion (4-2.1) we must show that if $F^{\mu\nu}$ satisfies these equations, then so does $F'^{\mu\nu}$ when these two fields are related by a Poincaré mapping. For the purposes of the discussion we will restrict ourselves to the subgroup of Poincaré mappings that can be reached from the identity mapping by a succession of infinitesimal mappings. It is then sufficient to consider only the infinitesimal mappings (1-4.5) in proving that these laws admit this subgroup as a covariance group. By making use of Eq. (1-8.3) we find that, for such mappings,

$$F^{\mu\nu}(x) \to F'^{\mu\nu}(x) = F^{\mu\nu}(x) + \bar{\delta}F^{\mu\nu}(x)$$

where

$$\bar{\delta}F^{\mu\nu} = \varepsilon_\rho{}^\mu F^{\rho\nu} + \varepsilon_\sigma{}^\nu F^{\mu\sigma} - F^{\mu\nu}{}_{,\rho}(\varepsilon_\sigma{}^\rho x^\sigma + \varepsilon^\sigma).$$

Since the equations of motion (4-2.1) are linear in the field $F^{\mu\nu}$ we need only show that $\bar{\delta}F^{\mu\nu}$ satisfies these equations when $F^{\mu\nu}$ satisfies them in order to complete the proof that they are covariant with respect to the subgroup of Poincaré mappings considered here. But this is indeed the case as can be shown by a simple calculation.

Problem 4.1. Show that the spatial reflection group, consisting of the identity mapping and the mapping

$$x^0 \to x'^0 = x^0$$

$$x^r \to x'^r = -x^r,$$

is a covariance group of Eqs. (4-2.1).

As a second example of the restrictions imposed on a physical theory by a covariance requirement, let us assume that the kpt of the theory are described by a symmetric tensor field $g_{\mu\nu}(x)$ and require that it be MMG-covariant.

Then one can show that there are only three essentially different theories whose dynamical laws are of second differential order and linear in the second derivatives of $g_{\mu\nu}$. We can take these dynamical laws to be

$$R_{\mu\nu\rho\sigma} = 0 \tag{4-2.2}$$

where $R_{\mu\nu\rho\sigma}$ is the Riemann-Christoffel tensor formed by treating $g_{\mu\nu}$ as a metric tensor. As a second possibility we can take the dynamical laws to be

$$C_{\mu\nu\rho\sigma} = 0 \qquad \text{and} \qquad R = 0 \tag{4-2.3}$$

where $C_{\mu\nu\rho\sigma}$ is the conformal tensor of Weyl given in terms of $g_{\mu\nu}$ by Eq. (3-4.10) and R is the curvature scalar formed from $g_{\mu\nu}$. Finally, we can take the dynamical laws to be

$$R_{\mu\nu} - \tfrac{1}{2}Rg_{\mu\nu} + \lambda g_{\mu\nu} = 0, \tag{4-2.4}$$

where $R_{\mu\nu}$ is the Ricci tensor formed from $g_{\mu\nu}$ and λ is some constant. (It is this latter set of equations, with $\lambda = 0$, that was finally adopted by Einstein as the empty-space gravitational field equations.) While we have used $g_{\mu\nu}$ as a metric tensor in forming the elements entering into the above equations of motion, we should emphasize that our assertions do not rest on $g_{\mu\nu}$ being a metric; given any symmetric tensor field, these are the only second-order MMG-covariant equations of motion that it can satisfy.

Like any other facet of physical theory, the existence of a covariance group associated with a physical system can only be verified by observations made on the system. If such a group exists, then, for each observed dpt of the system, there must exist all of the other dpt that are associated with this dpt by the elements of the covariance group. When one is in possession of a theory that is adequately supported by observations on the system, it is not necessary to check if all these other states actually exist in nature; one need only check to see if the theory that describes the system admits the group in question as a covariance group.

It often happens, however, that while we have a good idea of what mathematical quantities to associate with the elements of a physical system, we do not know the dynamical laws that govern the behavior of this system. In this case one can still determine experimentally if the system has associated with it any covariance groups. The possibilities are limited by the fact that the kpt of whatever theory describes the system must serve as the basis of a realization of such a group. Since we know the mathematical quantities that describe these kpt, we can find the groups for which they serve as the basis of a realization. One would then check to see if for each observed dpt there exists all of the others that are associated with it by elements of one or another of these groups. It was just such an investigation that lead to the discovery of the nonconservation of parity in the weak interactions. If parity was conserved in these interactions, then to every dpt of a weakly interacting system should correspond another, spatially reflected dpt. In the β-decay of cobalt-60 it was found that the electron is emitted preferentially in a direction opposite to the spin of the cobalt-60 nucleus. The spatially reflected dpt should correspond to one in which the electron is emitted parallel to the direction of the nuclear spin. However, no such dpt was observed[1] and it was concluded that the weak interactions do not admit the group of spatial reflections as a covariance group.

The problem of determining the covariance groups to be associated with a physical system is complicated in some cases by the fact that there exists a certain ambiguity as to what constitutes the observable elements of the system. Perhaps the most familiar example of such a situation arises in connection with the motion of a charged particle in a magnetic field. One can describe the kpt of this system solely in terms of the particle coordinates $\mathbf{x}(t)$. If one orients his coordinate axes so that the magnetic field points in the x^3-direction, the Newtonian equations of motion for the particle have the well-known form

$$\begin{aligned} m\ddot{x}^1 &= e\dot{x}^2 H_0 \\ m\ddot{x}^2 &= -e\dot{x}^1 H_0 \\ m\ddot{x}^3 &= 0 \end{aligned} \tag{4-2.5}$$

where e is the charge and m the mass of the particle and H_0 is the magnitude of the magnetic field. One can show that the covariance group of this theory contains elements corresponding to rotations about the x^3-axis but not to rotations about the x^1- and x^2-axes.

Alternately, one can characterize the kpt by the particle coordinates plus a vector field $\mathbf{H}(x, t)$. The equations of motion of this enlarged theory are then of the form

$$m\ddot{\mathbf{x}} = e\dot{\mathbf{x}} \times \mathbf{H} \tag{4-2.6a}$$

and

$$\dot{\mathbf{H}} = \nabla \cdot \mathbf{H} = \nabla \times \mathbf{H} = 0, \qquad \mathbf{H}^2 = H_0{}^2. \tag{4-2.6b}$$

The covariance group of this enlarged theory now contains, as a subgroup, the full three-dimensional rotation group which itself contains the group of rotations about a fixed axis as a subgroup, that is, a covariance group of the first theory. In the first theory the charged particle alone is considered to be observable while in the second theory both the particle and the field are

[1] C. S. Wu, E. Amber, R. W. Hayward, D. D. Hoppes, and P. R. Hudson, *Phys. Rev.*, **105**, 1413 (1957). The experiment of Wu, et. al., was based on a now famous suggestion of T. D. Lee and C. N. Yang, *Phys. Rev.*, **104**, 254 (1956).

considered to be observable. Nevertheless, it is clear that the two theories bear a direct relation to each other since we can always find a solution of Eqs. (4-2.6b) of the form $\mathbf{H} = (0, 0, H_0)$ and, for such solutions, Eq. (4-2.6a) reduces to Eq. (4-2.5). Furthermore, every solution of Eqs. (4-2.6b) can be transformed into this form by an element of the rotation group.

As a second example of two theories that are related to each other in a manner similar to the two theories discussed above, let us consider the theory in which the kpt are described by the antisymmetric tensor $F^{\mu\nu}$ with equations of motion (4-2.1) and a second theory in which the kpt are described by $F^{\mu\nu}$ and in addition a symmetric tensor field $g_{\mu\nu}$. We take, as the equations of motion of this second theory,

$$(\sqrt{-g}F^{\mu\nu})_{,\nu} = 0 \tag{4-2.7a}$$

$$F_{[\mu\nu,\rho]} = 0 \tag{4-2.7b}$$

with

$$F_{\mu\nu} = g_{\mu\rho} g_{\nu\sigma} F^{\rho\sigma}$$

and

$$R_{\mu\nu\rho\sigma} = 0 \tag{4-2.7c}$$

where g is the determinant of $g_{\mu\nu}$ and $R_{\mu\nu\rho\sigma}$ is the Riemann-Christoffel tensor formed from $g_{\mu\nu}$. Since the left-hand sides of these equations all transform as tensors or tensor densities under the MMG, it is clear that this second theory is MMG-covariant. These two theories again bear a direct relation to each other: We can always find a solution to Eq. (4-2.7c) of the form $g_{\mu\nu} = \eta_{\mu\nu}$ independent of $F^{\mu\nu}$ and for this solution Eqs. (4-2.7) reduce to Eqs. (4-2.1). Furthermore, it follows from the discussion of Section 3-4 that every solution of Eq. (4-2.7c) can be transformed by an element of the MMG into the form $g_{\mu\nu} = \eta_{\mu\nu}$.

Finally, let us consider a theory in which the kpt are described by a single scalar field ϕ and whose dynamical law has the form of the heat equation

$$\nabla^2\phi - \kappa \frac{\partial \phi}{\partial t} = 0. \tag{4-2.8}$$

The covariance group of this theory contains, as subgroups, the three-dimensional inhomogeneous rotation group and the time-translation group. A related theory can now be constructed in which the kpt are described by ϕ and a vector field n^μ and the dynamical laws have the form

$$(n^\mu n^\nu - \eta^{\mu\nu})\phi_{,\mu\nu} - \kappa n^\mu \phi_{,\mu} = 0 \tag{4-2.9a}$$

and

$$\eta_{\mu\nu} n^\mu n^\nu = 1, \qquad n^\mu{}_{,\nu} = 0. \tag{4-2.9b}$$

The covariance group of this second theory can then be shown to contain the Poincaré group as a subgroup which, in turn, contains the inhomogeneous rotation group and the time translation group as subgroups. A particular solution of Eqs. (4-2.9b) has the form $n^\mu = (1, 0, 0, 0)$ and, as we will prove later, every other solution of these equations can be transformed to this form by an element of the Poincaré group. For this solution Eq. (4-2.9a) reduces to Eq. (4-2.8) so that these two theories also bear a direct relation to each other.

Problem 4.2: Show that it is possible to construct still a third theory that is related to the previous two theories and that is MMG-covariant.

To make clear the relation between theories such as those discussed above we will introduce the notion of an *equivalence class of dpt*. If a physical theory admits a covariance group, it is possible to divide the totality of dpt of the theory into equivalence classes relative to this group. Two dpt are members of the same equivalence class if they are associated with each other by some element of the covariance group. All of the dpt belonging to a given equivalence class correspond to what we shall call the same *intrinsic state* of the physical system described by the theory. The individual dpt on the other hand correspond to *extrinsic states* of the system—that is, to the measurements made on the system by an observer. The existence of equivalence classes is thus a reflection of the fact that the relative "orientation" between a physical system and an observer can be varied without inducing at the same time a corresponding variation in the intrinsic state of the system.

Consider, for example, a many-particle system that has associated with it as a covariance group the three-dimensional rotation group and an observer who can measure the positions of the particles as functions of time. An observed set of values of these positions therefore corresponds to a dpt of the system. If now the system as a whole is rotated relative to the observer without disturbing the relative positions of the particles—that is, without disturbing the intrinsic state of the system—the observer will measure a new set of particle positions and hence a different dpt of the system. However, both of these dpt correspond to the same intrinsic state of the system. They also belong to the same equivalence class since they are related to each other by an element of the rotation group.

The question of whether two different extrinsic states of a system correspond to the same intrinsic state can be decided by finding an element of the covariance group associated with the system that transforms one of the corresponding dpt into the other. It can also be decided by calculating the *invariants* of the two dpt. Whenever a physical theory admits a covariance group, it is possible to construct invariants relative to this group from the mathematical quantities that describe the kpt of the theory. As the name implies, these invariants have the same values for two kpt that are associated with each other by an element of the covariance group. They have, therefore, the same values for all of the dpt belonging to the same equivalence class and hence corresponding to the same intrinsic state of the system described by the theory. In the above example the relative distances between the particles are invariants relative to the rotation group and hence any change in the system that does not affect these distances leaves unchanged the intrinsic state of the system.

Because it is the equivalence classes rather than the individual dpt of a theory that correspond to the various intrinsic states of a physical system, a knowledge of these states does not, by itself, determine completely the physical theory that describes the system. In fact, one can always construct a hierarchy of theories all of which have the same equivalence-class structure in the sense that the equivalence classes of these theories can be put into one-to-one correspondence with each other. Two theories of such a hierarchy will differ both with regard to the mathematical quantities that describe their respective kpt and their respective covariance groups. However, the set of mathematical quantities that describe the kpt of a given theory in such a hierarchy will contain, as subsets, those of each theory that precedes it in the hierarchy. Likewise, its covariance group will contain, as a subgroup, the covariance group of each preceding theory. It is clear that the pairs of theories discussed above belong to such hierarchies.

The question then arises as to which theory of a hierarchy one should use to describe a given physical system. The answer rests, of course, in the final analysis, on the measurements that one can make on the system. It is necessary that each quantity used to describe the kpt of a theory must, at least in principle, be measurable. This restriction will limit the choice of theories to the first few in a hierarchy. Which of these remaining theories one chooses to describe the system then depends on what measurements are made on the system. Thus, in the case of the particle moving in a magnetic field one could measure only the relative orientation between the particle trajectory and the field and hence use the first theory (the theory whose equations of motion are given by Eqs. (4-2.5)) to describe these measurements. Or, one could measure the orientation of both the particle trajectory and the field relative to some external reference frame, in which case one would use the second theory to describe these measurements. The fact that measurements of the second kind can be performed means that the second theory is physically acceptable since it is susceptible to experimental verification in all of its parts. It also shows that the first theory describes a subset of measurements that can be made on the system that is described by the second theory.

From the above discussion we see that not only can we rule out all but a few of the theories of a hierarchy on the grounds that they contain unobservable elements, but that of these remaining theories there is one that describes the maximum amount of information that can be obtained by means of measurements made on a system. This *maximal* theory describes the complete system under investigation; the remaining theories, in effect, describe only a part of the system. The existence of more than one description of a system is a reflection of the fact that a subset of elements of the system infiuence the behavior of the other elements but not vice versa. For certain types of measurements the values of the quantities associated with the elements of this subset are constants. By assigning these values to the corresponding quantities we can, in effect, reduce the maximal theory that describes the system to one of the theories that precedes it in the hierarchy. Thus, in the example of the particle moving in an external magnetic field, the components of the magnetic field remain unchanged during the course of the particle motion. When these components are substituted into the equations of motion (4-2.6), they reduce, as we have seen, to the equations of motion (4-2.5).

It is important to know what is the maximal theory that can describe a physical system since, in general, the constancy of a subset of the quantities that describe its kpt is only an approximation. We know, for example, that the description of the motion of a charged particle in an external magnetic field neglects the effect of this motion on the sources of the field. When we take account of this effect, we are led to the full theory of electrodynamics. Likewise, the constancy of $g_{\mu\nu}$ in the theory with equations of motion (4-2.7) suggests that if it is in fact a verifiable theory, it is also an approximation to another theory that takes account of the effect of the field described by $F^{\mu\nu}$ on the field described by $g_{\mu\nu}$. We will see later that such a theory does exist within the framework of general relativity.

So far we have made no assertions concerning the covariance groups of the physical theories that we shall deal with in this book. However, since all of these theories make use of the geometrical objects discussed in the previous chapters to describe their respective kpt, we will now put forth the hypothesis that the MMG must be a covariance group of all space-time theories. In succeeding chapters we will argue that the various dpt that constitute the equivalence classes of a MMG-covariant theory correspond to measurements made with one or another set of space-time measuring devices—that is, devices capable of assigning positions and times to physical events—and that the elements of this group correspond to the relations that exist between these various measurements. The hypothesis of MMG-covariance must, of course, have testable consequences before we can accept it. Among other things it leads to the conclusion that the field $g_{\mu\nu}$ which appears in the equations of motion (4-2.7) must be measurable. It also leads to the conclusion that the theories with equations of motion (4-2.6) and (4-2.9) are not maximal and that the quantities which must be introduced to make them maximal relative to the MMG must also be observable.

One final comment is in order concerning the covariance properties of physical theories. It has been argued by a number of authors that a covariance requirement is without physical content since it is always possible to achieve such covariance by employing a sufficient number of mathematical quantities in the description of the kpt of the theory.[2] However, unless all of these quantities can be measured, at least in principle, such a theory must be rejected on the grounds that it cannot be verified experimentally. Thus, far from being devoid of physical content, a covariance requirement has important physical consequences. It limits both the possible quantities that can describe the kpt and the forms of the dynamical laws of a theory.

4-3. Absolute and Dynamical Objects in Physical Theories

From the discussion of the previous section we see that there are essentially two different types of quantities that enter into the description of the kpt of a theory. Among other things, the determination of the values of one type for dpt is independent of the values of the other but not vice versa. Thus Eq. (4-2.7c) fixes the possible values of $g_{\mu\nu}$ independent of the values of $F^{\mu\nu}$. However, the values of $F^{\mu\nu}$ depend on the values of $g_{\mu\nu}$. Likewise, the possible values of n^{μ} are determined by Eq. (4-2.9b) independent of the values of ϕ. Furthermore, both $g_{\mu\nu}$ and n^{μ} have the same set of values for at least one dpt in every equivalence class and, hence, any allowed set of values of these quantities occurs in each equivalence class. In what follows we shall call quantities like $g_{\mu\nu}$ and n^{μ} *absolute objects* and quantities like $F^{\mu\nu}$ and ϕ, *dynamical objects*. We now turn to a precise definition of these quantities.

Let us assume that the kpt of a particular theory are characterized by a geometrical object y_A. It is possible that a subset of the components of y_A do not appear in the equations of motion for the remaining components and furthermore can be eliminated from the theory without altering the structure of its equivalence classes. Such a subset is obviously irrelevant to the theory. We shall assume, therefore, that no subset of the components of y_A is irrelevant in this sense. We now proceed to divide the components of y_A into two sets, ϕ_α and z_a where the ϕ_α have the following two properties:

(1) The ϕ_α constitute the basis of a faithful realization of the covariance group of the theory.

(2) Any ϕ_α that satisfies the equations of motion of the theory appears, together with all its transforms under the covariance group, in every equivalence class of dpt.

[2] Such an argument was developed in great detail by E. Kretschmann, *Ann. Physik*, **53**, 575 (1917).

The ϕ_α, if they exist, are the components of the absolute objects of the theory. The remaining part of y_A, the z_a, are then the components of the dynamical objects of the theory. It is therefore the z_a by themselves that distinguish between the various equivalence classes of dpt of the theory. In particular, if there is only one such equivalence class, then y_A is wholly absolute.

In most cases it is fairly evident which are the absolute and which the dynamical elements of the theory, although this is not always so. Thus, in the case of the enlarged set of equations (4-2.7), we see that $g_{\mu\nu}$ is an absolute object. Equation (4-2.7c) by itself determines $g_{\mu\nu}$ up to the arbitrariness that is a consequence of MMG-covariance. Furthermore, within each equivalence class, one can always find at least one element with $g_{\mu\nu} = \eta_{\mu\nu}$. $F^{\mu\nu}$ is therefore the dynamical object in this theory. In the case of the theory with dynamical laws given by Eq. (4-2.9) it is clear that the vector n^μ is the absolute element of the theory and ϕ is the dynamical element. In the case of Eqs. (4-2.3), the quantities $\tilde{g}_{\mu\nu} \equiv g_{\mu\nu}/\sqrt[4]{-g}$ constitute the components of an absolute object. This is so because the Weyl tensor $C_{\mu\nu\rho\sigma}$ can be expressed entirely in terms of the $\tilde{g}_{\mu\nu}$ and their derivatives, and the vanishing of this tensor is the necessary and sufficient condition that one can always find a mapping such that the transformed components of $\tilde{g}_{\mu\nu}$ take on the values $\eta_{\mu\nu}$. However, in this theory, g is a dynamical object, since the vanishing of the curvature scalar is not sufficient to determine this quantity without the imposition of boundary conditions. Finally, in the case of Eq. (4-2.8), we see that none of the components of $g_{\mu\nu}$ constitutes an absolute object.

4-4. Symmetry Groups

Having discussed the role played by covariance in the structure of physical theories, we now turn our attention to the concept of *symmetry* and its role in this structure. Like covariance, the notion of symmetry has been defined in a number of different ways, not all of which are equivalent. It has even been used interchangeably with the term covariance. For us, the words "covariance" and "symmetry" denote separate and distinct concepts.

Unlike the notion of covariance, the notion of symmetry is more immediately related to our intuitive notions concerning the structure of nature because we can speak of the symmetries of the individual dpt of a system, and it is the dpt that directly correspond to the observed states of physical systems. In simple cases one can see how the notion of symmetry arises. Consider, for instance, what we mean when we say that a ball is round. We ascribe the property of roundness to an object if its shape appears unaltered as we perform an arbitrary rotation of the object about some fixed point in the object.

In other words, we choose an arbitrary point on the surface of the object and compare the element of surface at that point before and after such a rotation. If the point still lies on the surface of the object and the two surface elements are identical for all points that initially lie on the surface, then we say that the object is round. This example suggests that we define a symmetry of a local geometrical object as follows: If $y_A(x)$ constitutes a realization of a group G and

$$y'_A(x) - y_A(x) = 0 \tag{4-4.1}$$

for some element of G, then that element is a symmetry of the object in question. Notice that we are comparing the transformed components $y'_A(x)$ with the original components $y_A(x)$ *at the same point* of the space-time manifold. (It would not do to require that $y'_A(x') - y_A(x) = 0$ because, among other things, this is always true for a scalar.) As a consequence, an infinitesimal element of G will be a symmetry of y_A if

$$\bar{\delta} y_A(x) = 0 \tag{4-4.2}$$

for this element. The totality of all elements of G that are symmetries of $y_A(x)$ constitute the *symmetry group* of this object.

Problem 4.3. Show that the symmetries of a geometrical object constitute a group.

As an example of a symmetry group let us consider a symmetric cotensor $g_{\mu\nu}$. For infinitesimal mappings of the MMG, Eq. (4-4.2) takes the form

$$\bar{\delta} g_{\mu\nu} \equiv -g_{\mu\rho}\xi^\rho{}_{,\nu} - g_{\rho\nu}\xi^\rho{}_{,\mu} - g_{\mu\nu,\rho}F^\rho = 0. \tag{4-4.3}$$

This equation, known as *Killing's equation*, will play an important role in later discussions. The descriptors ξ^ρ that satisfy Killing's equation are called *Killing vectors*, and serve to characterize important geometrical properties of various symmetric cotensors. Thus, if $g_{\mu\nu} = \eta_{\mu\nu}$, where $\eta_{\mu\nu}$ is the Minkowski matrix, at all points of the manifold, Eq. (4-4.3) reduces to

$$-\eta_{\mu\rho}\xi^\rho{}_{,\nu} - \eta_{\rho\nu}\xi^\rho{}_{,\mu} = 0.$$

The most general solution of this equation is of the form

$$\xi^\mu = \varepsilon^\mu + \varepsilon_\nu{}^\mu x^\nu, \tag{4-4.4}$$

where ε^μ and $\varepsilon_{\mu\nu} \equiv \eta_{\mu\rho}\varepsilon_\nu{}^\rho = -\varepsilon_{\nu\mu}$ are infinitesimal parameters. Hence the symmetry group of this particular cotensor is the Poincaré group.

Problem 4.4. Find the most general Killing vector for the two-dimensional cotensor $g_{\mu\nu}$ with components $g_{11} = 1, g_{22} = \sin^2(x^1), g_{12} = g_{21} = 0,$ and show that the symmetry group of this tensor is the three-dimensional rotation group. (Refer to the discussion at the end of Section 1-4.)

We can also make use of Eq. (4-4.2) to determine the most general functional dependence that a particular type of geometrical object can have and still admit a given group of mappings as a symmetry group. As a first example, let us determine the most general spherically symmetric scalar, that Is, the most general scalar that admits the three-dimensional rotation group as a symmetry group. A particular realization of this group is gotten by taking $\delta x^0 = 0$ and δx^r as given by

$$\delta x^r = \varepsilon^{rs} x^s, \tag{4-4.5}$$

where $\varepsilon^{rs} = -\varepsilon^{sr}$ are three arbitrary infinitesimal constants. For this particular form of δx^μ, Eq. (4-4.2) becomes, for a scalar ϕ,

$$\begin{aligned}\bar{\delta}\phi(x) &= -\phi_{,\mu}\delta x^\mu \\ &= -\phi_{,r}\varepsilon^{rs}x^s = 0.\end{aligned}$$

We see that this equation is satisfied by any arbitrary function of $r^2 \equiv (x^1)^2 + (x^2)^2 + (x^3)^2$ and x^0. Since we will need the result for our later work, let us next find the most general form of a spherically symmetric tensor $h_{\mu\nu}$. Equation (4-4.2) now takes the form

$$-h_{\mu\rho}\,\delta x^\rho{}_{,\nu} - h_{\rho\nu}\,\delta x^\rho{}_{,\mu} - h_{\mu\nu,\rho}\,\delta x^\rho = 0.$$

For the case when $\mu = \nu = 0$, this equation reduces to

$$-h_{00,r}\varepsilon^{rs}x^s = 0,$$

which implies that h_{00} is some arbitrary function of r^2 and x^0. For the case $\mu = 0$, $\nu \neq 0$, we have that

$$-h_{0r}\varepsilon^{rs} - h_{0s,u}\varepsilon^{ur}x^r = 0,$$

which implies that h_{0r} must be of the form

$$h_{0s} = \zeta(r^2, x^0)x^s,$$

where ζ is another arbitrary function of r^2 and x^0. Finally, for the case $\mu^0 \neq 0$, $\nu \neq 0$, we obtain the conditions

$$-h_{ru}\varepsilon^{us} - h_{us}\varepsilon^{ur} - h_{rs,u}\varepsilon^{uv}x^v = 0$$

which has, as a general solution,

$$h_{rs} = \psi(r^2, x^0)\delta_{rs} + \xi(r^2, x^0)x^r x^s,$$

where again ψ and ξ are arbitrary functions of r^2 and x^0 and where $\delta_{rs} = 1$ for $r = s$ and zero otherwise. In both the case of the scalar ϕ and the symmetric tensor $h_{\mu\nu}$, the general expressions found here agree with the usual expressions for these quantities when the δx^μ assume the Cartesian forms used here. We could, of course, have used any other set of expressions for the δx^μ that constituted a realization of the three-dimensional rotation group, thereby obtaining different forms for ϕ and $h_{\mu\nu}$. However, all such forms can be transformed into one another by means of mappings of the manifold onto itself, since symmetry under a group is an intrinsic property of a geometrical object.

Problem 4.5. Find the most general form for a symmetric tensor that admits the Galilean group as a symmetry group, using the Cartesian forms for δx^μ given by Eq. (1-4.6).

Problem 4.6. Find the most general symmetric tensor that is invariant under rotations about a fixed axis—that is, that has cylindrical symmetry.

Having defined the symmetry group of a geometrical object we can now define the *symmetry group of a physical system.* The symmetry group of this system is then defined to be the largest subgroup of the covariance group of this theory, which is simultaneously the symmetry group of its absolute objects. In particular, if the theory has no absolute objects, then the symmetry group of the physical system under consideration is just the covariance group of this theory. As we shall see presently, the requirement that a given system admits a particular group as a symmetry group restricts further the type of theories which can be used to describe the system. It also leads, in certain cases, to conservation laws for the system. However, before we discuss these consequences of the existence of a symmetry group, let us apply our definition to the theories considered in Section 4-2.

For the theory with the dynamical laws (4-2.1) there are no absolute objects. The Poincaré group, being a covariance group of this theory, is therefore a symmetry group of the system it describes. In the case of the related theory with dynamical laws (4-2.7), $g_{\mu\nu}$ is an absolute object whose symmetry group is, as we saw, just the Poincaré group. These two theories thus lead to the same symmetry group for the system they describe. There are no absolute objects in the theory with the dynamical law (4-2.8), so again the symmetry group of the system described by this theory coincides with the covariance group of this theory. The related theory, with dynamical laws (4-2.9), has n^μ as its absolute element. The covariance group of this theory is the Poincaré group, and n^μ transforms like a contravector under mappings of this group. Therefore

$$\bar{\delta} n^\mu = n^\nu \xi^\mu{}_{,\nu} - n^\mu{}_{,\nu}\xi^\nu,$$

with ξ^μ given by Eq. (4-4.4). To find the symmetry group of n^μ we need to know a solution of Eqs. (4-2.9b), which we can take to be given by $n^0 = 1$, $n^r = 0$. With n^μ of this form, the requirement that $\bar{\delta} n^\mu = 0$ leads to the condition that $\varepsilon_0{}^\mu = 0$ in Eq. (4-4.4). But in this case the ξ^μ reduce to the descriptors for the inhomogeneous rotation plus time translation group, which is just the covariance group of the theory with the dynamical law (4-2.8).

In the next section we shall discuss the relation between the symmetry group of a physical system and the conservation laws and continuity equations associated with the system. Here we will comment on the requirement that a given group be a symmetry group of a given physical system or a class of physical systems. Even more than the requirement that a given group be the covariance group of a theory, the requirement that a given group be a symmetry group limits the possible theories that satisfy this requirement. Thus the requirement that the Poincaré group be the symmetry group of a class of theories rules out the heat equation as a possible member of this class, even though it is possible to formulate the heat equation in a Poincaré covariant manner.

Likewise, the requirement that the MMG be a symmetry group of the system described by the symmetric tensor $g_{\mu\nu}$ rules out Eqs. (4-2.2) and (4-2.3) as dynamical laws for this object, even though they both admit the MMG as a covariance group. In fact, if we require that the dynamical laws for $g_{\mu\nu}$ be of second differential order, then only an equation of the form (4-2.4) has the MMG as a symmetry group. In what follows we shall speak of the *invariance* of a physical system under a group when that group is a symmetry group of the system. In our later discussions we will see that the totality of physical theories that go to make up Newtonian mechanics are invariant under the inhomogeneous Galilean group. Likewise, the theories of special relativity are characterized by invariance under the Poincaré group, while those of general relativity are invariant under the MMG.

4-5. Variational Principles and Conservation Laws

In addition to the differences between absolute and dynamical objects discussed in Section 4-3 there is another important difference that appears to be characteristic of these two types of objects. The equations of motion for the dynamical objects can often be derived from a variational principle, especially if these objects are fields. On the other hand, it appears to be the case, although we can give no proof of the assertion, that the equations of motion for the absolute objects do not have this property. In the case of the two examples considered in Section 4-2, neither Eq. (4-2.4c) nor Eq. (4-2.6b) can be derived from a variational principle, since in both cases there are more equations than independent components of the absolute objects. In the following discussion we will assume that the equations of motion for the dynamical objects of a theory follow from a variational principle and that those for the absolute elements do not.

In conformity with the above discussion, let us assume that the components of the geometrical object y_A, which describes the kpt of a given theory, can be divided into two groups according to the scheme

$$\{y_A\} \Rightarrow \{z_a, \phi_\alpha\}, \tag{4-5.1}$$

where the z_a represent the dynamical components and ϕ_α the absolute components of y_A. We assume further that there exists an action I of the form

$$I(z, \phi) = \int_R \mathfrak{L}(z_a, z_{a,\mu}, \phi_\alpha, \phi_{\alpha,\mu}; x^\mu)\, d^4x, \tag{4-5.2}$$

where R is some region of the space-time manifold. We have assumed that the Lagrangian $\mathfrak{L}$ does not depend on higher than the first derivatives of the components of y_A, since this condition is usually met in most cases of interest. This assumption does not in any way limit the generality of the conclusions to follow, and the whole analysis can be carried through with higher-order Lagrangians, although with considerably more labor. Under an arbitrary variation $\bar{\delta} y_A(x) \equiv y'_A(x) - y_A(x)$ of the y_A, where $y'_A(x)$ need not be related to $y_A(x)$ by means of a mapping, the change in the value of I is given by

$$\bar{\delta} I \equiv I(y + \bar{\delta} y) - I(y).$$

If we expand $I(y + \bar{\delta} y)$ in a Taylor series about $y_A(x)$ we have

$$\begin{aligned} I(y + \bar{\delta} y) &= \int_R \mathfrak{L}(y + \bar{\delta} y)\, d^4x \\ &= \int_R \mathfrak{L}(y)\, d^4x + \int_R \left\{ \frac{\partial \mathfrak{L}}{\partial y_A} \bar{\delta} y_A + \frac{\partial \mathfrak{L}}{\partial y_{A,\mu}} \bar{\delta} y_{A,\mu} \right\} d^4x \end{aligned}$$

correct to first order in $\bar{\delta} y_A$. Since $\bar{\delta} y_{A,\mu} = (\bar{\delta} y_A)_{,\mu}$, we can rewrite the last term as

$$\int_R \frac{\partial \mathfrak{L}}{\partial y_{A,\mu}} \bar{\delta} y_{A,\mu}\, d^4x = \int_R \left\{ -\left(\frac{\partial \mathfrak{L}}{\partial y_{A,\mu}} \right)_{,\mu} \bar{\delta} y_A + \left(\frac{\partial \mathfrak{L}}{\partial y_{A,\mu}} \bar{\delta} y_A \right)_{,\mu} \right\} d^4x.$$

Then, with the help of Gauss' law as expressed in Eq. (1-9.7), we can convert the last term on the right-hand side into the surface integral

$$\int_B \frac{\partial \mathfrak{L}}{\partial y_{A,\mu}} \bar{\delta} y_A\, dS_\mu,$$

where the integration is over the three-dimensional boundary B of the region R, and dS_μ is an element of area of this boundary with components given by

Eq. (1-9.8). Collecting terms we have

$$I(y + \bar{\delta} y) = \int_R \mathfrak{L}(y)\, d^4x + \int_R \frac{\partial \mathfrak{L}}{\delta y_A} \bar{\delta} y_A\, d^4x + \int_B \frac{\partial \mathfrak{L}}{\partial y_{A,\mu}} \bar{\delta} y_A\, dS_\mu .$$

The quantity $\delta\mathfrak{L}/\delta y_A$ appearing in the above equation is called the *variational derivative* of $\mathfrak{L}$ with respect to y_A and is given by

$$\frac{\delta \mathfrak{L}}{\delta y_A} \equiv \frac{\partial \mathfrak{L}}{\partial y_A} - \left\{\frac{\partial \mathfrak{L}}{\partial y_{A,\mu}}\right\}_{,\mu} . \tag{4-5.3}$$

Thus $\bar{\delta} I$ has the form

$$\bar{\delta} I = \int_R \frac{\delta \mathfrak{L}}{\delta y_A} \bar{\delta} y_A\, d^4x + \int_B \frac{\partial \mathfrak{L}}{\partial y_{A,\mu}} \bar{\delta} y_A\, dS_\mu . \tag{4-5.4}$$

Our fundamental assumption now is that the dynamical laws for the z_a follow from a variational principle using I as the action while those for the ϕ_α do not. This means that the dpt are such that I is stationary under arbitrary variational of the z_a that vanish on the boundary B. Thus $\bar{\delta} I$ given by Eq. (4-5.5) will vanish for such variations (we must, of course, not vary ϕ_α), and hence the dynamical laws of motion for the z_a will be just the Euler-Lagrange equations

$$\mathfrak{L}^a = \frac{\delta \mathfrak{L}}{\delta z_a} = \frac{\delta \mathfrak{L}}{\delta y_A} - \left\{\frac{\delta \mathfrak{L}}{\delta z_{a,\mu}}\right\}_{,\mu} = 0. \tag{4-5.5}$$

Since, however, the laws for the ϕ_α do not follow from a variational principle, it is not true for $\delta\mathfrak{L}/\delta\phi_\alpha = 0$ for a dpt.

In order to obtain the Lagrangian for the geometrical form of the equations of motion (4-2.4), we first introduce the potential covector A_μ and set

$$F_{\mu\nu} = A_{\nu,\mu} - A_{\mu,\nu} . \tag{4-5.6}$$

With this form for $F_{\mu\nu}$, Eq. (4-2.7b) is satisfied identically so that we need only to get Eq. (4-2.7a) from the variational principle. The Lagrangian that yields these equations upon variation is

$$\begin{aligned} \mathfrak{L} &= -\tfrac{1}{4}\sqrt{-g} F^{\mu\nu} F_{\mu\nu} \\ &= -\tfrac{1}{4}\sqrt{-g} g^{\mu\rho} g^{\nu\sigma} (A_{\nu,\mu} - A_{\mu,\nu})(A_{\sigma,\rho} - A_{\rho,\sigma}). \end{aligned} \tag{4-5.7}$$

Alternately one could treat the $F^{\mu\nu}$ and A_μ as independent objects and use the Lagrangian

$$\mathfrak{L} = \tfrac{1}{4}\sqrt{-g} F^{\mu\nu} F_{\mu\nu} - \tfrac{1}{2}\sqrt{-g} F^{\mu\nu}(A_{\nu,\mu} - A_{\mu,\nu}).$$

Upon varying $F^{\mu\nu}$, one obtains Eqs. (4-5.6), whereas variation of A_μ yields the equations

$$(\sqrt{-g}F^{\mu\nu})_{,\nu} = 0,$$

which are equivalent to Eqs. (4-2.7a).

Given an action of the form (4-5.2), let us now ask how the Lagrangian $\mathfrak{L}$ transforms under the covariance group of the theory with this action. The answer to this question will have important consequences if the system being described is invariant under a subgroup of this covariance group. To answer our question, we will make use of the basic requirement that under the action of the covariance group, dpt are transformed into dpt. This means first of all that the functional form of $\mathfrak{L}$ must not change under the action of the group. If this were not the case then varying the transformed action would lead to a set of dynamical laws that were different than those obtained by varying the original Lagrangian. This requirement, of course, does not assure us that the transform of a dpt is again a dpt. We now derive a sufficient condition for this to be the case.

Consider the action of an element of the covariance group on the action $I(y)$. Under the action of this element the region R of integration will get mapped onto a new region R' and the value $\mathfrak{L}'(x')$ of $\mathfrak{L}$ at the point x'^μ will, in general, be some function of $\mathfrak{L}(x)$ and the mapping function that maps x^μ onto x'^μ. Consequently the transformed action I' will be given by

$$I' = \int_{R'} \mathfrak{L}'(x)\, d^4x.$$

Consider now a kpt that makes I stationary for variations that vanish on the boundary of R. Under the action of the group, this kpt will get transformed into a new kpt, as will all variations of it. In particular, if the original variation vanishes on the boundary of R, then the transformed variation will vanish on the boundary of R' for all fields that we will have need to consider, that is, those with linear homogeneous transformation laws. Hence, if $I' = I$ for all kpt, the transform of a kpt that makes I stationary will make I' stationary. But $I' = I$ is just the condition that $\mathfrak{L}(x)$ transform like a scalar density of weight $+1$ under the action of the group. Actually this condition on $\mathfrak{L}(x)$ can be relaxed somewhat. All that is needed is that it differ from a scalar density by a complete divergence, that is, that there exist a set of four quantities Q^μ such that $\mathfrak{L} + Q^\mu{}_{,\mu}$ is a scalar density. The reason that we can relax the above restriction on $\mathfrak{L}$ to this extent is that the appearance of a complete divergence in an action will not contribute to the dynamical laws that follow from this action, since such a term can always be converted into a surface integral over the boundary of the region of integration and all allowed variations vanish on this boundary. Of course all that we have said above applies with equal force if the covariance group not only involves mappings of the manifold onto itself but also consists of internal transformations.

If the covariance group under consideration contains a part that can be generated from a set of infinitesimal elements, then the requirement that $\mathfrak{L} + Q^{\mu}{}_{,\mu}$ be a scalar density of weight $+1$ takes the form

$$\bar{\delta}(\mathfrak{L} + Q^{\mu}{}_{,\mu}) = -(\mathfrak{L} + Q^{\mu}{}_{,\mu})\xi^{\rho}{}_{,\rho} - (\mathfrak{L} + Q^{\mu}{}_{,\mu})_{,\rho}\xi^{\rho}$$

or

$$\bar{\delta}\mathfrak{L} = -(\mathfrak{L}\xi^{\rho} + Q^{\mu}{}_{,\mu}\xi^{\rho} + \bar{\delta}Q^{\rho})_{,\rho}. \tag{4-5.8}$$

Since the functional dependence of $\mathfrak{L}$ on y_A does not change under the action of the group, we can also calculate $\bar{\delta}\mathfrak{L}$ from a knowledge of $\bar{\delta}y_A$:

$$\begin{aligned}\bar{\delta}\mathfrak{L} &= \frac{\partial\mathfrak{L}}{\partial y_A}\bar{\delta}y_A + \frac{\partial\mathfrak{L}}{\partial y_{A,\mu}}\bar{\delta}y_{A,\mu} \\ &= \frac{\delta\mathfrak{L}}{\partial y_A}\bar{\delta}y_A + \left(\frac{\partial\mathfrak{L}}{\partial y_{A,\mu}}\bar{\delta}y_A\right)_{,\mu}\end{aligned} \tag{4-5.9}$$

for Lagrangians that depend on no higher than the first derivatives of the y_A. The two expressions for $\bar{\delta}\mathfrak{L}$ given by Eqs. (4-5.8) and (4-5.9) must be equal to each other for all kpt. Hence we obtain the important identity of Nöther:[3]

$$\bar{\delta}t^{\mu}{}_{,\mu} \equiv \frac{\delta\mathfrak{L}}{\delta y_A}\bar{\delta}y_A, \tag{4-5.10}$$

where

$$\bar{\delta}t^{\mu} = -\mathfrak{L}\xi^{\mu} - \frac{\partial\mathfrak{L}}{\partial y_{A,\mu}}\bar{\delta}y_A - Q^{\rho}{}_{,\rho}\xi^{\mu} - \bar{\delta}Q^{\mu}. \tag{4-5.11}$$

Consequently we conclude that if $\mathfrak{L}$ is such that $(\delta\mathfrak{L}/\delta y_A)\bar{\delta}y_A$ can be put into the form of a complete divergence, then the dpt of the dynamical laws that follow from this Lagrangian will be transformed into dpt under the action of the covariance group of the theory.

Let us now consider the situation where the y_A divide into two groups, as in Eq. (4-5.1). For a dpt we have Eq. (4-5.5) satisfied, so that Eq. (4-5.10) becomes

$$\bar{\delta}t^{\mu}{}_{,\mu} = \frac{\delta\mathfrak{L}}{\delta\phi_{\alpha}}\bar{\delta}\phi_{\alpha}. \tag{4-5.12}$$

[3] E. Nöther, *Nachr. Ges. Wiss. Göttingen*, 211 (1918).

In the special case that a particular mapping is such that $\bar{\delta}\phi_\alpha = 0$, we get an equation of continuity:

$$\bar{\delta}t^\mu{}_{,\mu} = 0 \tag{4-5.13}$$

which is satisfied for all dpt. Thus we arrive at an important conclusion concerning the relation between the symmetries of a physical system and the equations of continuity of that system. If the theory describing the system is such that the laws for the dynamical objects of the theory can be derived from a variational principle, then to each symmetry of the system there corresponds an equation of continuity, that is, a conservation law. When the symmetry group is either the inhomogeneous Galilean group or the Poincaré group, the corresponding conservation laws are those of energy, momentum, angular momentum, and center-of-mass motion.

If the symmetry group of a system is a Lie group, then the various quantities entering into the expression (4-5.11) for $\bar{\delta}t^\mu$ will have the form

$$\xi^\mu = f_i{}^\mu(x)\varepsilon^i, \tag{4-5.14a}$$

$$\bar{\delta}y_A = w_{Ai}(y(x), x)\varepsilon^i, \tag{4-5.14b}$$

and

$$\bar{\delta}Q^\mu = q_i{}^\mu(y(x), x)\varepsilon^i, \tag{4-5.14c}$$

where the $f_i{}^\mu$ and w_{Ai} are such that the y_A constitute the basis of a realization of the symmetry group and the $q_i{}^\mu$ are determined by the Lagrangian of the theory. Then we have the continuity equations

$$\bar{\delta}t^\mu{}_{,\mu} = t^\mu_{i\,,\mu}\varepsilon^i = 0, \tag{4-5.15}$$

where

$$\begin{aligned} \bar{\delta}t^\mu &= t_i{}^\mu\varepsilon^i \\ &= -\left\{(\mathfrak{L} + Q^\rho{}_{,\rho})f_i{}^\mu + \frac{\partial\mathfrak{L}}{\partial y_{A,\mu}} w_{Ai} + q_i{}^\mu\right\}\varepsilon^i \end{aligned} \tag{4-5.16}$$

is the defining equation for the quantity $t_i{}^\mu$. Since the ε^i are quite arbitrary, we can conclude from Eq. (4-5.15) that

$$t^\mu_{i\,,\mu} = 0. \tag{4-5.17}$$

Thus, if the symmetry group is an n parameter Lie group, we obtain n continuity equations. It should be kept in mind when evaluating $t_i{}^\mu$ that one must use the expressions for ϕ_α that led to the particular forms for ξ^μ, $\bar{\delta}y_A$, and $\bar{\delta}Q^\mu$, given by Eqs. (4-5.14). In the case of the equations of motion (4-2.1), where the symmetry group is the Poincaré group, if we take the ξ^μ to be given by Eq. (1-4.5) we must take $g_{\mu\nu} = \eta_{\mu\nu}$.

Problem 4.6. Show that the Lagrangian given by Eq. (4-5.7) is a scalar density of weight +1 under the MMG so that $Q^\mu = 0$. Work out the expressions for the various t_i^μ when $g_{\mu\nu} = \eta_{\mu\nu}$.

All that we have said here applies without change when we come to consider internal covariance groups. Again the symmetry group is defined by the condition that $\bar{\delta}\phi_\alpha = 0$ for the absolute objects, and again we obtain a continuity equation of the form (4-5.17) for each parameter of the symmetry group. In this regard the gauge group discussed in Sections 1-11 and 2-2 is of special interest. Under the action of this group a complex scalar ϕ transforms as

$$\bar{\delta}\phi = i\varepsilon\phi \quad \text{and} \quad \bar{\delta}\phi^* = -i\varepsilon\phi^* \tag{4-5.18}$$

while the vector potential A_μ transforms as

$$\bar{\delta}A_\mu = \varepsilon_{,\mu}. \tag{4-5.19}$$

If

$$F_{\mu\nu} = A_{\mu,\nu} - A_{\nu,\mu} = 0, \tag{4-5.20}$$

then in a simply connected domain of the manifold, A_μ can be expressed as the gradient of some scalar ψ:

$$A_\mu = \psi_{,\mu}.$$

If we now perform a gauge transformation with ψ as the gauge function, we find that $A'_\mu = A_\mu - \psi_{,\mu} = 0$. Thus, when $F_{\mu\nu} = 0$, A_μ is an absolute object. The symmetry group is then just the gauge transformations with ε a constant, as follows from Eq. (4-5.19). If the equations for ϕ and ϕ^* are derivable from a variational principle with a Lagrangian $\mathfrak{L}$ that is a density of weight +1, t_i^μ is given by

$$t^\mu = -i\frac{\partial\mathfrak{L}}{\partial\phi_{,\mu}}\phi + i\frac{\partial\mathfrak{L}}{\partial\phi^*_{,\mu}}\phi^* \tag{4-5.21}$$

and satisfies

$$t^\mu{}_{,\mu} = 0. \tag{4-5.22}$$

This continuity equation is usually interpreted as the continuity equation for electric charge.

The gauge group of electrodynamics differs from other covariance groups in one important respect. Being an Abelian group, it led to the transformation law (4-5.19) for the affinity A_μ. Consequently $\bar{\delta}A_\mu = 0$ when ε is a constant, whether or not A_μ is an absolute object. It then follows that t^μ, given by Eq.

(4-5.21) satisfies Eq. (4-5.22) in all cases, since the term $(\partial\mathfrak{L}/\partial A_{\mu,\nu})\bar{\delta}A_\nu$ in the expression (4-5.16) for $\bar{\delta}t^\mu$ is zero when ε is a constant. Physically this means that the electromagnetic field does not contribute to the electric current, a well-known fact.

Before we conclude this discussion of variational principles we must emphasize the fact that the requirement of invariance of the action I under a covariance transformation ($I' = I$) is a sufficient, but not necessary, condition that the dpt be transformed into dpt by the transformation. The Nöther identity (4-5.10) is therefore not a direct consequence of the covariance of the theory nor are the conservation laws (4-5.17) a consequence of the symmetry of the system being described. In both cases, to obtain these results, it is necessary to make the additional requirement that the action leading to the dynamical laws of the theory is an invariant. It is quite possible for a theory to possess a covariance group and to have dynamical laws that follow from a variational principle without the corresponding action being an invariant.

4-6. Gauge Groups

The MMG is a special example of a general class of covariance groups that is admitted by various physical theories, which have in common the property that the elements of such a group are characterized by one or more continuous space-time functions. Such groups will be called *gauge groups* after the gauge group of electrodynamics. Likewise, theories that admit gauge groups as covariance groups will be referred to as *gauge covariant theories*.

Whenever a theory admits a gauge group as a covariance group it is possible to impose *gauge conditions* on the dpt of the theory. These conditions consist of a set of relations between the objects that describe the kpt of the theory and that can be satisfied only by a subset of dpt in each equivalence class of dpt. The imposition of gauge conditions therefore has the effect of singling out from each such equivalence class a subclass of dpt that satisfy the gauge conditions. Usually these conditions take the form of local relations and are in general equal in number to the number of arbitrary functions that characterize the elements of the gauge group. The only restriction on the form of these conditions is that they can be satisfied by at least one member of every equivalence class of dpt of the theory. If we did not impose this restriction the use of a particular gauge condition could alter the physical content of the theory by eliminating the equivalence classes that did not contain dpt satisfying the condition.

Perhaps the most familiar example of a gauge condition is the Lorentz condition used in electrodynamics. We have already mentioned that when expressed in terms of the four-potential A_μ, Maxwell's equations admit the covariance group of gauge transformations $A_\mu \to A'_\mu = A_\mu + \Lambda_{,\mu}$, where Λ is an arbitrary space-time function. The Lorentz condition restricts the dpt of the theory to those that satisfy the equation

$$\eta^{\mu\nu} A_{\mu,\nu} = 0.$$

We will show later that every dpt of the Maxwell theory can be transformed into another dpt that satisfies the Lorentz condition by means of a suitable gauge transformation. The imposition of the Lorentz condition therefore does not alter the physical content of the Maxwell theory. In our study of general relativity we will also have occasion to make use of gauge conditions, more commonly called in this case *coordinate conditions.*

The imposition of gauge conditions is a mathematical device used mainly to simplify the equations of motion of a physical theory that admits a gauge group. Thus the imposition of the Lorentz condition reduces the Maxwell equations to a set of equations, each one of which involves only one component of A_μ. Also, the imposition of gauge conditions allows one to formulate an initial value problem for a gauge covariant theory. It should be emphasized, however, that gauge conditions by themselves have no physical content. While a particular condition may prove more convenient than others in certain applications, they all stand on an equal footing as far as the physical content of a theory is concerned. Only if kpt that satisfy the gauge condition were possible dpt would the condition have physical content; in fact, it would then have to be considered to be another dynamical law for specifying the dpt of the theory. But then, of course, we would be dealing with a different theory. In particular the observables of this new theory would be different from those of the original theory without the gauge condition.

The existence of a gauge group has far-reaching consequences for any theory that admits it as a covariance group. Hilbert[4] pointed out that the equations of motion of such a theory cannot all be independent of one another, but must satisfy a number of identities that are in general equal in number to the number of arbitrary functions that define an element of the group. These identities are sometimes called *Bianchi-type* identities, after the identities of this type that occur in general relativity. The reason for the existence of such identities is easy enough to see: Because of the freedom we have to impose gauge conditions on the dpt of a gauge covariant theory, the dpt of the theory that satisfy these conditions must, in effect, satisfy a larger number of equations than just the original equations of motion. It follows that these equations of motion cannot all be independent of one another, since if they were, they would in

[4] D. Hilbert, *Grundlagen der Physik, I, Nachr. Ges. Wiss. Göttingen,* 395 (1915).

general constitute, together with the gauge conditions, an overdetermined set of equations that would have no nontrivial solutions.

If the equations of motion of a gauge covariant theory follow from a variational principle, we can make use of the Nöther identity (4-5.10) to determine the form of the Bianchi identities of the theory. Let us assume that the basic variables of the theory are fields, $y_A(x)$, defined over the space-time manifold and that the action of the theory has the form

$$S = \int \mathfrak{L}\, d^4x.$$

The field equations are then

$$\mathfrak{L}^A \equiv \frac{\delta \mathfrak{L}}{\delta y_A(x)} = 0. \tag{4-6.1}$$

Consider now an infinitesimal gauge transformation under which

$$y_A(x) \rightarrow y'_A(x) = y_A(x) + \bar{\delta} y_A(x),$$

where $\bar{\delta} y_A(x)$ is assumed to have the form

$$\bar{\delta} y_A(x) = d^\mu_{Ai} \xi^i{}_{,\mu} + c_{Ai} \xi^i, \qquad i = 1, \cdots, n,$$

where d^μ_{Ai} and c_{Ai} are functions of the y_A. The arbitrary infinitesimal functions ξ^i appearing in the expression for $\bar{\delta} y_A$ are the descriptors of the transformation; they are equal in number to the number n of arbitrary functions that define an element of the gauge group. Although this expression is not the most general we could have assumed (we could have included higher derivatives of the ξ^i), it suffices for those gauge groups that we will have to deal with.

If we substitute the expression (4-6.1) for $\bar{\delta} y_A$ into the Nöther identity and perform an integration by parts, we obtain the identity

$$(\bar{\delta} t^\mu - d^\mu_{Ai} \mathfrak{L}^A \xi^i)_{,\mu} \equiv \{c_{Ai} \mathfrak{L}^A - (d^\mu_{Ai} \mathfrak{L}^A)_{,\mu}\} \xi^i. \tag{4-6.2}$$

Let us now integrate this identity over an arbitrary space-time region. With the help of Gauss' theorem the integral over the divergence appearing in this equation can be converted to a surface integral over the boundary of the region of integration. Since the ξ^i are arbitrary we can choose them so that they vanish on this boundary but are otherwise arbitrary. Consequently the surface integral can be made to vanish and we are left with the result that

$$\int d^A x \{c_{Ai} \mathfrak{L}^A - (d^\mu_{Ai} \mathfrak{L}^A)_{,\mu}\} \xi^i \equiv 0. \tag{4-6.3}$$

It then follows, since the ξ^i are arbitrary within the domain of integration, that

$$c_{Ai} \mathfrak{L}^A - (d^\mu_{Ai} \mathfrak{L}^A)_{,\mu} \equiv 0. \tag{4-6.4}$$

Finally, since the domain of integration is itself arbitrary, we can conclude that the above identities hold at all points of the space-time manifold. These are the Bianchi identities of the theory.

For us, the main consequence of the Bianchi identities is the way they control the dependence of $\mathfrak{L}^A$ on the highest derivatives of the y_A with respect to a given coordinate. Let us denote these derivatives by $y_A^{(n)}$. Then, if they appear linearly in $\mathfrak{L}^A$, we can write

$$\mathfrak{L}^A = \kappa^{AB} y_B^{(n)} + \mathfrak{J}^A \tag{4-6.5}$$

where κ^{AB} and $\mathfrak{J}^A$ depend on lower-order derivatives of the y_B with respect to this coordinate. If now we insert this expression for $\mathfrak{L}^A$ into the Bianchi identity and assume, for definiteness, that the coordinate in question is x^0, we obtain

$$d_{Ai}^0 \kappa^{AB} y_B^{(n+1)} + \cdots \equiv 0. \tag{4-6.6}$$

The exact form of the terms that are indicated by dots in this identity is not important; it suffices that they involve only derivatives of the y_A with respect to x^0 of order n or lower. Since all derivatives of the y_A at a point are arbitrary (there always will exist at least one kpt such that the derivatives of the y_A take preassigned values at a point), it follows that

$$d_{Ai}^0 \kappa^{AB} \equiv 0 \tag{4-6.7}$$

for all kpt. The d_{Ai}^0 are thus seen to be null vectors of the matrix κ^{AB}. It therefore follows that κ^{AB} is a singular matrix with n linearly independent null vectors. This has as a consequence that we cannot solve Eqs. (4-6.5) for all $y_A^{(n)}$ in terms of lower-order derivatives. As we shall see, our inability to solve the Eqs. (4-6.5) for all of the $y_A^{(n)}$ will have an effect on the initial value problem for theories with gauge groups not present in theories without such symmetry groups.

4-7. The Initial Value Problem

The initial value, or Cauchy, problem for a system of equations of motion can be stated in the following terms: Given a three-dimensional hypersurface imbedded in a space-time manifold and the values of the dynamical objects of a theory and a sufficient number of their normal derivatives on this hypersurface, construct a solution of the equations of motion of the theory that reduces to these values on the hypersurface. In the discussion to follow, the hypersurface on which the values of the dynamical objects and their normal derivatives are given will be referred to as the *initial value hypersurface* and the values themselves will be referred to as the *initial value data*.

In general the initial value problem will only possess solutions in a small but finite region surrounding any space-time point.

The exact nature of the initial value hypersurface and the type of initial value data given on it will, in general, depend on both the character of the dynamical objects and the equations of motion for these objects. Thus, if we are dealing with particle motion where the particle trajectories are given parametrically by

$$x^\mu = z_i{}^\mu(\lambda_i) \tag{4-7.1}$$

with λ_i a monotone increasing parameter along the ith trajectory, the initial value data might be specified by giving the values of $z_i{}^\mu$ and their first few derivatives at points where the trajectories intersect some hypersurface. Since we would want these points of intersection to be unique, the hypersurface must be such that no displacement along a trajectory lies in it. If, for instance, the tangents $\dot{z}_i{}^\mu = \partial z_i{}^\mu/\partial\lambda_i$ to the trajectories are required to satisfy the condition $g_{\mu\nu}(z_i)\dot{z}_i{}^\mu\dot{z}_i{}^\nu > 0$ then the normals n_μ to the hypersurface must also satisfy $g^{\mu\nu}n_\mu n_\nu > 0$ in order that the trajectories are guaranteed to intersect it in a point.

For a large class of theories that we shall be concerned with, the components y_A of the dynamical objects satisfy equations of the type

$$A^{\mu\nu}y_{A,\mu\nu} + B_A^{B\mu}y_{B,\mu} + C_A{}^B y_B = f_A \tag{4-7.2}$$

where the various coefficients and the f_A appearing here are functions defined on the manifold and where the signature of the matrix $A^{\mu\nu}$ is -3. Such equations are called *hyperbolic* partial differential equations. For such equations a suitable initial value surface is one whose normals satisfy $A^{\mu\nu}n_\mu n_\nu > 0$. Usually the components of $A^{\mu\nu}$ will be equal to the components of a metric $g^{\mu\nu}$. In this case such hypersurfaces are called spacelike hypersurfaces. The initial value data for such a surface would then consist of the values of the y_A and their first normal derivatives $\partial y_A/\partial n = g^{\mu\nu}n_\mu y_{A,\nu}$ on this hypersurface. It is also possible to give initial data on other types of hypersurfaces. In particular, considerable attention has been paid recently (see Section 13.2) to the initial value problem when the initial value data is given on characteristic surfaces.

In the case where the dynamical objects are fields, the key theorem of the initial value problem is due to Cauchy and Kowalewski.[5] It states that for a suitable initial value hypersurface the initial value problem admits a unique solution provided it is possible to solve the dynamical laws for the highest normal derivatives of the dynamical objects appearing therein. Consequently, equations of motion of a theory that admits a gauge group as a symmetry group will not, in general, be of the Cauchy-Kowalewski type.

[5] See R. Courant and D. Hilbert, *Methods of Mathematical Physics*, Vol. II (Wiley, New York, 1962) Chaps. 2 and 7 and references contained therein.

As an example of an application of the Cauchy-Kowalewski theorem let us consider the scalar wave equation

$$\eta^{\mu\nu}\phi_{,\mu\nu} = 0 \tag{4-7.3}$$

where $\eta^{\mu\nu} = \text{diag}(1, -1, -1, -1)$ and take, as an initial value hypersurface, the hypersurface $x^0 = 0$. The normals to this hypersurface, $n_\mu = (1, 0, 0, 0)$, clearly satisfy the condition $\eta^{\mu\nu}n_\mu n_\nu > 0$. Values of ϕ and its normal derivatives (in this case derivatives with respect to x^0, denoted by $\dot{\phi}$, $\ddot{\phi}$, etc.) on this hypersurface will be indicated by a subscript $_0$. Consider now a point in the neighborhood of the origin and assume that the value of ϕ at this point can be expanded in a power series in x^0 according to

$$\phi(x) = \phi_0(\mathbf{x}) + \dot{\phi}_0(\mathbf{x})x^0 + \tfrac{1}{2}\ddot{\phi}_0(\mathbf{x})(x^0)^2 + \cdots. \tag{4-7.4}$$

With the help of Eq. (4-7.3) we can express $\ddot{\phi}_0(\mathbf{x})$ and all higher derivatives in terms of the two functions $\phi_0(\mathbf{x})$ and $\dot{\phi}_0(\mathbf{x})$ so that, given these two functions on the $x^0 = 0$ hypersurface, we can compute all of the coefficients appearing in the expansion (4-7.4). If these functions are analytic, the series can be shown to converge in a small but finite region space-time region to a unique result. We see that the two functions $\phi_0(\mathbf{x})$ and $\dot{\phi}_0(\mathbf{x})$ constitute initial data for the wave equation.

The requirement that a set of equations of motion are of the Cauchy-Kowalewski type is not necessary either for their internal consistency or for the existence of solutions. As we shall see, the equations of motion for a system of particles acting via retarded or advanced action at a distance cannot be converted into the Cauchy-Kowalewski type. Nevertheless, such equations possess well-behaved, physically interpretable solutions. However, unless the equations of motion are of the Cauchy-Kowalewski type, it will not be possible to recast them in the framework of a canonical formalism. At the present time the canonical formulation of a set of equations of motion appears to be necessary for the construction of a quantum version of the theory with these equations of motion. Therefore, theories with equations of motion that are not of the Cauchy-Kowalewski type present a special problem in quantum physics.

Because of the Bianchi identities, the equations of motion associated with a gauge covariant theory are not of the Cauchy-Kowalewski type. If we take an initial value surface to be the submanifold $x^0 = \text{const}$, it follows from the discussion of the preceding section that one cannot solve these equations for the highest derivatives of the variables of the theory with respect to x^0. Therefore one cannot give sufficient initial value data on this surface for the determination of the coefficients in an expansion of the type employed in Eq. (4-7.4).

We can also give the following argument to show that the equations of motion of a gauge covariant theory are not of the Cauchy-Kowalewski type.

To do so, let us assume that the contrary is true—that is, that there exists an initial value surface and initial value data on this surface that correspond to a unique dpt of the theory. Consider now a gauge transformation that corresponds to the identity transformation in a finite region *R* of the space-time manifold that contains the initial value surface, but which differs from the identity transformation outside this region. When applied to the supposedly unique dpt, it yields another dpt that coincides with this dpt in the region *R*, but which differs from it outside of *R*. It follows that this second dpt must be related to the initial value data in the same way as the first dpt is related to it. We therefore have two different dpt associated with the same initial value data, contrary to assumption. The equations of motion of a gauge covariant theory thus require special treatment in their application to physical problems. Fortunately they usually can be converted to the Cauchy-Kowalewski type by the imposition of gauge conditions. Later we shall consider such a procedure in specific cases, such as electrodynamics and general relativity.

PART

II

ABSOLUTE SPACE-TIME THEORIES

5

Newtonian Mechanics

The first great modern scientific synthesis was achieved by Newton and was set forth in the *Principia*, first published in 1686. Many of the ideas developed in the *Principia* were not new, some having their genesis during the Renaissance and earlier. What was new was the presentation of these ideas as a single connected whole in what were essentially modern terms. The *Principia* contained many notable successes and also some notable failures; for example, Newton's inability to treat the motion of three or more bodies and continuum systems in general. But perhaps most important of all, Newton set forth in the *Principia* the ideas concerning the nature of space and time that lay at the foundations of all the advances of the next two hundred years and which were not seriously challenged until 1905, with the advent of special relativity.

In the Newtonian world picture space and time were separate and distinct absolutes and their symmetries were assumed to be the universal space-time symmetries of all physical systems. However, such systems appeared to possess an additional space-time symmetry over and above the Newtonian symmetries: They all seem to admit the special Galilean group as a symmetry group. But if these Galilean symmetries were indeed universal, it would mean that Newton's concept of absolute space could be eliminated from physics and with it the concept of absolute velocity, a conclusion known as the "principle of Galilean relativity."

Since the concept of absolute motion played a key role in the formulation of the famous three laws of motion, Newton and many of his predecessors were deeply troubled by the existence of the Galilean symmetries. While the Newtonian symmetries were assumed to be fundamental, the Galilean ones were not; it was assumed that at least some physical systems would not admit them. With the advent of electrodynamics physicists at last appeared to have at hand such a system. Thus were conceived the Michelson-Morley and Trouton-Noble experiments for determining the velocity of the earth through the "aether," that is, absolute space. It was the negative outcome of these and other such experiments that finally led Einstein to give up the Newtonian concepts in favor of those of special relativity. His work, together with that of Minkowski, led to the abandonment of both the concepts of absolute space and absolute time in favor of the single concept of absolute space-time. In this chapter we shall present those Newtonian concepts that are necessary for an understanding of the developments that led to the special theory of relativity.

5-1. Newtonian Space-Time

For Newton, space and time were conceived to be physical entities. Furthermore, they were absolute entities, unchanging and unchanged by the presence of other physical entities. Hence the definitions[1]:

1. Absolute, true, and mathematical time, of itself, and from its own nature, flows equably without relation to anything external.
2. Absolute space, in its own nature, without relation to anything external, remains always similar and immovable.

It is clear that in speaking of space and time in this way, Newton was referring to their geometrical properties. Furthermore the concept of a straight or right line was fundamental to the whole Newtonian picture, as was the concept of angles and distances between points. Since no other types of geometries were known at the time, there can be no doubt that when Newton spoke of space he meant the space of Euclid, that is, a three-dimensional manifold upon which was superimposed a flat, metric geometry. And by "absolute space" it is clear that he considered this geometry to be unaffected by any physical disturbances in this space. Likewise, Newtonian time must be understood as a one-dimensional manifold with a "distance" defined on it that is also unaffected by such physical disturbances.

The geometrical structure of Newtonian space-time requires, for its mathematical description, a number of geometrical objects. The notion of absolute time can be characterized by introducing onto the space-time manifold a one-parameter family of nonintersecting hypersurfaces defined by

$$t = t(x^\mu). \qquad (5\text{-}1.1)$$

[1] This and other quotes from the *Principia* are taken from I. Newton, *Principia*, edited by F. Cajori (Univ. of California Press, Berkeley, 1947).

The only other restriction placed on these hypersurfaces is that they be topologically equivalent to the three-dimensional Euclidean plane. Two points that lie in the same $t = \text{const}$ hypersurface are said to be *simultaneous* to each other. Each $t = \text{const}$ hypersurface can therefore be considered to correspond to a point of a one-dimensional manifold, the absolute time manifold of Newton. In what follows we shall sometimes refer to a $t = \text{const}$ hypersurface as a *plane of absolute simultaneity*.

In addition to the planes of absolute simultaneity we must also introduce objects that correspond to the absolute space of Newton. This is done by constructing a three-parameter congruence (space-time filling) of curves with the property that each curve of the congruence intersects a given $t = \text{const}$ hypersurface in only one point. Thus no two points on such a curve can be simultaneous to each other. We can therefore characterize the points along any one curve by a path parameter that coincides with the parameter associated with the planes of absolute simultaneity. In addition, the individual curves of the congruence are characterized by three parameters λ^r. The points on such a curve are thus given by the equations

$$\lambda^r = \lambda^r(x^\mu). \tag{5-1.2}$$

Each curve $\lambda^r = \text{const}$ can then be considered to be a point in a three-dimensional manifold corresponding to the absolute space of Newton. It is clear that one can always find a mapping that maps the planes of absolute simultaneity onto the hypersurfaces $x^0 = \text{const}$ and the congruence onto the curves $x^r = \text{const}$. The parameters t, λ^r associated with a point can then be so chosen as to be equal to the coordinates of the point, so that $t = x^0$, $\lambda^r = x^r$. In what follows we will assume that the parameters have been fixed in this way.

We solve Eqs. (5-1.1 and 5-1.2) for the coordinates of a point of the space-time manifold in terms of the four parameters t, λ^r associated with this point

$$x^\mu = x^\mu(t, \lambda^r). \tag{5-1.3}$$

It then follows that the normal, $n_\mu = t_{,\mu}$, to a $t = \text{const}$ hypersurface at a point of the hypersurface is related to the tangent $u^\mu = \partial x^\mu / \partial t$ of the member of the congruence passing through this point by

$$u^\mu n_\mu = 1. \tag{5-1.4}$$

Instead of starting with the Eqs. (5-1.1 and 5-1.2) for the planes of absolute simultaneity and the congruence of curves discussed above we could equally well have started out by introducing two vector fields n_μ and u^μ on the manifold and requiring them to satisfy Eq. (5-1.4). The congruence could then be constructed by requiring that the vector u^μ at a point be tangent to the curve passing through the point. Likewise, the planes of absolute simultaneity could be constructed so that a normal coincided with n_μ at each point of the manifold. In order that such hypersurfaces exist n_μ must be the gradient of a scalar field and, hence, in addition to Eq. (5-1.4), must satisfy the integrability condition

$$n_{\mu,\nu} - n_{\nu,\mu} = 0. \tag{5-1.5}$$

If we have mapped so that the parameters associated with a point coincide with the coordinates of the point, then n_μ (1, 0, 0, 0) and $u^\mu = (1, 0, 0, 0)$.

Problem 5.1. Show that there always exists a mapping such that the transformed components of two vectors u^μ and n_μ satisfying Eqs. (5-1.4) and (5-1.5) have the values (1, 0, 0, 0).

Having constructed the planes of absolute simultaneity and a congruence of curves on the space-time manifold we now proceed to introduce two distinct intervals between points of this manifold. Given two neighboring points x^μ and $x^\mu + dx^\mu$ we define the temporal interval, or time, between them to be

$$d\tau^2 = n_\mu n_\nu \, dx^\mu \, dx^\nu. \tag{5-1.6}$$

We see from this definition that the temporal interval between any two points lying in the same $t = \text{const}$ hypersurface is zero. Furthermore, since the interval defined above is integrable, the temporal interval between any two points of the manifold is seen to be equal to the difference between the values of the parameter t characterizing the planes of absolute simultaneity containing the points.

The second interval that we shall introduce will allow us to define a spatial distance between points of the manifold. If two such points lie on neighboring curves of the congruence characterized by the parameters λ^r and $\lambda^r + d\lambda^r$, this interval has the form

$$ds^2 = \delta_{rs} d\lambda^r \, d\lambda^s \tag{5-1.7}$$

corresponding to the Euclidean geometry of Newtonian space. Again this interval is integrable. The square of the distance between any two points of the manifold is given by

$$s_{12}^2 = \delta_{rs}(\lambda_1{}^r - \lambda_2{}^r)(\lambda_1{}^s - \lambda_2{}^s)$$

where $\lambda_1{}^r$ and $\lambda_2{}^r$ are the parameters associated with the curves passing through the two points. For neighboring points of the space-time manifold, $d\lambda^r = \lambda^r{}_{,\mu} \, dx^\mu$, so that the expression for ds^2 can be rewritten as

$$ds^2 = h_{\mu\nu}(x) \, dx^\mu \, dx^\nu \tag{5-1.8}$$

where

$$h_{\mu\nu} = \delta_{rs} \lambda^r{}_{,\mu} \lambda^s{}_{,\nu} \, . \tag{5-1.9}$$

When the coordinates of the points of the space-time manifold coincide in value with the parameters associated with these points, $h_{\mu\nu} = \text{diag}(0, 1, 1, 1)$.

It is possible to introduce a metric onto the space-time manifold that is equivalent to the metric $h_{\mu\nu}$ given by Eq. (5-1.9) in a way that avoids the use of the parameters t and λ^r. To do so we first introduce onto the manifold a flat, symmetric affinity $\mathring{\Gamma}^\rho_{\mu\nu}$. Later we will have need of this affinity when we give mathematical expression to Newton's three laws of motion. We will impose this affinity on the manifold in such a way that, when u^μ and n_μ have components (1, 0, 0, 0), the components of the affinity vanish. It follows then that, relative to this affinity, both u^μ and n_μ are covariantly constant

$$u^\mu{}_{;\nu} = 0 \tag{5-1.10}$$

and

$$n_{\mu;\nu} = 0. \tag{5-1.11}$$

The metric $h_{\mu\nu}$ is defined by the two conditions

$$h_{\mu\nu;\rho} = 0 \tag{5-1.12}$$

and

$$h_{\mu\nu} u^\nu = 0. \tag{5-1.13}$$

Thus $h_{\mu\nu}$ is covariantly constant and possesses a null vector.

Problem 5.2. Show that the two metrics discussed above are equivalent.

In characterizing the geometry of Newtonian space-time we have had to introduce a number of geometrical objects. If they are observable, they must appear in the description of the kpt of some physical theory. Since the possible values that they can assume for dpt of the theory are independent of the values of the remaining quantities needed to describe the kpt of the theory, it follows from our discussion of Section 4-3 that these objects are absolute. Thus the universal symmetry group of all Newtonian systems should be that subgroup of the MMG that leaves invariant these objects. To find this group we can either treat the planes of absolute simultaneity, the congruence of curves given by Eq. (5-1.2), and the metric $h_{\mu\nu}$ given by Eq. (5-1.9) as primary or else, equivalently, we can treat the vector fields n_μ and u^μ, the flat affinity $\mathring{\Gamma}^\rho_{\mu\nu}$, and the metric $h_{\mu\nu}$ which satisfies Eqs. (5-1.12) and (5-1.13) as primary; the result in either case is the same. If we consider the latter set and map so that all of these quantities take on their Cartesian values, that is, $n_\mu = u^\mu = (1, 0, 0, 0)$, $\mathring{\Gamma}^\rho_{\mu\nu} = 0$ and $h_{\mu\nu} = \text{diag}(0, 1, 1, 1)$, we see first of all that the requirement that $\mathring{\Gamma}^\rho_{\mu\nu} = 0$ leads to the condition that the symmetry mappings must be linear. If we consider the infinitesimal mappings of this group, that is, mappings of the form

$$x^\mu \to x'^\mu = x^\mu + \varepsilon_\nu{}^\mu x^\nu + \varepsilon^\mu, \tag{5-1.14}$$

the requirement that $\bar{\delta} h_{\mu\nu} = 0$ leads to the conditions

$$\varepsilon_0{}^r = 0 \tag{5-1.15}$$

and

$$\varepsilon_s{}^r + \varepsilon_r{}^s = 0. \tag{5-1.16}$$

The requirement that $\bar{\delta} n_\mu = 0$ leads to the further conditions

$$\varepsilon_\nu{}^0 = 0. \tag{5-1.17}$$

The additional requirement that $\bar{\delta} u^\mu = 0$ does not lead to any further conditions on $\varepsilon_\nu{}^\mu$. We see, therefore, that the mappings (5-1.14) with $\varepsilon_\nu{}^\mu$ restricted by the conditions (5-1.15) (5-1.16), and (5-1.17) are the infinitesimal mappings of the direct product of $0'_3$, the inhomogeneous rotation group, and T, the time translation group—that is, the group $0'_3 \times T$.

Problem 5.3. Show that the group of mappings that leaves invariant the planes of absolute simultaneity, the congruence given by Eq. (5-1.2), and the matrix $h_{\mu\nu}$ given by Eq. (5-1.9) is the group $0'_3 \times T$.

5-2. Newton's Laws of Motion

Newton did not feel it necessary to define time, space, place, and motion "as being well known to all." He did feel it necessary, however, to define certain other "less known" concepts. Among these was quantity of matter, defined by him as the "measure of the same, arising from its density and bulk conjointly," a circular definition at best. Nevertheless, it did serve to emphasize the distinction between weight and mass or "quantity of matter," a distinction that had not been clearly drawn before and was essential for an understanding of the gravitational force. He defined the quantity of motion as "the measure of the same, arising from the velocity and quantity of matter conjointly." If we take quantity of matter to be the mass of a body, then Newton's quantity of motion is nothing more than the momentum of that body.

He also defined the *vis insita*, or innate force, as the "power of resisting, by which every body, as much as in it lies, continues in its present state, whether it be of rest, or of moving uniformly forwards in a right line." This quantity was also called *inertia* by Newton.

Aside from a number of definitions having to do with centripetal force, which do not concern us here, Newton gave one other definition in this section of the *Principia* which was of importance for all that followed. Definition IV read: "An impressed force is an action exerted upon a body, in order to change its state, either of rest, or of uniform motion in a right line." He then went on to comment, that "this force consists in the action only, and remains no longer in the body when the action is over. For a body maintains every new state it acquires, by its inertia only. But impressed forces are of different origins, as from percussion, from pressure, from centripetal force." If nothing more, this was a clear statement of opposition to the Aristotelian view that a force was necessary to maintain any state of motion, even that of uniform motion.

Having discussed the various concepts of space, time, place, motion, force, etc., Newton then proceeded to set forth the famous axioms, or laws of motion. Law I states that:

> Every body continues in its state of rest, or of uniform motion in a right line, unless it is compelled to change that state by forces impressed upon it.

One should observe the difference between the definition of force given by Newton and his statement of the first law; they are not the same in content. While the definition of force tells us that a force is sufficient to change the state of rest or of uniform motion in a right line of a body, the first law tells us that a force is also necessary to effect this change.

Newton's first law in effect postulates the existence of a special class of motions which can be executed by material bodies and which are distinguishable from all other motions that such bodies may execute. Bodies that execute motions belonging to this class will in the future be referred to as *free* particles. It has been argued that the first law is, in effect, circular. According to the argument, the only way we have of knowing when no forces act on a body is that it moves as a free particle; that is, it does not supply us with an independent way of deciding when no forces act on a body. However, this argument misses the essential point of the first law, namely, that it postulates the existence of free particles. It is the job of the experimental physicist to decide which bodies can be taken to be free in just the same way as it is up to him to decide which systems correspond to harmonic oscillators, particles moving under the influence of an inverse square force, and so forth. Of course, it may be that there are no such things as free particles. But then the whole structure of Newtonian physics would have to be abandoned.

In order to give a mathematical formulation to the first law we must first find some way to characterize the kpt of a particle. One possibility is to give the coordinates of the points that lie on the trajectory of the particle as functions of a path parameter λ,

$$x^\mu = z^\mu(\lambda). \qquad (5\text{-}2.1)$$

Since two points on the trajectory of a particle can never be simultaneous to each other it must intersect each of the planes of simultaneity in just one point. Consequently, the path parameter can be taken to be the parameter t that characterizes these planes. The *velocity* of the particle then has components dz^μ/dt and is seen to satisfy

$$n_\mu \frac{dz^\mu}{dt} = 1. \qquad (5\text{-}2.2)$$

When the planes of simultaneity coincides with the $x^0 =$ const hypersurfaces, the velocity has the components $(1, \mathbf{v})$ where $\mathbf{v} = d\mathbf{z}/dt$.

To describe the motion of free bodies we will need an affinity, which we take to be the flat affinity introduced in the previous section. If Eq. (5-1.11) is satisfied, it follows that t will be an affine parameter along the geodesics of this affinity. The equations of motion for a free particle are then taken to be

$$\frac{\delta^2 z^\mu}{\delta t^2} \equiv \frac{d^2 z^\mu}{dt^2} + \mathring{\Gamma}^\mu_{\rho\sigma} \frac{dz^\rho}{dt}\frac{dz^\sigma}{dt} = 0 \qquad (5\text{-}2.3)$$

where $\mathring{\Gamma}^\mu_{\rho\sigma}$ is evaluated at the location of the particle and satisfies

$$\mathring{B}^\mu_{\nu\rho\sigma} = 0 \qquad (5\text{-}2.4)$$

with $\mathring{B}^\mu_{\nu\rho\sigma}$ the Riemann tensor. The vector $\delta^2 z^\mu/\delta t^2$ defined here will be taken to be the *acceleration* of the particle. (In some discussions $d^2 z^\mu/dt^2$ is taken to be the acceleration of the particle. However, it is not a geometrical object and we prefer to reserve generic names for such objects.)

The equations of motion (5–2.3) together with the ancillary equations (5-1.5), (5-1.11), and (5-2.4) constitute the dynamical laws of a theory whose kpt are described by n_μ, $\mathring{\Gamma}_{\rho\sigma}$ and z^μ. It is clear from the form of these equations that the theory is MMG-covariant. The individual dpt of an equivalence class of the theory correspond to measurements made by observers using various types of space-time measuring devices. In what follows we shall call a collection of such devices that is sufficient to enable the user to assign coordinates to any physical event that can occur in nature a *reference frame*. (We will defer a discussion of how one goes about constructing such frames until we come to discuss the special theory of relativity in Chapter 6.) Once one has identified a family of free particles one can then employ them to determine the components of the affinity in any particular reference frame with the help of Eq. (5-2.3). One can then check to see if these components satisfy

Eq. 5-2.4). If they do, then one can presume that the identification of the family of particles used to determine the affinity as free was justified. Of course, such measurements must be made over a region of space-time that is large enough to enable one to measure the derivatives of the affinity since a knowledge of the affinity itself is insufficient to determine whether or not Eq. (5-2.4) is satisfied.

Since every dpt predicted by a theory must be observable, it follows that there exists a class of reference frames for which the components of the affinity all vanish. Such frames are called *inertial frames*. In these frames the equations of motion (5-2.3) reduce to

$$\frac{d^2 z^\mu}{dt^2} = 0. \tag{5-2.5}$$

The dpt that satisfy these equations are therefore of the form

$$z^\mu = \alpha^\mu t + \beta^\mu \tag{5-2.6}$$

where α^μ and β^μ are constants. By an appropriate choice of the parameter t we can take $\alpha^0 = 1$ and $\beta^0 = 0$. In this case the α^r constitute the components of the velocity of the particle. If we use a noninertial frame, we must use the full equations (5-2.3) to describe the motion of free particles. Of course, the fact that we use a noninertial frame does not alter the intrinsic behavior of the particles; it only means that their trajectories will not be observed to satisfy the linear relation (5-2.5). When these other frames are used, the observed values of the affinity will no longer be zero and their explicit appearance in the equations of motion will correspond to fictitious Coriolis and centrifugal-type forces.

We now turn our attention to the second law of motion. As stated by Newton it is:

> The change of motion is proportional to the motive force impressed; and is made in the direction of the right line in which the force is impressed.

By the quantity of motion of a body, Newton meant the product of its mass with its velocity, that is, its momentum $p^\mu = m\, dz^\mu/dt$ where m is the mass of the body. The mathematical statement of the second law then has the form

$$\frac{\delta p^\mu}{\delta t} \equiv \frac{dp^\mu}{dt} + \mathring{\Gamma}^\mu_{\rho\sigma}\, p^\rho \frac{dz^\sigma}{dt} = f^\mu \tag{5-2.7}$$

where the force f^μ is at most a function of the position and momentum of the body. When m is constant along the trajectory, this equation reduces to

$$m \frac{\delta^2 z^\mu}{\delta t^2} = f^\mu. \tag{5-2.8}$$

Since $n_\mu\, \delta^2 z^\mu/\delta t^2 = 0$ it follows that f^μ must satisfy the relation

$$f^\mu n_\mu = 0. \tag{5-2.9}$$

Thus, in an inertial frame f^μ has components (0, **F**).

As they stand, the dynamical laws (5-2.7) are incomplete until the forces appearing therein are specified. They cannot, of course, be given arbitrarily but must be such functions of the particle coordinates and velocities and the absolute objects introduced to characterize the Newtonian geometry that the dynamical laws (5-2.8) are MMG-covariant. It is clear that in order to satisfy this requirement f^μ must transform like a vector as a consequence of its dependence on these quantities. It must therefore be a linear combination of whatever vectors are included among these quantities with coefficients that are functions of the scalars we can form from them. If we consider velocity independent forces, the only such vector is u^μ. However, since $n_\mu u^\mu = 1$, a force proportional to u^μ will not satisfy Eq. (5-2.9). If we admit velocity dependent forces then f^μ can be a linear combination of $\dot{z}^\mu = dz/dt$ and u^μ

$$f^\mu = \alpha \dot{z}^\mu + \beta u^\mu.$$

The coefficients α and β can depend on the single scalar $h_{\mu\nu}\dot{z}^\mu \dot{z}^\nu$ that we can form from the quantities available to us. In order that f^μ satisfy Eq. (5–2.9), however, we must have $\alpha + \beta = 0$ so that f^μ must be proportional to the vector $\dot{z}^\mu - u^\mu$.

If we consider many-particle systems or systems of particles and fields, then the number of possible forces that we can construct and that lead to MMG-covariant theories is greatly increased. For many-body systems this number can be restricted somewhat by taking account of Newton's third law

> To every action there is always opposed an equal reaction: or, the mutual actions of two bodies upon each other are always equal, and directed to contrary parts.

For systems of particles that interact via two-body forces, which were the only kind considered by Newton, the third law requires that

$$f^\mu_{ij}(t) = -f^\mu_{ji}(t) \tag{5-2.10}$$

where $f^\mu_{ij}(t)$ is the force that the jth particle exerts on the ith particle at time t. In general, this condition can only be satisfied if $f^\mu_{ij}(t)$ depends symmetrically on the parameters associated with the ith and jth particles. In particular, if f^μ_{ij} depends on the variables of the ith particle taken at the time t, then it can only depend on those of the jth particle at this same time.

When the forces between pairs of particles are restricted to be instantaneous, there are only a few quantities that they can depend upon. Again they must be linear combinations of the vectors at our disposal with coefficients that are functions of the scalars we can construct from the particle variables and the absolute quantities used to characterize the geometry of Newtonian space-time. Of particular importance is the scalar distance between the particles of the system. We could use the metric $h_{\mu\nu}$ for the purpose of calculating this distance. However, since we are only interested in computing the distance between points lying in the same $t = \text{const}$ hypersurface, it is unnecessary to do so.

Let us introduce a symmetric tensor $\gamma^{\mu\nu}$ and require it to satisfy

$$\gamma^{\mu\nu}{}_{;\rho} = 0 \tag{5-2.11}$$

and

$$\gamma^{\mu\nu} n_\nu = 0. \tag{5-2.12}$$

As a consequence of these conditions it follows that there always exists a mapping such that $\mathring{\Gamma}^{\mu}_{\rho\sigma} = 0$, $n_\mu = (1, 0, 0, 0)$ and $\gamma^{\mu\nu} = \text{diag}(0, 1, 1, 1)$. Since $\gamma^{\mu\nu}$ is singular it does not possess a proper inverse. However, there exists a class of tensors $\gamma_{\mu\nu}$ that satisfy

$$\gamma^{\mu\rho}\gamma_{\rho\sigma}\gamma^{\sigma\nu} = \gamma^{\mu\nu}; \tag{5-2.13}$$

such tensors are called "quasi-inverses." If $\mathring{\gamma}^{\mu\nu}$ is one such tensor then any other quasi-inverse will be of the form

$$\gamma_{\mu\nu} = \mathring{\gamma}_{\mu\nu} + b_{(\mu}n_{\nu)}$$

where b_μ is an arbitrary vector. In spite of the nonuniqueness of $\gamma_{\mu\nu}$ we can nevertheless use it to compute distances within a $t = \text{const}$ hypersurface since any displacement lying wholely within this hypersurface satisfies

$$n_\mu \, dx^\mu = 0.$$

Hence

$$\gamma_{\mu\nu} \, dx^\mu \, dx^\nu = \mathring{\gamma}_{\mu\nu} \, dx^\mu \, dx^\nu$$

for such a displacement. (Note, however, that $\gamma_{\mu\nu}\dot{z}^\mu\dot{z}^\nu$, for instance, is indeterminant.) The distance $s_{12}(t)$ between two points in a $t = \text{const}$ hypersurface can then be defined to be the greatest lower bound of $\int_1^2 \sqrt{\gamma_{\mu\nu} \, dx^\mu \, dx^\nu}$ for all curves joining these two points and lying in this hypersurface. When $\Gamma^{\mu}_{\rho\sigma}$, n_μ and $\gamma^{\mu\nu}$ assume their Cartesian values, a particular solution of Eq. (5-2.13) is given by $\gamma_{\mu\nu} = \text{diag}(0, 1, 1, 1)$ and hence in this case $s_{12}(t)$ is given by the Pythagorean formula

$$s_{12}^2(t) = \delta_{rs}(x_1{}^r(t) - x_2{}^r(t))(x_1{}^s(t) - x_2{}^s(t)) \tag{5-2.14}$$

Thus we can dispense with the vector u^μ and the metric $h_{\mu\nu}$ in forming a distance function in a $t = \text{const}$ hypersurface.

Let us now return to the problem of constructing two-body forces. If we restrict ourselves to velocity independent forces, then there is only one vector that we can construct from the quantities at our disposal, namely, $\gamma^{\mu\nu}(z_i)\partial s_{ij}/\partial z_i{}^\nu = -\gamma^{\mu\nu}(z_j)\partial s_{ij}/\partial z_j{}^\nu$. Hence f_{ij}^μ must be of the form

$$f_{ij}^\mu = \phi(s_{ij})\gamma^{\mu\nu}(z_i)\partial s_{ij}/\partial z_i{}^\nu \tag{5-2.15}$$

and is seen to satisfy Eqs. (5-2.9) and (5-2.10). By introducing a potential $U(s_{ij})$ such that $dU/ds_{ij} = -\phi$ we can rewrite this expression for f_{ij} as

$$f_{ij}^\mu = -\gamma^{\mu\nu}(z_i)\frac{\partial U(s_{ij})}{\partial z_i{}^\nu}. \tag{5-2.16}$$

If we admit the possibility of velocity dependent forces, then f_{ij} can, in addition, contain terms proportional to the vectors $\dot{z}_i{}^\mu$, $\dot{z}_j{}^\mu$ and u^μ. The coefficients of the vectors can now depend on the scalars s_{ij}, $\dot{s}_{ij}$, $h_{\mu\nu}(z_i)\dot{z}_i{}^\mu\dot{z}_i{}^\nu$, $h_{\mu\nu}(z_j)\dot{z}_j{}^\mu\dot{z}_j{}^\nu$, $h_{\mu\nu}(z_i)\dot{z}_i{}^\mu\dot{z}_j{}^\nu$, and $h_{\mu\nu}(z_j)\dot{z}_i{}^\mu\dot{z}_j{}^\nu$. We see that, in the case of velocity dependent forces, it is necessary to bring in the metric $h_{\mu\nu}$ in order to satisfy Eqs. (5-2.9) and (5-2.10). The only exception arises if f_{ij} contains a term equal to the product of $(\dot{z}_i{}^\mu - \dot{z}_j{}^\mu)$ times a function of s_{ij} and $\dot{s}_{ij}$.

Problem 5.4. Construct an action principle for the equations of motion of a system of particles with interaction forces given by Eq. (5-2.16) and determine the constants of the motion for such a system.

So far, the forces we have been considering lead to dpt that are transformed into one another by elements of the MMG. However, we often consider forces not having this property and so we must discuss how they fit into the Newtonian picture. As an example let us consider the harmonic oscillator. It is usually treated as a single particle moving under the influence of a force proportional to $\gamma^{\mu\nu}(z)\,\partial s_{i0}^2/\partial z^\nu$ where s_{i0} is the distance of the particle from some fixed point. Clearly the equations of motion for such a particle are not MMG-covariant. Another example is that of a particle moving in an electric field. The force in this case is taken to be a vector satisfying Eq. (5–2.9) that is a function of the particle coordinates. Again, the equations of motion for such a particle are not MMG-covariant. Nevertheless, both theories describe systems that are observed to occur in nature.

This apparent contradiction becomes resolved when one realizes that the equations of motion involving such forces are approximations of equations of motion for systems that are MMG-covariant. In the case of the harmonic oscillator these equations are those of a two-body system with an interaction potential proportional to s_{12}^2. The one-body harmonic oscillator equation is an approximation in which the mass of one of the particles is taken to be much larger than that of the other. Likewise, the equation of a single body moving in a given electric field is an approximation to the equations of motion of a system that takes account of the action of this body on the sources of the field.

5-3. The Law of Universal Gravitation and Action-at-a-Distance

One of Newton's greatest contributions to man's understanding of nature was his discovery of the universality of the gravitational force. With one stroke he did away with the thousand-year-old doctrine that held that celestial bodies were formed from the quintessence and that the laws governing their behavior were unrelated to those governing the behavior of terrestrial bodies formed from the four essences (earth, fire, air, and water). (A commonly held notion was that the planets were guided in their orbits by angels.)

The problem of a dynamical determination of the planetary orbits was the central problem of science of that day. Robert Hooke believed that the law of attraction between the heavenly bodies was the inverse square law, but was unable to give a convincing proof to support this belief. It was Newton who supplied such a proof and who demonstrated clearly that the terrestrial gravitational force was the same force that held the moon in its orbit. The question of the source of this attraction was left unanswered by Newton. He seriously considered an explanation based on an aether. However, Newton realized that an answer to this question was unnecessary for a description of the phenomena involved. And so he formulated his law of universal gravitation in terms of *action-at-a-distance*. This notion of a force acting through a distance became a central pillar in the classical edifice of theoretical mechanics, only to be superseded toward the end of the nineteenth century by the field concept.

Let us now turn to a mathematical formulation of the classical gravitational interaction of material bodies. The gravitational force of attraction was shown by Newton to depend on the inverse square of the instantaneous distance between pairs of bodies. This force can be derivable from a potential given by

$$U_{ij} = -G \frac{g_i g_j}{s_{ij}} \tag{5-3.1}$$

where G is the universal gravitational constant and g_i and g_j are the gravitational "charges" of the two bodies involved. It is an empirical fact that the gravitational charge is proportional to the inertial mass of a body and that the proportionality constant is independent of the composition or size of that body. With $G = 6.67 \times 10^{-8}$ dynes-cm^2-gm^{-2}, this proportionality constant is unity. Not until 1915 was the equivalence of gravitational charge and inertial mass recognized to be more than an accident of nature.[2]

[2] See Section 10-2 for a discussion of the experimental evidence of this equivalence.

The total force acting on the ith particle of an N-body system of gravitating particles is given by

$$f_i^{\mu} = -g_i \gamma^{\mu\nu}(z_i) \frac{\partial \phi_i}{\partial z_i^{\nu}} \tag{5-3.2}$$

where

$$\phi_i = -\sum_{j \neq i} G \frac{g_j}{s_{ij}} \tag{5-3.3}$$

is the *gravitational potential* seen by the ith particle, sometimes called the ith *partial potential*. This potential differs from the total gravitational potential $\phi(x)$ given by

$$\phi(x) = -\sum_i G \frac{g_i}{s_{xi}} \tag{5-3.4}$$

where s_{xi} is the instantaneous distance between the point x^{μ} and the ith particle. The *gravitational field* $\mathscr{G}^{\mu}(x)$ is then defined by

$$\mathscr{G}^{\mu}(x) = -\gamma^{\mu\nu}(x)\phi(x)_{,\nu} \tag{5-3.5}$$

We see that both the gravitational potential and the gravitational field diverge for values of x^{μ} that coincide with the coordinates of a source of the field. For this reason their introduction leads to a number of difficulties common to all field-theory descriptions of interacting particles. Fortunately, the introduction of the field concept is unnecessary for the description of the gravitational interaction within the framework of Newtonian mechanics. For future reference we note that when the absolute objects of the theory take on their Cartesian values, $\phi(x)$ satisfies the Poisson equation

$$\nabla^2 \phi(x) = 4\pi G \rho(x) \tag{5-3.6}$$

where

$$\rho(x) = \sum_i g_i \int \delta^4(x - z_i(t))\, dt \tag{5-3.7}$$

with $\delta^4(x)$ the four-dimensional Dirac delta function.

Once one takes account of the gravitational interaction in Newtonian physics one encounters difficulties of the kind that ultimately led Einstein to propose the general theory of relativity. Suppose that we want to determine the components of the affinity of Newtonian space-time when gravitating bodies are present. For this purpose we would want to use test bodies whose masses were sufficiently small that we could neglect their action on the gravitating masses present. Because of the equivalence of inertial mass and gravitational charge, however, all bodies possessing an inertial mass also possess an equal gravitational charge, that is, $m_i = g_i$ in appropriate units. Consequently we cannot neglect the action of these gravitating bodies on the test body. The equations of motion for a test body must therefore include this action and, hence, after dividing by the mass of the test body, have the form

$$\frac{d^2 z^\mu}{dt^2} + \mathring{\Gamma}^\mu_{\rho\sigma}(z) \frac{dz^\rho}{dt} \frac{dz^\sigma}{dt} = -\gamma^{\mu\nu}(z)\phi(z)_{,\nu} \tag{5-3.8}$$

where $\phi(x)$ is the gravitational potential of whatever gravitating bodies, excluding the test body, are present and $\mathring{\Gamma}^\mu_{\rho\sigma}$ is the flat affinity of Newtonian space-time.

Following Trautman[3] we now introduce a new affinity $\Gamma^\mu_{\rho\sigma}$ by

$$\Gamma^\mu_{\rho\sigma} = \mathring{\Gamma}^\mu_{\rho\sigma} + \gamma^{\mu\nu}\phi_{,\nu} n_\rho n_\sigma \tag{5-3.9}$$

and rewrite Eq. (5-3.8) in the form

$$\frac{d^2 z^\mu}{dt^2} + \Gamma^\mu_{\rho\sigma}(z) \frac{dz^\rho}{dt} \frac{dz^\sigma}{dt} = 0. \tag{5-3.10}$$

We see, therefore, that all we can determine by local measurements—that is, measurements made in a finite region of space-time—using test bodies is the affinity $\Gamma^\mu_{\rho\sigma}$. If there were test bodies with different g/m ratios, it would be possible to determine the values of the components of the flat affinity from a knowledge of the $\Gamma^\mu_{\rho\sigma}$ obtained by using test bodies with different values of this ratio. Since this ratio is the same for all bodies found in nature, however, we will always measure the same value of $\Gamma^\mu_{\rho\sigma}$ in a given region of space-time. This difficulty is, of course, directly related to the fact that once we take account of gravitational effects we can no longer unambiguously and in all cases identify a class of particles as free. Only to the extent that we can ignore these effects can such an identification be made.

We might still hope to measure the components of the flat affinity by decomposing the measured components of $\Gamma^\mu_{\rho\sigma}$ as in Eq. (5-3.9). However, Trautman has shown that this decomposition is not unique; if ψ is any solution of

$$n_\mu \psi_{;\rho\sigma} - n_\rho \psi_{;\mu\sigma} = 0$$

then $\mathring{\Gamma}^\mu_{\rho\sigma} - \gamma^{\mu\nu}\psi_{,\nu} n_\rho n_\sigma$ is also a flat affinity. The best that one can do if one is dealing with bounded sources of the gravitational field is to normalize the gravitational potential ϕ by requiring that it vanish at large distances from these sources. We see that the gravitational potential given by Eq. (5-3.4) has this property. However, if we are dealing with gravitational fields of the type

[3] A. Trautman, *Comptes Rendus* (Paris), **257**, 617 (1963); see also K. Friedrichs, *Math. Ann.*, **98**, 566 (1927).

that occur in cosmology or when gravitational radiation is present, no such normalization is, in general, possible. Even when we can so normalize the gravitational potential, we see that the components of the flat affinity can only be determined by nonlocal measurements, that is, by measurements that include observations on test bodies that are spatially far removed from whatever sources of the gravitational field are present.

From the above discussion we see that when real gravitational fields are present, only the total affinity $\Gamma^{\mu}_{\rho\sigma}$ is observable by means of local observations made with test bodies. It may be, of course, that there are other physical systems existing in nature that could be used to measure the flat Newtonian affinity. These systems would either not be affected by a gravitational field or else depend on parameters that in effect change the interaction as the parameters are changed. Later we shall see that the assumption of the nonexistence of such systems constitutes one of the fundamental principles of general relativity—the principle of equivalence. But if this assumption is valid, the flat affinity, and with it one of the key elements of Newtonian space-time, can be eliminated from our space-time descriptions. Consequently, there being fewer absolute elements without the flat affinity, the symmetry group of all space-time theories will be larger than the inhomogeneous Galilean group since, in general, the fewer the number of absolute elements, the bigger the symmetry group of a theory. We shall examine the consequences of these conclusions later when we come to discuss the general theory of relativity.

5-4. Galilean Relativity and Absolute Motion

The unobservability of the flat affinity in the presence of a gravitational field discussed in the last section raised the question of whether the group of symmetries of Newtonian space-time was in fact the largest universal space-time symmetry group. Because of the weakness of the gravitational interaction, the Newtonian symmetries would still retain their privileged character, at least to a high degree of approximation, were it not for the apparent unobservability of another element used to describe the geometry of Newtonian space-time. We saw in Section 5-2 that only if one considered velocity dependent interparticle forces was it necessary to introduce the vector field u^{μ} and the associated metric $h_{\mu\nu}$. Except in the case of a charged particle moving in a magnetic field, such velocity dependent forces did not seem to exist in nature. But if in fact u^{μ} was unobservable, it would mean that there were fewer absolute objects in nature than were supposed by Newton. In particular, his whole concept of an absolute space would have to be discarded.

If u^{μ} and $h_{\mu\nu}$ were not observable, then the universal space-time symmetry group would have to be determined by the remaining absolute objects of

Newtonian space-time: $\mathring{\Gamma}^{\mu}_{\rho\sigma}$, n_{μ}, and $\gamma^{\mu\nu}$. When these objects assume their Cartesian values, the infinitesimal elements of their symmetry group can easily be determined in a manner analogous to that used in Section 5-1 to determine the symmetries of Newtonian space-time. We now find that these infinitesimal mappings have the form

$$x^{\mu} \to x'^{\mu} = \varepsilon_{\nu}{}^{\mu} x^{\nu} + \varepsilon^{\mu} \tag{5-4.1}$$

where the parameters $\varepsilon_{\nu}{}^{\mu}$ are restricted by the conditions

$$\varepsilon_{s}{}^{r} = -\varepsilon_{r}{}^{s} \qquad \text{and} \qquad \varepsilon_{\nu}{}^{0} = 0. \tag{5-4.2}$$

These infinitesimal mappings clearly form a group—the *inhomogeneous Galilean group*. This group contains, as a subgroup, the Newtonian symmetry group $0'_3 \times T$. But in addition it contains the group of mappings

$$\begin{aligned} x^{0} &\to x'^{0} = x^{0} \\ x^{r} &\to x'^{r} = x^{r} + v^{r} x^{0} \end{aligned} \tag{5-4.3}$$

where the v^r are three arbitrary constants. These mappings by themself constitute a group—the group of *special Galilean* or *velocity* mappings. The parameters v^r clearly represent the relative velocity between inertial frames that are related to each other by the mapping (5-4.3). The existence of these additional special Galilean symmetries has had a profound influence on the development of physics. The assertion that they constitute a universal symmetry group of all physical systems is known as the principle of *Galilean* relativity.

The most important consequence of the principle of Galilean relativity is that it is impossible to distinguish between the various inertial frames moving relative to each other. If u^{μ} were observable, it would be possible to single out a subclass of such frames, namely, those in which it would have the components (1, 0, 0, 0). The frames of this subclass would differ from each other only with regard to the orientation of axes and the choice of the spatial and temporal origins. These frames could then be considered to be absolutely at rest in space and the velocity of a particle in such a frame would then be its absolute velocity. However, if there is no way to distinguish between the various inertial frames, it is impossible to distinguish between the states of absolute rest and uniform motion referred to by Newton in his first law of motion. Furthermore, since the mappings (5-4.1) and (5-4.2) do not leave invariant the vector field u^{μ}, they do not map the congruence of curves associated with the absolute space of Newton onto themselves. Therefore the whole concept of an absolute space as envisaged by Newton loses its meaning and can be eliminated from our space-time descriptions.

In formulating the first law, Newton clearly had in mind the absolute motion of the particle—the motion of the particle in moving from one absolute place to another. He nevertheless realized that just as the absolute place of an object was unobservable because of the homogeneity and isotropy of space, so also the absolute velocity of an object was unobservable. Only relative place and velocity were observable; hence the relativity of uniform or Galilean motion. What was absolute for Newton was the acceleration of a body. It was just this distinction between uniform motion as being relative and nonuniform motion as being absolute that led Newton to the second law of motion. If he had conceived of all motion as being relative, he could never have arrived at this law.

Newton went to great lengths to distinguish between absolute or, as he called them, "true motions and relative motions." He says:

> "The causes by which true and relative motions are distinguished, one from the other, are the forces impressed upon bodies to generate motion. True motion is neither generated nor altered, but by some force impressed upon the body moved; but relative motion may be generated or altered without any force impressed on the body. For it is sufficient only to impress some force on other bodies with which the former is compared, that by their giving way, that relation may be changed, in which the relative rest or motion of this other body did consist."

Thus, since force was an absolute concept in the Newtonian picture, its application must produce an absolute effect. It is this relation between the two absolutes, force and acceleration, that underlies the second law of motion.

The problem of deciding observationally whether a given motion was true or apparent was also discussed by Newton. According to him, "The effects which distinguish absolute from relative motion are the forces of receding from the axis of circular motion." Most readers will be familiar with Newton's discussion of the rotating pail filled with water. He discussed another example that is not so generally known but in which he very clearly drew a distinction between absolute and relative motion and (more important for us) their relation to absolute space. This example is important in that it bears directly on the later arguments of Mach, which in turn so greatly influenced Einstein in his formulation of the general theory of relativity. Newton considered two globes rotating about their common center of gravity and held together by a connecting cord. He then proposed to discover how much of this rotation is absolute and how much is relative, by measuring the tension in the cord. He went on to make the following crucial assertion: "And thus we might find both the quantity and the determination of this circular motion, even in an immense vacuum, where there was nothing external or sensible with which the globes could be compared."

Newton did not stop at this point, however. He then brought up the problem of the visual or kinematic relativity of motion, first discussed by Nicole Oresme in 1377 in his commentaries on Aristotle's *De Caelo et Mundi*. Newton wrote:

> "But now, if in that space some remote bodies were placed that kept always a given position one to another, as the fixed stars do in our regions, we could not indeed determine from the relative translation of the globes among those bodies whether the motion did belong to the globes or to the bodies. But if we observed the cord, and found that its tension was that very tension which the motions of the globes required, we might conclude the motion to be in the globes, and the bodies to be at rest."

The only alternative to this conclusion would be to say that the motion of the stars produced the tension in the cord. It was just this alternative that was proposed by Mach. However, no force known to Newton could have produced such a tension, and so he was perfectly justified in his conclusion. The only long range force known at the time was the gravitational force, and it depended only on the separation and not the velocities of the bodies involved.

We will return to this question of the observation of absolute motion when we come to study the general theory of relativity. In that theory the gravitational force is velocity-dependent, and hence the possibility arises of adopting the alternate view that the globes are stationary and the stars are rotating, and the tension in the cord connecting the two globes is due to this rotation. For the moment, however, we merely point out that the visual relativity of motion implies nothing whatsoever concerning the dynamic relativity of that motion.

Before we conclude this discussion of Galilean relativity we would like to contrast the attitude of the physicists who followed Newton, with regard to the two transformation groups $O'_3 \times T$ and the group of special Galilean transformations (5-4.3). The former was accepted without question as reflecting the fundamental symmetry of space and time. The invariance of mechanical laws under the group of special Galilean transformations was considered to be an accident and did not reflect any fundamental symmetry of nature. It was thought, and indeed for a while appeared to be the case, that other, nonmechanical, systems would not possess this invariance property. If such systems were found to exist in nature, then one could determine their absolute velocity and hence the absolute velocity of all systems in space. It therefore became a pressing problem of the nineteenth century physicist to discover such systems.

5-5. Maxwell's Electrodynamics and the Galilean Symmetries; the Aether

As we have seen, Newtonian mechanics was unable to furnish us with a dynamical means of distinguishing between the various states of uniform motion of a body. However, Maxwell's electrodynamics did at first appear to offer just such a possibility. Among other things, it predicted that a uniformly moving charge q would produce a magnetic field B, given by

$$\mathbf{B} = \frac{\mu_0 q}{4\pi} \frac{\mathbf{v} \times \mathbf{r}}{r^3}, \tag{5-5.1}$$

where $\mathbf{r}$ is the radius vector from the charge to the point where $\mathbf{B}$ is measured. The velocity $\mathbf{v}$ of the charge appearing in this expression was assumed to be its velocity with respect to absolute space. In addition, Maxwell's theory also

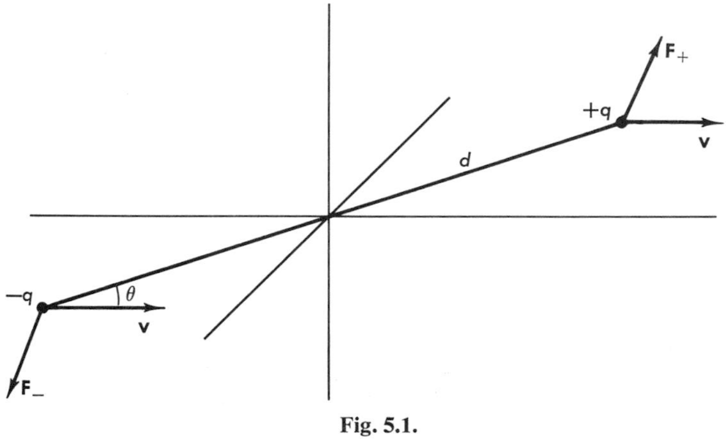

Fig. 5.1.

predicted that a charge moving with a velocity $\mathbf{u}$ (again assumed to be with respect to absolute space) in a magnetic field $\mathbf{B}$ would experience a force $\mathbf{F}$, given by

$$\mathbf{F} = q\mathbf{u} \times \mathbf{B} \tag{5-5.2}$$

Now imagine equal and opposite charges placed at the ends of a rigid insulating rod, the whole assumed to be moving uniformly with a velocity $\mathbf{v}$ with respect to absolute space. According to the above discussion each charge should feel a force acting on it owing to the presence of the other, given by

$$F = \frac{\mu_0 q^2}{4\pi} \frac{\mathbf{v} \times (\mathbf{v} \times \mathbf{d})}{d^3}, \tag{5-5.3}$$

where **d** is the vector from the charge producing the field to the charge being acted upon. Although the two forces on the charges are oppositely directed, they act at different points of the system and so produce a net torque of magnitude

$$T = \frac{1}{4\pi\varepsilon_0} \frac{q^2 v^2 \sin\theta \cos\theta}{c^2 d}, \tag{5-5.4}$$

where θ is the angle between **v** and **d**. Thus, unless such a system of charges was absolutely at rest, or $\theta = 0$ or $\pi/2$, it should experience a net torque, causing it to rotate. In principle, then, one could determine the absolute velocity of such a system and hence by reference to it could find the absolute velocity of any other system. Of course it would be a difficult determination, since the effect depends upon $(v/c)^2$. An investigation based on this idea was actually carried out by Trouton and Noble (see Section 6-1) to determine the velocity of the earth through absolute space, but their work led to a null result.

Another, and far-reaching, conclusion to be drawn from Maxwell's equations was that electromagnetic disturbances could be propagated as waves in otherwise empty space with a characteristic velocity $c = 1/\sqrt{\varepsilon_0 \mu_0} = 3 \times 10^8$ meters/sec. Again it was assumed that this was the velocity of propagation with respect to absolute space. If this were the case then one could assume that the apparatus measuring this value for the velocity of light was in a state of absolute rest.

It is clear that the vacuum Maxwell equations cannot admit the full Galilean group as a symmetry group, since the only invariant velocity under this group is one with an infinite magnitude, while Maxwell's equations contain the characteristic finite velocity c. One can also show directly that they do not admit the Galilean group as a symmetry group. In order for them to do so, one must be able to construct for the fields **E** and **B** a transformation law that is a realization of the Galilean group and which would also transform solutions of these equations into other solutions. Furthermore this realization must in fact be a representation, that is, the transformed fields must be linear functions of the original fields in order that the sum of two transformed solutions would be equal to the transform of the sum, as required by the linearity of the Maxwell equations. It is then an easy matter to show that no such representation exists. For this reason, physicists of the time were of the opinion that the symmetry group of Maxwell's equations was just $O'_3 \times T$.

The failure of Maxwell's equations to admit the special Galilean symmetries was regarded by physicists of the day as proof of the existence of an underlying substratum of the physical world in a state of absolute rest. This substratum was assumed to be the long-sought aether of the nineteenth century physicists.[4] Originally introduced into contemporary science by Descartes as a mechanism for the transmission of force, it was endowed through the years with many and marvelous properties. The concept of an aether was employed by Fresnel with great success in explaining the propagation properties of light. Later, when it was realized that light was but a manifestation of the electromagnetic field and when Maxwell's theory of this field was developed, much effort was expended on constructing mechanical models based on an aether that yielded the results of Maxwell's theory. Even after attempts to give a mechanistic interpretation to electromagnetic phenomena were abandoned, the aether was retained, notably by Lorentz, in connection with questions relating to the electrodynamics of moving bodies. A central consequence of these considerations was that it should be possible, with the help of electromagnetic phenomena, to devise experiments that would enable an observer to determine his velocity with respect to the aether, that is, his absolute velocity. In the next chapter we will discuss several such experiments and how their failure led to the special theory of relativity.

[4] For a history of the concept of an aether, see E. T. Whittaker, *A History of the Theories of Aether and Electricity*, Vol. I (Thomas Nelson and Sons, Ltd., Edinburgh, 1951).

6

Special Relativity

The early years of this century marked one of the turning points in the development of physics. All attempts to determine the absolute velocity of a physical system by electromagnetic means had failed. Many physicists at the time advocated a strict adherence to the concepts of absolute space and time of the Newtonian world-picture and attempted to explain these failures by introducing new, but essentially classic, laws governing the behavior of matter in motion. However, while these proposals were able to explain some of these failures, they were unable to explain others. Furthermore, they all suffered from the logical defect that while they employed the concept of absolute motion, they precluded if true, the possibility of ever observing such motion.

A much simpler but at the same time more profound solution to the problem was put forth independently by Poincaré and Einstein and constitutes what is known today as the special theory of relativity. Both men advocated an even more drastic revision of the Newtonian ideas of space and time than that implied by the principle of Galilean relativity. While the latter, as we have seen, deprived space of its absolute status, the proposals of Poincaré and Einstein ultimately led to the realization that neither space nor time by themselves were absolute and that only when taken together did they form an absolute entity.

In this chapter we first discuss some of the key experiments that led to the crisis in physics at the beginning of the present century and briefly describe attempts to accommodate these experiments within the framework of the Newtonian world-picture. We then examine the changes wrought in this picture by the ideas of special relativity. However, we do not attempt to arrive at the special theory by an inductive method based on experimental results. Rather we follow a deductive procedure based on the fundamental assumption that the underlying symmetry of all physical systems is the Poincaré or inhomogeneous Lorentz group. Our main concern, then, is to discover simple physical theories that are in accord with this assumption and whose elements can be made to correspond to physical systems found in nature. With the help of these theories we can develop models for space-time measuring devices. In particular we consider free particles and light rays and with their help deduce the properties of a light clock constructed from these objects. Our reason for proceeding in this way rests on the fact that all measurements, space-time or otherwise, are in fact comparisons between different physical systems. Therefore it must be possible (and, we hold, necessary) to describe both the systems being measured and the measuring device within the framework of the same world-picture. Consequently the principles underlying a given world-picture must precede, in our view, a theory of measurement, rather than the other way around.

Once we have constructed simple models of space-time measuring devices it will then be possible to deduce the familiar kinematic consequences of special relativity, such as length contraction and time dilation. In doing so we will see that measurements made with such devices can be made to correspond to the space-time coordinates employed in the theories of special relativity and that the relations between such measurements correspond to the Poincaré symmetry of these theories.

6-1. The Search for Absolute Velocity

Actual attempts to determine the earth's velocity through the aether might be said to have begun with the first careful measurements of the velocity of light by Fizeau[1] in 1894 and Foucault[2] in 1850. It was found to be impossible, on the basis of the results, to detect any effect due to the earth's motion. However, a detailed analysis of these measurements show that such effects would be only corrections, of the order of $(v_{\text{earth}}/c)^2$, to results one would predict if the earth were absolutely at rest. Within the accuracy of Fizeau and Foucault's observations, then, it was impossible to rule out reasonable values of v_{earth}, and thus the notion of absolute motion.

Shortly after Fizeau's measurement, Weber and Kohlrausch[3] determined the value of $1/\sqrt{\varepsilon_0\mu_0}$ by measuring the charge of a Leyden jar both electrostatically and electromagnetically. They obtained a value of 3.1×10^8

[1] H. Fizeau, *Compt. Rend.*, **29**, 90 (1849).

[2] L. Foucault, *Compt. Rend.*, **30**, 551 (1850).

[3] W. Weber and R. Kohlrausch, *Ann. der Phys.*, **99**, 10 (1856).

meters/sec that was in agreement, within experimental error, with that obtained by Fizeau. Maxwell took this amazing coincidence as conclusive evidence that "light consists in the transverse undulations of the same medium which is the cause of electric and magnetic phenomena."

The next attempts to observe the earth's absolute motion were those of Fizeau[4] in 1851 and Hoek[5] in 1868, who measured the velocity of light in transparent substances. The interpretation of their results was complicated by the fact that one had to make an assumption with regard to the effective phase velocity of the light in the moving media. If the medium were taken to be absolutely at rest then the velocity of light in the medium would be, according to Maxwell's phenomenological electrodynamics, just c/n, where n is the index of refraction of the medium. If an observer moving with absolute velocity $\mathbf{v}$ were to measure the velocity of light in the medium, he would find a velocity c' given by

$$c' = \frac{c}{n} - \boldsymbol{\alpha} \cdot \mathbf{v}, \tag{6-1.1}$$

where $\boldsymbol{\alpha}$ is a unit vector in the direction of the wave normal. The question to be answered, however, was: What would be the observed velocity if both medium and observer were at rest with respect to each other but moving with the same absolute velocity? The simplest hypothesis regarding this velocity was that it was independent of the state of motion of the medium and depended only on the absolute velocity of the observer. In this case the observed velocity would again be given by Eq. (6-1.1). Alternately, Stokes[6] assumed that it depended only on the relative velocity of medium and observer. In this case c' would be given by

$$c' = \frac{c}{n}. \tag{6-1.2}$$

Midway between these two hypotheses was the one put forth by Fresnel[7] and based on an elastic aether theory. According to him, c' was given by

$$c' = \frac{c}{n} - \frac{1}{n^2} \boldsymbol{\alpha} \cdot \mathbf{v}. \tag{6-1.3}$$

If we use either expression (6-1.1) or (6-1.2) for c', then the effect of the earth's motion would contribute a correction of order of magnitude v/c. However, no such effects were observed. If, on the other hand, one uses Fresnel's expression (6-1.3) for the velocity c', then the corrections due to the earth's motion are of order $(v/c)^2$ and are not detectable on the basis of Fizeau and

[4] H. Fizeau, *Compt. Rend.*, **33**, 349 (1851).

[5] M. Hoek, *Archives Néerlandaises des Sciences Exactes et Naturalles*, **3**, 180 (1868).

[6] G. G. Stokes, *Phil. Mag.*, (3), **27**, 9 (1849).

[7] A. J. Fresnel, *Ann. de Chim. et de Phys.* **9** 57 (1818).

Hoek's results. Thus their results did not rule out absolute motion but rather verified the correctness of Fresnel's equation (6-1.3), at least to first order, in v/c.

There was a serious objection to Fresnel's derivation of Eq. (6-1.3). According to him, the aether was partly dragged along by the moving body with a velocity $(1 - 1/n^2)\mathbf{v}$, where $\mathbf{v}$ is the velocity of the body relative to absolute space. Since, in general, n is a function of the frequency of the light wave, this would mean in effect that there had to be a separate aether for each such frequency. A more satisfactory discussion of the phase velocity of light in a moving medium was given by Lorentz,[8] based on his theory of electrons. We will not go into the details of this theory except to say that it makes use of the notion of absolute rest; in a medium in a state of absolute rest the phase velocity of propagation of light is c/n. On the basis of this assumption Lorentz was then able to show that for a moving medium, the phase velocity was given to a first approximation by Fresnel's expression (6-1.3). As a consequence the interpretation of Hoek and Frizeau's experiments on the basis of the Lorentz theory showed that one could not detect any effects of the earth's motion to first order in v_{earth}/c. In fact Lorentz' theory yielded results for all optical phenomena that were in agreement with the principle of relativity up to first order in v_{earth}/c.[9]

It thus became clear that if one were ever to detect deviations of Maxwell's equations due to the earth's absolute motion by optical means, one had to devise an experiment that was sensitive enough to detect effects of the order $(v_{\text{earth}}c/)^2$.

The first experimental attempt to detect second-order effects in optical phenomena due to the earth's motion was carried out by Michelson[10] in the year 1881 and later refined by Michelson and Morley[11] in 1887. The essential idea of the experiment was to split a beam of light, allow the two resultant beams to travel paths at right angles to each other, and finally to recombine them to produce a series of interference fringes. The apparatus was then rotated through 90 deg. If the velocity of light in the apparatus depended upon the absolute velocity of the earth then the velocities of light along the two paths should change when the instrument is rotated, and hence one should observe a shift of the fringes.

In Fig. 6.1 we have schematically indicated a simplified version of the Michelson-Morley experiment. Light from the source L is split into two beams

[8] H. A. Lorentz, *The Theory of Electrons* (2nd ed., Teubner, Leipzig, 1916) (Dover reprint, 1952).

[9] For a detailed analysis of the various experimental determinations of the velocity of light, see C. Møller, *The Theory of Relativity* (Clarendon Press, Oxford, 1952).

[10] A. A. Michelson, *Am. J. Sci.*, **22**, 120 (1881).

[11] A. A. Michelson and E. W. Morley, *Am. J. Sci.*, **34**, 333 (1887).

at right angles to each other by the half-silvered mirror. One of these beams is reflected by a mirror M_1; the other, by a mirror M_2. The reflected beams are then recombined with the help of the half-silvered mirror and observed

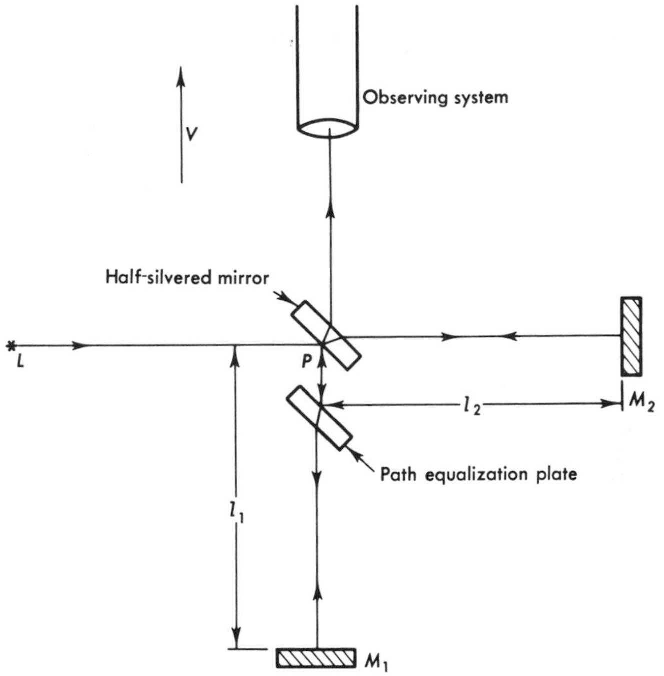

Fig. 6.1.

with a telescope. For an on-axis beam the time to travel from the point P on the half-silvered mirror down to M_1 and back again to P is given by

$$t_1 = l_1\left(\frac{1}{c - v} + \frac{1}{c + v}\right) = \frac{2l_1}{c(1 - \beta^2)}, \qquad \beta = \frac{v}{c}, \tag{6-1.4}$$

if the apparatus is assumed to be moving with a velocity v in the direction indicated. During the time $t_2/2$ taken for the light to go from P to M_2 the apparatus as a whole will have moved through a distance δ, given by

$$\frac{v}{c} = \frac{\delta}{\sqrt{\delta^2 + l_2{}^2}},$$

so that

$$\delta = \frac{\beta l_2}{\sqrt{1 - \beta^2}}. \tag{6-1.5}$$

Consequently t_2 is given by

$$t_2 = \frac{2}{c}\sqrt{l_2{}^2 + \delta^2} = \frac{2l_2}{c\sqrt{1-\beta^2}}. \qquad (6\text{-}1.6)$$

The difference in optical path is, then, neglecting the thickness of the half-silvered mirror,

$$\begin{aligned}\Delta &= c(t_1 - t_2) \\ &= \frac{2}{\sqrt{1-\beta^2}}\left(\frac{l_1}{\sqrt{1-\beta^2}} - l_2\right) \qquad (6\text{-}1.7)\end{aligned}$$

If the interferometer is now rotated through 90 deg., l_1 and l_2 interchange roles and

$$t'_1 = \frac{2l_1/c}{\sqrt{1-\beta^2}}, \qquad t'_2 = \frac{2l_2/c}{\sqrt{1-\beta^2}},$$

so that the difference in optical path is given by

$$\Delta' = c(t'_1 - t'_2) = \frac{2}{\sqrt{1-\beta^2}}\left(l_1 - \frac{l_2}{\sqrt{1-\beta^2}}\right) \qquad (6\text{-}1.8)$$

Thus, upon rotation, the interference pattern will shift by n fringes, given by

$$\begin{aligned}n &= \frac{\Delta' - \Delta}{\lambda} = \frac{2(l_1 + l_2)}{\lambda\sqrt{1-\beta^2}}\left(1 - \frac{1}{\sqrt{1-\beta^2}}\right) \\ &\cong -\frac{(l_1 + l_2)}{\lambda}\beta^2. \qquad (6\text{-}1.9)\end{aligned}$$

Within the accuracy of the Michelson-Morley experiment it should have been possible to detect absolute velocities of the earth greater than about 10 km sec. No such shift was observed, even though the velocity of the earth in its orbit with respect to the background of the "fixed" stars is about 30 km/sec. The experiment has been repeated from time to time with varying accuracies, but in no case has any effect due to the earth's motion ever been detected.[12] The most recent Michelson-Morley type of experiment has been carried out by Cedarholm, Bland, Havens, and Townes,[13] who used maser beams in place

[12] For a discussion of these experiments, see G. W. Stokes, "Michelson-Morley Experiment," in *The Encyclopaedic Dictionary of Physics*, Vol. 4 (Pergamon Press, Oxford, 1961), p. 635.

[13] J. P. Cedarholm, G. F. Bland, B. L. Havens, and C. H. Townes, *Phys. Rev. Letters*, **1**, 342 (1958).

of the usual light sources. With their experimental arrangement they were able to obtain accuracy of 1 part in 10^{12} and placed an upper limit of less than 1 part in 10^3 on the effect for an earth velocity of 30 km/sec.

Various other experiments have been proposed from time to time to detect the absolute motion of the earth.[14] While a detailed treatment of these experiments is beyond the scope of the present discussion, we mention the experiment of Trouton and Noble[15] who, using an analysis similar to that leading to Eq. (5-5.4), attempted to detect the torque on a suspended parallel plate condenser, with decisive negative results. Similarly negative were the results of Ives and Stillwell[16] on the transverse Doppler shift (see Section 8-3).

Several attempts were made to explain the negative results of the Michelson-Morley experiment. One of these was the Fitzgerald-Lorentz contraction hypothesis,[17] which postulated that material bodies were contracted as they moved through absolute space. The factor of contraction, $\sqrt{1-\beta^2}$, in the direction of motion was just sufficient to explain the negative results of the Michelson-Morley and Trouton-Noble experiments. Thus in the expression (6-1.7), one should replace l_1 by $l_1{}^0\sqrt{1-\beta^2}$, and l_2 by $l_2{}^0\sqrt{1-\beta^2}$ in Eq. (6-1.8). As a consequence

$$\Delta = \frac{2}{\sqrt{1-\beta^2}}(l_1{}^0 - l_2{}^0) = \Delta' \tag{6-1.10}$$

and hence there will be no fringe shift.

Lorentz[18] justified the contraction hypothesis on the grounds that the molecular forces that hold material bodies together might be affected by translational motion through absolute space. On the assumption that molecules at rest interact only electrostatically, Lorentz was able to show on the basis of his theory that a moving material object will be in equilibrium if all dimensions in the direction of motion are shortened by just the Fitzgerald-Lorentz contraction factor.

Although the Fitzgerald-Lorentz hypothesis was dropped after the advent of special relativity, it is interesting to note that it was subjected to a direct experiment test in 1932 by Kennedy and Thorndike.[19] Their test was based on the fact that a Michelson-Morley type of apparatus, in which the two beams of light travel appreciably different distances, should be able to detect

[14] A detailed list of references to these experiments is to be found in W. Pauli, *Theory of Relativity* (Pergamon Press, New York, 1958).

[15] F. T. Trouton, *Trans. Roy. Soc.* (London), **202**, 165 (1903); *Proc. Roy. Soc.* (London), **72**, 132, (1903).

[16] H. E. Ives and C. R. Stillwell, *J. Optical Soc. Amer.*, **28**, 215 (1938); **31**, 369 (1941).

[17] See O. Lodge, *Nature*, **46**, 165 (1892), where Fitgerald's conjecture is credited to him.

[18] H. A. Lorentz, *Amst. Verh. Akad. van Wetenschappen*, **1**, 74 (1892).

[19] R. J. Kennedy and E. W. Thorndike, *Phys. Rev.*, **42**, 400 (1932).

second-order effects due to diurnal and yearly changes in the absolute motion of the earth, even assuming the validity of the Fitzgerald-Lorentz contraction hypothesis. Thus in Eq. (6-1.10) for the difference in optical path length, if β were to change to β' one would have a fringe shift, given approximately by

$$n \cong \frac{l_1^{\,0} - l_2^{\,0}}{\lambda} (\beta'^2 - \beta^2).$$

Their results were null to within $\pm$ 10 km/sec. As in the case of the Michelson-Morley experiment, this null result could be explained by an additional ad hoc hypothesis. In addition to assuming a Fitgerald-Lorentz contraction one had now to assume, as did Larmor,[20] that a clock moving with a velocity v relative to absolute space ran slower by a factor $1/\sqrt{1 - \beta^2}$ than did one in a state of absolute rest.

Problem 6.1. Show that the Larmor hypothesis, together with the Fitzgerald-Lorentz hypothesis, would explain the null result of the Kennedy-Thorndike experiment.

A second attempt to explain the negative results of the Michelson-Morley experiment was based on the so-called aether drag hypothesis.[21] It was assumed that the medium which supported the electromagnetic oscillations that give rise to light was actually carried or dragged along with the earth. Such a hypothesis would automatically explain all null results of interference experiments performed on the earth's surface. However, if the hypothesis were correct, there should be no aberration of starlight, whereas we observe an aberration that would be predicted without an aether drag by using the known orbital velocity of the earth, Furthermore the velocity of light in a moving medium would no longer be given by the Fresnel formula of Eq. (6-1.3) but rather by that given in Eq. (6-1.1). This, as we saw, would lead to first-order effects due to the earth's motion, but these are not observed.

The emission theories of light represented a third attempt to explain the negative results of all attempts to use the laws of electrodynamics to observe a motion of the earth through absolute space. In all these theories it was assumed that the velocity of light was c/n with *respect to its source* and was independent of the state of motion of the transmitting medium of refractive index n. All these theories had difficulties with the question of what happened

[20] J. Larmor, *Aether and Matter* (Cambridge Univ. Press, Cambridge, 1900), pp. 167–177.

[21] A. A. Michelson, *Am. J. Sci.*, **4**, 3, 475 (1897). In this paper Michelson describes a repetition of his experiment on a high mountain in order to see if there is an aether wind, again with null result.

to the light reflected from a mirror moving with respect to the source. There were three alternatives proposed: The velocity

1. Remained c/n relative to the original source.
2. Changed to c/n relative to the mirror.
3. Changed to c/n relative to the image of the source in the mirror.

Of these three, the latter two lead to coherence difficulties with the reflected light. The first alternative was proposed by Ritz in 1910 in an attempt to obviate the conclusions on the nature of simultaneity reached by Einstein five years earlier. Ritz retained the two Maxwell equations $\nabla \cdot \mathbf{B} = 0$ and $\nabla \times \mathbf{E} = -(1/c)(\partial \mathbf{B}/\partial t)$, but replaced the other two equations involving source terms by

$$\phi = \frac{1}{4\pi\varepsilon_0} \int \rho\left(\mathbf{x}', t - \frac{r}{c + v_r}\right) \frac{d^3x'}{r}$$

and

$$\mathbf{A} = \frac{\mu_0}{4\pi} \int \mathbf{j}\left(\mathbf{x}', t - \frac{r}{c + v'_r}\right) \frac{d^3x'}{r},$$

where v_r is the component of the velocity of the source in the direction pointing from the source to the point of observation. All emission theories were in disagreement with observations on double stars. If the velocity of light depended on the velocity of the source, then one should expect to see ghost images at times. de Sitter[22] showed that if $v_{\text{light}} = c + kv_{\text{star}}$, k would have to be less than 0.002. Also, Tomaschek[23] repeated the Michelson-Morley experiment, using sunlight. The Ritz theory predicted complications in the interference pattern due to the sun's rotation, but they were not observed.

The failure of all attempts to measure the absolute velocity of the earth by electromagnetic means presented the scientists at the turn of the century with a profound mystery. It appeared that electrodynamics, like Newtonian mechanics, admitted an additional symmetry over and above the symmetries that arise as a consequence of the homogeneity and isotropy of space and the homogeneity of time. This by itself would not have been so serious; physicists had lived with such a situation ever since Newton first set down his laws of motion. What was serious was that the symmetry group of Maxwell's equations was different from that of Newtonian mechanics. Scientists could therefore no longer turn their backs on the question as they had up to then. The fact that the symmetry group was different for the two theories meant that a combined electromagnetic-mechanical system should allow one to determine

[22] W. de Sitter, *Proc. Acad. Sci., Amsterdam*, **15**, 1297 (1913); **16**, 395 (1913).
[23] R. Tomaschek, *Ann. der Phys.*, **73**, 105 (1924).

the earth's velocity by dynamical means. A satisfactory solution to the problem was set forth in 1905 by Einstein. It was embodied in the theory we now know as special relativity, to which we now turn our attention.

6-2. The Development of Special Relativity[24]

The realization that the Poincaré rather than the Galilean group was the fundamental symmetry group of nature was arrived at independently by Poincaré and Einstein in 1905. As early as 1887 Voight[25] had considered special Lorentz transformations in connection with the wave equation. A transformation similar to the one used by Voight was introduced by Lorentz[26] in 1892. It was Larmor[27] who first set down in 1900 the expression we now use for the Lorentz transformation. Lorentz[28] in 1904, was the first to show that the vacuum Maxwell equations admitted the Lorentz group as a symmetry group. However, in his paper, he did not consider this group as representing a fundamental symmetry of nature. He still employed the concept of absolute velocity and tried to understand length contraction in terms of absolute motion. Poincaré[29] showed that the electrodynamics of Lorentz was covariant with respect to the Lorentz group and asserted that this group was the fundamental symmetry group of all physical systems. Einstein's epochal paper,[30] *Zur Elektrodynamik bewegter Körper*, was submitted for publication at about the same time as was Poincaré's without prior knowledge of Lorentz' paper. At the very beginning of this paper Einstein sets forth the two basic postulates of special relativity. They are, as stated by Einstein:

(i) *The Principle of Relativity.* The laws of electrodynamics and optics are valid for all frames of reference for which the equations of mechanics hold good.

(ii) *The Principle of the Constancy of the Velocity of Light.* Light is always propagated in empty space with a definite velocity c, which is independent of the state of motion of the emitting body.

[24] An account of the history of the development of special relativity is contained in E. T. Whittaker, *A History of the Theories of Aether and Electricity*, Vol. II (Thomas Nelson and Sons, Ltd., Edinburgh, 1951). His view that the special theory is due in the main to Lorentz and Poincaré is not generally held. For an alternative account, see G. Holton, *Am. J. Phys.*, 28, 627 (1960).

[25] W. Voight, *Nachr. Ges. Wiss. Göttingen*, 41 (1887).

[26] H. A. Lorentz, *Arch. néerl. Sci.*, **25**, 363 (1892).

[27] J. Larmor, *loc. cit.*

[28] H. A. Lorentz, *Proc. Acad. Sci. Amsterdam*, **6**, 809 (1904).

[29] H. Poincaré, *C. R. Acad. Sci., Paris*, **140**, 1504 (1905).

[30] A. Einstein, *Ann. Phys., Leipzig*, **17**, 891 (1905).

On the basis of these two postulates and an analysis of the concept of simultaneity Einstein was able to derive all results contained in the papers of Lorentz and Poincaré without having to make any special assumptions concerning the nature of the forces responsible for the stability of rigid bodies. Above all, it was Einstein's analysis of space-time measurements and especially the concept of temporal simultaneity that gave a justification to the principle of relativity and made it acceptable to most physicists of the time, with a few notable exceptions (for example, Ritz). Shortly before he died, James Frank mentioned to the author that on the morning he first read Einstein's paper, he knew that Einstein was right in all essential details.

6-3. Relation Between Theory and Experiment

Given a physical theory, the problem that confronts the physicist is to make a connection between the mathematical elements of the theory and the elements of the physical world. The particular theory from which the physicist starts may in part be suggested by the results of observations, but in the final analysis it is a free invention of the human mind, as Einstein pointed out. Thus, while the negative outcomes of the Michelson-Morley and other such experiments may have suggested that a universal relativity principle was operative, they did not force one to accept this principle, as witness the various attempts to explain the experiments that did not make use of such a principle. It is therefore not necessary to justify *ab initio* the basic postulates of a physical theory. Such justification can come only after the theory has been compared with experiment. This fact is often overlooked by critics of relativity theory, who usually try to find discrepancies in the foundations of the theory. Unfortunately it usually does little good to point out that what they are objecting to has nothing to do with the observable predictions of the theory or that the predictions of relativity theory have been amply borne out by observations.

A good deal of the confusion that arises out of relativity theory is associated with the measurements of those quantities we usually call "space" and "time," and more explicitly the coordinates that appear, for instance, in the usual formulation of Maxwell's laws. It was Einstein's analysis of space-time measurements more than anything else that led to the acceptance of the relativity theory. It was clear from the start that Einstein considered these measurements in terms of comparisons between different physical systems. He says: "It might appear possible to overcome all the difficulties attending the definition of 'time' by substituting 'the position of the small hand of my watch' for 'time'. And in fact, such a definition is satisfactory when we are concerned with defining a time exclusively for the place where the watch is located." Likewise he considered spatial measurements to be comparisons between rigid bodies and other physical systems. He puts it thus: "The theory to be developed is based—like all electrodynamics—on the kinematics of the rigid body, since the assertions of any such theory have to do with relationships between the rigid bodies (systems of coordinates), clocks, and electromagnetic processes." In this he was very close in spirit to Newton, who wrote in the preface to the first edition of the *Principia*: "Therefore geometry is founded in mechanical practice and is nothing but that part of universal mechanics which accurately proposes and demonstrates the art of measuring." Thus we should not demand that what the physicist calls a space measurement or a time measurement bear much relation to our psychologically conditioned senses of these concepts. All that we must require is that such measurements should be realizable by real physical systems. The great success of relativity theory lies, in the final analysis, in the fact that space-time measurements carried out with real physical systems are more correctly described in that theory than in Newtonian mechanics.

In his discussion of space-time measurements Einstein employed the notion of rigid bodies, standard clocks, and light signals to synchronize these measurements. Unfortunately clocks are not simple physical systems. The notion of a rigid rod is even more complicated. We now know from the modern theory of solids that quantum phenomena play an essential role in the internal dynamics of such systems. But even on a classical level we shall see later (Section 9-3) that the concept of a rigid rod is, strictly speaking, inadmissible in relativity theory. Therefore, to associate the coordinates appearing in special forms of the laws of nature, with reference frames constructed with the help of such objects, is a questionable procedure at best, as Einstein himself was later to point out. As long as one restricts oneself to special relativity it is possible to find systems in nature that correlate with special coordinate systems. However, in general relativity such a procedure is wholly impracticable. Consequently we will not follow Einstein's original development in our approach to relativity theory. In no way is this meant to imply that Einstein's treatment was wrong or less of an achievement than it is regarded to be. Our final conclusions will be the same as his. But at the present time, with the whole of relativity theory behind us, it is possible to give a unified discussion of the measurement process, one that is applicable to both the special and the general theories of relativity.

In a certain sense the position we take here is very much like that adopted by Newton. In Section 5-2 we saw that he distinguished between absolute space, which could not be immediately observed, and relative space, which is directly observable by means of physical apparatus whose behavior is governed by the laws of Newtonian mechanics. Newton makes this distinction even more clearly in the case of time. In addition to absolute time he introduces the notion of relative, apparent, or common time as "some sensible and external (whether accurate or unequable) measure of duration by the means of motion." Somewhat later, Newton makes a very penetrating comment regarding the relation of the two times (an analogous comment can be made for the relation between absolute and relative space). He says: "It may be that there is no such thing as an equable motion, whereby time may be measured. All motions may be accelerated and retarded, but the flow of absolute time is not liable to any change." For Newton, then, the question of the existence of an absolute space and an absolute time was independent of whether or not they were directly observable; the question rested, in the final analysis, on the success of the system based on their existence in describing the physical world. And exactly the same can be said concerning the existence of the absolute space-time of special relativity. Consequently the only real objection that one can raise to Newton is that his system did not describe the physical world as accurately as did later theories. But within the context of the known facts at that time it afforded a complete description of the world. Newton could have been correct, the world could have been as he described it, and no amount of philosophical disquisition could have proved him wrong.

The fundamental requirement we place on any physical theory is that it be complete. By this we mean that it must contain a description of all physical systems with which it purports to deal, including those systems employed in the measurement process. If we wish to discuss the temporal behavior of some physical system, we must do so by correlating this behavior with another system we take to be a clock. The dynamics of both the system *and the clock* must be describable within the framework of the class of physical theories under consideration, as must also the method of correlating their configurations.

If we are to adhere to our fundamental requirements it is necessary to set forth the properties of the class of physical theories that we have to deal with in a manner that is not directly related to any measurement process, and then to examine the various systems encompassed by those theories so as to find those that are suitable as models for clocks, meter sticks, and other measuring devices. In Newtonian mechanics it is fairly easy to construct models of clocks; for example, two masses coupled by a spring will do nicely. It is also easy to describe how one would correlate the motion of such a clock with some other Newtonian system. Our ability to do both things hinges on the fact that absolute simultaneity of events is a concept inherent in the Newtonian world-picture. Since this concept is not available to us in special relativity, it will be harder to discuss the measurement process in this theory, and we can expect to find that the relationship between space-time measurements will differ in the two theories.

6-4. Symmetries of Special Relativistic Systems: Absolute Space-Time

Einstein's hypothesis of the principle of relativity forces us to the conclusion that the laws of nature possess a symmetry over and above the symmetries associated with the homogeneity and isotropy of space and the homogeneity of time. In effect he deduced the nature of this symmetry group by coupling his first hypothesis with that proclaiming the independence of the velocity of light on the motion of the source. The group he found was just the inhomogeneous Lorentz group or, as it is now called, the Poincaré group.

Today there is no need to justify the Poincaré group as there was in Einstein's time, nor to derive it from a minimal set of hypotheses. Rather we simply assert that the Poincaré group is the symmetry group (at least to a high degree of approximation) of all physical systems and then explore the consequences of such an assertion. The justification for this assertion will then rest on the degree to which these consequences are borne out in nature.

Of course it is a simple matter for us to adopt the straightforward procedure of taking the Poincaré group to be a universal symmetry group, since we already know that it works. Einstein's great achievement was to realize that a relativity principle was operative and then to discover that it was the Poincaré group which encompassed this principle. This required, among other things, that he be willing to modify Newtonian mechanics, since the Poincaré group was not the symmetry group of Newtonian systems. But even more, it meant that he had to be willing to give up the Newtonian concepts of absolute space and absolute time, which had been generally accepted for more than two hundred years.

To see what is involved in accepting the Poincaré group as the symmetry group of all physical systems, let us begin by recalling that it is the motion group of a flat, four-dimensional metric of signature -2. Such a metric satisfies

$$R_{\mu\nu\rho\sigma}(g) = 0 \tag{6-4.1}$$

and can always be mapped so that

$$g_{\mu\nu}(x) = \eta_{\mu\nu}. \tag{6-4.2}$$

The infinitesimal mappings that leave this form of the metric invariant, that is, those mappings for which

$$\bar{\delta}\eta_{\mu\nu} = -\eta_{\mu\rho}\xi^{\rho}{}_{,\nu} - \eta_{\nu}\xi^{\rho}{}_{,\mu} = 0, \tag{6-4.3}$$

are of the form

$$x^{\mu} \to x'^{\mu} = x^{\mu} + \varepsilon_{\nu}{}^{\mu}x^{\nu} + \varepsilon^{\mu}, \tag{6-4.4}$$

where $\varepsilon_{\mu\nu} \equiv \eta_{\mu\rho}\varepsilon_{\nu}{}^{\rho} = -\varepsilon_{\nu\mu}$ and ε^{μ} are infinitesimal parameters.

The essential feature of the Poincaré mappings is that they inextricably mix the space and time coordinates of a point. We saw that the special Galilean mappings left invariant the $t = \text{const}$ surfaces of Newton's absolute time, but not the $\lambda^r = \text{const}$ curves of his absolute space. Under the Poincaré mappings neither the $t = \text{const}$ surfaces or the $\lambda^r = \text{const}$ curves are invariant; neither space by itself or time by itself can be considered as being absolute in the Newtonian sense. Only space-time as a whole can be said to be absolute in the sense that the metric tensor $g_{\mu\nu}$ satisfies Eq. (6-4.1), and hence is an absolute object in all Poincaré invariant theories. H. Minkowski[31] put it aptly when he said, "Henceforth space by itself, and time by itself, are doomed to fade away into mere shadows, and only a kind of union of the two will preserve an independent reality."

6-5. The Poincaré Group

In this section we will collect a number of useful facts concerning the Lorentz and Poincaré groups, which follow from their definitions as groups of mappings that leave invariant the flat metric. To simplify matters we will use the Minkowski form of this metric as given in Eq. (6-4.2).

The Lorentz mappings are those linear homogeneous mappings

$$x^\mu \to x'^\mu = \Lambda_\nu{}^\mu x^\nu \tag{6-5.1}$$

that leave invariant the Minkowski metric $\eta_{\mu\nu}$ and therefore satisfy

$$\Lambda_\rho{}^\mu \Lambda_\sigma{}^\nu \eta_{\mu\nu} = \eta_{\rho\sigma} . \tag{6-5.2}$$

Equation (6-4.3) is a special case of the above condition when the mapping is infinitesimal, that is, when $\Lambda_\nu{}^\mu$ is of the form

$$\Lambda_\nu{}^\mu = \delta_\nu{}^\mu + \varepsilon_\nu{}^\mu . \tag{6-5.3}$$

Problem 6.2. Show that the product of two finite Lorentz transformations is again a Lorentz transformation.

If we take the determinant of both sides of Eq. (6-5.2) we find, by the well-known rule that the determinant of a product of matrices is equal to the product of the determinants of the matrices that

$$(\det \Lambda)^2 = 1,$$

so that

$$\det \Lambda = \pm 1. \tag{6-5.4}$$

[31] H. Minkowski, *Space and Time*. Address delivered at the 80th Assembly of German Natural Scientists and Physicians, Cologne, Sept. 21, 1908.

Let us further set $\rho = \sigma = 0$ in Eq. (6-5.2). We then have

$$(\Lambda_0{}^0)^2 - \sum_{i=1}^{3} (\Lambda_0{}^i)^2 = 1,$$

from which it follows that

$$\Lambda_0{}^0 \geqq 1 \tag{6-5.5a}$$

or

$$\Lambda_0{}^0 \leqq -1. \tag{6-5.5b}$$

Consequently the set of all Lorentz matrices Λ can be divided into four subsets according to whether $\det \Lambda = \pm 1$ and $\Lambda_0{}^0$ is greater than or equal to 1 or less than -1. If $\det \Lambda = +1$ we say that Λ is *proper*, while if $\Lambda_0{}^0 \geqq 1$, Λ is *orthochronous*. The proper orthochronous Lorentz transformations form a subgroup, the group of restricted Lorentz mappings. It is a six-parameter continuous group, so that any Λ belonging to this subgroup can be obtained by starting from the identity and integrating a series of infinitesimal mappings of the form given in Eq. (6-5.3). This is clearly not the case for the other mappings, since for the identity mapping $\Lambda_\nu{}^\mu = \delta_\nu{}^\mu$, and so $\det \Lambda = +1$ and $\Lambda_0{}^0 = +1$. Since $\det \Lambda$ can be only ± 1 and $\Lambda_0{}^0 \geqq 1$ or $\leqq -1$, starting with the value $+1$, $\det \Lambda$ can never be made to change sign, nor can $\Lambda_0{}^0$ become $\leqq -1$ if originally it is $\geqq 1$ by a continuous process. However, the improper or nonorthochronous Λ, or both, can all be obtained by multiplying the restricted Λ by one of the following three finite Λ:

1. For $\det \Lambda = -1$, $\Lambda_0{}^0 \geqq 1$, multiply restricted matrices by the space inversion matrix.

$$\Lambda(i_s) = \operatorname{diag}(1, -1, -1, -1),$$

which sends

$$x^0 \to x^0, \qquad x^r \to -x^r.$$

2. For $\det \Lambda = -1$, $\Lambda_0{}^0 \leqq -1$ multiply restricted matrices by the time inversion matrix.

$$\Lambda(i_t) = \operatorname{diag}(-1, 1, 1, 1),$$

which sends

$$x^0 \to -x^0, \qquad x^r \to x^r.$$

3. For $\det \Lambda = +1$, $\Lambda_0{}^0 \leqq -1$, multiply restricted matrices by the space-time inversion

$$\begin{aligned} \Lambda(i_{st}) &= \Lambda(i_s)\Lambda(i_t) \\ &= \operatorname{diag}(-1, -1, -1, -1), \end{aligned}$$

which sends

$$x^\mu \to -x^\mu.$$

In order to obtain expressions for the finite restricted Λ one can try to find Λ with $\det \Lambda = +1$ and $\Lambda_0{}^0 \geqq 1$ that satisfy Eq. (6–5.2). Actually it is easier to integrate the infinitesimal mappings. For such a mapping let us write

$$x^\mu \to x'^\mu = x^\mu + \varepsilon_i I^\mu_{i\nu} x^\nu, \tag{6-5.6}$$

where the new parameters ε_i are related to the $\varepsilon_\nu{}^\mu$ by

$$\varepsilon_i I^\mu_{i\nu} = \varepsilon_\nu{}^\mu. \tag{6-5.7}$$

Since, for an arbitrary infinitesimal mapping there are six $\varepsilon^{\mu\nu}$, there will be six ε_i and hence six generators $I^\mu_{i\nu}$. Then, since $\varepsilon^{\mu\nu} = -\varepsilon^{\nu\mu}$ we have

$$\varepsilon_i I^\mu_{i\rho} \eta^{\rho\nu} = -\varepsilon_i I^\nu_{i\rho} \eta^{\rho\mu},$$

or since the ε_i are arbitrary and independent,

$$I^\mu_{i\rho} \eta^{\rho\nu} = -I^\nu_{i\rho} \eta^{\rho\mu}. \tag{6-5.8}$$

A set of six independent generators that satisfy these conditions are given by

$$I^\rho_{1\sigma} = \delta_3{}^\rho \delta_\sigma{}^2 - \delta_2{}^\rho \delta_\sigma{}^3, \tag{6-5.8a}$$

$$I^\rho_{2\sigma} = \delta_1{}^\rho \delta_\sigma{}^3 - \delta_3{}^\rho \delta_\sigma{}^3, \tag{6-5.8b}$$

$$I^\rho_{3\sigma} = \delta_2{}^\rho \delta_\sigma{}^1 - \delta_1{}^\rho \delta_\sigma{}^2, \tag{6-5.8c}$$

$$I^\rho_{4\sigma} = \delta_0{}^\rho \delta_\sigma{}^1 + \delta_1{}^\rho \delta_\sigma{}^0, \tag{6-5.8d}$$

$$I^\rho_{5\sigma} = \delta_0{}^\rho \delta_\sigma{}^2 + \delta_2{}^\rho \delta_\sigma{}^0, \tag{6-5.8e}$$

$$I^\rho_{6\sigma} = \delta_0{}^\rho \delta_\sigma{}^3 + \delta_3{}^\rho \delta_\sigma{}^0. \tag{6-5.8f}$$

In order to examine further the structure of the Lorentz group, let us determine the commutation relations among the above six generators. If we introduce the notation

$$\{I_r\} = \{I_1, I_2, I_3\} \quad \text{and} \quad \{J_r\} = \{I_4, I_5, I_6\},$$

we find that

$$[I_i, I_j]_\nu{}^\mu = \varepsilon_{ijk} I^\mu_{k\nu}, \tag{6-5.9}$$

$$[J_i, J_j]_\nu{}^\mu = -\varepsilon_{ijk} I^\mu_{k\nu}, \tag{6-5.10}$$

and

$$[I_i, J_j]_\nu{}^\mu = \varepsilon_{ijk} J^\mu_{k\nu}, \tag{6-5.11}$$

where ε_{ijk} is the three-dimensional alternating symbol. From the commutation relations (6-5.9) it follows that the I_k by themselves constitute the generators of a subgroup of the Lorentz group. It further follows from these relations that this subgroup is just the three-dimensional rotation group O_3 whose generators obey the same commutation relations. However, the J_i are not the generators of any subgroup of the Lorentz group.

Because of the fact that the rotation group is a subgroup, it is sometimes convenient to make use of the so-called (1 + 3) notation. Since a rotation generated by an I_r leaves invariant the o component of a contravector and transforms the r components among themselves, we can in effect treat the o component like a scalar and the r components like a three-vector in a three-dimensional flat space with a Euclidean metric, as far as the rotations are concerned. Then, as long as we consider only rotations, we can make use of the conventional results of vector analysis. Thus we will decompose the vector A^μ as $A^\mu = \{A^0, \mathbf{A}\}$. In this notation,

$$\eta_{\mu\nu}A^\mu A^\nu = (A^0)^2 - \mathbf{A}^2. \tag{6-5.12}$$

With the help of the ∇ operator of vector analysis we can write $A^\mu{}_{,\mu}$ as

$$A^\mu{}_{,\mu} = A^0{}_{,0} - \nabla \cdot \mathbf{A}. \tag{6-5.13}$$

We can also write second-rank tensors in the (1 + 3) notation. If $F^{\mu\nu} = -F^{\nu\mu}$, set

$$F^{\mu\nu} = \left\{\begin{array}{c|c} O & -\mathbf{F} \\ \hline \mathbf{F} & F^{rs} \end{array}\right\}, \tag{6-5.14}$$

were $\mathbf{F}$ transforms like a polar vector under a rotation and F^{rs} transforms like a skew contra-three tensor. If we introduce the vector $\mathfrak{F}$ with components $\mathfrak{F}_r = \frac{1}{2}\varepsilon_{rst}F^{st}$, where ε_{rst} is the three-dimensional alternating symbol, then $\mathfrak{F}$ transforms like an axial vector under rotations. With the help of these quantities and the ∇ operator we can write $F^{\mu\nu}{}_{,\nu}$ as

$$F^{\mu\nu}{}_{,\nu} = \{-\nabla \cdot \mathbf{F}, \nabla \times \mathfrak{F} + \dot{\mathbf{F}}\}. \tag{6-5.15}$$

where a dot over a quantity denotes differentiation with respect to x^0. Also one has

$$\tfrac{1}{2}\varepsilon^{\mu\nu\rho\sigma}F_{\nu\rho,\sigma} = \{\nabla \cdot \mathfrak{F}, \nabla \times \mathbf{F} - \dot{\mathfrak{F}}\} \tag{6-5.16}$$

It is also possible to extend the (1 + 3) notation to other tensors, although we will not do so here, since the results are not often used.

Problem 6.3. Prove Eqs. (6-5.15) and (6-5.16).

Let us now consider finite transformations of the group. Form the Lorentz matrix

$$R(\theta,\mathbf{u}) = e^{\theta u \cdot \mathbf{I}} \tag{6-5.17}$$

where $\mathbf{u}\cdot\mathbf{I} = u_1I_1 + u_2I_2 + u_3I_3$ and where, for convenience in writing, we have suppressed all indices. As usual the exponential of a matrix is defined in terms of the series expansion of the exponential, that is,

$$\exp\{A_\nu{}^\mu\} = \delta_\nu{}^\mu + A_\nu{}^\mu + \frac{1}{2}A_\rho{}^\mu A_\nu{}^\rho + \frac{1}{3!}A_\rho{}^\mu A_\sigma{}^\rho A_\nu{}^\sigma + \cdots.$$

If $\mathbf{u}\cdot\mathbf{u} = 1$ it can be shown that

$$(\mathbf{u}\cdot\mathbf{I})^3 = -\mathbf{u}\cdot\mathbf{I}.$$

Making use of this fact, it then follows that

$$R(\theta,\mathbf{u}) = [1 + (\mathbf{u}\cdot\mathbf{I})^2] - (\mathbf{u}\cdot I)^2\cos\theta + (\mathbf{u}\cdot\mathbf{I})\sin\theta. \tag{6-5.18}$$

Since $R_0{}^0(\theta, \mathbf{u}) = 1$ we can interpret $R(\theta, \mathbf{u})$ as a rotation about $\mathbf{u}$ through an angle θ.

Next consider the Lorentz matrix

$$V(\phi,\mathbf{w}) \equiv e^{\phi\mathbf{w}\cdot\mathbf{J}}, \tag{6-5.19}$$

where we use the same notation as before. If $\mathbf{w}\cdot\mathbf{w} = 1$ then

$$(\mathbf{w}\cdot\boldsymbol{J})^3 = \mathbf{w}\cdot\boldsymbol{J},$$

and so

$$V(\phi,\mathbf{w}) = [1 - (\mathbf{w}\cdot\boldsymbol{J})^2] + (\mathbf{w}\cdot\boldsymbol{J})^2\cosh\phi + (\mathbf{w}\cdot\boldsymbol{J})\sinh\phi. \tag{6-5.20}$$

The mapping formed by using $V(\phi, \mathbf{w})$ will be called a *velocity mapping*. Let us introduce a new quantity $\mathbf{v}$ given by

$$\mathbf{v} = \mathbf{w}\tanh\phi. \tag{6-5.21}$$

Then it follows that

$$\frac{\mathbf{v}}{\sqrt{1-\mathbf{v}^2}} = \mathbf{w}\sinh\phi \quad\text{and}\quad \frac{1}{\sqrt{1-\mathbf{v}^2}} = \cosh\phi,$$

where $\mathbf{v}^2 = \mathbf{v}\cdot\mathbf{v}$. From these relations we see that

$$\sqrt{\mathbf{v}^2} < 1.$$

In the future we shall designate a velocity mapping by $V(\mathbf{v})$ instead of $V(\theta, \mathbf{w})$.

Problem 6.4. Show that $V(v)V(-v) = 1$.

To interpret the velocity mappings, consider the curve $x^r = 0$. Under a velocity mapping $(V(v)$ this curve gets mapped onto the curve $x^r = v^r x^0$.

Later we will see that such curves will represent the trajectories of free particles and that, if the velocity of the particle associated with the curve $x^r = 0$ is zero, the particle moving along the curve $x^r = v^r x^0$ will have a velocity v^r.

Problem 6.5. Show that if $A^\mu = \{0, \mathbf{A}\}$ and $\mathbf{A}\cdot\mathbf{v} = 0$, then under the velocity mapping $V(\mathbf{v})$, $A^\mu \to A'^\mu = A^\mu$.

Although the matrix of an arbitrary velocity mapping is rather complicated, its action on a covector A^μ can be written in a simple form. If under the action of $V(\mathbf{v})$, $A^\mu \to A^\mu$, then it follows from Eq. (6-5.20) that

$$A'^\mu = \left\{\gamma(A^0 + \mathbf{A}\cdot\mathbf{v}), \mathbf{A} + \frac{(\mathbf{A}\cdot\mathbf{v})\mathbf{v}}{\mathbf{v}^2}(\gamma - 1) + \gamma\mathbf{v}A^0\right\}, \tag{6-5.22}$$

where $\gamma = (1 - \mathbf{v}^2)^{-1/2}$. If, in particular, $\mathbf{v} = \{v, 0, 0\}$, then $V(\mathbf{v})$ has the simple form

$$V(v) = \begin{Bmatrix} \gamma & v\gamma & 0 & 0 \\ v\gamma & \gamma & 0 & 0 \\ 0 & 0 & 1 & 0 \\ 0 & 0 & 0 & 1 \end{Bmatrix}. \tag{6-5.23}$$

Problem 6.6. Show that when an antisymmetric contratensor $F^{\mu\nu}$ is decomposed into its (1 + 3) components according to Eq. (6-5.14) and the discussion following it, then under a velocity mapping $V(\mathbf{v})$,

$$\mathbf{F}' = \gamma\mathbf{F} + \frac{\mathbf{v}}{\mathbf{v}^2}(\mathbf{v}\cdot\mathbf{F})(1-\gamma) + \gamma(\mathbf{v}\times\mathfrak{F}) \tag{6-5.24a}$$

and

$$\mathfrak{F}' = \gamma\mathfrak{F} + \frac{\mathbf{v}}{\mathbf{v}^2}(\mathbf{v}\cdot\mathfrak{F})(1-\gamma) + \gamma(\mathbf{v}\times\mathbf{F}). \tag{6-5.24b}$$

Since the velocity mappings do not form a group it is not surprising that the product of two such mappings is no longer a velocity mapping but involves a rotation as well. This property of the Lorentz group leads to the well-known Thomas precession.[32] (See also Section 7-20.) Only when the two velocities involved are parallel do we again get a velocity transformation. Let $\mathbf{v}_1 = (v_1, 0, 0)$ and $\mathbf{v}_2 = (v_2, 0, 0)$ be two such velocities. Then, from Eq. (6-5.21) and the fact that

$$\tanh(\phi_1 + \phi_2) = \frac{\tanh\phi_1 + \tanh\phi_2}{1 + \tanh\phi_1\tanh\phi_2},$$

[32] L. H. Thomas, *Nature*, **117**, 514 (1926); *Phil. Mag.*, **7**, 3, 1 (1927).

it follows that

$$v_3 = \frac{v_1 + v_2}{1 + v_1 v_2} \tag{6-5.25}$$

for $V(\mathbf{v}_3) \equiv V(\mathbf{v}_1)V(\mathbf{v}_2)$. We see that $\mathbf{v}_3$ is not simply the sum of $\mathbf{v}_1 + \mathbf{v}_2$, as would be the case for the Galilean group.

We now prove a fundamental property of the Lorentz group, namely, that any Lorentz matrix Λ can be factored in two different ways:

$$\Lambda = V(\mathbf{v})R \quad \text{and} \quad \Lambda = R'V(\mathbf{v}'),$$

where the velocities $\mathbf{v}$ and $\mathbf{v}'$ and the rotations R and R' are unique. To prove this theorem choose $\mathbf{v}$ to have the components $v^r = \Lambda_0{}^r/\Lambda_0{}^0$. As a consequence of Eq. (6-5.2) it follows that $\mathbf{v}^2 < 1$, so $\mathbf{v}$ is an admissible velocity. It then follows from Eq. (6-5.20) that

$$V_0{}^0(-\mathbf{v}) = \Lambda_0{}^0, \qquad V_r{}^0(-\mathbf{v}) = -\Lambda_0{}^r.$$

Now define a matrix $R = V(-\mathbf{v})\Lambda$. Then, again making use of Eq. (6-5.2), we find that $R_0{}^0 = 1$. Since R is a Lorentz matrix it must be a rotation. Finally, since $V(\mathbf{v})V(-\mathbf{v}) = 1$ we have

$$\Lambda = V(\mathbf{v})V(-\mathbf{v})\Lambda = V(\mathbf{v})R.$$

To prove the uniqueness of the decomposition, let $V(\mathbf{v})R = V(\bar{\mathbf{v}})\bar{R}$, where $\bar{R}$ is some other rotation. Then, by multiplying on the left by $V(-\mathbf{v})$ and on the right by $\bar{R}^{-1}$, we get the result

$$R\bar{R}^{-1} = V(-\mathbf{v})V(\bar{\mathbf{v}}),$$

so that $V(-\mathbf{v})V(\bar{\mathbf{v}})$ is a proper rotation. But, from Eq. (6-5.20) it follows that

$$(V(-\mathbf{v})V(\bar{\mathbf{v}}))_0{}^0 = \frac{1 - \mathbf{v}\cdot\bar{\mathbf{v}}}{\sqrt{1-\mathbf{v}^2}\sqrt{1-\bar{\mathbf{v}}^2}}$$

and the right side equals $+1$ if and only if $\mathbf{v} = \bar{\mathbf{v}}$, which proves the uniqueness. The proof of the uniqueness of the second factorization follows along similar lines.

The problem of classifying all finite dimensional irreducible representations of the Lorentz group is straightforward once we know the irreducible representations of the rotation group. These are characterized by a single number j which can take on all integer and half-integer values from 0 to ∞. The basis space for the jth representation has dimensions $2j + 1$. In order to find irreducible representations of the Lorentz group, that is, irreducible matrices that have the same multiplication properties as the Lorentz matrices and whose infinitesimal elements satisfy the commutation relations (6-5.9, 6-5.10 and 6-5.11), we form the matrices

$$M_k = \tfrac{1}{2}i(I_k + iJ_k) \quad \text{and} \quad N_k = \tfrac{1}{2}i(I_k - iJ_k).$$

These matrices satisfy the commutation relations

$$[M_i, M_j] = i\varepsilon_{ijk}M_k, \quad [N_i, N_j] = i\varepsilon_{ijk}N_k, \quad [M_i, N_j] = 0,$$

Thus the M by themselves and the N by themselves satisfy the standard angular momentum commutation relations. Therefore the finite dimensional representations of the Lorentz group are characterized by two integers or half-integers (j, j'), which can take on all values from 0 to ∞, and the corresponding basis space is $(2j + 1)(2j + 1)$ dimensional. In addition to the finite dimensional representations there are also infinite dimensional representations. They have not found a use in physics to date and will not be discussed here.[33]

The (0, 0) representation acts on the scalars and the Lorentz matrices $\Lambda^{(00)}$ are all equal to unity. The $(\frac{1}{2}, \frac{1}{2})$ representation acts on the four component vectors and the corresponding Lorentz matrices $\Lambda^{(\frac{1}{2},\frac{1}{2})}$ are just the usual 4×4 Lorentz matrices. However, there are also the $(\frac{1}{2}, 0)$ and $(0, \frac{1}{2})$ representations. The objects that transform under these representations are called *two-component spinors*. These are a new type of object that we have not yet met with. We will discuss their significance when we come to discuss the Dirac equation. The M and N matrices for these representations are given by

$$M_k^{(\frac{1}{2},0)} = \tfrac{1}{2}i\sigma_k, \qquad M_k^{(0,\frac{1}{2})} = \tfrac{1}{2}i\sigma_k,$$

$$N_k^{(\frac{1}{2},0)} = \tfrac{1}{2}\sigma_k, \qquad N_j^{(0,\frac{1}{2})} = -\tfrac{1}{2}\sigma_k,$$

where the three σ_k are the 2×2 Pauli spin matrices.

The Poincaré mappings also leave invariant the Minkowski tensor $\eta_{\mu\nu}$. They consist of a Lorentz mapping plus an arbitrary translation and are of the form

$$x^\mu \to x'^\mu = \Lambda_\nu{}^\mu x^\nu + a^\mu. \tag{6-5.26}$$

We will denote such a mapping by the symbol $P(\Lambda, a)$.

We can get a faithful representation of the Poincaré group in the following way: Let $P_B{}^A(\Lambda,a)$ be a five-dimensional matrix with A, B running from 0 to 4. We take $P_\nu{}^\mu(\Lambda,a) = \Lambda_\nu{}^\mu$, $P_4{}^\mu(\Lambda,a) = a^\mu$, $P_\mu{}^4(\Lambda,a) = 0$, and $P_4{}^4(\Lambda,a) = 1$, so that

$$P_B{}^A(\Lambda,a) = \left\{\begin{array}{c|c} \Lambda & a \\ \hline 0 & 1 \end{array}\right\}. \tag{6-5.27}$$

[33] For a unified derivation of all irreducible representations of the Lorentz group, see M. A. Naimark, "Linear Representations of the Lorentz Group," *Amer. Math. Soc.*, Transl., Ser. 2, Vol. 6 (Secs. 2b, 2c).

Further, let y^A be a column matrix with $y^\mu = x^\mu$, $y^4 = 1$. Then Eq. (6-5.26) can be written as

$$y^A \to y'^A = P_B{}^A(\Lambda,a)y^B. \tag{6-5.28}$$

The Poincaré group is thus the group of all matrices of the form (6-5.27), where Λ is a Lorentz matrix and a is a column matrix.

Within the Poincaré group the Lorentz group is a proper subgroup, and we have

$$P(\Lambda',0)P(\Lambda,0) = P(\Lambda'\Lambda,0) \tag{6-5.29}$$

Likewise the translations by themselves form a subgroup. A pure translation is characterized by $P(1, a)$ and we have

$$P(1,a)P(1,a') = P(1, a + a'). \tag{6-5.30}$$

In addition to Eqs. (6-5.29) and (6-5.30), one has that

$$P^{-1}(\Lambda,a) = P(\Lambda^{-1}, -\Lambda^{-1}a), \tag{6-5.31}$$

$$P(\Lambda,a)P(\Lambda',a') = P(\Lambda\Lambda',a + \Lambda a'), \tag{6-5.32}$$

and

$$P(\Lambda,a)P(1,a')P^{-1}(\Lambda,a) = P(1,\Lambda a'). \tag{6-5.33}$$

From Eq. (6-5.32) it follows that $P(\Lambda, a)$ can be factored according to

$$P(\Lambda,a) = P(1,a)P(\Lambda,0) \tag{6-5.34}$$

and that this factorization is unique.

As a consequence of Eq. (6-5.33) we see that the subgroup of translations is an invariant subgroup of the Poincaré group. The quotient or factor group of the Poincaré group P by the translation group T is just the Lorentz group L, that is, $L \sim P/T$.

We now construct the infinitesimal generators of the Poincaré group. Since this group is a ten-parameter group (six $\varepsilon_\nu{}^\mu$, four ε^μ in the infinitesimal case), there will be ten generators. Six of these are gotten immediately from the I of Eq. (6-5.8). We define the matrices $(J_\nu{}^\mu)_B{}^A$ by the conditions

$$(J_\nu{}^\mu)_1{}^0 = I^\mu_{4\nu}, \quad (J_\nu{}^\mu)_2{}^0 = I^\mu_{5\nu},$$

$$(J_\nu{}^\mu)_3{}^0 = I^\mu_{6\nu}, \quad (J_\nu{}^\mu)_2{}^1 = I^\mu_{3\nu},$$

$$(J_\nu{}^\mu)_3{}^2 = I^\mu_{1\nu}, \quad (J_\nu{}^\mu)_1{}^3 = I^\mu_{2\nu},$$

and

$$(J_\nu{}^\mu)_A{}^4 = (J_\nu{}^\mu)_4{}^A = 0.$$

These matrices satisfy the condition $J^{\mu\nu} = J_\rho{}^\mu \eta^{\rho\nu} = -J^{\nu\mu}$, and are the generators of the Lorentz part of the mapping. The generators of the translations are given by four matrices, $(P_\mu)_B{}^A$, with nonvanishing components

$$(P_0)_4{}^4 = (P_1)_4{}^1 = (P_2)_4{}^2 = (P_3)_4{}^3 = -1.$$

In terms of these generators the infinitesimal form of the mapping (6-5.28) can be written as

$$y^A \to y'^A = y^A + \{\varepsilon_\mu{}^\nu(J_\nu{}^\mu)_B{}^A + \varepsilon^\mu(P_\mu)_B{}^A\}y^B. \tag{6-5.35}$$

The commutation relations between the various generators can be worked out with some labor. As matrix equations they are

$$[J^{\mu\nu}, J^{\rho\sigma}] = \eta^{\nu\rho}J^{\mu\sigma} - \eta^{\mu\rho}J^{\nu\sigma} + \eta^{\nu\sigma}J^{\rho\mu} - \eta^{\mu\sigma}J^{\rho\nu}, \tag{6-5.36}$$

$$[J^{\mu\nu}, P^\rho] = \eta^{\nu\rho}P^\mu - \eta^{\mu\rho}P^\nu, \tag{6-5.37}$$

and

$$[P^\rho, P^\sigma] = 0, \tag{6-5.38}$$

where

$$P^\rho = \eta^{\rho\sigma}P_\sigma. \tag{6-5.39}$$

Dirac[34] has made use of these commutation relations to formulate a set of conditions for a quantum field theory to be Poincaré invariant.

As in the case of the Lorentz group we can ask for representations of the Poincaré group, that is, a collection of matrices that satisfies Eqs. (6-5.31), (6-5.32), and (6-5.33), and whose infinitesimal elements satisfy the commutation relations (6-5.36), (6-5.37), and (6-5.38). The unitary representations have been studied extensively by Wigner.[35] Their principal application lies in the field of elementary-particle physics, where they are used to classify these particles.[36] The relevance of the Poincaré group lies in the fact that in any irreducible representation, the matrices

$$P = P^\mu P_\mu \tag{6-5.40}$$

and

$$W = \tfrac{1}{2}J_{\mu\nu}J^{\mu\nu}P^\rho P_\rho - J_{\mu\rho}J^{\nu\rho}P^\mu P_\nu \tag{6-5.41}$$

are multiples of the unit matrix. The possible values of P and W therefore characterize the various irreducible representations of the Poincaré group. The unitary representations are all infinite dimensional. When $P > 0$ they are characterized by two real numbers m and s, with m any number > 0 and

[34] P. A. M. Dirac, *Rev. Mod. Phys.*, **34**, 592 (1962).

[35] E. P. Wigner, *Ann. Math.*, **40**, No. 1 (1939).

[36] For a discussion of the relevance of the Poincaré group to physical applications, see E. P. Wigner, *Nuovo Cimento*, **3**, 517 (1956).

s any integer or half-integer $\geqq 0$. The values of P and W are given in terms of m and s by

$$P = m^2 \quad \text{and} \quad W = m^2 s(s+1). \tag{6-5.42}$$

The representation spaces in these cases are Hilbert spaces and are taken as the Hilbert spaces of a relativistic particle with mass m and spin s. The case of $P = 0$ is more complicated and will not be discussed here other than to say that among the unitary representations are those for zero rest-mass particles of integer or half-integer spin.

6-6. Geometry of Minkowski Space

In what follows we shall refer to a four-dimensional manifold with a flat metric of signature -2 defined on it as a *Minkowski space.* In many respects Minkowski space is very much like *Euclidean space*, with which we are more familiar. The fact that Euclidean space is usually taken to be three-dimensional while Minkowski space is four-dimensional does not make a great deal of difference for the purpose of this discussion. What is important is that the flat metric of Euclidean space has a signature $+4$ if the Euclidean space is also taken to be four-dimensional. The main difference between these two spaces arises from the difference in signature of the two metrics involved.

In order to explore the similarities and differences between the two types of spaces, it is convenient to study their two-dimensional analogues. We further assume that in both cases, the metric has been mapped onto the manifold in such a way that it is diagonal and everywhere constant. If then, x^1 and x^2 are the coordinates of the Euclidean two-space and x^0 and x^1 are those of the Minkowski two-space, the interval ds between neighboring points is given in the two cases by

$$ds_E{}^2 = (dx^1)^2 + d(x^2)^2 \tag{6-6.1}$$

for the Euclidean space, and

$$ds_M{}^2 = (dx^0)^2 - (dx^1)^2 \tag{6-6.2}$$

for the Minkowski space. One immediately sees an essential difference between the two cases. The vanishing of ds_E implies that $dx^1 = dx^2 = 0$. However, the vanishing of ds_M implies only that $dx^0 = dx^1$. Thus, in Minkowski space, two separate and distinct neighboring points can have a vanishing interval. The mappings that leave invariant the distance ds_E between two neighboring points in Euclidean space are just the rotations through an angle θ:

$$\begin{aligned} x^1 \to x'^1 &= x^1 \cos\theta + x^2 \sin\theta, \\ x^2 \to x'^2 &= -x^1 \sin\theta + x^2 \cos\theta, \end{aligned} \tag{6-6.3}$$

while the mappings that leave invariant the Minkowski distance ds_M are the "pseudo-rotations"

$$\begin{aligned} x^0 \to x'^0 &= x^0 \cosh \phi + x^1 \sinh \phi, \\ x^1 \to x'^1 &= x^0 \sinh \phi + x^1 \cosh \phi. \end{aligned} \tag{6-6.4}$$

The difference in the form of these mappings is again due to the essential difference between Euclidean and Minkowski space.

There are other oddities of Minkowski space that have no analogue in Euclidean space. In both cases we can define the length A of a vector with components A^r by the equation $A^2 = g_{rs}A^rA^s$. In Euclidean space a vector of zero length has *a fortiori* all its components equal to zero, whereas this is not true in Minkowski space.

In Euclidean space, two vectors with nonvanishing components A^r and B^r are said to be perpendicular if $g_{rs}A^rB^s = 0$. We can carry over this definition of perpendicularity to Minkowski space, but then we must be prepared to accept the fact that a vector with nonvanishing components can be perpendicular to itself. Also, two vectors that are not perpendicular in Euclidean space may be perpendicular in Minkowski space. Two such vectors are those with components (2, 5) and (5, 2). However, the affine properties of the two spaces are the same, and in particular all properties of parallel lines are identical in the two spaces. Thus, since in both spaces the affinity is integrable, it follows that through a given point one can draw only one straight line parallel to a given straight line in the two spaces.

If we take the integral of ds along a curve to be the length of that curve in both cases, we can again define a straight line as that curve for which $\int ds$ is stationary. With ds given by Eq. (6-6.1) or (6-6.2), the equation of a straight line is then a linear relation between the coordinates. The difference in the two cases is that in the Euclidean case, the straight line is the curve of least length connecting two points, while in the Minkowski case the straight line is the curve of greatest length connecting two points.

The differences between Euclidean and Minkowski space are especially troublesome when we come to depict the geometry of the latter by means of pictures. It has often been said that the hard part about relativity theory is that it is a four-dimensional theory (popularly, time is the fourth dimension) and that it is hard to represent four-dimensional structures by means of two-dimensional drawings. But really the same thing can be said of Newtonian mechanics, which is also a four-dimensional theory with, if anything, a much more difficult underlying geometrical structure to visualize. The real difficulty lies in drawing pictures that involve metrical properties when the geometry has a nondefinite metric. This difficulty arises because the paper on which one makes such drawings is already endowed with a Euclidean metric. Thus, in a drawing that represents two mutually perpendicular vectors in a Euclidean geometry they "look" perpendicular, whereas in a drawing of two perpendicular vectors in a Minkowski geometry, the two lines may look anything but perpendicular. Thus the vectors with components (1, 2) and (2, 1) are perpendicular in a Minkowski geometry but do not appear to be so in a diagrammatic representation. (See Fig. 6.2.)

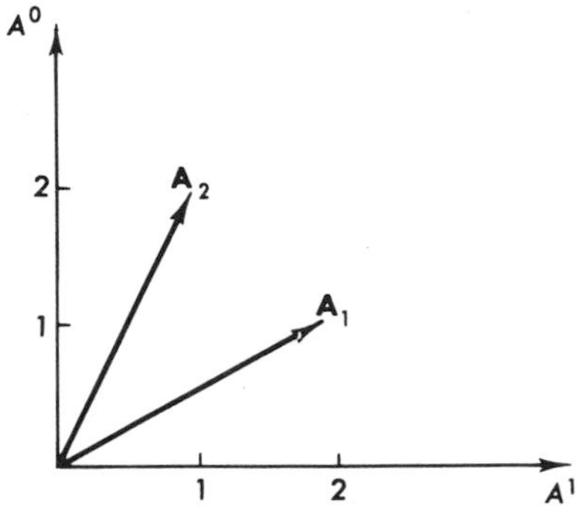

Fig. 6.2.

In Euclidean geometry one can, as we saw, define the distance between nonneighboring points as the length of the straight line (geodesic) connecting points. If $\{x_1{}^1, x_1{}^2\}$ and $\{x_2{}^1, x_2{}^2\}$ are two such points and ds_E is given by Eq. (6-6.1), this distance, S_{E21}, is given by the usual Pythagorean expression

$$S_{E21}^2 = (x_2{}^1 - x_1{}^1)^2 + (x_2{}^2 - x_1{}^2)^2.$$

We can likewise define a distance in Minkowski space. Now, however, if $\{x_1{}^0, x_1{}^1\}$ and $\{x_2{}^0, x_2{}^1\}$ are two points and ds_M is given by Eq. (6-6.2), then S_{M21} is given by

$$S_{M21}^2 = (x_2{}^0 - x_1{}^0)^2 - (x_2{}^1 - x_1{}^1)^2$$

While the locus of points equidistant from a given point in Euclidean space is a circle, the locus of such points in Minkowski space appears as an hyperbola in a drawing.

Of particular importance to us is the degenerate hyperbola $x^1 = \pm x^0$. Points located on this surface are all at a zero distance from the origin. Furthermore the lines $x^1 = \pm x^0$ are null geodesics. As we shall see in the next section, they represent the path of propagation of light-signals. Because of this the degenerate hyperbola constitutes what is called a *light cone*. In the four-dimensional case the totality of all null geodesics passing through a given point form the light cone at that point. The light cone at a point serves to divide space-time into distinct regions with respect to that point, as indicated in Fig. 6.3, where we have taken the point to be at the origin of coordinates. If $S_{M21}^2 > 0$ and $x_2{}^0 - x_1{}^0 > 0$, then 2 is said to lie in the *future* of 1, while if $x_2{}^0 - x_1{}^0 > 0$, 2 is said to lie in the *past* of 1. If $S_{M12}^2 < 0$, 2 is sometimes said to lie in the *elsewhere* of 1. The significance of this division of space-time with respect to a point lies in the fact that under a Lorentz mapping, light cones are mapped onto light cones as follows from the invariance

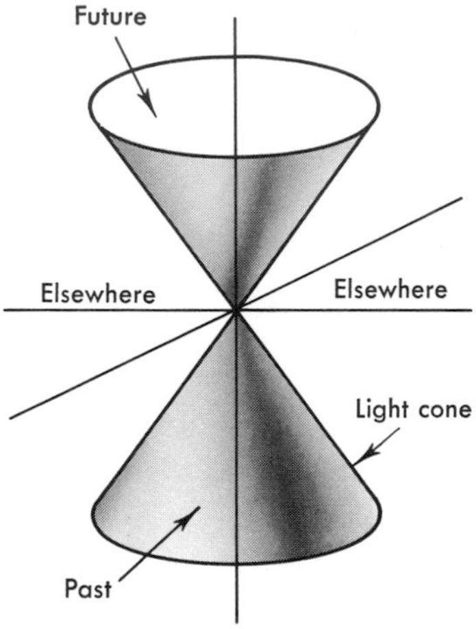

Fig. 6.3.

of the Minkowski metric under such a mapping. Consequently, under an orthochronous Lorentz mapping, the points that lie in the future (past) of a given point are mapped onto points that lie in the future (past) of the point onto which the given point is mapped. Under a nonorthochronous Lorentz mapping, past and future are reversed, but in all cases the elsewhere is mapped onto the elsewhere. In a like manner one can characterize a vector according as $A^2 = g_{\mu\nu}A^\mu A^\nu \gtreqless 0$. If $A^2 > 0$ the vector is said to be *timelike*; if $A^2 < 0$ it is said to be *spacelike*; and if $A^2 = 0$, it is a *null* vector. Under a Lorentz mapping the characterization of a vector does not change.

Problem 6.7. Show that if a vector is timelike, then there always exists a Lorentz mapping such that its transformed spatial components all vanish.

Problem 6.8. Show that if a vector is spacelike, then there always exists a Lorentz mapping such that its transformed temporal component vanishes.

Problem 6.9. Show that if A^μ is timelike, then any other vector orthogonal to it is spacelike.

If the tangents to a given curve are everywhere timelike, the curve is said to be a *timelike world-line* and, as we will see, it can represent the trajectory of a material particle. If, in particular, the world-line is a geodesic, we have a timelike geodesic. Under a Lorentz mapping a timelike world-line is transformed into a timelike world-line, and in particular a timelike geodesic is transformed into a timelike geodesic. The timelike geodesics are of particular importance because they will be seen to represent the world-lines of free particles.

In addition to the notion of a timelike world-line we will also have occasion to use that of a *spacelike hypersurface.* A hypersurface is spacelike, provided any displacement lying wholly within the surface is spacelike. If

$$\phi(x) = 0 \tag{6-6.5}$$

is the equation of the surface, then a displacement with components dx^μ will lie within the surface, provided

$$\phi_{,\mu}\, dx^\mu = 0. \tag{6-6.6}$$

Hence ϕ will be spacelike, provided

$$g_{\mu\nu}\, dx^\mu\, dx^\nu < 0 \tag{6-6.7}$$

for all dx^μ satisfying Eq. (6-6.6). Alternatively, one could require that the normal to the surface at any point be timelike. The components of the normal are

$$n^\mu = g^{\mu\nu}\phi_{,\nu}, \tag{6-6.8}$$

and so we must have

$$g_{\mu\nu} n^\mu n^\nu > 0. \tag{6-6.9}$$

Problem 6.10. Show that if the normal to a surface is timelike, then all displacements lying in the surface are spacelike.

There is nothing in Euclidean geometry to correspond to the light cone, since distinct points are always at a finite distance from one another. The mappings that leave invariant the Euclidean metric δ_{rs} are just the translations and rotations about the origin. The nearest thing to a light cone is the plane of absolute simultaneity of Newtonian mechanics. In fact, if light propagated with infinite velocity, the light cone would collapse into a plane of absolute simultaneity. To see this, let us introduce a coordinate t related to x^0 by $x^0 = ct$, where c is the velocity of light. (Its numerical value is unimportant. For the moment all that matters is that c is finite.) In terms of t, the equation of the light cone through the origin can be written as

$$t^2 - \frac{1}{c^2}\{(x^1)^2 + (x^2)^2 + (x^3)^2\} = 0.$$

Thus, in the limit $c \to \infty$, the light cone collapses onto the surface $t = 0$, that is, on the plane of simultaneity through the origin. Furthermore this surface is invariant under the group of mappings that are the limit of the Lorentz mappings when $c \to \infty$. To see what these mappings are, let us consider a velocity mapping given by Eq. (6-5.22), with A^μ replaced by $x^\mu = \{x^0, \mathbf{x}\}$, and $\mathbf{v}$ replaced by $\mathbf{v}/c$. In the limit $c \to \infty$ the mapping reduces to

$$t' = t \quad \text{and} \quad \mathbf{x}' = \mathbf{x} + \mathbf{v}t$$

which is just the velocity mapping of the Galilean group as given in Eq. (5-4.3). This reduction also justifies our calling the surface $t = 0$ a plane of absolute simultaneity.

The method developed here for obtaining the velocity mappings of the Galilean group from those of the Poincaré group by replacing x^0 by ct and v by v/c and taking the limit $c \to \infty$ will be employed later (see Section 7-11) to obtain the nonrelativistic limit of a set of relativistic equations. The equations obtained thereby can be properly considered to be this limit since their symmetry group will be the inhomogeneous Galilean group if the Poincaré group was the symmetry group of the original relativistic equations.

Before concluding this section we would like to comment briefly on the use of the so-called imaginary time in special relativity. It was first introduced by Minkowski as a simplifying device. If one introduces a new coordinate $x^4 = ix^0$ and $d\bar{s} = i\, ds_M$, then Eq. (6–6.2) gets rewritten as

$$d\bar{s}^2 = (dx^1)^2 + (dx^4)^2, \tag{6-6.10}$$

which is formally the same as the expression for ds_E. In terms of this new coordinate the metric has been made Euclidean, but at the expense of introducing an imaginary coordinate. Now the locus of points at a constant distance from a given point appears as a circle in a drawing instead of a hyperbola. Furthermore, if we set $\phi = \theta/i$ in Eq. (6-6.4), we obtain

$$x'^1 = x^1 \cos\theta + x^4 \sin\theta,$$

$$x'^4 = -x^1 \sin\theta + x^4 \cos\theta,$$

which has the form of an ordinary rotation.

While the use of an imaginary coordinate simplifies the writing of some equations (the covariant components of a vector are now respectively equal to its contravariant components), it is unphysical and obscures the essential distinction between Minkowski and Euclidean geometries. Furthermore, although widely used in quantum field theory, one runs into trouble when taking the complex conjugate of a tensor, since the imaginary appearing in the expression for the coordinate does not change sign under this operation. Consequently one cannot take the complex conjugate of a tensor directly, but must proceed term by term, thereby destroying the manifest covariance of the formalism. Finally, nothing is gained by the use of the imaginary coordinate in general relativity, since the $g_{\mu\nu}$ must appear explicitly in the various equations of that theory. Consequently we will not make use of the imaginary coordinate in this work.

6-7. Light Signals

From the totality of geodesics associated with a metric we can separate a special subclass, namely, the null geodesics. According to the discussion of Section 3-2, these null geodesics with points $x^\mu = z^\mu(\lambda)$ are determined by the equations

$$\ddot{z}^\mu + \begin{Bmatrix} \mu \\ \rho\sigma \end{Bmatrix} \dot{z}^\rho \dot{z}^\sigma = 0 \tag{6-7.1}$$

together with the supplementary condition

$$g_{\mu\nu}\dot{z}^\mu \dot{z}^\nu = 0 \tag{6-7.2}$$

for a suitable choice of the path parameter λ. (Here a dot over a quantity denotes differentiation with respect to the path parameter.) With $g_{\mu\nu} = \eta_{\mu\nu}$, Eq. (6-7.1) reduces to $\ddot{z}^\mu = 0$, which has as its general solution,

$$z^\mu - z_0{}^\mu = k^\mu \lambda, \tag{6-7.3}$$

where $z_0{}^\mu$ and k^μ are constants of integration. As a consequence of Eq. (6-7.2), k^μ is seen to satisfy

$$\eta_{\mu\nu} k^\mu k^\nu = 0 \tag{6-7.4}$$

and is therefore a null vector.

If we define the three-velocity $\mathbf{u}$ along a geodesic to have components $u^r = dz^r/dz^0$, we see that for null geodesics, $u^r = k^r/k^0$, and hence it follows from Eq. (6-7.3) that

$$\mathbf{u}^2 = \delta_{rs} u^r u^s = 1.$$

Furthermore, under a Poincaré mapping, the geodesic (6-7.3) is mapped onto the trajectory

$$z^\mu - z'_0{}^\mu = k'^\mu \lambda, \tag{6-7.5}$$

where

$$k'^\mu = \Lambda_\nu{}^\mu k^\nu \quad \text{and} \quad z'_0{}^\mu = \Lambda_\nu{}^\mu z_0{}^\nu + a^\mu.$$

Since $\Lambda_\nu{}^\mu$ is a Lorentz matrix it follows from Eqs. (6-5.2) and (6-7.4) that

$$\eta_{\mu\nu}k'^\mu k'^\nu = 0,$$

so that the trajectory (6-7.5) is also a null geodesic. It therefore follows that the magnitude of the velocity **u** associated with a null geodesic is an invariant under a Poincaré mapping. As we will see, this is true only for the null geodesics. We note finally that the surface generated by all null geodesics (6-7.3) through a point $x_0{}^\mu$ is determined by the equation

$$\eta_{\mu\nu}(x^\mu - x_0{}^\mu)(x^\nu - x_0{}^\nu) = 0,$$

which by definition is the light cone at the point $x_0{}^\mu$.

We are now going to make our first association between a physical quantity and the elements of our theory developed here. There is, of course, nothing in the theory that tells us how to make such associations. Their justification must rest on how well the predictions of the theory accord with the properties of the physical quantities that are associated with the elements of the theory. We will assume that the null geodesics correspond to the trajectories of light signals or rays. This is a reasonable association to make because, as we saw, light appears to be propagated with a characteristic velocity that is independent of its source, and the null geodesics all have associated with them a characteristic velocity. Later, when we come to study the electromagnetic field, we will discuss the motion of wave packets and show that they propagate along null geodesics. Since a light signal can be considered to be an electromagnetic wave packet, we will obtain additional justification for associating null geodesics with the trajectories of light signals. Of course we have not said how we are to measure the velocity of light. What we shall do in fact is to use light rays and free particles (to be discussed in the next section) as primitive physical elements from which we will construct space-time measuring devices.

6-8. Free Particles in Special Relativity

Having made the association of the trajectories of light signals with null geodesics of the Minkowski metric, we now look for elements of our theory that we can associate with the trajectories of free particles. Again we are free to make this choice in whatever way we wish. It is nevertheless reasonable to associate the trajectories of free particles with other geodesics of the Minkowski metric.

In addition to the null geodesics there are two other types of geodesics available to us in special relativity for the description of unaccelerated particle motion, the timelike geodesics and the spacelike geodesics. The latter, by analogy with the former, are defined as those geodesics whose tangents are all spacelike. The equation for either type can, by a suitable choice of path parameter τ be written in the form

$$\ddot{z}^{\mu} = 0 \tag{6-8.1}$$

when $g_{\mu\nu}$ takes on its Minkowski values, Equation (6-8.1) has, as an integral of the motion,

$$\eta_{\mu\nu}\dot{z}^{\mu}\dot{z}^{\nu} = k, \tag{6-8.2}$$

where k is a nonzero constant. If $k > 0$, the geodesic is timelike, and if $k < 0$, it is spacelike. At this point the choice of the type of geodesic to represent the motion of free particles is arbitrary. There is nothing in the formalism of special relativity as so far developed that rules out one or the other possibilities. Our choice, namely, to use the timelike geodesics, will be justified later, after we have developed a theory of space-time measurements, by causality conditions.

By adjusting the path parameter the constant k appearing in Eq. (6-8.2) can be made to take on the value $+1$ for a timelike geodesic. In this case the path parameter, which we will denote by τ, is called the *proper time* along the trajectory. It is also convenient for some purposes to re-express Eq. (6-8.1) in terms of the three-velocity $\mathbf{u}$ defined by

$$\mathbf{u} \equiv \frac{d\mathbf{z}}{dz^0}. \tag{6-8.3}$$

From Eq. (6-8.2), with $k = +1$, it follows that

$$\dot{z}^0 = \frac{1}{\sqrt{1 - \mathbf{u}^2}}, \qquad \dot{\mathbf{z}} = \frac{\mathbf{u}}{\sqrt{1 - \mathbf{u}^2}}. \tag{6-8.4}$$

By making use of this result Eq. (6-8.1) can be rewritten in the form

$$\frac{d}{dz^0}\frac{\mathbf{u}}{\sqrt{1 - \mathbf{u}^2}} = 0 \tag{6-8.5a}$$

and

$$\frac{d}{dz^0}\frac{1}{\sqrt{1 - \mathbf{u}^2}} = 0. \tag{6-8.5b}$$

6-9. A Light Clock

With the help of a system of free particles and light signals we can now proceed to construct a simple model of a clock. As are all models, it is an idealization, but one that we can control because we know how it is constructed. The particular idealization we will make is to assume that our free particles can scatter light without suffering any recoil. At least within the framework of classical physics, such an idealization is possible. The problem of making space-time measurements in the quantum domain is much more complicated and will not be discussed here.

In our construction of a light clock we shall follow a construction due to Wheeler and Marzke.[37] The essential idea is to find two free particles that are moving parallel to each other in space-time and to bounce a light signal between them. The time interval between successive "ticks" on such a clock is then the time needed for a complete circuit of the light signal. The main difficulty in the construction is to assure oneself that the two particles are indeed moving parallel to each other. For this purpose we need a gas of free particles moving with various velocities in different directions. For simplicity we shall confine ourselves to the case of one spatial dimension. To proceed with the construction (see Fig. 6.4) one chooses a given free particle

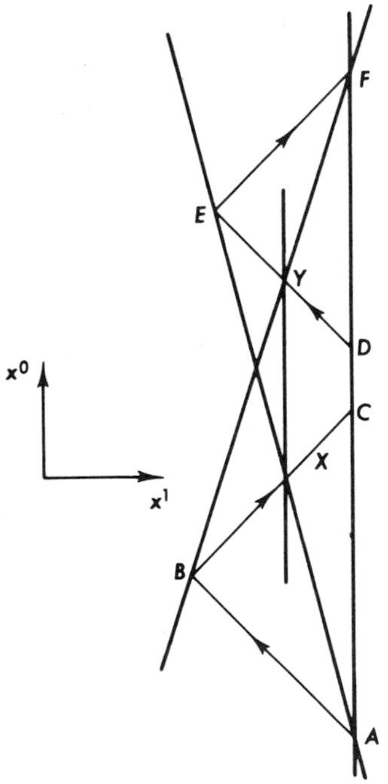

Fig. 6.4.

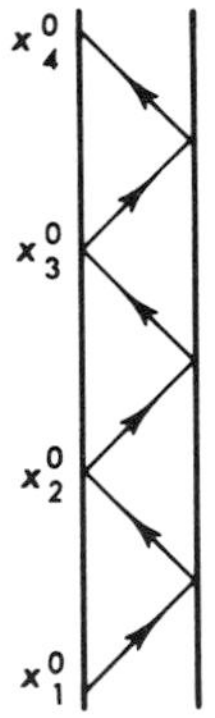

Fig. 6.5.

[37] Robert F. Marzke and J. A. Wheeler, article in *Gravitation and Relativity* (Benjamin, New York, 1964), Chap. 3.

moving along the trajectory AF and another free particle with trajectory BF that will collide with the first particle at F. Let the first particle emit a light flash at A. It will strike the second particle at B and is subsequently reflected back to C. Now consider a third free particle whose trajectory intersects that of the first at A and the light ray BC at X. By trial and error determine the point D and the rays DE and EF. Then the free particle with trajectory XY is moving parallel to the original trajectory $ACDF$.

In Table 6.1 we give the Cartesian analysis of the Marzke-Wheeler construction. The system is so mapped that $g_{\mu\nu} = \eta_{\mu\nu}$, the equation of $ACDF$, is $x^1 = 0$, the event A has coordinates (0, 0), and the event at C coordinates (2a, 0). It follows, therefore, that XY is parallel to $ACDF$ as desired.

TABLE 6.1

CARTESIAN ANALYSIS OF MARZKE-WHEELER CONSTRUCTION

Event or Geodesic	Coordinates or Equation	Reason
A	$(0, 0)$	By definition and choice of mapping
C	$(2a, 0)$	Same
B	$(a, -a)$	B connected by rays to A and C
E	$(c, -b)$	Definition of b and c
D	$(c - b, 0)$	On axis and connected by ray to E
F	$(0, c + b)$	Same
AE	$bx^0 + cx^1 = 0$	Checks for A and E
BC	$x^0 - x^1 = 2a$	Light ray through C
X	$\left(\dfrac{2ac}{b+c}, \dfrac{-2ab}{b+c}\right)$	Intersection of AE and BC
BF	$ax^0 + (a - b - c)x^1 = 0$	Checks for B and F
DE	$x^0 + x^1 = c - b$	Checks for D and E
Y	$\left(c - b + \dfrac{2a}{b+c}, \dfrac{-2a}{b+c}\right)$	Intersection of DE and BF
XY	$x^1 = \dfrac{-2ab}{b+c}$	Checks for X and Y

Using the two parallel-moving particles, we now construct our light clock by reflecting a light signal back and forth between them. Such a clock can then be used to define the time interval between two events lying on its world-line by the number of reflections taking place between the two events. In computing this interval we can make use of the fact that if the world-line of one of the particles comprising the clock is $x^r = 0$, then the coordinate differences $x_2{}^0 - x_1{}^0$, $x_3{}^0 - x_2{}^0$, etc. (see Fig. 6.5), are equal to each other.

Problem 6.11. Show that $x_2{}^0 - x_1{}^0 = x_3{}^0 - x_2{}^0$ in Fig. 6.5.

6-10. Measurements of Space and Time Intervals with Light Clocks; Inertial Frames

Once we have constructed a light clock we can use it, in conjunction with a light source attached to it, to measure spatial and temporal intervals between events in space-time. To see how this can be done, let us first consider the case of one spatial dimension. Consider an event A (see Fig. 6.6) that can be

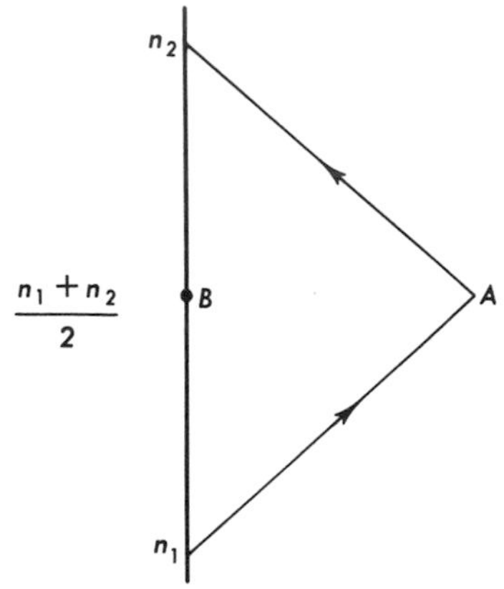

Fig. 6.6.

seen by the clock, that is, one that is capable of reflecting a light ray sent out by the clock. The point A might, for instance, be a point on the trajectory of some other particle, the point of collision of two particles, etc. The clock now emits a series of light signals, and note is made of the number of ticks that elapse between the time of emission of a signal and some arbitrary origin of time. If the particular signal that is reflected from A is emitted at the point 1 after n_1 ticks and received back from the origin at the point 2 after n_2 ticks, then the distance d_A of the event A from the clock is taken to be

$$d_A = \frac{n_2 - n_1}{2}. \tag{6-10.1}$$

Likewise the time t_A at which the event occurs is taken to be

$$t_A = \frac{n_2 + n_1}{2} \tag{6-10.2}$$

At first sight the above definitions of d_A and t_A appear quite arbitrary. While the values of n_1 and n_2 are directly observable, d_A and t_A are not; they must be computed from n_1 and n_2 by means of Eqs. (6-10.1) and (6-10.2).

Actually, the definitions of d_A and t_A are formulated so as to maintain certain properties of what one normally takes to be space and time measurements. Thus, in order that d_A be independent of the origin of time, it must be a function only of $n_2 - n_1$. Likewise the requirement that the spatial and temporal intervals $d_{AB} \equiv d_B - d_A$ and $t_{AB} \equiv t_B - t_A$ be independent of this origin leads to the result that d_A and t_A must be linear functions of n_1 and n_2. Finally, if we require that the distance of a free particle from a light clock be a linear function of the time when the particle is at that distance, we are led unambiguously to the expressions for d_A and t_A given in Eqs. (6-10.1) and (6-10.2).

Problem 6.12. Discuss the measurement of the velocity of a particle, using a light clock. In particular, show that the velocity so determined does not depend upon the separation between the particles comprising the light clock.

Problem 6.13. Show that if the velocity of the particles comprising clock II is measured by clock I is v, then the velocity of the particles comprising clock I will be measured by clock II to be $-v$.

In giving a mathematical description of the measurements depicted in Fig. 6.6 it is convenient, though by no means necessary, to take the flat metric to have its Minkowski values $(+1, -1, -1, -1,)$, to take the work line of the light clock to be $x^1 = 0$, and to choose a time coordinate so that the coordinates of the point 1 are $(n_1, 0)$ and those of point 2 are $(n_2, 0)$. With this placement of events one sees that the coordinates of point A are then (t_A, d_A). Furthermore, if a free particle is observed to be moving with a velocity u with respect to the clock, its trajectory will be the curve $x^1 = ux^0 + x_0{}^1$. Note that with our definitions of time and distance, the velocity of light is unity and all velocities are dimensionless.

The extension of the above procedure to the case of three spatial dimensions is quite straightforward. Rather than one light clock, one employs three light clocks, all having one of their two particles in common. In addition, the planes that contain the various clocks are perpendicular to each other. Finally, the rates of the three clocks are adjusted so that all are alike. This adjustment is accomplished by requiring that a light signal emitted by the particle common to all three light clocks, upon reflection by the other three particles of the clocks, be received back at the common particle at the same point. Such a configuration of clocks will be referred to as an *inertial reference frame*, or *inertial frame* for short.

One can define the distance d_A and the time t_A associated with the event A exactly as in the one-dimensional case by using Eqs. (6-10.1) and (6-10.2). In addition one can now measure the direction cosines $\alpha_A{}^1$, $\alpha_A{}^2$, $\alpha_A{}^3$, which the light ray from the clock to A makes with the orthogonal axes determined by the three clocks, and so define the orthogonal distances $d_A{}^1 = \alpha_A{}^1 d_A$, etc. Now take the trajectory of the particle common to all three light clocks to be $x^1 = x^2 = x^3 = 0$ and the trajectories of the other three particles comprising the inertial frame to be $x^1 = 1, x^2 = x^3 = 0$; $x^2 = 1, x^1 = x^3 = 0$, and $x^3 = 1, x^1 = x^2 = 0$. Then, as in the one-dimensional case, the coordinates of the point A are $\{t_A, d_A{}^1, d_A{}^2, d_A{}^3\}$.

6-11. Simultaneity

In the construction of Fig. 6.4 the temporal coordinate of the point A is the same as that of point B. In this case the point A is *simultaneous* with the point B. Furthermore, if we imagine a light clock whose trajectory is parallel to the original clock and passes through A, by the same construction B would be simultaneous with A. Also, any point that is simultaneous with A will be simultaneous with B, and vice versa. As defined, then, simultaneity is an equivalence relation: The totality of points simultaneous with a given point as determined by a light clock whose trajectory passes through that point constitutes a plane of simultaneity whose normal is tangent to the trajectory of the light clock used for the construction. However, there are an infinity of different light clocks whose trajectories pass through a given point, and it is clear that each one defines a *different* plane of simultaneity containing the point. Thus simultaneity becomes a relative concept with our above definition.

Problem 6.14. Show that if A and B cannot be connected by a light ray, then there always exists a light clock passing through A which makes B simultaneous with A.

While our definition of simultaneity might appear deficient at first sight, it is clear that it is the only one possible with the systems at hand. Only if the invariant velocity for the transmission of information were infinite, as is the case in Newtonian mechanics, would it be possible to define an absolute concept of simultaneity that would enable all observers to decide whether two events were or were not simultaneous. It was just this realization of the relativity of simultaneity and an analysis of its consequences that led Einstein to the special theory of relativity.

6-12. Comparison of Clock Rates, Time Dilation, Length Contraction

So far we have dealt with only one clock and have used it to set up a standard of time (and also distance). We now turn to the question of when two different clocks can be considered as identical—that is, when they serve to define the same unit of time. As long as the two clocks are at rest with respect to each other, the problem is straightforward. One sends out two light signals from one of them, n ticks apart. Then, if the other clock observes the two signals to arrive n ticks apart, the two clocks will be said to run at the same rate. With this definition, it is clear that it is immaterial which clock sends out the signals and which receives it. Furthermore, if two clocks at rest with respect to each other run at the same rate, then a third clock at rest with respect to one of them and running at the same rate as it, will be at rest with respect to the other clock and run at the same rate as it. We can therefore group all clocks at rest with respect to one another into equivalence classes of clocks running at the same rate. In a like manner we can group all reference frames at rest with respect to one another into equivalence classes, the clocks of any two equivalent frames running at the same rate and the 1-, 2-, and 3-directions coinciding. With this definition of equivalent clocks and reference frames the time interval $t_A - t_B$ and the spatial intervals $d_A{}^r - d_B{}^r$ between two events is independent of which clock or reference frame of an equivalence class is used to determine these intervals. This would not have been true if we had used definitions of these quantities other than those given by Eqs. (6-10.1) and (6-10.2).

Suppose now that two clocks are moving with respect to one another. Suppose further that when one clock sends out two signals n ticks apart, the other receives them also n ticks apart. Then it is no longer true that if the second clock sends out signals n ticks apart, the first one will receive them n ticks apart. One can see this directly by referring to Fig. 6.7. Here clock A

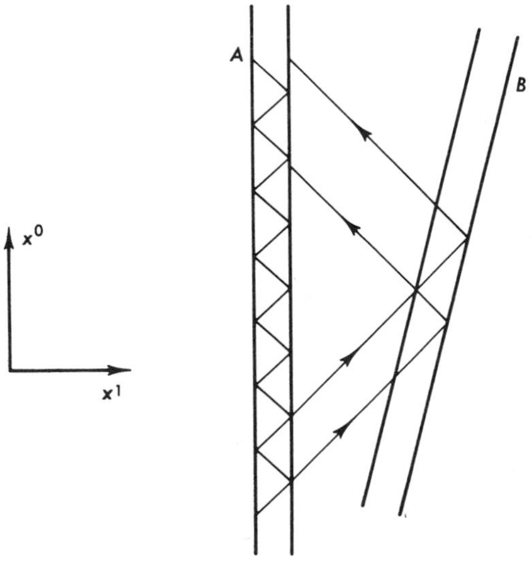

Fig. 6.7.

sends out a signal after each tick and clock B receives them one tick apart. However, if B sends out signals one tick apart, A will receive them more than one tick apart. Consequently this method of comparing rates does not allow one to set up an equivalence class of relatively moving identical clocks, and the standard time would depend on which clock sends out the signals.

In order to construct equivalent clocks that are moving relative to one another, we will take advantage of the fact that the space-time manifold is four-dimensional and that the orientation in space of the two particles constituting a light clock has no effect on the rate of the clock. The validity of this latter conclusion can be seen in the following way: Consider three free particles at rest with respect to each other and oriented so that their world-lines are not coplanar. We can use the particles in pairs to form light clocks with the property that if their rates agree at some point on one of their world-lines, it will continue to agree along the whole world-line. Furthermore, if the rate of two of the three light clocks agree, the rate of all three must agree.

Consider now two light clocks moving with a relative velocity v with respect to each other and assume that the world-lines of one particle of each clock cross at some point. Through this point one constructs the two planes of simultaneity defined by the two clocks. These planes will intersect in a two-dimensional plane. The particles of the two light clocks are then adjusted so that the two world-lines of the particles of one clock intersect the world-lines of the particles of the other clock in pairs, as illustrated in Fig. 6.8. The two

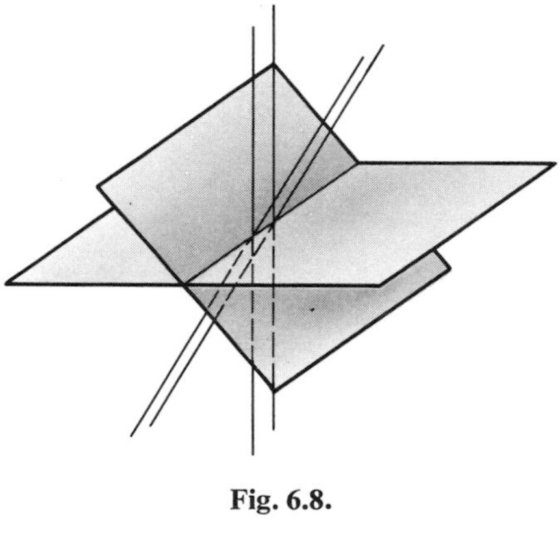

Fig. 6.8.

clocks are then defined to run at the same rate. One can extend this adjustment procedure to clocks with nonintersecting world-lines by using a third clock whose world-line crosses the other two. The reader can convince himself with the help of a few space-time diagrams that, starting with a given clock, one can construct an equivalence class of moving identical clocks, and thus set up a universal time standard for all clocks.

Having set up a universal time standard we can ask how the rate of a clock moving with respect to another identical clock would appear to this latter clock. For simplicity we shall consider two spatial dimensions. Let the particles of one clock move along the curves (see Fig. 6.9) OC and PA, while those of the other move along the curves OD and PB. If a light signal is

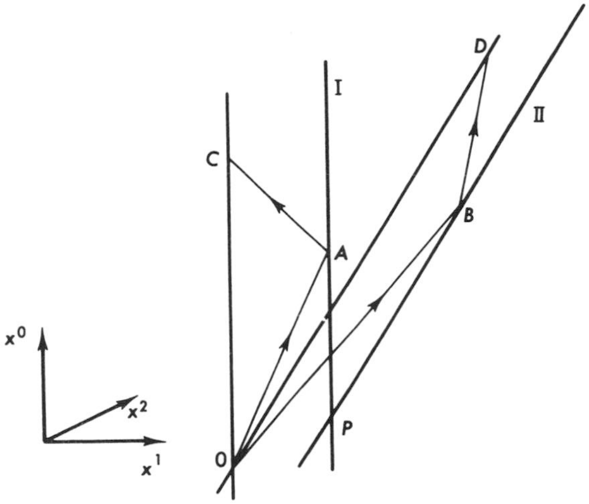

Fig. 6.9.

emitted at 0, clock I will have ticked once when the signal reaches C, while clock II will have ticked once when the signal reaches D. The Cartesian analysis of the process is given in Table 6.2 on page 168.

From the analysis of Table 6.2 we see that the point on OC that is simultaneous with the point D relative to I has coordinates $\left(\frac{2a}{\sqrt{1-v^2}}, 0, 0\right)$, which is clearly later than the point C by the factor $1/\sqrt{1-v^2}$. As a consequence, if two events lying on OD are measured to be separated by a time interval Δt_{II} as measured by clock II, they will be observed by clock I to have a time interval

$$\Delta t_{\text{I}} = \frac{\Delta t_{\text{II}}}{\sqrt{1-v^2}} \tag{6-12.1}$$

This is the famous *time dilation effect*. We see that it arises as a consequence of our definitional procedure for measuring time intervals.

From our above discussion we see that clock I would see clock II running at a slower rate than itself. But since there is nothing to distinguish between the

TABLE 6.2

CARTESIAN ANALYSIS OF TIME DILATION

Event or Geodesic	Coordinates or Equation	Reason
O	$(0, 0, 0)$	Achieved by mapping
P	$(0, 0, a)$	Same
OC	$x^1 = x^2 = 0$	Same
OD	$x^1 = vx^0,\ x^2 = 0$	Clock II moving with velocity v with respect to clock I
PA	$x^1 = 0,\ x^2 = a$	PA parallel to OC
PB	$x^1 = vx^0,\ x^2 = a$	PB parallel to OD and definition of identical clocks
OA, OB	$x^{02} - x^{12} - x^{22} = 0$	Light cone at 0
A	$(a, 0, a)$	Intersection of PA and OA
B	$\left(\frac{a}{\sqrt{1-v^2}}, \frac{av}{\sqrt{1-v^2}}, a\right)$	Intersection of PB and OB
AC	$(x^0 - a)^2 - x^{12} - (x^2 - a)^2 = 0$	Light cone at A
C	$(2a, 0, 0)$	Intersection of OC and AC
BD	$\left(x^0 - \frac{a}{\sqrt{1-v^2}}\right)^2 - \left(x^1 - \frac{av}{\sqrt{1-v^2}}\right)^2 - (x^2 - a)^2 = 0$	Light cone at B
D	$\left(\frac{2a}{\sqrt{1-v^2}}, \frac{2av}{\sqrt{1-v^2}}, 0\right)$	Intersection of OD and BD

two clocks, it follows that clock II must see clock I running at a slower rate also. Again by drawing a space-time diagram the reader can convince himself that the point on OD that is simultaneous with the event C as determined by clock II will be above point D, and that if two events lying on OC are observed by clock I to be separated by a time interval Δt_{I} they will be separated by a time interval

$$\Delta t_{\mathrm{II}} = \frac{\Delta t_{\mathrm{I}}}{\sqrt{1 - v^2}}$$

as determined by clock II.

We see from the above discussion that the time interval between two events depends on which of a set of equivalent clocks determines the interval, and hence it is not an invariant for those events. Likewise, the measurement of distance will depend on the clock used to make the measurement. In Fig. 6.10 we have redrawn Fig. 6.9, but now have drawn the trajectories of only one particle of each clock, OC for clock I and OD for clock II. The same letters label corresponding points in the two diagrams. In Fig. 6.10 we have added the

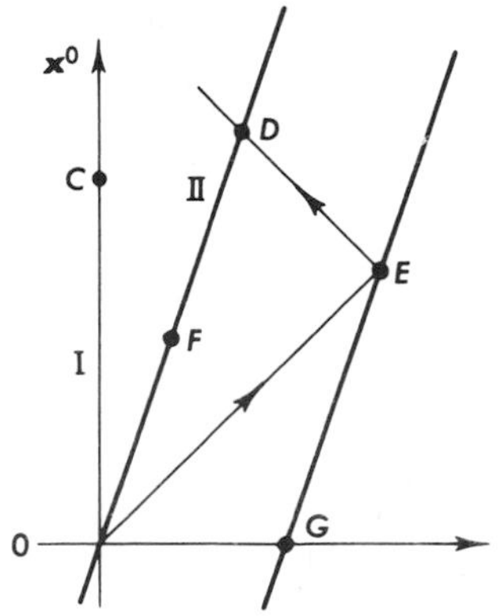

Fig. 6.10.

additional trajectory GE parallel to OD and passing through point E. This latter point is determined by the intersection of light rays through O and D. Furthermore both clocks are assumed to adopt a standard of length that makes the velocity of light unity. The Cartesian analysis is given in Table 6.3.

From our definition of distance measurements and the choice of units, clock II would determine that the particle moving along the trajectory GE is at a distance $l_{\text{II}} = a$ (distance between simultaneous points F and E). On the other hand, clock I would determine the instantaneous distance between this particle and clock II to be $l_{\text{I}} = a\sqrt{1 - v^2}$ (distance between simultaneous points O and G). Consequently one has

$$l_{\text{I}} = \sqrt{1 - v^2}\, l_{\text{II}} \tag{6-12.2}$$

Equation (6-12.2) expresses the *length contraction* of distance. One could reverse the procedure and construct a trajectory parallel to OC and passing through the point of intersection of light rays through O and C. Then clock I would determine the distance between itself and the particle moving along this trajectory to be a, while clock I would determine it to be $a\sqrt{1 - v^2}$. The relation between these two distances would then be given by Eq. (6-12.2), with l_{I} and l_{II} reversed.

TABLE 6.3

CARTESIAN ANALYSIS OF LENGTH CONTRACTION

Event or Geodesic	Coordinate or Equation	Reason
OC	$x^1 = 0$	See Table 6.2
OD	$x^1 = vx^0$	Same
O	$(0, 0)$	Same
D	$\left(\frac{2a}{\sqrt{1-v^2}}, \frac{2av}{\sqrt{1-v^2}}\right)$	Same
OE	$x^0 - x^1 = 0$	Light ray through O
ED	$x^0 + x^1 = 2a\sqrt{\frac{1+v}{1-v}}$	Light ray through D
E	$\left(a\sqrt{\frac{1+v}{1-v}}, a\sqrt{\frac{1+v}{1-v}}\right)$	Intersection of OE and ED
GE	$x^1 = vx^0 + a\sqrt{1-v^2}$	Trajectory through E and parallel to OD
F	$\left(\frac{a}{\sqrt{1-v^2}}, \frac{av}{\sqrt{1-v^2}}\right)$	Point on OD simultaneous with E as determined by clock II
G	$(0, a\sqrt{1-v^2}$	Point on GE simultaneous with O as determined by clock I

Because of the contraction and dilation factors that enter into the relation between time and distance intervals as determined by reference frames moving uniformly with respect to one another, the space and time intervals between two events will depend on the reference system that is used for their determination. However, by analyzing the distance and time measurements made with equivalent reference frames, one can show that the *proper interval* τ_{AB} between two events A and B, given by

$$\tau_{AB}^2 = (t_A - t_B)^2 - ({d_A}^1 - {d_B}^1)^2 - ({d_A}^2 - {d_B}^2)^2 - ({d_A}^3 - {d_B}^3)^2 \tag{6-12.3}$$

has the same value in all equivalent reference frames.

Problem 6.15. By using the methods developed in this section show that, in the case of one spatial dimension, τ_{AB} has the same value as determined by two clocks moving uniformly with respect to one another.

The invariance of the proper interval is sometimes useful in comparing the space-time intervals measured by two different clocks or reference frames.

As an example consider two events lying along the world-line of clock II. The value of τ_{AB} as determined by this clock will be given by

$$\tau_{AB}^2 = (t_A - t_B)_{\mathrm{II}}^2 = \Delta t_{\mathrm{II}}^2. \tag{6-12.4}$$

If clock II is moving with velocity v with respect to clock I, then the distances d_{I} and times t_{I} of points lying on the trajectory of clock II as measured by clock I will be related by

$$d_{\mathrm{I}} = vt_{\mathrm{I}} + d_{\mathrm{I}_0}.$$

Consequently the interval τ_{AB} as determined by clock I will be given by

$$\begin{aligned} \tau_{AB}^2 &= (t_A - t_B)_{\mathrm{I}}{}^2 - (d_A - d_B)_{\mathrm{I}}{}^2 \\ &= (t_A - t_B)_{\mathrm{I}}{}^2 - v^2(t_A - t_B)_{\mathrm{I}}{}^2 \\ &= (1 - v^2)\, \Delta t_{\mathrm{I}}{}^2. \end{aligned} \tag{6-12.5}$$

Since the expressions (6-12.4) and (6-12.5) for τ_{AB} must be equal, we again have the result that

$$\Delta t_{\mathrm{I}} = \frac{\Delta t_{\mathrm{II}}^2}{\sqrt{1 - v^2}}.$$

Problem 6.16. Use the invariance of τ_{AB} to derive the length contraction given by Eq. (6-12.2).

6-13. Other Clocks and Measuring Rods

So far we have dealt with a very special type of clock constructed from free particles and light rays. In order that our discussion be meaningful, we have assumed that such simple systems exist in nature with the properties ascribed to them. Of course the theory does not tell us how to recognize and identify these systems. However, once we have tentatively made the identification, these systems should behave according to the predictions of theory. If they do not, then either the theory is incorrect, that is, does not describe the identified systems, as would be the case if we had used Newtonian mechanics instead of relativistic mechanics, or the identification is incorrect. If we are unsuccessful there is no way of deciding between the two alternatives. In order to verify the predictions of the theory, however, it is necessary that we find some systems that behave as predicted, but failure to find such systems does not rule out a given theory. Thus the singular lack of success in finding a system whose dynamics depended on its overall velocity did not deter people from using Newtonian mechanics. It was only after they became convinced that such systems did not exist that they were willing to accept the new ideas of special relativity.

Even though systems approximating the behavior of free particles and light rays exist in nature, it would be awkward if we had to build all of our measuring devices from them. Fortunately nature, with the help of man, has provided us with other dynamical systems that can be used to make space-time measurements. The motion of the earth, atomic clocks, and pocket watches all serve as devices to measure time. We have not made use of such clocks in our discussion so far because they are rather complicated systems, obeying more or less involved dynamical laws. However, if these laws admit the Poincaré group as a symmetry group then we can make certain predictions concerning their behavior without a detailed knowledge of these laws. Suppose that a system (for example, an atomic clock) is observed to be synchronous with a light clock. Then, if we compare the rates of two identical atomic clocks, moving with respect to one another, with the help of light signals we will observe a time dilation effect as predicted by Eq. (6-12.1). Let us emphasize that this prediction is a consequence solely of the assumed Poincaré symmetry of the dynamical laws of the clocks involved.

Perhaps the most spectacular verification of the time dilation effect occurs in the case of μ meson decay.[38] The μ meson is observed to decay with a half-life of $\sim 10^{-8}$ sec as measured by a standard clock at rest with respect to the meson. Nevertheless μ mesons produced in the upper atmosphere by cosmic ray particles are observed to reach the surface of the earth. Since the total distance they must traverse in such a journey is of the order of 100 km, even if they traveled with essentially the velocity of light they could not traverse this distance if they lived for only 10^{-8} sec. However, this apparent contradiction is removed if the decay rate of a moving μ meson, as measured by a clock at rest on the earth, is increased as predicted by Eq. (6-12.1).[39] Alternately one can describe the observed effect by saying that the total length of atmosphere traversed by the meson, as measured by a clock moving with the meson, is only a few meters, as predicted by Eq. (6-12.2).

All that we have said concerning clocks can also be applied without change to systems that can serve as measuring rods, for example, a metal ruler. Such an object has the property that, when unaccelerated, its end points move along parallel, timelike geodesics. In other words, two free particles moving with the same velocity as the rod and coincident with its end points remain in coincidence with these points. Thus in Fig. 6.10 the lines OD and GE can also be taken to be the trajectories of a measuring rod. The straight line

[38] B. Rossi and D. B. Hall, *Phys. Rev.* **59**, 223 (1941).

[39] R. Durbin, H. H. Loar, and W. W. Havens, *Phys. Rev.*, **88**, 179 (1952), determined the half-life of 73 MeV mesons and found agreement to within 10 percent of the predicted time-dilation factor of 1.5.

joining points E and F then represents the points simultaneously occupied by the intermediate parts of the rod, as determined by clock II, while the straight line joining the points O and G represent the points simultaneously occupied by the intermediate parts of the rod, as determined by clock I. An additional property of a measuring rod is that its length is independent of orientation, as measured by a clock at rest with respect to the rod. Systems with these properties will be referred to as *rigid rods*. From this point of view the null results of the Michelson-Morley experiment can be looked on as verifying that the material out of which the arms of the interferometer have been constructed behaves like a rigid rod.

While the above predictions concerning the behavior of unaccelerated clocks and rods could be made without a detailed knowledge of the dynamics of these systems, this is no longer true when these systems are accelerated. Furthermore it is clear that one must also take account of the accelerative process. Thus consider two rods whose end points are coincident when the rods are unaccelerated. One rod is made of soft lead and the other of hardened steel. They are now accelerated by pushing in such a way that the end points where the accelerative forces are applied remain coincident and the forces are in the same direction. Then, for sufficiently large accelerations, the lead rod will be physically shortened in comparison to the steel rod. Likewise, if we pull on one rod and push on the other so that one pair of end points remains in coincidence, then the other pair will not remain so.

The same state of affairs exists for clocks. Thus, if an atom is accelerated by means of a strong electric field, there will be a shift in frequency in its spectrum due to the Stark effect, and if the field is too large, the atom will be ionized and can no longer function as a clock. One can appreciate the problem more easily for the case of a light clock. If it is to be moved along an arbitrary trajectory in space, one might arrange matters so that one of the particles which constitute the clock moves along this trajectory. But then, how does the second particle move?

One sometimes defines an *ideal clock* as being one that ticks off intervals proportional to $\int d\tau$ along its trajectory. And provided the accelerations are not too great, most real clocks will behave as an ideal clock. However, if one drops his watch on the floor it will usually stop running although it will continue to trace out a trajectory in space-time. While this is an extreme example it does illustrate the point. In effect, the assumption that an ideal clock measures $\int d\tau$ along its trajectory is equivalent to the assumption that one can ignore the accelerative effects on its internal workings. One sees this by noting that any trajectory can be approximated by a series of straight-line segments and that an unaccelerated light clock does measure $\int d\tau$ along its trajectory. Thus an ideal clock is equivalent to a series of unaccelerated light clocks moving tangent to the trajectory of the ideal clock. However, in the discussions to follow we will not need to introduce the notion of an ideal clock.

6-14. The Twin Paradox

As an illustration of the ideas set forth above we will now discuss the so-called twin or clock paradox of special relativity. Perhaps no other problem in special relativity has had so many papers written about it, nor called forth more nonsense. The discussion here closely follows that given by A. Schild.[40]

Consider two identical twins I and II. Twin I spends his life on earth. At birth, twin II is put into a rocket ship that then accelerates away from the earth until it reaches a velocity v with respect to the earth. After some time the rocket ship slows down, turns around, and returns to earth again with a velocity v. Upon reaching the vicinity of the earth the ship slows down and comes to rest. We are now asked to compare the biological ages of the two twins.

In Fig. 6.11 we have drawn the world-lines of the two twins. The determination of the ages of the two twins upon reaching C can be carried out in a number of ways. As a model we can replace the twins by identical light clocks. The age of twin I will be proportional to the number of ticks experienced by a clock in moving along AC from A to C. Likewise the age of twin II will be proportional (same constant of proportionality) to the number of ticks experienced by an identical clock in moving from A to B plus the number of ticks experienced in moving from B to C. According to the results of Section 6-12, if $T_{\text{II}}/2$ is the time recorded by the clock in going from A to B, then

$$\frac{T_{\text{I}}}{2} = \frac{T_{\text{II}}/2}{\sqrt{1 - v^2}} > \frac{T_{\text{II}}}{2}$$

will be the time recorded by clock I in going from A to B', since B', the midpoint of AC, is simultaneous with B for clock I. A similar argument holds for the second part of the journey, and one concludes that the age T_{I} of twin I is greater by the factor $1/\sqrt{1 - v^2}$ than the age of T_{II} of twin II.

The difference in ages is a real difference and not apparent because the two twins are contiguous at the beginning and end points of their journeys. If $v^2 = 0.99$ and T_{II} is one year, then T_{I} will be ten years. While it is unlikely that such large velocities can be achieved for macroscopic objects as large as rockets, the effect has been already observed in the decay of μ mesons.

At this point a number of objections can be raised. One might argue that biological time has nothing to do with time as discussed in special relativity.

[40] A. Schild, *The Amer. Math. Monthly*, **66**, 1 (1959).

But the heart, of course, in every real sense is a clock, and we know that it remains synchronous with other clocks and in particular would remain synchronous with a light clock. Thus, if biological age is measured by the number of heart beats elapsed, then biological time and the time special relativity talks about are in fact one and the same. One might also object that we have completely neglected the effects of acceleration on twin II. Strictly speaking, the path of twin II should be as indicated in Fig. 6.12, where

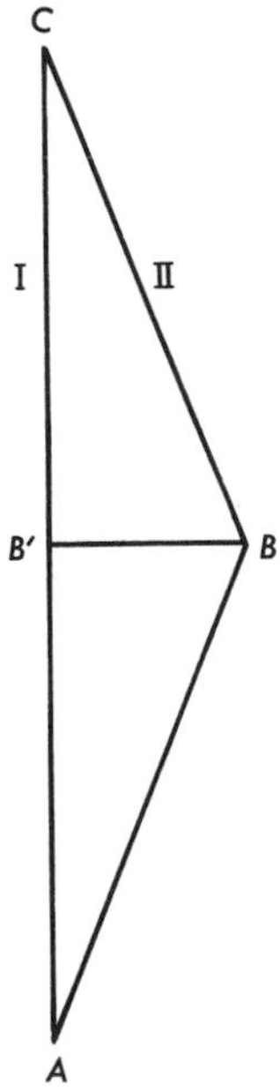

Fig. 6.11.

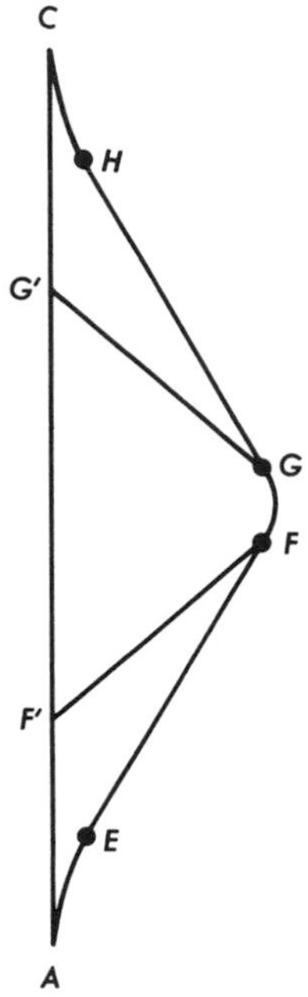

Fig. 6.12.

the curved portions between *AE*, *FG*, and *HC* represent the accelerative parts of the journey of twin II. Certainly, if the accelerations are too large, twin II will be killed and will arrive at *C* even younger than calculated above, as measured by the total number of his heart beats. Unless we know the effect of the accelerative process on twin II we cannot calculate how many heart beats occur during the accelerative parts of the journey. However, whatever the effect, we can disregard it by making the accelerative periods short compared with the unaccelerated period in his journey.

Another objection is raised, however, which bears more directly on the problem at hand. It is argued that twin II can use equally well the time dilation to calculate that twin I should be the younger when both come together at *C*. At first sight the objection seems valid and gives rise to what has been called in the literature the *twin paradox*. To support the argument it is further argued that one could redraw the two trajectories so that the curve ABC of Fig. 6.11 becomes a straight line while the curve $AB'C$ is bent at its midpoint. In other words, twin II is considered to be at rest while twin I is considered to be doing the traveling. Such a procedure is justified by arguing that all motion is relative. But of course all motion is not relative in special relativity. Acceleration is just as absolute in this description as in Newtonian mechanics. A free particle can follow the trajectory $AB'C$ but not ABC. While either twin could conclude visually that he was at rest and his sibling was doing the traveling, only twin I could consider himself dynamically to be at rest.

The error that arises in applying the dilation factor to the age of twin I can be understood by referring to Fig. 6-12. Here the point F' is simultaneous to the point F and the point G' is simultaneous to the point G as determined by twin II. Now it is quite true that if twin II ages by an amount $T_{\text{II}}/2$ in going from A to F; then, neglecting accelerations,

$$\frac{T'_{\text{I}}}{2} = \sqrt{1 - v^2}\,\frac{T_{\text{II}}}{2} < \frac{T_{\text{II}}}{2}$$

will be the amount twin I ages in going from A to F'. This is also the relation between the amounts of aging of the two twins in going from GC to $G'C$, respectively. But of course $T_{\text{I}} \neq T'_{\text{I}}$, since twin I also ages in going from F' to G'. It has often been asserted that one needs general relativity to calculate how much twin I ages in going from F' to G'. Since, however, we have assumed that the laws of special relativity hold throughout the space-time region in question, there is no reason for bringing general relativity into the argument. In fact, based on the above calculations, it follows directly that

$$t''_{\text{I}} = t_{\text{I}} - t'_{\text{I}} = \frac{v^2 T_{\text{II}}}{\sqrt{1 - v^2}}.$$

However, we cannot conclude from this result that twin II would suddenly see twin I age by the amount t''_{I} as he moves from F to G. It is in fact instructive to discuss just what each twin would actually see if he viewed his sibling through a telescope.

In Fig. 6.13 we have drawn a picture of what twin I would see, neglecting the accelerative parts of the journey of twin II. Let D be a point on AB with coordinates (T, vT). A light signal leaving D will travel along the ray $D\bar{D}$ with equation

$$x^0 + x^1 = T(1 + v),$$

and hence will meet the world-line of twin I at the point $\bar{D}$, with coordinates

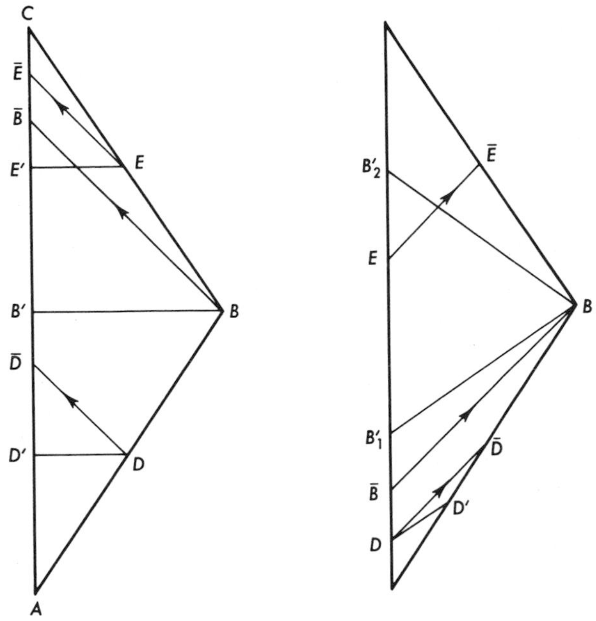

Fig. 6.13.

$(T(1 + v), 0)$. The point $\bar{D}$ is thus later than the point D' which is simultaneous with D relative to twin I. At D' the biological age of twin I will be T, while that of twin II at the point D is T'. The two ages are related by Eq. (6-12.1):

$$T = \frac{T'}{\sqrt{1 - v^2}}.$$

However, twin I will see twin II only at the age T' when he arrives at $\bar{D}$, at which time twin I will be $T(1 + v)$ years old. Consequently he will see twin II age at a rate $((1 - v)/(1 + v))^{1/2}$ slower than he is aging during the first part of the journey. If twin II is $T_{\text{II}}/2$ years old when he reaches the midpoint B of his journey, twin I will be $((1 + v)/(1 - v))^{1/2}(T_{\text{II}}/2)$ years old when he sees twin II there and will be at point $\bar{B}$. Since $\bar{B}$ is much later than B', the point simultaneous with B relative to twin I, twin I sees twin II at age $T_{\text{II}}/2$, much later than $T_{\text{I}}/2$. If one repeats the argument for the homeward portion of the journey of twin II, he finds that twin II will appear to age at a rate $(1 + v)/(1 - v))^{1/2}$ faster than twin I. However, twin II appears to age more rapidly than twin I for only a short time,

$$T_{\text{I}} - \left(\frac{1 + v}{1 - v}\right)^{1/2} \frac{T_{\text{II}}}{2}.$$

It then follows that

$$T_{\mathrm{I}} - \left(\frac{1+v}{1-v}\right)^{1/2}\frac{T_{\mathrm{II}}}{2} = \left(\frac{1-v}{1+v}\right)^{1/2}\frac{T_{\mathrm{II}}}{2},$$

and hence

$$T_{\mathrm{I}} = \frac{T_{\mathrm{II}}}{\sqrt{1-v^2}} > T_{\mathrm{II}}, \tag{6-14.1}$$

as before.

We can employ a line of reasoning similar to that above, to discuss what twin II sees. Referring to Fig. 6.13(b), let D be a point lying on the world-line of twin I, with coordinates $(T, 0)$, when twin I is viewed by twin II. The point $\bar{D}$ of observation on AB has coordinates $((T/(1-v)), (vT/(1-v))$ and is later than the point D', which is simultaneous with D relative to twin II. If twin I is T years old at D, then twin II will be $T\sqrt{1-v^2}$ years old at D' and

$$\left(\frac{T}{1-v}\right)\sqrt{1-v^2} = T\left(\frac{1+v}{1-v}\right)^{1/2}$$

years old at $\bar{D}$, and so twin I will appear to be aging less rapidly during the first half of the journey. However, from point B onward, twin II sees twin I aging more rapidly by a factor $((1+v)/(1-v))^{1/2}$. Since the time taken for twin I to go from $\bar{B}$ to C is

$$T_{\mathrm{I}} - \left(\frac{1-v}{1+v}\right)^{1/2}\frac{T_{\mathrm{II}}}{2},$$

it follows that

$$\frac{T_{\mathrm{II}}}{2} = \left(\frac{1-v}{1+v}\right)^{1/2}\left(T_{\mathrm{I}} - \left(\frac{1-v}{1+v}\right)^{1/2}\frac{T_{\mathrm{II}}}{2}\right),$$

and again one has the result of Eq. (6-14.1).

From the above discussion we see that each twin will first see the other aging less rapidly than himself, and then at some point (B for twin II in Fig. 6.13(b), and $\bar{B}$ for twin I in Fig. 6.13(a) the other abruptly appears to age more rapidly. While the observed aging ratios

$$\left(\frac{1-v}{1+v}\right)^{1/2} \quad \text{and} \quad \left(\frac{1+v}{1-v}\right)^{1/2}$$

are the same for each twin, the periods of less and more rapid aging are, for each twin just such as to give Eq. (6-14.1) as the relation between their final ages. If twin II had been allowed to accelerate smoothly at the far end of his journey, then even that rate of aging would have changed continuously.

6-15. Relation Between Space-Time Measurements in Different Inertial Frames

In Section 6-10 we discussed how an observer could, with the help of an inertial reference frame, assign a time and location to an event occurring in space-time. Then, in Section 6-12, we developed relations between certain types of space-time measurements made by different observers moving uniformly with respect to each other; for example, Eqs. (6-12.1) and (6-12.2). We now wish to derive a general relation between the space-time measurements made by different observers using inertial frames to make these measurements. In doing so, we will obtain an interpretation of the Poincaré mappings.

Suppose that in an inertial frame I, the event A is observed to occur at a time t_{I} and a location determined by the three orthogonal distances $d_{\mathrm{I}}{}^r$. If the origin of frame I (the particle common to the three clocks comprising this frame) is taken to move along the trajectory $x^r = 0$ and the three directions determined in this frame are taken to lie in the direction of the three coordinate axes, then by a suitable choice of units the coordinates of event A will be $(t_{\mathrm{I}}, d_{\mathrm{I}}{}^r)$. Consider now a second inertial frame II. In general, ten parameters will be needed to fix the relation of this second frame to frame I. Three of these parameters will determine the velocity of frame II relative to frame I, three will determine its orientation, and four will be needed to locate its origin. In this second frame the event A will be observed to occur at a time t_{II} and a location determined by the distances $d_{\mathrm{II}}{}^r$. We now want to find the relation between t_{I}, $d_{\mathrm{I}}{}^r$, and t_{II}, $d_{\mathrm{II}}{}^r$.

Let us perform a Poincaré mapping that brings frame II into coincidence with the coordinate system being employed on our manifold. Since the most general Poincaré mapping involves ten arbitrary parameters, such a mapping can always be found. Under this mapping the point A, with coordinates $x^\mu = (t_{\mathrm{I}}, d_{\mathrm{I}}{}^r)$, gets mapped onto a new point A', with coordinates x'^μ. But now, since II coincides with the coordinate system, the coordinates x'^μ will be numerically equal to $(t_{\mathrm{II}}, d_{\mathrm{II}}{}^r)$ because the values of these latter quantities do not change under a mapping. Consequently we have the desired relation that

$$t_{\mathrm{II}} = \Lambda_0{}^0 t_{\mathrm{I}} + \Lambda_r{}^0 d_{\mathrm{I}}{}^r + a^0 \tag{6-15.1a}$$

and

$$d_{\mathrm{II}}^r = \Lambda_0{}^r t_{\mathrm{I}} + \Lambda_s{}^r d_{\mathrm{I}}{}^s + a^r. \tag{6-15.1b}$$

From these results it is clear that τ_{AB} as given by Eq. (6-12.3) has the same value in all reference frames.

As an example consider the case when II is moving in the positive x' direction with a velocity v with respect to I. Further, let the origins of the two systems coincide so that the points lying on the trajectory of II as determined by I satisfy

$$d_{\mathrm{I}} = vT_{\mathrm{I}}.$$

One sees from Eq. (6-5.22) that the Poincaré mapping that maps the trajectory of II onto the x^0 axis is the velocity mapping with $\mathbf{v} = (-v, 0, 0)$; that is, the Lorentz mapping with Lorentz matrix $\Lambda_\nu{}^\mu$ with nonvanishing components given by

$$\Lambda_0{}^0 = \Lambda_1{}^1 = \frac{1}{\sqrt{1 - v^2}} \tag{6-15.2a}$$

$$\Lambda_1{}^0 = \Lambda_0{}^1 = \frac{v}{\sqrt{1 - v^2}} \tag{6-15.2b}$$

For this mapping, Eq. (6-15.1) reduces to

$$t_{\mathrm{II}} = \frac{t_{\mathrm{I}} - v\, d_{\mathrm{I}}{}^1}{\sqrt{1-v^2}} \tag{6-15.3a}$$

and

$$d_{\mathrm{II}}^1 = \frac{d_{\mathrm{I}}{}^1 - vt_{\mathrm{I}}}{\sqrt{1 - v^2}}. \tag{6-15.3b}$$

We can then invert these equations to obtain the more familiar form of the relation between times and distances as measured in the two systems

$$t_{\mathrm{I}} = \frac{t_{\mathrm{II}} + v\, d_{\mathrm{II}}^1}{\sqrt{1 - v^2}} \tag{6-15.4a}$$

and

$$d_{\mathrm{I}}{}^1 = \frac{d_{\mathrm{II}}^1 + vt_{\mathrm{II}}}{\sqrt{1 - v^2}}, \qquad d_{\mathrm{I}}{}^2 = d_{\mathrm{II}}^2, \quad d_{\mathrm{I}}{}^3 = d_{\mathrm{II}}^3. \tag{6-15.4b}$$

One can make use of Eqs. (6-15.4) to derive the Lorentz contraction of lengths. Let a rod be at rest in I and let its end points be at distances d_{I} and d'_{I} from the origin of I in the direction of the x' axis at some time t_{I}. The length of the rod in I, its rest frame, is thus $\Delta l_{\mathrm{I}} = d'_{\mathrm{I}} - d_{\mathrm{I}}$. To find its length in II one would have to determine the simultaneous distances of the two end points in II. In other words, one must find the values d_{II} and d'_{II} when $t_{\mathrm{II}} = t'_{\mathrm{II}}$. Since the rod is at rest in I, the distances of the end points are independent of time, it follows from Eq. (6-15.4b) that the one end of the rod is at a distance d_{II} from the origin of II at t_{II}, given by

$$d_{\text{I}} = \frac{d_{\text{II}} + vt_{\text{II}}}{\sqrt{1 - v^2}}$$

while the other end is at a distance d'_{II} at this time, given by

$$d'_{\text{I}} = \frac{d'_{\text{II}} + vt_{\text{II}}}{\sqrt{1 - v^2}}.$$

Subtracting, one has the expected result that

$$\Delta l_{\text{II}} = d'_{\text{II}} - d_{\text{II}} = \sqrt{1 - v^2}\, \Delta d'_{\text{I}}.$$

Of course the times t_{I} and t'_{I}, as determined in I, when II observes the simultaneous (relative to II) distances of the end points of the rod, are not equal.

Problem 6.17. Derive the time dilation from the transformation Eqs. (6-15.4).

In Section 6-10 we saw that when the flat metric takes on its Minkowski values, the space-time coordinates correspond to space the time measurements made with the help of an inertial frame. The Poincaré mappings that leave invariant this form of the metric are now seen to correspond to the relations between space and time measurements made in two different inertial frames. In the next section we will argue that the other manifold mappings correspond to the relations between such measurements made in noninertial frames.

6-16. Noninertial Reference Systems

So far in our discussion of space and time measurements, we have considered classes of measuring devices that go to make up inertial reference frames. However, one often has occasion to make measurements with noninertial devices; for example, measurements made in a laboratory rotating with the earth. In this section we will discuss such measurements and show how they can be incorporated into a space-time description.

As a model of noninertial reference frame let us consider a uniformly accelerated light clock, one reflector of which moves in the plane $x^2 = x^3 = 0$ and the other in the plane $x^2 = \Delta$, $x^3 = 0$, where Δ is the separation between the reflectors. By using a radar technique in conjunction with such a clock, one could proceed to make space-time measurements in the same way as discussed in Section 6–10. Let n_1 be the number of ticks recorded by the light clock, counting from some arbitrary point on its trajectory, and n_2 the number of ticks recorded when the light signal returns after being reflected from some space-time event; for example, a particle occupying a point on its trajectory. Then, as before, we can take the distance between the clock at a point on its trajectory when it has recorded $(n_1 + n_2)/2$ ticks and the point where the light is reflected to be

$$s_a = \frac{n_2 - n_1}{2}. \tag{6-16.1}$$

Also, the time of reflection is taken to be

$$t_a = \frac{n_2 + n_1}{2}. \tag{6-16.2}$$

Suppose now that the trajectory of the reflector of the light clock moving in the $x^2 = x^3 = 0$ plane is given by

$$\left(x^1 + \frac{1}{a^2}\right)^2 - x^{0^2} = \frac{1}{a^2}, \qquad a = \text{const}, \tag{6-16.3}$$

and a free particle moving in the $x^2 = x^3 = 0$ plane parallel to the x^0 axis is observed by the accelerated light clock. Then it can be shown[41] that s_a and t_a corresponding to points on the trajectory of this particle are related by

$$1 - as_0 = e^{-as_a} \cosh at_a, \tag{6-16.4}$$

where s_0 is the initial distance of the free particle from the accelerated light clock. Note in particular that this observed trajectory does not have the same form as the trajectory of the accelerated light clock as observed by a free light clock. This latter trajectory is given by Eq. (6-16.3), with x^1 replaced by s and x^0 replaced by t.

Let us now perform a mapping such that the trajectory of the accelerated light clock is parallel to the x^0 axis, and equal time intervals of this clock correspond to equal intervals along the x^0 axis and are such that an event in the $x^2 = x^3 = 0$ plane to which the accelerated light clock assigns a distance s_a and a time t_a has coordinates $(t_a, s_a, 0, 0)$. Such a mapping can be found, and has the form

$$x'^0 = \frac{1}{2a} \ln[\{1 + a(x^0 + x^1 - s_0)\}\{1 + a(-x^0 + x^1 - s_0)\}],$$

$$x'^1 = \frac{1}{2a} \ln\left[\frac{1 + a(x^0 + x^1 - s_0)}{1 + a(-x^0 + x^1 - s_0)}\right], \qquad x'^2 = x^2, \quad x'^3 = x^3.$$

[41] J. L. Anderson and R. Gautreau, unpublished.

Under this mapping the metric gets transformed to

$$g'_{\mu\nu} = e^{2ax'}\eta_{\mu\nu}, \tag{6-16.5}$$

so that the light cones are still given by

$$\eta_{\mu\nu}(x^\mu - x_0{}^\mu)(x^\nu - x_0{}^\nu) = 0.$$

Furthermore the trajectory of a free particle, given by Eq. (6-16.4) with s replaced by x^1 and t by x^0, is again a geodesic of this transformed metric. Finally, since $g'_{\mu\nu}$ is just the transform of a flat metric, it too must be flat, that is, $R_{\mu\nu\rho\sigma}(g') = 0$.

On the basis of such examples we are led to conclude that, in general, distance and time measurements made with some arbitrary set of measuring devices can always be made to correspond to the coordinates of the space-time manifold by means of a suitable space-time mapping. Under such a mapping the space-time metric will in general no longer have its Minkowski values but will nevertheless still be flat. The mappings of the space-time manifold onto itself are thus seen to correspond to the relations between the space and time measurements made with two different sets of space-time measuring devices. This same interpretation can be maintained even when the space-time geometry is no longer flat. It follows that we can always make space-time measurements in such a way that the corresponding metric can be made to take on its Minkowski values and its derivatives can be made to vanish at any one space-time point.

Finally we should point out that only when the flat metric does take on its Minkowski values everywhere on the manifold do the space-time coordinates correspond to *a priori* identifiable space-time measurements, that is, measurements made with unaccelerated space-time measuring devices. However, when the metric is not Minkowskian everywhere, or at least in a finite region of space-time, there is in general no *a priori* way of deciding what measuring devices correspond to the space-time coordinates nor what is the mapping that relates the measurements made with two different sets of devices.

It is customary in Newtonian mechanics to relate the space and time measurements made in a rotating reference frame to those in an inertial frame by a mapping of the form

$$\begin{aligned}
x'^0 &= x^0,\\
x'^1 &= x^1 \cos \omega x^0 + x^2 \sin \omega x^0,\\
x'^2 &= -x^1 \sin \omega x^0 + x^2 \cos \omega x^0,\\
x'^3 &= x^3.
\end{aligned}$$

Much effort has been expended on trying to construct the counterpart of these mappings in special relativity. For us, such effort is pointless unless the details of how the rotating reference frame is constructed are given. As we discussed in Section 6-13, only in the case of unaccelerated measuring devices can we predict *ab initio* the relations between the measurements made with them. Similarly, given an arbitrary flat metric, we have no *a priori* way of deciding what are the space-time measurements to which the coordinates correspond. In the general case we are forced to perform actual experiments in order to correlate one set of space-time measurements with another. Thus the transformation law quoted above must be considered to be either the result of observations or of special assumptions concerning the behavior of the rotating measuring devices. We certainly cannot expect it to be valid for arbitrarily large values of ω, if for no other reason than that all physical systems experience strains when forces are exerted on them and hence will tend to become distorted. Also, for sufficiently large ω, parts of the rotating reference frame would have to be moving faster than the speed of light. But if this happened, the usual causal relations between space-time measurements would be destroyed (see Section 6-19).

6-17. Derivation of the Poincaré Mappings

So far, our approach to special relativity was to assume that the Poincaré group is a universal symmetry group of all physical systems. We then investigated the interrelations between two simple systems satisfying this requirement, free particles and light rays. In this section we will reverse the prccedure and assume the existence of a system of free particles and light rays, and from their assumed properties will derive the fact that the Poincaré group is the symmetry group of this system. This derivation will then tell us the minimum assumptions needed to define this group. In this we will give a derivation that closely follows Einstein's original derivation.

Our fundamental assumption is that there exist simple systems in nature with the property that the mathematical description of their trajectories in a space-time manifold can be so mapped as to take the form

$$x^\mu = x_0{}^\mu + \zeta^\mu \lambda \tag{6-17.1}$$

where the $x_0{}^\mu$ and ζ^μ are constants and λ is a suitable path parameter. We further assume that the magnitude of the velocity $dx^r/dx^0 = \zeta^r/\zeta^0$ is less than or equal to unity:

$$\sum_r \left(\frac{\zeta^r}{\zeta^0}\right)^2 \leqq 1$$

or

$$(\zeta^0)^2 - \sum_r (\zeta^r)^2 = \eta_{\mu\nu}\zeta^\mu\zeta^\nu \geqq 0. \tag{6-17.2}$$

When the inequality holds, the trajectories are taken to be those of free particles, and when the equality holds, the trajectories are those of light rays.

We now ask for those mappings that map this system of trajectories onto itself, with the property that light rays are mapped onto light rays. Thus we require that under the mapping

$$x^{\mu} \rightarrow x'^{\mu} = x'^{\mu}(x), \tag{6-17.3}$$

the trajectories described by Eq. (6-17.1) are mapped onto the trajectories

$$x^{\mu} = x'_0{}^{\mu} + \zeta'^{\mu}\lambda \tag{6-17.4}$$

with

$$\eta_{\mu\nu}\zeta'^{\mu}\zeta'^{\nu} \geqq 0. \tag{6-17.5}$$

We further require that if the equality in Eq. (6-17.2) holds, it also holds in Eq. (6-17.5). It therefore follows that

$$\frac{dx'^{\mu}}{d\lambda} = \zeta'^{\mu} \tag{6-17.6}$$

and

$$\frac{d^2x'^{\mu}}{d\lambda^2} = 0. \tag{6-17.7}$$

But it also follows from Eq. (6-17.1) and (6-17.3) that

$$\frac{dx'^{\mu}}{d\lambda} = \frac{\partial x'^{\mu}}{\partial x^{\rho}}\zeta_{\rho} \tag{6-17.8}$$

and

$$\frac{\partial^2 x'^{\mu}}{\partial\lambda^2} = \frac{\partial^2 x'^{\mu}}{\partial x^{\rho}\,\partial x^{\sigma}}\zeta^{\rho}\zeta^{\sigma}, \tag{6-17.9}$$

so that

$$\frac{\partial^2 x'^{\mu}}{\partial x^{\rho}\,\partial x^{\sigma}}\zeta^{\rho}\zeta^{\sigma} = 0. \tag{6-17.10}$$

Since, however, the ζ^{ρ} need only satisfy the inequality of Eq. (6-17.2) we have enough freedom in their choice to conclude that

$$\frac{\partial^2 x'^{\mu}}{\partial x^{\rho}\,\partial x^{\sigma}} = 0. \tag{6-17.11}$$

It therefore follows that the mappings in question are linear, that is, of the form

$$x'^{\mu} = a_{\nu}{}^{\mu}x^{\nu} + b^{\mu}. \tag{6-17.12}$$

In his derivation Einstein assumed the linearity of the mappings, giving as his reason the homogeneity of space and time. We see from our derivation that our requirements represent the minimum expression of this homogeneity. Note, however, that it was necessary to use the trajectories of both free particles and light rays to arrive at our conclusion of linearity. If we had used only light rays the ζ^μ would have had to satisfy the equality of Eq. (6-17.2), and hence would not be independent of each other. Under this circumstance we could not have obtained Eq. (6-17.11) from Eq. (6-17.10).

To conclude the derivation we now bring in our requirement that light rays are mapped into light rays. From Eqs. (6-17.6) and (6-17.8) it follows that

$$\zeta'^\mu = \frac{\partial x'^\mu}{\partial x^\nu}\zeta^\nu$$
$$= a_\nu{}^\mu \zeta^\nu. \tag{6-17.13}$$

Then, since the equality of Eq. (6-17.5) holds, we have

$$(\eta^{\rho\sigma} a_\mu{}^\rho a_\nu{}^\sigma)\zeta^\mu\zeta^\nu = 0 \tag{6-17.14}$$

We cannot, of course, conclude that the bracketed coefficients in this equation are individually zero, since the ζ^μ satisfy Eq. (6-17.2) with the equality holding. To proceed, we make use of Lagrange's method of undetermined multipliers. Let us multiply Eq. (6-17.2) by an undetermined multiplier γ and add the result to Eq. (6-17.14) to obtain the equation

$$\{\eta_{\rho\sigma} a_\mu{}^\rho a_\nu{}^\sigma - \gamma\eta_{\mu\nu}\}\zeta^\mu\zeta^\nu = 0. \tag{6-17.15}$$

We now choose γ so that one of the coefficients in Eq. (6–17.15) is zero. Then there is enough arbitrariness in the ζ^μ to conclude that the remaining coefficients also vanish. Thus we have the result that the a^μ must satisfy

$$\eta_{\rho\sigma} a_\mu{}^\rho a_\nu{}^\sigma = \gamma\eta_{\mu\nu}. \tag{6-17.16}$$

The mapping coefficients $a_\nu{}^\mu$ are thus seen to form a matrix that is proportional to a Lorentz matrix so that

$$a_\nu{}^\mu = \gamma\Lambda_\nu{}^\mu, \tag{6-17.17}$$

where $\Lambda_\nu{}^\mu$ is a Lorentz matrix and γ is some function of the parameters appearing in $\Lambda_\nu{}^\mu$. We rule out the possibility that γ is independent of these parameters, since in that case it would correspond to a pure scale change, as can be seen by taking $\Lambda_\nu{}^\mu = \delta_\nu{}^\mu$ in Eq. (6-17.17).

To fix the coefficient γ appearing in Eq. (6-17.16) we will require that the matrices $a_\nu{}^\mu$ constitute the representation of a group, that is, that the product of two such matrices is again a matrix of this form. Thus, if $a^\mu_{1\nu} = \gamma_1\Lambda^\mu_{1\nu}$, $a^\mu_{2\nu} = \gamma_2\Lambda^\mu_{2\nu}$, and $a^\mu_{3\nu} = \gamma_3\Lambda^\mu_{3\nu}$, where $a^\mu_{3\nu} = a^\mu_{2\rho}a^\rho_{1\nu}$, we must have $\gamma_3 = \gamma_2\gamma_1$. Now, according to the factorization theorem proved in Section 6-15, an arbitrary Lorentz matrix can be written as the product of a pure velocity matrix times a rotation. It follows, therefore, than any γ can be factored into the product of two terms, one of which depends only on the rotation parameters that arise in the decomposition of its associated $\Lambda_\nu{}^\mu$ matrix and another that depends only on the velocity parameters involved in the decomposition. Since, under a rotation, time and distance scales do not change, the factor of γ associated with a rotation must have the value unity. Furthermore, since the inverse of $a_\nu{}^\mu$ has the form $\gamma^{-1}\Lambda_\nu^{-1\mu}$, it follows that

$$\gamma(\mathbf{v})\gamma(-\mathbf{v}) = 1.$$

But since the factor $\gamma(\mathbf{v})$ has the effect of only changing the distance scales in directions normal to $\mathbf{v}$, $\gamma(\mathbf{v})$ must be independent of the direction of $\mathbf{v}$, that is, $\gamma(\mathbf{v}) = \gamma(-\mathbf{v})$. Consequently $\gamma^2(\mathbf{v}) = \pm 1$, and since $\gamma(0) = 1$ we have finally that $\gamma = +1$ for all mappings.

When we originally defined a Poincaré mapping, we did so by requiring it to be a symmetry mapping of the flat Minkowski metric. Now we see that it can be equally well defined as a symmetry mapping of a system of free particles and light rays without any reference to a metric. Thus we see that the introduction of a metric is merely a convenient way of expressing the properties of free particles and light rays.

With the help of this new view of the origin of the Poincaré group we can better understand our original requirement that this group must be a universal symmetry group of all physical systems. Suppose that the behavior of some physical system can be correlated with the behavior of a light clock or system of light clocks that are taken to be at rest with respect to the system. Now imagine that we have a replica of this system, which is moving uniformly with respect to the original system, and a light clock or clocks running at the same rate as the other light clocks and at rest with respect to this second system. (Here the phrase "at rest with respect to" is to taken have the same sense for both systems.) Then, if the principle of the relativity of uniform motion is valid, the correlation of this second system with these latter clocks must be the same as the correlation of the first system with the clocks that are at rest with respect to it. But this will be the case only if the Poincaré group is a symmetry group of the system in question. Thus it follows that any system whose behavior can be correlated with light clocks must possess the Poincaré symmetry if the relativity of uniform motion is to hold.

6-18. The Einstein Addition Theorem for Velocities; the Fresnel Coefficient

The procedure developed for relating the spatial and temporal intervals between events as determined in two different inertial frames can be extended to other types of space-time measurements. Consider the motion of a particle (or light ray). From the observed locations of points on the trajectory as a function of time, as determined by frame I, it is possible to construct the parametric equations of this trajectory on the manifold when I coincides with the coordinate systems on the manifold. Let these equations be

$$x^\mu = z^\mu(\tau), \tag{6-18.1}$$

where the parameter is chosen to be the proper time (or affine parameter in the case of a light ray) along the trajectory.

Consider now a Poincaré mapping that sends a second inertial frame II onto the coordinate axes. Under this mapping the trajectory of the particle gets mapped onto another curve:

$$x^\mu = z'^\mu(\tau) = \Lambda_\nu{}^\mu z^\nu(\tau) + a^\mu.$$

The corresponding four-velocities, $\dot{z}^\mu$ and $\dot{z}'^\mu$, are related by

$$\dot{z}'^\mu(\tau) = \Lambda_\nu{}^\mu \dot{z}^\nu(\tau). \tag{6-18.2}$$

For the case when II is moving uniformly with respect to I with a velocity $\mathbf{v}$, this mapping is a velocity mapping, and we can make use of Eq. (6-5.22) to relate the three-velocity $\mathbf{u}_{\mathrm{II}}$ with components $\dot{z}'^r/\dot{z}'^0$, as measured in II, to the three-velocity $\mathbf{u}_{\mathrm{I}}$ with components $\dot{z}^r/\dot{z}^0$, as measured in I. According to Eq. (6-8.4), $\dot{z}^\mu = \delta_{\mathrm{I}}(1, \mathbf{u}_{\mathrm{I}})$ and $\dot{z}'^\mu = \delta_{\mathrm{II}}(1, \mathbf{u}_{\mathrm{II}})$, where $\delta = (1 - \mathbf{u}^2)^{-1/2}$. Hence we have

$$\delta_{\mathrm{II}} = \gamma\delta_{\mathrm{I}}[1 - \mathbf{u}_{\mathrm{I}} \cdot \mathbf{v}] \tag{6-18.3}$$

and

$$\mathbf{u}_{\mathrm{II}} = \frac{\gamma^{-1}\mathbf{u}_{\mathrm{I}} + (\mathbf{u}_{\mathrm{I}} \cdot \mathbf{v}/\mathbf{v}^2)\mathbf{v}(1 - \gamma^{-1}) - \mathbf{v}}{1 - \mathbf{u}_{\mathrm{I}} \cdot \mathbf{v}}, \tag{6-18.4}$$

where $\gamma = (1 - \mathbf{v}^2)^{-1/2}$.

The inverse relations giving δ_{I} and $\mathbf{u}_{\mathrm{I}}$ in terms of δ_{II} and $\mathbf{u}_{\mathrm{II}}$ are obtained from the above equations by interchanging these quantities and replacing $\mathbf{v}$ by $-\mathbf{v}$.

In the special case that $\mathbf{u}_{\mathrm{II}} \cdot \mathbf{v} = 0$, the velocity transformation equations reduce to

$$\mathbf{u}_{\mathrm{I}} = \gamma^{-1}\mathbf{u}_{\mathrm{II}} + \mathbf{v}, \tag{6-18.5}$$

while if $\mathbf{u}_{\mathrm{II}}$ and $\mathbf{v}$ are parallel, they reduce to

$$\mathbf{u}_{\mathrm{I}} = \frac{\mathbf{u}_{\mathrm{II}} + \mathbf{v}}{1 + \mathbf{u}_{\mathrm{II}} \cdot \mathbf{v}}. \tag{6-18.6}$$

Finally, when $V(\mathbf{v}) = V(v, 0, 0)$, we have

$$u_{\text{II}}^1 = \frac{u_{\text{I}}^{\ 1} - v}{1 - u_{\text{I}}^{\ 1} v}, \quad u_{\text{II}}^2 = \gamma^{-1} \frac{u_{\text{I}}^{\ 2}}{1 - u_{\text{I}}^{\ 1} v}, \quad u_{\text{II}}^3 = \gamma^{-1} \frac{u_{\text{I}}^{\ 3}}{1 - u_{\text{I}}^{\ 1} v}. \tag{6-18.7}$$

These are the Einstein addition formulas for velocities.

As an example of the application of the transformation law for velocities we will now derive the velocity of propagation of light in a moving medium. If one knew the detailed mechanism of propagation in the medium, one could derive an expression for this velocity directly. However, the detailed dynamical description of the propagation of light in a material medium consistent with the requirements of relativity theory would be extremely complicated and hard to come by. Fortunately, the knowledge that, whatever these laws may be, they must be consistent with the requirement of relativity theory, allows us to apply to Einstein velocity transformation formulas directly.

Let us assume that a homogeneous isotropic material capable of transmitting electromagnetic signals is moving with a velocity $\mathbf{v}$ in the 1-direction with respect to a reference frame I. In a reference frame II at rest with respect to the material the velocity of propagation of light is just $1/n$ in all directions, n being the index of refraction of the material as determined by measurements in II. Now consider a signal being propagated through the medium in the 1-direction. Taking $u_{\text{II}}^1 = 1/n$ and using the inverse velocity transformation, one finds that its velocity $u_{\text{I}}^{\ 1}$ as measured in I is given by

$$u_{\text{I}}^{\ 1} = \frac{v + 1/n}{1 + v/n} \simeq \frac{1}{n} + v\left(1 - \frac{1}{n^2}\right) + O(v^2) \tag{6-18.8}$$

This result agrees, to first order in v with that of Fresnel. Note that if the signal is transmitted transverse to the motion of the material (for example, with $u_{\text{II}} = 1/n$), the observed velocity in I has components $u_{\text{I}}^{\ 1}$ and $u_{\text{I}}^{\ 2}$, given by

$$u_{\text{I}}^{\ 1} = v, \qquad u_{\text{I}}^{\ 2} = \sqrt{1 - \mathbf{v}^2}\left(\frac{1}{n}\right) \simeq \frac{1}{n} - \frac{1}{2}\frac{v^2}{n} + O(v^4).$$

There is thus a transverse second-order correction to the velocity in this case.

Problem 6.18. Derive an expression for the aberration of light coming from distant stars, using the Einstein velocity transformation formulas.

Problem 6.19. The relative velocity between two particles can be defined as the velocity of one of them, as measured in a reference frame in which the other particle is at rest. Is this definition reflexive, that is, if we reverse the roles of the two particles, do we get the same value for the relative velocity (with an appropriate change in sign)?

Problem 6.20. If $\mathbf{u}_1$ and $\mathbf{u}_2$ are the velocities of two particles as measured in a given frame and $\mathbf{u}_{12}$ is their relative velocity in this frame, show that

$$\mathbf{u}_{12}^2 = \frac{(\mathbf{u}_1 - \mathbf{u}_2)^2 - (\mathbf{u}_1 \times \mathbf{u}_2)^2}{(1 - \mathbf{u}_1 \cdot \mathbf{u}_2)^2}$$

where the dot and cross product have their usual meanings.

Problem 6.21. In the previous problem let the two velocities be $\mathbf{u}$ and $\mathbf{u} + d\mathbf{u}$. Show that the magnitude, dw, of the relative velocity is given in terms of u, the magnitude of $\mathbf{u}$ and θ, ϕ, the polar angle and azimuthal angle of $\mathbf{u}$ by

$$dw^2 = \frac{du^2}{\sqrt{1-u^2}} + \frac{u^2}{\sqrt{1-u^2}}(d\theta^2 + \sin^2\theta\, d\phi^2).$$

Note that if we introduce the variable $\psi = \tanh^{-1}\mathbf{u}$, du^2 can be written in the form

$$dw^2 = d\psi^2 + \sinh^2\psi\{d\theta^2 + \sin^2\theta\, d\phi^2\}.$$

Geometrically this expression has the form of the expression for ds^2 in a three-dimensional Lobachevski space, that is, a space of constant negative curvature.

Problem 6.22. Show that the components of the four-acceleration $\ddot{z}^u$ are given in terms of the three-velocity $\mathbf{u}$ and the three-acceleration $\mathbf{a} = d\mathbf{u}/dz^0$ by

$$\ddot{z}^\mu = \frac{1}{1-\mathbf{u}^2}\left\{\frac{\mathbf{u}\cdot\boldsymbol{a}}{1-\mathbf{u}^2},\quad \boldsymbol{a} + \frac{\mathbf{u}(\mathbf{u}\cdot\boldsymbol{a})}{1-\mathbf{u}^2}\right\}.$$

From this result derive the transformation law for $\mathbf{a}$ analogous to the transformation law (6-18.4) for $\mathbf{u}$.

6-19. Causality

So far, the only requirement we have placed on a set of dynamical laws is that they admit the Poincaré group as a symmetry group. As we said, this requirement is directly related to the existence of certain simple structures that can exist in space-time. One of these structures consists of the light cones at each space-time point. The other consists of the totality of timelike geodesics through each point of space-time, with the property that the totality of such geodesics is mapped into itself under a Lorentz mapping. We interpreted the elements of this latter structure as the trajectories of free particles. However, there is a third invariant structure that has not been assigned a physical interpretation. This structure consists of all spacelike geodesics through a point. For such trajectories the magnitude of the three-velocity is greater than unity. *A priori*, there is no reason why the spacelike trajectories should not have their physical counterparts. It is only when we bring in the requirement of causality that we can rule out such trajectories.

The timelike trajectories (as well as the null trajectories that lie in a light cone) have the property that if two points specified by parameter values τ and $\tau + d\tau$ are such that $\dot{z}^0 > 0$, then under any orthochronous mapping, $\dot{z}'^0 > 0$. Indeed we have that

$$\dot{z}'^0 = \frac{1}{\sqrt{1 - v^2}} \{\dot{z}^0 - v\dot{z}^1\} = \frac{\dot{z}^0}{\sqrt{1 - v^2}} \{1 - vu^1\},$$

for the special Lorentz mapping with velocity v in the x^1 direction. Since $\mathbf{u}^2 < 1$, it follows that $|u^1| < 1$. Also $|v| < 1$. Thus if $\dot{z}^0 > 0$, then $\dot{z}'^0 > 0$. We can interpret this result by saying that if the point $\tau + d\tau$ is later than τ as observed in one reference frame, it will be later as observed in all other frames that do not reverse the overall sense of time.[42] Thus the point $\tau + d\tau$ can be said to be absolutely later than the point τ.

Let us now consider a spacelike trajectory. Since $\mathbf{u}^2 > 1$ for such trajectories it follows that $|u^1| > 1$ for some spacelike trajectory as measured in a particular reference frame. Let us suppose that $\dot{z}^0 > 0$ in this frame. Then there will exist other frames in which $\dot{z}'^0 < 0$, namely, those frames moving with a velocity v relative to the initial frame such that $vu^1 > 1$. In these frames the point $\tau + d\tau$ will appear earlier than τ. Of course, there will be other spacelike trajectories for which the earlier-than–later-than relation is preserved.

If we say that the arrival of a particle at $\tau + d\tau$ is caused by its having previously been at the point τ on its trajectory, then the sequence of cause and effect is not preserved for a spacelike trajectory by all Lorentz mappings and hence is not an invariant concept for these trajectories. The requirement of the absolute invariance of the cause-effect relation for all physical systems thus forces us to rule out all spacelike trajectories from physical consideration. We are thus led to the requirement, logically independent of the requirements of the relativity principle, that the velocity of propagation of any physical effect that connects cause and effect must be less than or equal to the velocity of light. The velocity of light is thus a limiting velocity for the propagation of all physical effects and hence for the propagation of information. We will take this requirement to be the second requirement on all physical laws. It will be referred to as the requirement of *causality*. We should point out that this requirement is not entirely free of ambiguity in its formulation. Especially in relativistic quantum field theory, a completely satisfactory formulation has still to be achieved.

[42] A nonorthochronous Lorentz mapping would, of course, invert the earlier-than–later-than relation between points on a timelike trajectory. But such a mapping will invert this relation for *all* such trajectories.

7

Relativistic Particle Dynamics

The problem of constructing a theory of interacting particles is a relatively simple matter in the Newtonian world-picture, the action-at-a-distance description of the gravitational interaction being the most famous example. Because of the structure of these equations it is possible to derive laws of conservation of energy, of momentum, etc., associated with the Galilean symmetries that are local in time. One can also formulate a well-defined initial value problem, express the theory in Hamiltonian form, develop a statistical mechanics, and quantize such a theory. All of this changes when we come to consider a relativistic system of interacting particles. The essential reason for this change is that the planes of absolute simultaneity of the Newtonian world-picture get replaced by light cones in special relativity. Consequently particles can no longer interact with each other instantaneously except on direct contact and must interact via retarded or advanced interactions, or both. It is still possible to derive conservation laws for such a system corresponding to the Poincaré symmetries, but they will be nonlocal in time. Furthermore, to date, no well-defined initial value problem has been posed for such a system nor does a statistical mechanics or a quantum mechanics exist for it. Nevertheless an action-at-a-distance description of interacting particles has certain advantages over the more customary field description that make its study worth while. In particular it does not require a particle to interact with itself and thereby avoids the self-energy problems of the field description.

As an example of a relativistic system of interacting particles we will consider the action-at-a-distance theory of the electromagnetic interaction of charged particles. The resulting equations of motion for a two-body system will be shown to possess at least one class of well-behaved solutions and the theory as a whole will be shown to reproduce most of the results of Maxwell-Lorentz electrodynamics. However, since a particle does not interact with itself, the problem of radiation and radiation reaction requires additional, essentially cosmological, assumptions. We will also consider the problem of describing the gravitational interaction by an action-at-a-distance theory within the framework of special relativity.

7-1. The Dynamics of a Free Particle

We saw in Section 6-8 that the trajectory of a free particle can be associated with a timelike geodesic. Equation (6-8.1) can therefore be derived from a variational principle, with an action S given by

$$S = -\alpha \int_{\lambda_1}^{\lambda_2} ds, \tag{7-1.1}$$

where α is a constant to be determined and

$$ds^2 = g_{\mu\nu}\dot{z}^\mu \dot{z}^\nu \, d\lambda^2$$

At this point the path parameter λ is left unspecified. We cannot take it to be the proper time along a trajectory, since to do so would limit the variations to those that are compatible with the condition $g_{\mu\nu}\dot{z}^\mu\dot{z}^\nu = 1$. The equations of motion associated with the action S are obtained by requiring that $\delta S = 0$ for variations of the dpt that vanish at the end points 1 and 2. For such variations, with $g_{\mu\nu} = \eta_{\mu\nu}$,

$$\delta S = 2\alpha \int_{\lambda_1}^{\lambda_2} \frac{d}{d\lambda}\left\{\frac{\eta_{\mu\nu}\dot{z}^\mu}{\sqrt{\eta_{\alpha\beta}\dot{z}^\alpha\dot{z}^\beta}}\right\} \delta z^\nu \, d\lambda.$$

Hence δS will be zero if

$$\frac{d}{d\lambda}\left\{\frac{\dot{z}^\mu}{\sqrt{\eta_{\alpha\beta}\dot{z}^\alpha\dot{z}^\beta}}\right\} = 0.$$

At this point we are now free to take the path parameter along a dpt to be the proper time τ, in which case these equations reduce to

$$\ddot{z}^\mu = 0,$$

in agreement with Eqs. (6-8.1).

Let us now express the action (7-1.1) in terms of the three-velocity $\mathbf{u}$, with components dz^r/dz^0. With $g_{\mu\nu} = \eta_{\mu\nu}$, it becomes

$$S = -\alpha \int_{z_1^0}^{z_2^0} \sqrt{1 - \mathbf{u}^2} \, dz^0. \tag{7-1.2}$$

Setting $S = \int L\, dz^0$, we see that the free-particle Lagrangian is

$$L = -\alpha\sqrt{1 - \mathbf{u}^2}, \tag{7-1.3}$$

so that, in the limit of small velocities,

$$L \simeq -\alpha + \frac{\alpha \mathbf{u}^2}{2} + \cdots.$$

If we compare this expression with the classical expression for the Lagrangian of a free particle, that is,

$$L_c = \tfrac{1}{2} m \mathbf{u}^2,$$

we see that, aside from a constant factor which is immaterial in an action principle, the approximation relativistic Lagrangian of the particle agrees with the classical Lagrangian, provided we identify α with the mass of the particle. Actually this identification is meaningful only in the limit $\mathbf{u} \to 0$, so we shall identify α with the *rest mass* of the particle, that is, its observed mass when the particle is at rest. Thus, writing m_0 for the rest mass, the relativistic Lagrangian is given by

$$L = -m_0\sqrt{1 - \mathbf{u}^2}.$$

The three-momentum, $\mathbf{p}$, is defined as usual by

$$\begin{aligned} \mathbf{p} &= \frac{\partial L}{\partial \mathbf{u}} \\ &= \frac{m_0 \mathbf{u}}{\sqrt{1 - \mathbf{u}^2}}. \end{aligned} \tag{7-1.4}$$

If we define the mass of the particle in the usual way as being the ratio of the momentum of the particle to its velocity, we see that (as in classical mechanics) the definition leads to a unique result, since here also the momentum is parallel to the velocity. However, we now have as the expression for the relativistic mass m,

$$m = \frac{p^r}{u^r} = \frac{m_0}{\sqrt{1 - \mathbf{u}^2}} \tag{7-1.5}$$

This expression for the relativistic mass of the particle, unlike its classical counterpart $m = m_0$, is velocity-dependent and increases with increasing velocity.

The energy is again defined to be equal to the Hamiltonian of the system so that

$$\mathscr{E} = H = \mathbf{p} \cdot \mathbf{u} - L$$
$$= \frac{m_0}{\sqrt{1 - \mathbf{u}^2}}, \tag{7-1.6}$$

which, for small velocities, reduces to

$$\mathscr{E} \simeq m_0 + \tfrac{1}{2} m_0 \mathbf{u}^2 + \cdots$$

and differs from the usual classical expression for the energy of a free particle by the term m_0. This term corresponds to the rest energy of the particle. However, one should not assign too much significance to this term at present, since the energy is defined only to within an additive constant, both in Newtonian mechanics and in special relativity. Some authors define the kinetic energy T to be

$$T = \mathscr{E} - m_0 .$$

We now derive an important relation between the relativistic energy and momentum of a free particle. Since

$$\mathscr{E}^2 = \frac{m_0^{\,2}}{1 - \mathbf{u}^2} \quad \text{and} \quad \mathbf{p}^2 = \frac{m_0^{\,2}\mathbf{u}^2}{1 - \mathbf{u}^2},$$

we have

$$\mathscr{E}^2 = \mathbf{p}^2 + m_0^{\,2}. \tag{7-1.7}$$

Also we have directly

$$\mathbf{p} = \mathscr{E}\mathbf{u}.$$

For a finite rest-mass particle, both $\mathbf{p}$ and $\mathscr{E} \to \infty$ as $\mathbf{u}^2 \to 1$, so that the velocity of a material particle can never exceed the velocity of light. This result is directly related to the fact that a timelike trajectory can never be mapped onto a spacelike trajectory by a Lorentz mapping.

It is possible, in special relativity, to also deal with zero rest-mass particles, for example, photons or neutrinos. Zero rest-mass particles travel along lightlike trajectories with finite momenta and energy. For such particles the expressions (7-1.4) and (7-1.6) for $\mathbf{p}$ and $\mathscr{E}$ in terms of $\mathbf{u}$ cannot be used. Nevertheless, relation (7-1.7) is still assumed to be valid, with $m_0 = 0$.

Expressed in terms of the momenta and for arbitrary rest mass, the Hamiltonian for a free particle is

$$H = \sqrt{\mathbf{p}^2 + m_0^{\,2}}. \tag{7-1.8}$$

For momenta $\mathbf{p}^2 \ll m_0{}^2$, corresponding to velocities $\mathbf{u}^2 \ll 1$, H approximates the classical expression for the Hamiltonian of a free particle plus a rest-mass term:

$$H \simeq m_0 + \frac{\mathbf{p}^2}{2m_0} + \cdots.$$

7-2. Conservation Laws for a Free Particle

The action for a free particle is, with $g_{\mu\nu} = \eta_{\mu\nu}$,

$$S = -m_0 \int_{\lambda_1}^{\lambda_2} \sqrt{\eta_{\mu\nu}\dot{z}^\mu \dot{z}^\nu}\, d\lambda. \tag{7-2.1}$$

For a symmetry mapping in which

$$z'^\mu(\lambda) = \Lambda_\nu{}^\mu z^\nu(\lambda) + a^\mu,$$

we must have

$$S' = -m_0 \int_{\lambda_1}^{\lambda_2} \sqrt{\eta_{\mu\nu}\dot{z}'^\mu \dot{z}'^\nu}\, d\lambda = S$$

if, when $z^\mu(\lambda)$ is a dpt, $z'(\lambda)$ is also a dpt. This requirement leads in the case of infinitesimal mapping to the result that

$$\frac{\partial L}{\partial \dot{z}^\mu}\, \bar{\delta} z^\mu \Big|_{\lambda_1}^{\lambda_2} = 0, \tag{7-2.2}$$

where

$$\bar{\delta} z^\mu = \varepsilon_\nu{}^\mu z^\nu + \varepsilon^\mu.$$

If we define

$$p_\mu \equiv -\frac{\partial L}{\partial \dot{z}^\mu} = \frac{m_0 \eta_{\mu\nu}\dot{z}^\nu}{\sqrt{\eta_{\alpha\beta}\dot{z}^\alpha \dot{z}^\beta}},$$

Eq. (7-2.2) takes the form

$$\{p_\mu z^\nu \varepsilon_\nu{}^\mu + p_\mu \varepsilon^\mu\} \Big|_{\lambda_1}^{\lambda_2} = 0. \tag{7-2.3}$$

Corresponding to the translations, the quantities p_μ are conserved along the trajectory. The four-momentum $p^\mu = \eta_{\mu\nu} p_\nu$ has as its spatial components the three components of linear momentum, and the fourth component is the energy of the particle. Thus

$$p^\mu = (\mathscr{E}, \mathbf{p}); \tag{7-2.4}$$

the energy and momentum of a particle together form the components of a four-vector, and hence we know how they transform under a Lorentz mapping. Note that the scalar $\eta_{\mu\nu}p^\mu p^\nu$ is just

$$\eta_{\mu\nu}p^\mu p^\nu = m_0{}^2\eta_{\mu\nu}\dot{z}^\mu\dot{z}^\nu = m_0{}^2,$$

which is another way of writing $\mathscr{E}^2 - \mathbf{p}^2 = m_0{}^2$.

Corresponding to the "rotations," the quantities

$$l^{\mu\nu} = (z^\mu p^\nu - z^\nu p^\mu)$$

are conserved along a trajectory. The spatial components l^{rs} are just the components of the three-dimensional angular-momentum vector

$$\mathbf{l} = \mathbf{z} \times \mathbf{p}, \qquad \mathbf{z} = \{z^1, z^2, z^3\},$$

so that

$$l^{12} = (z^1p^2 - z^2p^1) = l^3, \quad \text{etc.}$$

The mixed component l^{r0} is given by

$$l^{r0} = (p^0z^r - p^rz^0) = z_0{}^rp^0,$$

where $z_0{}^r$ is the value of z^r for $z^0 = 0$.

7-3. Relativistic Force

If a particle deviates from a straight path, we can conclude that it is being acted upon by a force. As in Newtonian mechanics, we can take as a measure of this force the change in $\mathbf{p}$ with respect to z^0 along the particle's trajectory. Thus at any point on this trajectory we will have acting a force $\mathbf{F}$, given by

$$\begin{aligned}\mathbf{F} &= \frac{d}{dz^0}\left\{\frac{m_0\mathbf{u}}{\sqrt{1-\mathbf{u}^2}}\right\} \\ &= \frac{m_0}{\sqrt{1-\mathbf{u}^2}}\frac{d\mathbf{u}}{dz^0} + \frac{m_0\mathbf{u}\cdot(d\mathbf{u}/dz^0)}{(1-\mathbf{u}^2)^{3/2}}\,\mathbf{u}. \end{aligned} \tag{7-3.1}$$

If the velocity changes in direction only and not in magnitude, the second term is zero, since

$$\frac{d}{dz^0}\mathbf{u}^2 = 2\mathbf{u}\cdot\frac{d\mathbf{u}}{dz^0}.$$

In this case, then $\mathbf{F}\cdot\mathbf{u} = 0$; that is, the force is perpendicular to the velocity. For this reason $m_0/\sqrt{1-\mathbf{u}^2}$ is sometimes referred to as the "transverse" mass of the particle. If, on the other hand, the velocity changes in magnitude only so that the force is parallel to the velocity, then

$$\mathbf{F}\cdot\mathbf{u} = \frac{m_0}{\sqrt{1-\mathbf{u}^2}}\frac{d\mathbf{u}}{dz^0}\cdot\mathbf{u} + \frac{m_0}{(1-\mathbf{u}^2)^{3/2}}\mathbf{u}^2\mathbf{u}\cdot\frac{d\mathbf{u}}{dz^0}$$

$$= \frac{m_0}{(1-\mathbf{u}^2)^{3/2}}\frac{d\mathbf{u}}{dz^0}\cdot\mathbf{u},$$

so that

$$\mathbf{F} = \frac{m_0}{(1-\mathbf{u}^2)^{3/2}}\frac{d\mathbf{u}}{dz^0}.$$

For this reason $m_0/(1-\mathbf{u}^2)^{3/2}$ is sometimes called the "longitudinal" mass of the particle.

As might be expected, the three-force $\mathbf{F}$ has no simple transformation properties. One can, however, introduce a related four-force f^μ, the Minkowski force, defined as

$$f^\mu = \frac{d}{d\tau}p^\mu. \tag{7-3.2}$$

Now

$$\frac{d}{d\tau}p^\mu = \frac{1}{\sqrt{1-\mathbf{u}^2}}\frac{d}{dz^0}p^\mu,$$

so

$$f^\mu = \frac{1}{\sqrt{1-\mathbf{u}^2}}\left(\frac{d}{dz^0}\mathscr{E}, \frac{d}{dz^0}\mathbf{p}\right). \tag{7-3.3}$$

But $(d/dz^0)\mathbf{p} = \mathbf{F}$, while

$$\frac{d}{dz^0}\mathscr{E} = \mathbf{u}\cdot\frac{d}{dz^0}\mathbf{p} = \mathbf{u}\cdot\mathbf{F}, \tag{7-3.4}$$

and hence

$$f^\mu = \frac{1}{\sqrt{1-\mathbf{u}^2}}(\mathbf{u}\cdot\mathbf{F}, \mathbf{F}). \tag{7-3.5}$$

The quantity $\mathbf{u}\cdot\mathbf{F}$ appearing as the zeroth component of the Minkowski four-force can thus be interpreted as the rate at which work is being done on the particle. This interpretation agrees with our interpretation of $\mathscr{E}$ as the energy of the particle, since by Eq. (7-3.4), the increase in $\mathscr{E}$ is just equal to $\mathbf{u}\cdot\mathbf{F}$.

Problem 7.1. Show that f^μ, given by Eq. (7-3.5) satisfies

$$\eta_{\mu\nu} f^\mu \dot{z}^\nu = 0.$$

Problem 7.2. By making use of Eq. (6-5.22), show that for a general velocity mapping $V(\mathbf{v})$,

$$\mathbf{F} \to \mathbf{F}' = \frac{\gamma^{-1}\mathbf{F} + \dfrac{\mathbf{v}(\mathbf{v}\cdot\mathbf{F})}{\mathbf{v}^2}[1 - \gamma^{-1}] + \mathbf{v}(\mathbf{u}\cdot\mathbf{F})}{1 + \mathbf{u}\cdot\mathbf{v}}$$

In Section 7-1 we defined the relativistic mass of a particle as the ratio of its momentum to its velocity. We can maintain this definition even if the particle is acted upon by external forces, and so obtain the same expression for m as before. We also saw that the energy of a free particle was, to within an additive constant, numerically equal to this relativistic mass. While the value of this constant has no significance, we see that a change in the energy of a particle due to the action of force is numerically equal to the change in its relativistic mass:

$$\Delta\mathscr{E} = \Delta m,$$

where $\Delta\mathscr{E} = \mathscr{E}_2 - \mathscr{E}_1$, the difference in energy of the particle before and after a force acts on it. Likewise $\Delta m = m_2 - m_1$ is the difference in mass of the particle due to the action of the force. Thus we see that there exists in relativistic mechanics a very close relation between the mechanical energy of a particle and its mass. We shall return to this relation when we come to consider a system of particles.

Experimental evidence for the increase of mass with energy was first presented by Kaufmann.[1] However, his results were not precise enough to establish the exact dependence on velocity. Later experiments by Bucherer and others[2] verified the velocity dependence of m on $\mathbf{u}$ as given by Eq. (7-1.5) for values of u in the range 0.38 to 0.69. Recently Rogers, McReynolds, and Rogers[3] checked the mass formula for electrons with velocities up to 0.8. The mass formula has also been checked indirectly in high-energy accelerators. As the velocity of the particle approaches unity, one must alter the synchronism mechanism of an accelerator to take account of the mass increase. Since accelerators in the BeV region have been designed to take account

[1] W. Kaufmann, *Nachr. Ges. Wiss. Göttingen*, 143 (1901), 291 (1902), 90 (1903); *Ber. Sitz.*, **45**, 949 (1905); *Ann. der Phys.*, **19**, 487 (1906), and **20**, 639 (1906).

[2] A. H. Bucherer, *Verh. d. Deutschen Phys. Ges.*, **6**, 688 (1908); *Phys. Zeits.*, **9**, 755 (1908). Also E. Hupka, *Ann. der Phys.*, **31**, 169 (1910); G. Neumann, *ibid.*, **45**, 529 (1914); C. Schäfer, *ibid.*, **49**, 934 (1916); Ch. E. Guye and Ch. Lavanchy, *Arch. des Sci. Phys. natur., Geneva*, **41**, 286, 353, 441 (1916).

[3] M. M. Rogers, A. W. McReynolds, and F. T. Rogers, Jr., *Phys. Rev.*, **57**, 379 (1940).

of this mass increase and operate satisfactorily, we can take it that the relativistic dependence of mass on velocity has been established.

Problem 7.3. What is the maximum energy one could get out of a fixed-frequency electron cyclotron?

As in the nonrelativistic case, the introduction of an external force must be considered as an approximation in which, for the particle under consideration, one neglects the action of the particle on the sources producing this force. The question of when a given force has this property is a much more difficult problem in relativistic mechanics and will be deferred until after we discuss the problem of interacting particles. For the present we will assume that such is the case. Then Eq. (7-3.1) or equivalently Eq. (7-3.2), can be considered to be the relativistic analogue of Newton's second law when $\mathbf{F}$ or f_μ are given.

In general the relativistic equations of motion with a given force are more difficult to solve than the corresponding nonrelativistic problem with the same force because of the factors $(1 - \mathbf{u}^2)^{-1/2}$ that occur in these equations. One problem that can be solved and which is of some interest is the case of so-called *hyperbolic motion*. This is the motion produced by a constant three-force. Restricting ourselves to one-dimensional motion we can write Eq. (7-3.1) as

$$\frac{d}{dz^0}\frac{u}{\sqrt{1-u^2}} = k,$$

so that

$$\frac{u}{\sqrt{1-u^2}} = kz^0 + \text{const.}$$

We will fix the constant of integration by taking $u = 0$ for $z^0 = 0$. Then, solving for u, we get

$$u = \frac{kz^0}{\sqrt{1-k^2z^{02}}}.$$

A final integration then yields the result that

$$z^1 = \frac{1}{k}\sqrt{1+k^2z^{02}},$$

so that

$$z^{12} - z^{02} = \frac{1}{k^2},$$

that is, the trajectory of the particle is a hyperbola in space-time.

We see in the above example that the velocity of the particle asymptotically approaches unity, in contrast to the corresponding Newtonian case where it diverges. One can attribute this behavior to the increase in mass of the particle with velocity. As u increases, m increases, so that the force, being constant, is less effective in producing a given change per unit time in u at high velocities than at low velocities, although it always produces the same change in the momentum per unit time.

Problem 7.4. Solve the equations of motion of the one-dimensional "relativistic" oscillator; that is, take $F = -kz^1$ in Eq. (7-3.1).

Problem 7.5. Discuss the relativistic Kepler problem (see Section 7-12). Show in particular that the orbits are precessing ellipses. Derive an expression for the rate of precession of the perihelion of the orbit.

7-4. Relativistic Systems of Particles

Within the framework of Newtonian mechanics it is a relatively simple matter to formulate a comprehensive action principle for a system of interacting particles and, with the help of Nöther's identity, to derive conservation laws associated with the symmetries of the system. In the final analysis, this simplicity is due to the existence of the planes of absolute simultaneity in the Newtonian world picture. These planes, in turn, make possible the instantaneous interactions between particles dealt with in that picture. In special relativity the problem of interacting particles is much more complicated. There are no planes of absolute simultaneity and consequently no possibility of instantaneous interactions in this world picture. In fact, from our discussion of Section 6-19 we can conclude that all interaction effects are propagated through space with a finite velocity less than or equal to the speed of light.

In spite of the above-mentioned difficulties it is possible to formulate an action principle for a relativistic system of interacting particles. The action S for the system is taken to be of the form

$$S = -\sum_i m_{0i} \int_{\sigma_1}^{\sigma_2} \sqrt{\eta_{\mu\nu} \dot{z}_i^{\mu} \dot{z}_i^{\nu}}\, d\lambda_i + \sum_{i \neq j}\sum \iint_{(\sigma_1)}^{(\sigma_2)} V_{ij}\, d\lambda_i\, d\lambda_j . \tag{7-4.1}$$

The first term in this expression for S is just the sum of free-particle actions of the form (7-1.1), one for each particle of the system. The integrals in this term are over those parts of the particle trajectories that lie between two nonintersecting spacelike surfaces σ_1 and σ_2 (see Fig. 7.1). The second term in the expression for S represents the interaction part of the action: for simplicity we have restricted ourselves to two-body interactions. The integrals in this interaction term extend far enough beyond σ_1 and σ_2 to include all effects of any one particle on the portions of the trajectories of the other particles lying between σ_1 and σ_2. The limits of these integrals will therefore depend not only on σ_1 and σ_2, but also on the form of the interaction Lagrangian V_{ij}. We have indicated this dependence by the notation (σ_1) and (σ_2).

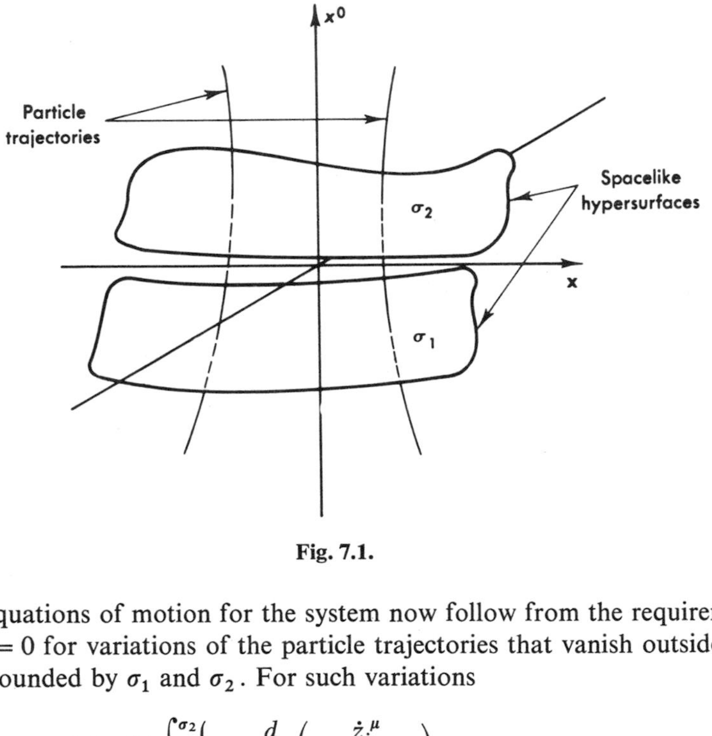

Fig. 7.1.

The equations of motion for the system now follow from the requirement that $\delta S = 0$ for variations of the particle trajectories that vanish outside the region bounded by σ_1 and σ_2. For such variations

$$\delta S = \sum_i \int_{\sigma_1}^{\sigma_2} \left\{ m_{0i} \frac{d}{d\lambda_i} \left(\frac{\dot{z}_i^\mu}{\sqrt{\eta_{\alpha\beta}\dot{z}_i^\alpha \dot{z}_i^\beta}} \right) \right.$$

$$\left. + \sum_{j \neq i} \int_{(\sigma_1)}^{(\sigma_2)} \left(\frac{\partial V_{ij}}{\partial z_i^\mu} - \frac{d}{d\lambda_i} \left(\frac{\partial V_{ij}}{\partial \dot{z}_i^\mu} \right) \right) d\lambda_j \right\} \delta z_{i\mu} \, d\lambda_i . \tag{7-4.2}$$

Consequently, δS will be zero, provided

$$m_{0i}\ddot{z}_i^\mu + \sum_{j \neq i} \int_{(\sigma_1)}^{(\sigma_2)} \left\{ \frac{\partial V_{ij}}{\partial z_i^\mu} - \frac{d}{d\tau_i} \left(\frac{\partial V_{ij}}{\partial \dot{z}_i^\mu} \right) \right\} d\tau_i = 0, \tag{7-4.3}$$

where now we have taken the path parameters to be the proper times τ_i along the trajectories.

Equations (7-4.3) also follow from the requirement that the *partial* action S_i, defined by

$$S_i = \int_{\sigma_1}^{\sigma_2} L_i(\lambda_i)\, d\lambda_i$$

$$= \int_{\sigma_1}^{\sigma_2} \left\{ -m_{0i}\sqrt{\eta_{\mu\nu}\dot{z}_i^{\mu}\dot{z}_i^{\nu}} + \sum_{j\neq i} \int_{(\sigma_1)}^{(\sigma_2)} V_{ij}\, d\lambda_j \right\} d\lambda_i \qquad (7\text{-}4.4)$$

be stationary for variations of $\dot{z}_i^{\mu}$ that vanish outside the region bounded by σ_1 and σ_2. This requirement leads to the equations

$$\frac{\delta L_i}{\delta z_i^{\mu}} \equiv \frac{\partial L_i}{\partial z_i^{\mu}} - \frac{d}{d\tau_i}\left(\frac{\partial L_i}{\partial \dot{z}_i^{\mu}}\right) = 0, \qquad (7\text{-}4.5)$$

which are equivalent to Eqs. (7-4.3). Equations of this type are known as *Fokker-type equations of motion*, after the equations developed by Fokker to describe the electromagnetic interaction.

The Fokker-type equations of motion (7-4.5) are the relativistic analogues of the Newtonian action-at-a-distance equations. However, they differ fundamentally in their structure. The Newtonian equations are ordinary differential equations for which there is a well-defined initial value problem. Given the values of the particle positions and velocities on one plane of absolute simultaneity, it is possible to find their values on a later plane by an integration process. The relativistic equations of motion (7-4.5), however, are integro-differential equations because of the integrals appearing in the partial Lagrangian L_i defined by Eq. (7-4.4). As a consequence it does not appear possible to formulate a well-defined initial value problem for these equations. In Fig. 7.2 we have pictured the trajectories of two particles and a light cone whose apex lies between them. Since physical effects can propagate with at most the speed of light, it is clear that the motion of particle 1 between the points 1a and 1b, where its trajectory intersects the light cone, can have no effect on particle 2 as it moves between the points 2a and 2b, and vice versa. It appears, therefore, that these two portions of the trajectories of the two particles can be completely arbitrary. Once they are specified, one can then make use of the equations of motion to determine the remaining portions of the two trajectories by an integration process. It is also clear from Fig. 7.2 that a knowledge of the positions and velocities of the particles on an $x^0 = \text{const}$ or any other spacelike surface is insufficient to determine the forces action on the particles on such a surface, and hence it is difficult to calculate their accelerations on this surface.

Thus, far from being able to characterize the dpt of a system of particles by a finite number of parameters, it appears that we need to specify finite segments of each of their trajectories. Because of this property of the equations of motion for a relativistic system of particles, it has not been possible, to date, to recast these equations in Hamiltonian form. As a consequence, one encounters as yet unsolved problems in constructing a relativistic statistical mechanics

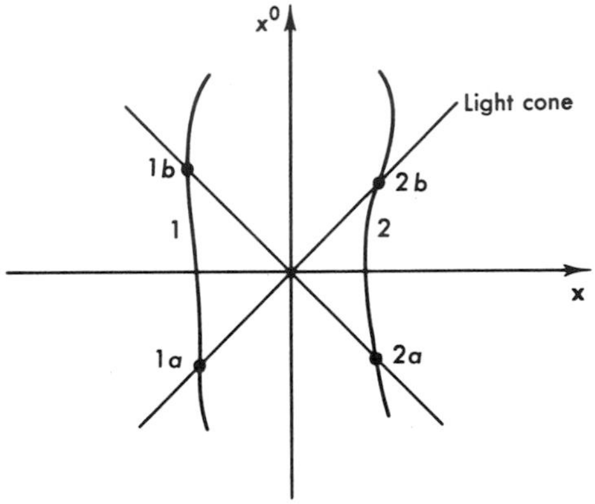

Fig. 7.2.

or a quantum theory for such a system, both of which appear to require equations of motion in Hamiltonian form. In spite of these difficulties it is possible to formulate an action-at-a-distance theory of the electromagnetic interactions between charged particles and to find solutions to the equations of motion of this theory. The great advantage of this and all action-at-a-distance theories over the more customary field-theory descriptions of particle interactions is that in these theories a given particle interacts only with the other particles in the system and not with itself. One thereby avoids the usual self-energy problems of the field-theory descriptions.

7-5. Conservation Laws for Relativistic Systems of Particles

From our previous discussions we can expect that the requirement of Poincaré invariance for the system with equations of motion (7-4.3) will lead to conservation laws. However, since S is a nonlocal action due to the occurrence of the double integrals appearing in Eq. (7-4.1), we cannot make direct use of the Nöther identity to obtain these laws as we did in Newtonian mechanics but must proceed anew. The conservation laws will again follow from the requirement that the value of S for a given kpt be equal to its value for the transform of this kpt under a Poincaré mapping. For infinitesimal mappings $x^\mu \to x'^\mu = x^\mu + \varepsilon_\nu{}^\mu x^\nu + \varepsilon^\mu$, the difference $\delta S = S' - S$ must therefore be zero for all kpt. This requirement on S then leads, for the dpt, to a set of conservation laws of the form

$$\sum_i (\mathfrak{P}_{i\mu}\varepsilon^\mu + \mathfrak{L}_{i\mu\nu}\varepsilon^{\mu\nu})\Big|_{\sigma_1}^{\sigma_2} = 0. \tag{7-5.1}$$

In general the quantities $\mathfrak{P}_{i\mu}$ and $\mathfrak{L}_{i\mu\nu}$ are complicated functions of the dpt involving integrals over these trajectories, analogous to those appearing in the expression (7-4.1) for S, and are not simply functions of the particle coordinates and velocities on the surface σ_1 and σ_2, as in the Newtonian case. They have the general form

$$\mathfrak{P}_{i\mu} = p_{i\mu} + \text{nonlocal term} \tag{7-5.2}$$

and

$$\mathfrak{L}_{i\mu\nu} = l_{i\mu\nu} + \text{nonlocal terms}, \tag{7-5.3}$$

where

$$l_{i\mu\nu} = z_{i\mu} p_{i\nu} - z_{i\nu} p_{i\mu}$$

and where the nonlocal (in time) terms indicated in these equations arise from the interaction terms in S.

Later (see Section 7-9), we will give explicit expressions for these nonlocal terms for the case of the electromagnetic interaction. However, here we will content ourselves with the application of the conservation laws (7-5.1) to the case of scattering problems, where we will not need their explicit form. We consider first the case of elastic scattering.

In a typical elastic scattering problem, two or more particles approach each other, interact for a period, and then recede from each other, while their individual identities remain constant. In particular, the parameters associated with the particles (for example, charge, rest mass, and spin), remain unchanged during the scattering process. To apply the conservation laws (7-5.1) to this situation we assume that σ_1 is a spacelike surface on which the particles involved can be considered to be noninteracting and lying wholly in the past of the region of interaction (see Fig. 7.3). Likewise σ_2 is another spacelike surface on which the particles are noninteracting and lying wholly in the future of the region of interaction. On these surfaces the contribution to $\mathfrak{P}_{i\mu}$ coming from the interaction part of the action S is assumed to be negligible compared with the contribution from the free-particle part. In this case $\mathfrak{P}_{i\mu}$ is equal to $p_{i\mu}$ and we have, corresponding to the translational invariance of the system, the conservation laws

$$\sum_{\substack{i \\ \text{initial}}} p_{i\mu} = \sum_{\substack{i \\ \text{final}}} p_{i\mu}, \tag{7-5.4}$$

where "initial" and "final" refer to the values of $p_{i\mu}$ on σ_1 and σ_2, respectively, They represent the conservation of linear momentum ($\mu = 1, 2, 3$) and energy ($\mu = 0$) for the system, Equation (7-5.4) is obviously an invariant statement to the extent that, under a Poincaré mapping, the region of interaction does not intersect σ_1 or σ_2. Furthermore the surfaces σ_1 and σ_2 are themselves restricted only by the requirement that they be spacelike and not pass through the region of interaction.

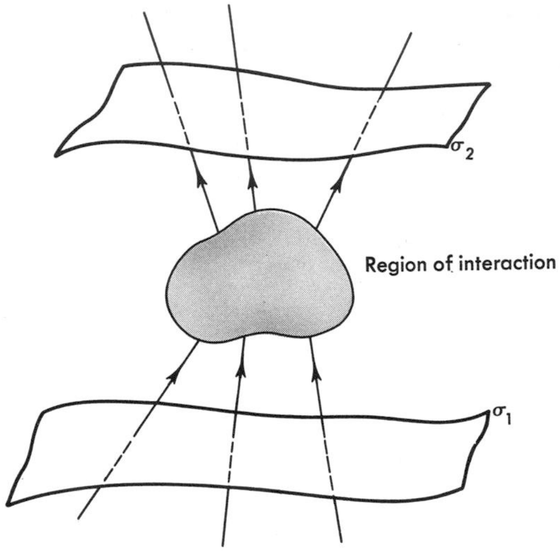

Fig. 7.3.

In describing a collision process it is sometimes convenient, as in the theory of dispersion relations, to work with the scalars that can be formed from the momenta involved. In particular, one is interested in constructing a set of algebraically independent scalars that can be varied independently and hence characterize a given collision process. We state without proof the following theorem:

> If n is the number of initial plus final particles involved in the collision, then the number of such scalars is $(3n - 10)$. In particular it is zero for $n = 2$ and 3.

Problem 7.6. Construct the two scalars that characterize a two-body scattering experiment ($n = 4$).

In addition to the conservation laws (7-5.4) for the energy and momentum of a system of particles, there is also a conservation law for angular momentum associated with the Lorentz invariance of the system. Under the same conditions for which Eqs. (7-5.4) are valid we therefore have

$$\sum_{\substack{i \\ \text{initial}}} l_i^{\mu\nu} = \sum_{\substack{i \\ \text{final}}} l_i^{\mu\nu}. \tag{7-5.5}$$

For μ, $\nu \neq 0$, Eqs. (7-5.5) are just the conservation laws for angular momentum. To interpret the r, 0 equations let us evaluate the two sums appearing therein on two spacelike surfaces $x^0 = x_1{}^0$ and $x^0 = x_2{}^0$. For these surfaces, $z_i{}^0 = x_1{}^0$ initially and $z_i{}^0 = x_2{}^0$ finally. Therefore, with this choice of spacelike surfaces, the r, 0 equations take the form

$$L_1^{0r} = L_2^{0r} = \text{const}, \tag{7-5.6}$$

where

$$L_1^{0r} = \sum_{\substack{i \\ \text{initial}}} (p_i{}^r x_1{}^0 - p_i{}^0 z_i{}^r),$$

and

$$L_2^{0r} = \sum_{\substack{i \\ \text{final}}} (p_i{}^r x_2{}^0 - p_i{}^0 z_i{}^r).$$

Likewise defining

$$P_1{}^\mu = \sum_{\substack{i \\ \text{initial}}} p_i{}^\mu \quad \text{and} \quad P_2{}^\mu = \sum_{\substack{i \\ \text{final}}} p_i{}^\mu,$$

we have, from Eq. (7-5.4), $P_1{}^\mu = P_2{}^\mu = P^\mu$.

With the help of P^μ we can rewrite Eq. (7-5.6) in the following form:

$$0 = L_2^{0r} - L_1^{0r} = (x_2{}^0 - x_1{}^0)P^r - \sum_{\substack{i \\ \text{final}}} z_i{}^r p_i{}^0 - \sum_{\substack{i \\ \text{initial}}} z_i{}^r p_i{}^0.$$

If we now divide this equation by P^0 we finally obtain the result that

$$\mathbf{Z}_{\text{final}} - \mathbf{Z}_{\text{initial}} = \mathbf{V}(x_2{}^0 - x_1{}^0), \tag{7-5.7}$$

where

$$\mathbf{Z}_{\text{final}} = \sum_{\substack{i \\ \text{final}}} \frac{\mathbf{z}_i p_i{}^0}{P^0}, \qquad \mathbf{Z}_{\text{initial}} = \sum_{\substack{i \\ \text{initial}}} \frac{\mathbf{z}_i p_i{}^0}{P^0} \tag{7-5.8}$$

and

$$\mathbf{V} = \mathbf{P}/P^0. \tag{7-5.9}$$

We can call $\mathbf{Z}$ the center of energy of the system of particles, in analogy to the center of mass as defined in Newtonian mechanics. According to Eq.

(7-5.7), this point moves like a free particle with a velocity $\mathbf{V}$, given by Eq. (7-5.9). If $P^\mu P_\mu = M^2 \geqq 0$, we can always perform a mapping so that $P'^\mu = (M, 0)$. The corresponding reference frame is called variously, the center of mass, or center of momentum, or center of energy frame. We prefer the latter terminology. In this frame $\mathbf{V} = 0$.

Problem 7.7. Show that

$$M \geqq \sum_{\substack{i \\ \text{initial}}} m_{0i} \quad \text{and} \quad M \geqq \sum_{\substack{i \\ \text{final}}} m_{0i}.$$

7-6. Elastic Scattering

Let us now consider the elastic scattering of two particles. Initially the particles are assumed to have energies and momenta $\mathscr{E}_{10}$, $\mathscr{E}_{20}$, $\mathbf{p}_{10}$, $\mathbf{p}_{20}$. With the help of the conservation laws we can then find the values of the energy and momenta of the particles after a collision has occurred in terms of these initial values and the scattering angle. As in the Newtonian case it is convenient to perform a mapping such that the total momentum of the two particles is zero. This can be accomplished by a Lorentz mapping $V(-\mathbf{v})$ with a velocity $\mathbf{v}$ given by Eq. (7-5.9). In this case

$$\mathbf{v} = \frac{\mathbf{p}_{10} + \mathbf{p}_{20}}{\mathscr{E}_{10} + \mathscr{E}_{20}}.$$

From the fact that $p_1{}^\mu$ and $p_2{}^\mu$ transform under such a mapping as four vectors, one can then calculate the transformed values of these quantities.

Because of the conservation laws for the total energy and momentum of the two particles we can conclude that since $\mathbf{p}'_{10} + \mathbf{p}'_{20} = 0$ before the collision, the same must be true after the collision. Consequently, after the collision, the two momenta must also be equal and opposite. The most they can do is to rotate about the center of energy. Furthermore, since the collision is assumed to be elastic, the magnitude of the momentum of each particle must remain unchanged. Once the angle of rotation about the center of energy has been determined from a knowledge of the collision process, one can perform an inverse Lorentz mapping to obtain the final energies and momenta in terms of their initial values and this scattering angle.

As an example consider a collision process in which $\mathbf{p}_{20} = 0$ and $\mathbf{p}_{10}$, which is in the direction of the x^1 axis, has the magnitude p_{10}. Also let the rest masses of the two particles be m_{0_1} and m_{0_2}. After the mapping, particle 2 will have an initial velocity $-v$ in the x^1-direction, given by

$$v = \frac{p_{10}}{\mathscr{E}_{10} + \mathscr{E}_{20}} \tag{7-6.1}$$

and a corresponding momentum

$$\mathbf{p}'_{20} = \left\{\frac{-m_{0_2}v}{\sqrt{1-v^2}}, 0,0\right\}. \tag{7-6.2}$$

The momentum of particle 1 is then just $-\mathbf{p}'_{20}$. If χ is the angle of scattering about the center of energy, then after the scattering,

$$p'_2{}^1 = p'_{20} \cos \chi, \tag{7-6.3}$$

where p'_{20} is the magnitude of $\mathbf{p}'_{20}$. Furthermore the energy of this particle is

$$\mathscr{E}'_2 = \mathscr{E}'_{20} = \frac{m_{0_2}}{\sqrt{1-v^2}}. \tag{7-6.4}$$

Hence, performing the inverse Lorentz mapping $V(\mathbf{v})$, we find that the final energy of particle 2 is

$$\mathscr{E}_2 = \frac{\mathscr{E}'_2 + p'_2{}^1 v}{\sqrt{1-v^2}} = \frac{m_{0_2}(1 - v^2 \cos \chi)}{1-v^2}.$$

Then, since

$$p_{10} = \sqrt{\mathscr{E}_{10}^2 - m_{0_1}^2}, \qquad \mathscr{E}_{20} = m_{0_2},$$

we have

$$v = \frac{\sqrt{\mathscr{E}_{10}^2 - m_{0_1}^2}}{\mathscr{E}_{10} + m_{0_2}}, \tag{7-6.5}$$

and finally, after a short calculation,

$$\mathscr{E}_2 = m_{0_2} + \frac{m_{0_2}(\mathscr{E}_{10}^2 - m_{0_1}^2)}{m_{0_1}^2 + m_{0_2}^2 + 2m_{0_2}\mathscr{E}_{10}}(1 - \cos \chi). \tag{7-6.6}$$

Since the total energy is conserved during the collision,

$$\mathscr{E}_1 + \mathscr{E}_2 = \mathscr{E}_{10} + m_{0_2};$$

hence

$$\mathscr{E}_1 = \mathscr{E}_{10} - \frac{m_{0_2}(\mathscr{E}_{10}^2 - m_{0_1}^2)}{m_{0_1}^2 + m_{0_2}^2 + 2m_{0_2}\mathscr{E}_{10}}(1 - \cos \chi). \tag{7-6.7}$$

The actual scattering angles θ_1 and θ_2 can be determined from a knowledge of the final momenta. Since the transformed momenta are equal and opposite, both before and after scattering, we can use the initial and final momenta of either particle to define a plane of scattering. The two planes obviously coincide. Consequently the actual initial and final momenta of the two particles must be coplanar and lie in the same plane which, without loss of generality, we can take to be the x^1x^2-plane. Then, in the center-of-energy system,

$$p'_2{}^1 = p'_2 \cos\chi = p'_{20} \cos\chi,$$

$$p'_2{}^2 = p'_2 \sin\chi = p'_{20} \sin\chi.$$

By applying the inverse Lorentz mapping and making use of the expressions (7-6.2) and (7-6.4) for p'_{20} and $\mathscr{E}'_2$, we obtain for $p_2{}^1$ and $p_2{}^2$ the expressions

$$p_2{}^1 = \frac{v\mathscr{E}'_2 + p'_2{}^1}{\sqrt{1-v^2}} = m_{0_2} v \frac{1-\cos\chi}{1-v^2}$$

and

$$p_2{}^2 = p'_1{}^2 = -m_{0_2} v \frac{\sin\chi}{\sqrt{1-v^2}}.$$

Consequently

$$\tan\theta_2 = \frac{-p_2{}^2}{p_2{}^1} = \sqrt{1-v^2}\,\frac{\sin\chi}{1-\cos\chi} \tag{7-6.8}$$

which if we wish, we can express in terms of $\mathscr{E}_{10}$ by means of Eq. (7-6.5).

To find θ_2 we proceed in the same manner. Making use of the fact that $p'_{10} = -p'_{20}$ we have that

$$p'_1{}^1 = -p'_{20} \cos\chi,$$

$$p'_1{}^2 = -p'_{20} \sin\chi,$$

and

$$\mathscr{E}'_1 = \sqrt{(p'_1)^2 + m_{0_1}^2} = \sqrt{(p'_{20})^2 + m_{0_1}^2}.$$

Consequently

$$p_1{}^1 = \frac{v\mathscr{E}'_1 + p'_1{}^1}{\sqrt{1-v^2}} = m_{0_2} v \frac{\sqrt{v^2 + (m_{0_1}/m_{0_2})^2(1-v^2)} + \cos\chi}{1-v^2},$$

$$p_1{}^2 = p'_1{}^2 = m_{0_2} v \frac{\sin\chi}{\sqrt{1-v^2}}. \tag{7-6.9}$$

In the case where $m_{0_1} = m_{0_2} = m_0$ these expressions simplify somewhat. In particular

$$\tan\theta_1 = \sqrt{1-v^2}\,\frac{\sin\chi}{1+\cos\chi}. \tag{7-6.10}$$

Now

$$\tan(\theta_1 + \theta_2) = \frac{\tan\theta_1 + \tan\theta_2}{1 - \tan\theta_1 \tan\theta_2},$$

and from Eqs. (7-6.8) and (7-6.10) it follows that

$$\tan \theta_1 \tan \theta_2 = 1 - v^2. \tag{7-6.11}$$

Hence in the nonrelativistic limit $v \ll 1$,

$$\tan(\theta_1 + \theta_2) = \infty \quad \text{or} \quad \theta_1 + \theta_2 = \frac{\pi}{2}.$$

This is a well-known result of Newtonian mechanics; the directions of motion of two like particles after a collision are perpendicular to each other. However, when the velocity of the incident particle approaches unity, $p_{10} \gg m_{0_1}$ and $v \approx 1$. In this case, $\tan \theta_1 \tan \theta_2 < 1$, and hence $\theta_1 + \theta_2 < \pi/2$. This effect has been observed experimentally by Champion[4] in the collisions of β particles with electrons at rest in a cloud chamber. These measurements confirmed the relation (7-6.11) and indirectly the variation of mass of a particle with velocity.

Problem 7.8. A photon can be considered as a particle of rest-mass zero. Its energy is $h\nu$, where ν is the frequency associated with the photon and the magnitude of its momentum is $h\nu/c$. Consider the scattering of a photon by an electron at rest and derive the Compton formula for the change in wavelength:

$$\lambda = \lambda_0 + \frac{2h}{m_e} \sin^2 \frac{\theta}{2},$$

where m_e is the rest mass of the electron and θ is the angle through which the photon is scattered.

7-7. Inelastic Scattering

The problem of obtaining simple conservation laws for inelastic scattering processes from the general conservation laws (7-5.1), analogous to the simple laws (7-5.4) and (7-5.5) for the case of elastic scattering will now be considered. The chief difficulty arising in the discussion of an inelastic collision is that one or more of the particles involved must be considered to be a composite system composed of interacting particles forming a bound state. (For a discussion of such a system, see Section 7-10.) Thus, a typical inelastic collision involves two initially noninteracting particles that come together to form such a bound state. Likewise, a composite system can disintegrate into a number of constituent parts, which themselves may be either simple or composite. Finally, a composite system can interact with another system in such a way that its internal state is altered.

[4] F. C. Champion, *Proc. Roy. Soc.* **A136**, 630 (1932).

In each of the above cases it is impossible to find an initial surface, a final surface, or an initial and final surface on which all particles involved can be considered to be noninteracting. Therefore not all of the $\mathfrak{P}_{i\mu}$ that appear in Eq. (7-5.1) can be taken to be equal to just $p_{i\mu}$. In order to deal with such situations we will make the assumption that the sum of the $\mathfrak{P}_{i\mu}$ for a composite particle on an initial or a final surface is equal to the energy and momentum of that particle as determined by measurements that do not disturb the internal state of the particle. Since such measurements are made by observing the respose of the particle to external forces, these forces must be small compared with the forces that hold the particle together. It would be desirable, and should be possible, to prove this assumption within the framework of the formalism developed in Section 7-4. Since no such proof is known to the author, we are forced to bring in this additional assumption in our discussion of inelastic scattering processes. Its justification will rest, therefore, on the experimental verification of its consequences. If we make this assumption, the conservation laws (7-5.1) again reduce to the simple forms (7-5.4) and (7-5.5) where the $p_{i\mu}$ and z_i^{μ} represent the observed values of these quantities for both the simple and composite particles of the system and where σ_1 and σ_2 are again spacelike surfaces on which these particles can be taken to be noninteracting.

Let us now compare the relativistic conservation laws (7-5.4) with their Newtonian counterparts as applied to inelastic collision processes. In terms of three-velocities $\mathbf{u}_i$, the total linear momentum of a system of particles is given by

$$\mathbf{P} = \sum_i \frac{m_{0_i}\mathbf{u}_i}{\sqrt{1 - \mathbf{u}_i^2}}.$$

The conservation law $\mathbf{P}_1 = \mathbf{P}_2$ is the relativistic analogue of the Newtonian law of conservation of momentum. Since, however,

$$\mathscr{E} = \sum_i \frac{m_{0_i}}{\sqrt{1 - \mathbf{u}_i^2}},$$

the conservation law $\mathscr{E}_1 = \mathscr{E}_2$ has no simple Newtonian analogue. The individual terms in the expression for $\mathscr{E}$ can be represented as the sum of the rest energy m_{0_1} of the particle plus the kinetic energy T_i, given by

$$T_i = \frac{m_{0_i}}{\sqrt{1 - \mathbf{u}_1^2}} - m_{0_i},$$

so that

$$\mathscr{E} = \sum_i m_{0_i} + \sum_i T_i.$$

The conservation of $\mathscr{E}$ thus says that the total rest mass of a system plus its total kinetic energy is conserved.

In Newtonian mechanics the assumption was always made that the total mass of a system was conserved in any interaction. On the other hand, the total kinetic energy of the system need not be conserved. The net change in the total kinetic energy during an interaction was assumed to go into the internal energy of the composite particles taking part in the interaction, in order to conserve overall energy of the system. In relativistic mechanics we see that the two laws of conservation of mass and energy are combined in a single conservation law. But if this law is correct, we arrive at a very profound conclusion, first stated in all generality by Einstein[5] in 1905. If the total kinetic energy of a system of particles increases during an interaction, then the rest mass of the system must decrease, and vice versa. If again we think of the change in kinetic energy as going into the interval energy of the particles, we can say that an increase $\Delta\mathscr{E}$ in the internal energy of a particle produces an increase Δm_0 in its rest mass, where $\Delta\mathscr{E} = \Delta m_0$.

This is the famous Einstein relation (popularly written as $\mathscr{E} = mc^2$) between energy and mass. At the time it was put forth by Einstein there was no way of directly testing this relation. In all reactions where a system gained or lost internal energy the net change in its rest mass was vastly smaller than the original rest mass. Consequently, the rest mass of the system remained constant to a high degree of accuracy. However, with the advent of nuclear physics it became possible to make an experimental test of general equivalence of mass and energy. In 1932 Cockcroft and Walton[6] studied the nuclear reaction

$$ {}_1^1\mathrm{H} + {}_3^7\mathrm{Li} \rightarrow {}_2^4\mathrm{He} + {}_2^4\mathrm{He}. $$

The mass of ${}_1^1\mathrm{H} + {}_3^7\mathrm{Li}$ is $1.0076 + 7.0166$ mass units, and that of two ${}_2^4\mathrm{He}$ nuclei is 2×4.0028, giving a rest-mass difference of 0.0186 mass units, or 0.309×10^{-25} gram. This corresponds to an energy equivalent of 27.7×10^{-6} erg. Precise measurements of the difference between the total kinetic energy of the α particles and kinetic energy of the incident proton gave a value of $(27.6 \pm 0.05) \times 10^{-6}$ erg. Since the uncertainty in the determination of the mass of ${}_3^7\mathrm{Li}$ is $\pm 0.2 \times 10^{-6}$ erg, the Einstein relation can be considered as being verified in this instance to an accuracy of better than 1 part in 100.

An even more startling verification of the equivalence of energy and mass came about with the discovery of the positron. A positron and an electron, both of rest mass m_0, can annihilate each other, and in the process they produce two photons of zero rest mass.[7] Measurements of the energy of these

[5] A. Einstein, *Ann. der Phys.*, **18**, 639 (1905). See also H. Poincaré, *Arch. Néerland*, **5**, 252 (1900).

[6] J. D. Cockcroft and E. T. S. Walton, *Proc. Roy. Soc.* (London), **A137**, 229 (1932).

[7] C. D. Anderson and S. H. Neddermeyer, *Phys. Rev.*, **43**, 1034 (1933); F. Rasetti, L. Meitner, and K. Phillip, *Naturw.*, **21**, 286 (1933).

photons show that it is always $\geqq 2m_0$. The inverse process is also observed to take place, provided the energy of the photons is at least $2m_0$.[8] Recently the antiproton has been produced[9] and is observed to annihilate the proton.

When a composite body of rest mass M_0 breaks up into its constituent parts (it may do this in several different ways), each of rest mass m_{0_i}, the difference $M_0 - \sum_i m_{0_i}$ is called its *mass defect*. The equivalent energy is taken to be the binding energy of the body. Let us consider the case of a composite body at rest, which breaks up spontaneously into two parts with masses m_{0_1} and m_{0_2} and velocities u_1 and u_2. The conservation of energy then requires that

$$m_0 = \frac{m_{0_1}}{\sqrt{1 - {u_1}^2}} + \frac{m_{0_2}}{\sqrt{1 - {u_2}^2}}. \tag{7-7.1}$$

Clearly $m_0 > m_{0_1} + m_{0_2}$ if this equation is to hold. The mass defect $m_0 - m_{0_1} - m_{0_2}$ is thus positive, as is its binding energy. Thus a body can disintegrate, if its mass defect is positive, and in so doing liberate usable energy. If it is negative, no spontaneous breakup is possible and the body is stable. Thus, knowing the mass of a given nucleus, one can predict whether or not it is stable against such breakup. In all cases these predictions are borne out by observation. If the mass defect is negative the nucleus can disintegrate only if it is supplied from the outside with an energy equal to its mass defect.

In the case of the spontaneous disintegration of a body of mass m_0 at rest, the combined laws of conservation of energy and momentum yield values for the velocities u_1 and u_2 of the constituent particles. If the composite particle is initially at rest the momenta of the two constituents must be equal and opposite. Thus we have, in addition to Eq. (7-7.1),

$$\frac{m_{0_1} u_1}{\sqrt{1 - {u_1}^2}} = \frac{m_{0_2} u_2}{\sqrt{1 - {u_2}^2}} \tag{7-7.2}$$

Equations (7-7.1) and (7-7.2) together then determine u_1 and u_2.

As a second example of the application of the conservation laws (7-5.1), consider a particle of rest mass m_{0_1} and velocity u_1 which collides with another particle of rest mass m_{0_2} at rest. After the collision the particles coalesce to form a composite particle of mass m_0 moving with a velocity u. Conservation of energy requires that

$$\frac{m_{0_1}}{\sqrt{1 - {u_1}^2}} + m_{0_2} = \frac{m_0}{\sqrt{1 - u^2}},$$

[8] C. D. Anderson, *Phys. Rev.*, **41**, 405 (1932); P. M. S. Blackett and G. P. S. Occhialini, *Proc. Roy. Soc.* (London), **A139**, 699 (1933).

[9] O. Chamberlain, E. G. Segré, C. E. Wiegand, and T. Ypsilantis, *Phys. Rev.*, **100**, 947 (1955).

while conservation of momentum requires that

$$\frac{m_{0_1}u_1}{\sqrt{1-u_1^2}} = \frac{m_0 u}{\sqrt{1-u^2}}.$$

Solving these two equations for m_0 and u we find that

$$u = \frac{m_{0_1}u_1}{(m_{0_1}+m_{0_2})\sqrt{1-u_1^2}} \quad \text{and} \quad m_0^2 = m_{0_1}^2 + m_{0_2}^2 + \frac{2m_{0_1}m_{0_2}}{\sqrt{1-u_1^2}}.$$

On the basis of our above discussion we can conclude that the relativistic conservation laws, as embodied in Eqs. (7-5.4) and (7-5.5), have been well verified by experiment. So strong is our belief in their universal validity that when (in 1931) they appeared not to hold for beta decay, Pauli invented the neutrino to save them.

7-8. Relativistic Statistical Mechanics

As might be expected on the basis of our previous discussion, a fully developed relativistic statistical mechanics is still to be achieved. Without a Hamiltonian formalism one cannot prove Liouville's theorem in general, and hence the usual deductions that follow from that theorem are not available to us. Only in the case of a system of point particles that interact via point contact (that is, an ideal gas) is a relativistic formulation possible. The consequences of such a formulation were first worked out by Jüttner[10] in 1911.

Between collisions the energy of a free particle is given by Eq. (7-1.6). Since the particles interact only upon contact, the total partition function Z for the system is just equal to the product of the individual particle partition functions z_i. Thus, for an N particle system of equal mass particles $Z = z^N$, where

$$z = V\int d^3p \exp\{-\beta(m_0^2 + p^2)^{1/2}\}, \qquad (\beta = 1/kT).$$

The integral can be evaluated to give

$$z = Vm_0^3 \cdot 2\pi^2(-i)\frac{H_2^{(1)}(im_0\beta)}{m_0\beta},$$

where $H_2^{(1)}$ is the second-order Hankel function of the first kind and V is the volume of the container of the particles.

[10] F. Jüttner, *Ann. der Phys. Lpz.*, **34**, 856 (1911).

The free energy F is related to the partition function by the usual relation

$$F = -\frac{1}{\beta} \ln Z$$

which in our case gives

$$F = -\frac{N}{\beta}\left\{\ln V + \ln\left(-\frac{iH_2^{(1)}(im_0\beta)}{m_0\beta}\right)\right\} + \text{const.}$$

From F one can obtain all other thermodynamic quantities of interest. Thus p and E, the pressure and internal energy, respectively, are given by

$$p = -\frac{\partial F}{\partial V} \quad \text{and} \quad E = F - T\frac{\partial F}{\partial T} = -T^2 \frac{\partial}{\partial T}\left(\frac{F}{T}\right).$$

Thus we find as the thermal equation of state:

$$pV = NkT, \tag{7-8.1}$$

which is the same as in the nonrelativistic case. On the other hand, the caloric equation of state is different in the two cases. One has

$$E = \frac{N}{\beta}\left\{1 - \frac{iH'^{(1)}_2(im_0\beta)}{H_2^{(1)}(im_0\beta)}\, m_0\beta\right\}. \tag{7-8.2}$$

In the nonrelativistic limit, $m_0 \gg \beta^{-1}$, and so, in this limit, we can replace the Hankel function by its asymptotic value

$$-iH_2^{(1)}(ix) \simeq \frac{e^{-x}}{\sqrt{\frac{1}{2}\pi x}}. \tag{7-8.3}$$

Using this asymptotic form one obtains for E the expression

$$E \simeq \frac{N}{\beta}\left(m_{0_\beta} + \frac{3}{2}\right) = Nm_0 + \frac{3}{2}NkT, \tag{7-8.4}$$

which, except for the rest-energy term, is the same as the nonrelativistic result.

In the extreme relativistic case, $m \ll \beta^{-1}$, and we can make use of the series expansion of $H_2^{(1)}$ in the neighborhood of zero to obtain a limiting expression for E in this case. For small x,

$$H_2^{(1)}(ix) \simeq -\frac{1}{\pi}\left(\frac{2}{ix}\right)^2,$$

and therefore

$$E = 3NkT. \tag{7-8.5}$$

Hence, in the extreme relativistic limit, pressure and total energy density ε are related by

$$p = \tfrac{1}{3}\varepsilon. \tag{7-8.6}$$

Although we have derived this last result for an ideal gas, we can expect that it will have a general validity for all systems when the thermal energy of the system is large compared with rest mass and interaction energy.

Problem 7.9. A photon gas may be considered to be an ideal case of massless particles and to correspond to black-body radiation confined to a cavity. Calculate the pressure, internal energy, and entropy of such a gas. What assumption must be made concerning N, to obtain the usual black-body expressions for p and E?

7-9. Action-at-a-Distance Electrodynamics

An action-at-a-distance formulation of electrodynamics has been discussed by a number of authors.[11] It is usually referred to as the Fokker or Wheeler-Feynman *action principle*. These authors take V_{ik} in Eq. (7-4.1) to be given by

$$V_{ik} = -e_i e_k \delta((z_i - z_k)^2)\dot{z}_i^{\mu}\dot{z}_k^{\nu}\eta_{\mu\nu}, \tag{7-9.1}$$

where $(z_i - z_k)^2 = \eta_{\mu\nu}(z_i^{\mu} - z_k^{\mu})(z_i^{\nu} - z_k^{\nu})$. The parameter e_i appearing in this expression for V_{ik} is taken to be the charge of the ith particle. The partial Lagrangian L_i is thus

$$L_i = -m_{0i}\sqrt{\dot{z}_i^2} - e_i \sum_{k\neq i} e_k \int_{(\sigma_1)}^{(\sigma_2)} \dot{z}_i^{\mu}\dot{z}_k^{\nu}\eta_{\mu\nu}\delta((z_i - z_k)^2)\, d\lambda_k. \tag{7-9.2}$$

When the parameter λ_i is taken to be the proper time along the ith trajectory so that $\dot{z}_i^2 = 1$, the equations of motion (7-4.5) take the form

$$m_{0i}\ddot{z}_i^{\mu} = e_i F_{\nu}{}^{\mu}(z_i)\dot{z}_i^{\nu}, \tag{7-9.3}$$

where

$$F_{\mu\nu}(z_i) = \frac{\partial A_{\nu}(z_i)}{\partial z_i^{\mu}} - \frac{\partial A_{\mu}(z_i)}{\partial z_i^{\nu}} \tag{7-9.4}$$

and

$$A^{\mu}(z_i) = \sum_{k\neq i} e_k \int_{(\sigma_1)}^{(\sigma_2)} \delta((z_i - z_k)^2)\dot{z}_k^{\mu}\, d\lambda_k. \tag{7-9.5}$$

[11] K. Schwarzschild, *Nachr. Akad. Wiss. Göttingen Math. Physik. Kl.*, IIa, 1903; 128, 132, 245 (1903); H. Tetrode, *Z. Physik*, **10**, 317 (1922); A. D. Fokker, *ibid*, **58**, 386 (1929), *Physica*, **9**, 33 (1929); **12**, 145 (1932); J. A. Wheeler and R. P. Feynman, *Rev. Mod. Phys.*, **17** 157 (1945); **21**, 425 (1949).

As we will see later, these equations have the form of the Lorentz equations of motion of a particle in an electromagnetic field described by the potential $A_\mu(x)$. In their analysis of these equations Wheeler and Feynman have shown that it is possible to derive from them all the well-known results of conventional Maxwell electrodynamics.

Because of the special properties of the δ function, it is possible to evaluate the integrals appearing in Eq. (7-9.5) directly. The only contributions to the integral come from the points where the light cone emanating from the point z_i^μ intersects the trajectories of the other particles. The light cone will intersect each trajectory at just two points, the advanced and retarded points, determined by the equation

$$(z_i - z_k)^2 = 0.$$

For a given z_i^μ, just two values of the parameter λ_k will satisfy these equations, corresponding to the advanced and retarded points on the kth trajectory (see Fig. 7.4). Consequently the motion of the ith particle is determined not only by the past motion of the other particles but by their future motion as well.

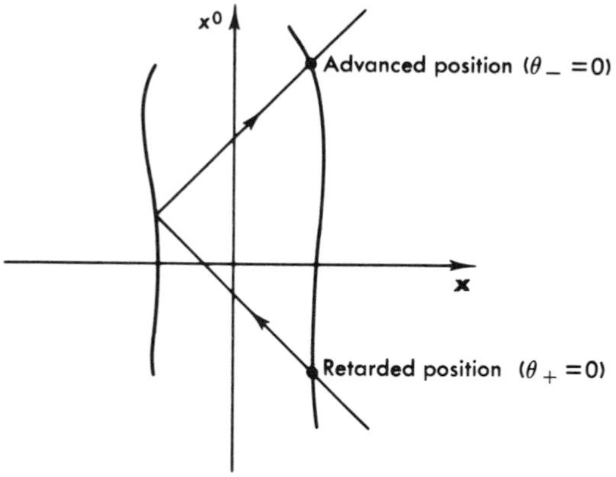

Fig. 7.4.

While this type of interaction violates our sense of causality, it is completely consistent with the requirements of special relativity. It corresponds in the field theoretic description to using half-advanced plus half-retarded solutions of the field equations.

In evaluating $A_\mu(z_i)$ we make use of the following property of the δ function:[12]

$$\delta(x^2) = \frac{1}{2|\mathbf{x}|}\,[\delta(|\mathbf{x}| - x^0) + \delta(|\mathbf{x}| + x^0)], \qquad |\mathbf{x}| = \sqrt{\delta_{rs} x^r x^s}. \qquad (7\text{-}9.6)$$

[12] See Appendix.

Thus the contribution to A_μ from the trajectory $z(\lambda)$ at the point x^μ is given by

$$A_\mu(x) = e\int_{-\infty}^{+\infty} \delta((x-z)^2)\dot{z}_\mu \, d\lambda$$

$$= \frac{e}{2}\int_{-\infty}^{+\infty} \frac{1}{|\mathbf{x}-\mathbf{z}|}\{\delta(|\mathbf{x}-\mathbf{z}| - (x^0 - z^0)) + \delta(|\mathbf{x}-\mathbf{z}| + (x^0 - z^0))\}\dot{z}_\mu \, d\lambda.$$

To evaluate the integrals appearing in the above expression for $A_\mu(x)$, it is convenient to introduce new variables of integration θ_+ and θ_- defined by

$$\theta_\pm = |\mathbf{x}-\mathbf{z}| \mp (x^0 - z^0),$$

θ_+ being used for the first integral and θ_- for the second. Since the trajectory of the source is assumed to be everywhere timelike, it follows that $\theta_+ \to \pm\infty$, while $\theta_- \to \mp\infty$ as $\lambda \to \pm\infty$, resulting in a change of sign in the second integral. To complete the change of variables we need to express $d\lambda$ in terms of $d\theta_\pm$ in the two cases. By straightforward differentiation we have

$$d\theta_\pm = \left(\pm\dot{z}^0 - \frac{\mathbf{x}-\mathbf{z}}{|\mathbf{x}-\mathbf{z}|}\cdot\dot{\mathbf{z}}\right) d\lambda.$$

However, instead of these expressions for $d\theta_\pm$, we can equally well take

$$d\theta_\pm = \left(\frac{x^0 - z^0}{|\mathbf{x}-\mathbf{z}|}\dot{z}^0 - \frac{\mathbf{x}-\mathbf{z}}{|\mathbf{x}-\mathbf{z}|}\cdot\dot{\mathbf{z}}\right) d\lambda$$

$$= \frac{x^\mu - z^\mu}{|\mathbf{x}-\mathbf{z}|}\dot{z}_\mu \, d\lambda,$$

since the two sets of expressions agree with each other when $\theta_\pm = 0$, and it is only for this value that the integrands in the respective integrals are nonzero. Combining these results, then, we have

$$A_\mu(x) = \frac{e}{2}\int_{-\infty}^{+\infty} \frac{\dot{z}_\mu}{(x^\rho - z^\rho)\dot{z}_\rho}\{\delta(\theta_+)\,d\theta_+ - \delta(\theta_-)\,d\theta_-\}$$

$$= \tfrac{1}{2}\{A_\mu{}^+(x) + A_\mu{}^-(x)\}, \tag{7-9.7}$$

where

$$A_\mu{}^\pm(x) = \pm\left.\frac{e\dot{z}_\mu}{(x^\rho - z^\rho)\dot{z}_\rho}\right|_{\theta_\pm = 0}. \tag{7-9.8}$$

These are the well-known retarded and advanced potential of Lienard-Wiechert, in whose evaluation z^μ and $\dot{z}^\mu$ are to be evaluated for values of the parameter λ for which $\theta_\pm = 0$. Thus $A_\mu{}^\pm(x)$ depends upon x^μ, not only explicitly in the denominator but implicitly through the dependence of z^μ and $\dot{z}^\mu$ on the retarded and advanced values of λ, which depend on x^μ.

In order to evaluate the "field strength" $F_{\mu\nu}$ we must compute the derivatives of A_μ. Because of the implicit dependence of A_μ on x, this computation requires some care. In particular we need to know the derivatives of λ with respect to x^μ. To compute these derivatives we note that both the retarded and advanced values of λ are determined by the condition

$$(x^\mu - z^\mu(\lambda))(x^\nu - z^\nu(\lambda))\eta_{\mu\nu} = 0.$$

If we differentiate this equation with respect to x^ρ we obtain the result that

$$(\delta_\rho{}^\mu - \dot{z}^\mu \lambda_{,\rho}) r_\mu = 0, \qquad r_\mu = \eta_{\mu\nu}(x^\nu - z^\nu),$$

so that

$$\lambda_{,\mu} = \frac{r_\mu}{\rho},$$

where $\rho = r_\mu \dot{z}^\mu$. The remainder of the calculation is now straightforward, and one finds that when λ is taken to be the proper time τ along a trajectory, that is, when $\dot{z}^\mu \dot{z}_\mu = 1$,

$$\begin{aligned} F^{\pm}_{\mu\nu} &= A^{\pm}_{\nu,\mu} - A^{\pm}_{\mu,\nu} \\ &= \pm 2e \left\{ \frac{1}{\rho^3} r_{[\mu}\dot{z}_{\nu]} + \frac{1}{\rho^2} r_{[\mu}\ddot{z}_{\nu]} - \frac{1}{\rho^3} r_{[\mu}\dot{z}_{\nu]} r^\alpha \ddot{z}_\alpha \right\} \Bigg|_{\theta_\pm = 0} \end{aligned} \tag{7-9.9}$$

One can now substitute this result into Eq. (7-9.3) to obtain the equations of the motion of the system of charges.

Problem 7.10. Show that Eq. (7-9.9) can be written in three-vector form as

$$\begin{aligned} \mathbf{E}^{\pm}(x^0,\mathbf{x}) = \pm \frac{e}{(\pm r - \mathbf{r}\cdot\mathbf{u})^3} \{&(1 - \mathbf{u}^2)(\mathbf{r} \mp r\mathbf{u}) - \mathbf{a}r(r \mp \mathbf{r}\cdot\mathbf{u}) \\ &+ (\mathbf{r} \mp r\mathbf{u})\mathbf{r}\cdot\mathbf{a}\}|_{\theta^\pm = 0} \end{aligned} \tag{7-9.10}$$

and

$$\mathbf{B}^{\pm}(x^0,\mathbf{x}) = \frac{\mathbf{r}}{r}\Bigg|_{\theta^\pm = 0} \times \mathbf{E}^{\pm}(x^0,\mathbf{x}), \tag{7-9.11}$$

where $\mathbf{r} = \mathbf{x} - \mathbf{z}$, $E_r = F_{0r}$, and $B_r = -\frac{1}{2}\varepsilon_{ruv}F^{uv}$.

As an application of the above results let us calculate the retarded field of a charge e moving uniformly with a velocity $\mathbf{v}$. Taking the coordinates of the point at which we wish to evaluate the field to be $\{x_0{}^0, \mathbf{x}_0\}$ and the position of the charge at the time $x_0{}^0$ to be $\mathbf{z}_0$, we seek the retarder position $\mathbf{z}_+$ of the charge in terms of these quantities. The trajectory of the charge is given by

$$\mathbf{z} = \mathbf{z}_0 + \mathbf{v}(x^0 - x_0{}^0),$$

while the points on the backward light cone emanating from the point $\{x_0{}^0, \mathbf{x}_0\}$ are determined by the equation

$$|\mathbf{x}_0 - \mathbf{x}| - (x_0{}^0 - x^0) = 0.$$

Referring to Fig. 7.5, we see that the retarded position is determined by the equation

$$\mathbf{z}_+ = \mathbf{z}_0 - \mathbf{v}|\mathbf{x}_0 - \mathbf{z}_+|.$$

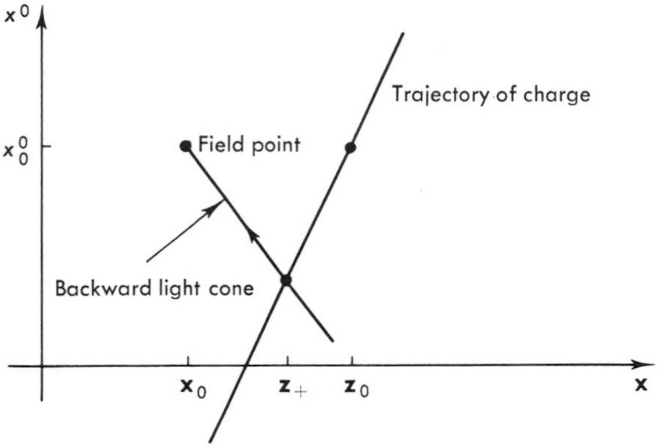

Fig. 7.5.

Subtracting $\mathbf{x}_0$ from both sides of this equation we have the result

$$\mathbf{r}_+ - \mathbf{v}r_+ = \mathbf{r}_0, \tag{7-9.12}$$

where $\mathbf{r}_+ = \mathbf{x}_0 - \mathbf{z}_+$ and $\mathbf{r}_0 = \mathbf{x}_0 - \mathbf{z}_0$. We see that $\mathbf{r}_0 \times \mathbf{v} = \mathbf{r}_+ \times \mathbf{v}$, from which it follows after a short calculation that

$$s \equiv r_+ - \mathbf{r}_+ \cdot \mathbf{v} = \sqrt{\mathbf{r}_0{}^2 - (\mathbf{r}_0 \times \mathbf{v})^2}. \tag{7-9.13}$$

We can now make use of Eqs. (7-9.12) and (7-9.13) to evaluate $\mathrm{E}^+(x_0{}^0, \mathbf{x}_0)$, as given by Eq. (7-9.10). Since a $= 0$ in the present case, we have

$$\mathbf{E}^+(x_0{}^0,\mathbf{x}_0) = \frac{e}{s^3}(1 - \mathbf{v}^2)\mathbf{r}_0. \tag{7-9.14a}$$

Furthermore, since $\mathbf{r}_+ \times \mathbf{r}_0 = r_+\mathbf{v} \times \mathbf{r}_0$, it follows that

$$\mathbf{B}^+(x_0{}^0,\mathbf{x}_0) = \mathbf{v} \times \mathbf{E}^+(x_0{}^0,\mathbf{x}_0). \tag{7-9.14b}$$

The advanced fields can be also calculated in a similar manner.

Wheeler and Feynman[13] discussed the equivalence between the action-at-a-distance formulation of the electromagnetic interaction between particles and the more conventional description of this interaction in terms of an electromagnetic field satisfying Maxwell's equations. They were able to show that, to the extent that one ignores the self-interaction of charges in the Maxwell theory, the two formalisms are equivalent. All well-known consequences of the Maxwell theory (for example, Ampère's law) can, when properly interpreted, be obtained from the action-at-a-distance formalism. We will show later that the fields $F^{\pm}_{\mu\nu}$ given by Eq. (7-9.9) satisfy Maxwell's equations, thereby demonstrating further the equivalence of the two formalisms.

Let us now evaluate the quantities $\mathfrak{P}_{i\mu}$ and $\mathfrak{L}_{i\mu\nu}$ that appear in the conservation laws (7-5.1) for a system of charged particles. To do so, we must calculate $\bar{\delta}S = S' - S$. With V_{ij} given by Eq. (7-9.1), one has, after performing a number of integrations by parts,

$$\begin{aligned}\bar{\delta}S = \sum_i \int_{\sigma_1}^{\sigma_2} \frac{d}{d\tau_i}(m_{0_i}\dot{z}_{i\mu})\bar{\delta}z_i^{\mu}\, d\tau_i - \left\{\sum_i m_{0_i}\dot{z}_{i\mu}\bar{\delta}z_i^{\mu}\right\}\Bigg|_{\sigma_1}^{\sigma_2} \\ - \sum_{i\neq j}\sum e_i e_j \iint_{(\sigma_1)}^{(\sigma_2)} \{\partial_{\mu_i}\delta(z_i - z_j)^2 \dot{z}_i^{\rho}\dot{z}_{j\rho} - \partial_{\rho i}\delta(z_i - z_j)^2\dot{z}_{i\rho}\dot{z}_{j\mu}\}\bar{\delta}z_i^{\mu}\, d\tau_j\, d\tau_i \\ - \sum_{i\neq j}\sum e_i e_j \left\{\int_{(\sigma_1)}^{(\sigma_2)} \delta(z_i - z_{j\mu})^2 \dot{z}_j^{\sigma}\bar{\delta}z_i^{\mu}\, d\tau_j\right\}\Bigg|_{(\sigma_1)}^{(\sigma_2)},\end{aligned} \qquad (7\text{-}9.15)$$

where $\partial_{\mu_i} \equiv \partial/\partial z_i^{\mu}$ and where we have taken the path parameters to be the proper times τ_i. If we now make use of the equations of motion (7-9.3), we have

$$\begin{aligned}\bar{\delta}S = -\sum_{i\neq j}\sum e_i e_j\left(\int_{\sigma_2}^{(\sigma_2)} + \int_{(\sigma_1)}^{\sigma_1}\right)\int_{(\sigma_1)}^{(\sigma_2)} \{\partial_{\mu_i}\delta(z_i - z_j)^2\dot{z}_i^{\rho}\dot{z}_{j\rho} - \partial_{\rho_i}\delta(z_i - z_j)^2\dot{z}_i^{\rho}\dot{z}_{j\mu}\} \\ \times\, \bar{\delta}z_i^{\mu}\, d\tau_j\, d\tau_i \\ -\left\{\sum_i m_{0_i}\dot{z}_{i\mu}\bar{\delta}z_i^{\mu}\right\}\Bigg|_{\sigma_1}^{\sigma_2} - \sum_{i\neq j}\sum e_i e_j\left\{\int_{(\sigma_1)}^{(\sigma_2)} \delta(z_i - z_j)^2\dot{z}_{j\mu}\bar{\delta}z_i^{\mu}\, d\tau_j\right\}\Bigg|_{(\sigma_1)}^{(\sigma_2)}.\end{aligned}$$

Now setting $\bar{\delta}z_i^{\mu} = \varepsilon_{\nu}{}^{\mu}z_i^{\nu} + \varepsilon^{\mu}$, one finds that $\mathfrak{P}_{i\mu}$ in Eq. (7-5.1) has the form

$$\begin{aligned}\mathfrak{P}_i^{\mu}(\sigma) = m_{0_i}\dot{z}_i^{\mu}(\sigma) + e_i A^{\mu}(z_i(\sigma)) \\ + \sum_{j\neq i} e_i e_j\left(\int_{\sigma}^{(\sigma)}\int_{(\sigma)}^{\sigma} - \int_{(\sigma)}^{\sigma}\int_{\sigma}^{(\sigma)}\right)\delta'(z_i - z_j)^2(z_i^{\mu} - z_j^{\mu})\dot{z}_i^{\nu}\dot{z}_{j\nu}\, d\tau_i\, d\tau_j,\end{aligned} \qquad (7\text{-}9.16)$$

[13] J. A. Wheeler and R. P. Feynman, *op. cit.*

while $\mathfrak{L}_{i\mu\nu}$ has the form

$$\begin{aligned}\mathfrak{L}_{i\mu\nu}(\sigma) = {} & z_i^{\mu}(\sigma)\{m_{0_i}\dot{z}_i^{\nu}(\sigma) + e_i A^{\nu}(z_i(\sigma)\} \\ & - z_i^{\nu}(\sigma)\{m_{0_i}\dot{z}_i^{\mu}(\sigma) + e_i A^{\mu}(z_i(\sigma))\} \\ & + \sum_{j \neq i} e_i e_j \left(\int_{\sigma}^{(\sigma)} \int_{(\sigma)}^{\sigma} - \int_{(\sigma)}^{\sigma} \int_{\sigma}^{(\sigma)} \right) \{\delta'(z_i - z_j)^2 (z_i^{\nu} z_j^{\mu} - z_i^{\mu} z_j^{\nu}) \dot{z}_i^{\rho} \dot{z}_{i\rho} \\ & - \tfrac{1}{2}\delta(z_i - z_j)^2 (\dot{z}_i^{\nu} \dot{z}_j^{\mu} - \dot{z}_i^{\mu} \dot{z}_j^{\nu})\} \, d\tau_i \, d\tau_j ,\end{aligned} \tag{7-9.17}$$

where

$$A^{\mu}(z_i) = \sum_{j \neq i} \left\{ \frac{e_j \dot{z}_j^{\mu}}{(z_i^{\rho} - z_j^{\rho})\dot{z}_{j\rho}} \bigg|_{\theta_{+ij}=0} - \frac{e_j \dot{z}_j^{\mu}}{(z_i^{\rho} - z_j^{\rho}) z_{j\rho}} \bigg|_{\theta_{-ij}=0} \right\},$$

with

$$\theta_{\pm ij} = |\mathbf{z}_i - \mathbf{z}_j| \mp (z_i^0 - z_j^0).$$

From these results we see that in the limit when the spatial separation between particles on an $x^0 = \text{const}$ surface tends to infinity, $\mathfrak{P}_{i\mu}$ reduces to $p_{i\mu}$, and $\mathfrak{L}_{i\mu\nu}$ reduces to $l_{i\mu\nu}$.

7-10. The Relativistic Two-Body Problem

As we mentioned before, even the simplest relativistic two-body interaction leads to integro-differential equations of motion. In spite of this it is possible to find solutions to the two-body problem.[14] The simplest solution consists of two opposite charges (for example, an electron and a proton) revolving about a common origin. The position of the proton is given by

$$z_p^{\mu} = \{x^0, -a \cos \omega x^0, a \sin \omega x^0, 0\} \tag{7-10.1}$$

and that of the electron by

$$z_e^{\mu} = \{x^0, b \cos \omega x^0, -b \sin \omega x^0, 0\} \tag{7-10.2}$$

If the proton is at the point $z_p^{\mu} = \{0, a, 0, 0\}$, then the advanced position of the electron will be $z_{e-}^{\mu} = \{\tau, -b \cos \omega\tau, -b \sin \omega t, 0\}$, and the retarded position will be $z_{e+}^{\mu} = \{-\tau, -b \cos \omega t, b \sin \omega t, 0\}$, where τ is the positive solution of the equation

$$\tau^2 - (b \cos \omega\tau + a)^2 - b^2 \sin^2 \omega\tau = 0. \tag{7-10.3}$$

(It is clear that if τ is a solution, then so is $-\tau$.) Rather than work with τ it is more convenient to work with the retardation angle θ, defined as the angle

[14] A. Schild, *Phys. Rev.*, **131**, 2762 (1963).

through which one particle turns during the time $\tau = \theta/\omega$ that it takes light from the other particle to reach it (see Fig. 7.6). It follows from Eq. (7-10.3) that θ satisfies

$$\theta^2 - v_p^{\,2} - v_e^{\,2} - 2v_p v_e \cos\theta = 0, \tag{7-10.4}$$

where

$$v_p = a\omega, \qquad v_e = b\omega.$$

Upon substituting the solutions (7-10.1) and (7-10.2) into the radial equations of motion for the proton, one obtains the condition

$$\frac{m_p}{\omega}\frac{v_p}{\sqrt{1-v_p^{\,2}}} = \frac{e^2}{(\theta + v_p v_e \sin\theta)^3}\,[(v_e + v_p\cos\theta)(1 - v_p^{\,2})(1 - v_e^{\,2}) + (v_p\theta + v_e\sin\theta)(\theta + v_p v_e \sin\theta)], \tag{7-10.5}$$

and similarly from the equation of motion of the electron, one obtains the condition

$$\frac{m_e}{\omega}\frac{v_e}{\sqrt{1-v_e^{\,2}}} = \frac{e^2}{(\theta + v_e v_p \sin\theta)^3}\,[(v_e + v_p\cos\theta)(1 - v_e^{\,2})(1 - v_p^{\,2}) + (v_e\theta + v_p\sin\theta)(\theta + v_e v_p \sin\theta)]. \tag{7-10.6}$$

Because we have used the symmetric interaction rather than just the retarded interaction, the tangential and temporal parts of the equations of motion are satisfied identically. Equations (7-10.4), (7-10.5), and (7-10.6) then serve to determine v_p, v_e, and ω, and hence the radii a and b of the orbits. To the best of the author's knowledge this is the only rigorous solution of a two-body problem in special relativity.

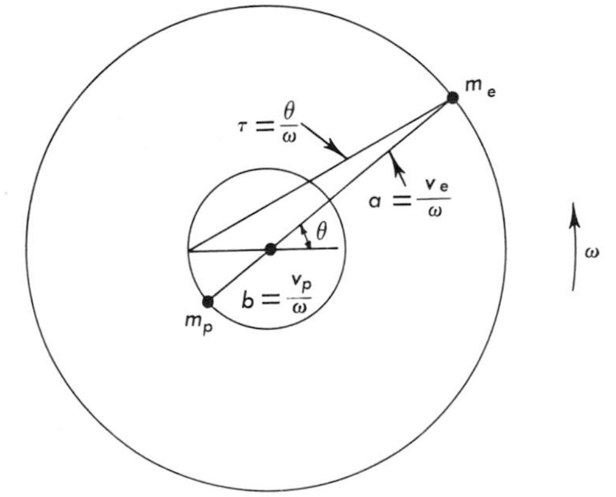

Fig. 7.6.

Schild has evaluated the energy, momentum, and angular momentum for the above solution of the two-body problem. Taking the spacelike surface σ to be the surface $x^0 = 0$, he finds that

$$P^r = \sum_i \mathfrak{P}_i{}^r = 0$$

and

$$\mathscr{E} = \sum_i \mathfrak{P}_i{}^0 = m_p(1 - v_p{}^2)^{+1/2} + m_e(1 - v_e{}^2)^{+1/2}.$$

Since $\mathscr{E} < m_p + m_e$, the mass defect for this dpt is negative, and so we are dealing with a bound state. The only nonvanishing component of the angular momentum tensor is the 12-component, which has the value

$$\begin{aligned} L^{12} &= \sum_i \mathfrak{L}_i^{12} \\ &= e^2 \frac{1 + v_p v_e \cos\theta}{\theta + v_p v_e \sin\theta}. \end{aligned}$$

7-11. Slow-Motion Approximation to Electrodynamics

In Section (6-6) we saw that the Galilean group could be obtained from the Lorentz group as the limiting case when the invariant velocity that defines the light cone becomes infinite. We will make use of this fact to derive an approximation to the fully relativistic equations (7-9.3), which can be considered to be the nonrelativistic analogue of these equations. To do this we will introduce the light velocity c explicitly into our equations by replacing z^0 by the variable ct, recalling that just this change of variable led to the Galilean group when $c \to \infty$. Having introduced c explicitly, we can obtain the desired results by expanding the equations of motion in powers of c^{-1}, retaining only the lowest powers in this expansion. Actually, it is easier to work with the partial Lagrangian L_i, given by Eq. (7-9.2), than with the equations of motion themselves.

When we include the light velocity explicitly, the ith partial action S_i is given by

$$\begin{aligned} S_i = \int_1^2 L_i\, d\lambda_i = &-\int_1^2 m_{0_i} c^2 \sqrt{1 - \frac{\mathbf{u}_i^2}{c^2}}\, dt_i \\ &- e_i \sum_{k \neq i} e_k \int_1^2 dt_i \int_{-\infty}^{+\infty} dt_k \left(1 - \frac{\mathbf{u}_i \cdot \mathbf{u}_k}{c^2}\right) \frac{1}{2r_{ij}} \left[\delta\left(\frac{1}{c} r_{ik} - t_{ik}\right) \right. \\ &\left. + \delta\left(\frac{1}{c} r_{ik} + t_{ik}\right)\right], \end{aligned} \tag{7-11.1}$$

where now $\mathbf{u}_i = d\mathbf{z}_i/dt_i$, $t_{ik} = t_i - t_k$, and $\mathbf{r}_{ik} = (\mathbf{z}_i - \mathbf{z}_k)$. The c are most easily inserted by requiring that S_i have the overall dimensions of an action. In the limit $c \to \infty$, S_i reduces to the Galilean form (after subtracting the constant $-m_{0_i}c^2\int_1^2 dt_i$)

$$S_i = \int_1^2 \left\{\frac{1}{2} m_{0_i}\mathbf{u}_i^2 - \sum_{k \neq i} \frac{e_i e_k}{r_{ik}}\right\} dt_i, \tag{7-11.2}$$

and hence leads to the equations of motion of a system of nonrelativistic particles interacting via an instantaneous Coulomb force.

For some applications (for example, positronium) one needs to know the first-order corrections to the nonrelativistic action due to relativistic effects. These can be obtained by retaining the first nonvanishing terms in c^{-1} in the fully relativistic action. To find these terms, one expands the delta functions in Eq. (7-11.1) about the point $(1/c)r_{ij} = 0$. Thus we have

$$\delta\left(\frac{1}{c} r_{ij} - t_{ij}\right) = \delta(t_{ij}) - \delta'(t_{ij}) \frac{1}{c} r_{ij} + \delta''(t_{ij}) \frac{1}{2c^2} r_{ij}^2 + \cdots,$$

and similarly for the other delta function appearing in Eq. (7-11.1). When one inserts these expansions into Eq. (7-11.1) and retains terms up to c^{-2}, one obtains after a number of integrations by parts,

$$S_i = \int_1^2 \left[\frac{1}{2} m_{0_i}\mathbf{u}_i^2 - \sum_{k \neq i} \frac{e_i e_k}{r_{ik}} + \frac{1}{2c^2}\left\{\frac{1}{4} m_{0_i}\mathbf{u}_i^4 \right.\right.$$
$$\left.\left. + \sum_{k \neq i} \frac{e_i e_k}{r_{ik}} \left(\mathbf{u}_i \cdot \mathbf{u}_k + \frac{(\mathbf{u}_i \cdot \mathbf{r}_{ik})(\mathbf{u}_k \cdot \mathbf{r}_{ik})}{r_{ik}^2}\right)\right\}\right] dt_i .$$

Since all interactions in S_i are instantaneous and symmetric in the particle variables, we can add the partial actions to obtain an overall action S, given by

$$S = \int_1^2 dt \left[\sum_i \frac{1}{2} m_{0_i}\mathbf{u}_i^2 - \frac{1}{2}\sum_{i \neq k}\sum \frac{e_i e_k}{r_{ik}} + \frac{1}{2c^2}\left\{\frac{1}{4}\sum_i m_{0_i}\mathbf{u}_i^4 \right.\right.$$
$$\left.\left. + \frac{1}{2}\sum_{i \neq k}\sum \left(\mathbf{u}_i \cdot \mathbf{u}_k + \frac{(\mathbf{u}_i \cdot \mathbf{r}_{ik})(\mathbf{u}_k \cdot \mathbf{r}_{ik})}{r_{ik}^2}\right)\right\}\right]. \tag{7-11.3}$$

This action was first obtained by Darwin[15] in 1920. It leads to a set of equations of motion that are approximations to the fully relativistic equations (7-9.3). From the form of S one sees that the approximation being made is in effect a slow-motion approximation in which one includes, at each stage of the

[15] C. G. Darwin, *Phil. Mag.*, **39**, 537 (1920).

approximation, only a finite number of derivatives of the position variables.[16] Higher derivatives do not arise, since in effect each derivative with respect to t brings in an additional power of c^{-1}.

Problem 7.11. Derive an expression for the Hamiltonian corresponding to the action of Eq. (7-11.3).

Problem 7.12. Find the next order in the expansion of S_i, that is, the term of order c^{-3}, both when S_i is given by Eq. (7-11.1) and when S_i involves only a retarded term. Discuss the significance of these terms in these two cases.

7-12. The Motion of Charges in Fields

The Eqs. (7-9.3) are the equations of motion of the ith particle of a system of charged particles, all interacting with one another. While the force

$$f_i^{\mu} = e_i F_{\nu}{}^{\mu}(z_i)\dot{z}_i{}^{\nu}$$

acting on the ith particle is determined by the motion of all other particles, the motions of the other particles are affected by the ith particle. If we neglect this latter effect, then we can, at least in principle, determine the motion of the ith particle from a knowledge of its position and velocity at some point on its world-line, provided we know $F_{\nu}{}^{\mu}(z_i)$. We will call this object the electromagnetic field of the remaining particles. In order to compute $F_{\nu}{}^{\mu}$ we would first have to find a consistent set of motions of these remaining trajectories and then compute $F_{\nu}{}^{\mu}$ from its definition (7-9.4) or from Eq. (7-9.9). In general this task is beyond our abilities, so we must look for other means of determining $F_{\nu}{}^{\mu}$. We will turn to this task in the next chapter. Here we will merely assume that $F_{\nu}{}^{\mu}$ is a given function of $z_i{}^{\mu}$ and will not worry about whether it is a field that could be produced by a real system of moving charges.

With the above assumptions the equations of motion of a charged particle in a field can be derived from a variational principle, with a Lagrangian L given by Eq. (7-9.2). Written in terms of the potential A_{μ}, L takes the form

$$L = -m_0\sqrt{\dot{z}^2} - eA_{\mu}(z)\dot{z}^{\mu}. \tag{7-12.1}$$

The requirement that

$$\delta \int_{\lambda_1}^{\lambda_2} L \, d\lambda = 0$$

[16] See G. Breit, *Phys. Rev.*, **51**, 248 (1937) for a discussion of the quantum analogue of the slow-motion approximation as applied to nuclear physics.

for variations of a dpt that vanish at the end points then leads, as we saw, to the equations of motion

$$m_0 \ddot{z}^\mu = eF_\nu{}^\mu \dot{z}^\nu \tag{7-12.2}$$

when the parameter along the path is chosen such that $\dot{z}^2 = 1$. For an arbitrary parameter the equations have the form

$$\frac{d}{d\lambda}\left(m_0 \frac{\dot{z}^\mu}{\sqrt{\dot{z}^2}}\right) = eF_\nu{}^\mu \dot{z}^\nu. \tag{7-12.3}$$

For many applications it is useful to let z^0 itself be the parameter. In this case the spatial part of the equations of motion reads

$$\frac{d}{dx^0} p^r = eF_s{}^r u^s + eF_0{}^r. \tag{7-12.4}$$

The right-hand side represents the force on the particle due to the field described by $F_\nu{}^\mu$. The velocity-independent part of this force is said to be due to the electric field.

$$E_r = F_{0r}. \tag{7-12.5}$$

The velocity-dependent part of the force can be written, with the help of the magnetic induction

$$B_r = -\tfrac{1}{2}\varepsilon_{ruv} F^{uv}, \tag{7-12.6}$$

as

$$eF_s{}^r u^s = e\varepsilon^{rst} u^s B_t.$$

Thus in the usual three-vector notation we have the equations

$$\frac{d}{dx^0}\,\mathbf{p} = e(\mathbf{E} + \mathbf{u} \times \mathbf{B}). \tag{7-12.7}$$

The three-force appearing on the right is just the usual Lorentz force on a charged particle moving in a combined electric and magnetic field.

The fourth of the equations (7-12.3), where $\mu = 0$, is not an independent equation. If we multiply it by $\dot{z}_\mu$ we obtain

$$\dot{z}_\mu \frac{d}{d\lambda}\left(m_0 \frac{\dot{z}^\mu}{\sqrt{\dot{z}^2}}\right) = eF^{\mu\nu} \dot{z}_\mu \dot{z}_\nu.$$

The left-hand side is zero because the four-acceleration is orthogonal to the four-velocity, and the right-hand side is zero because of the antisymmetry of $F^{\mu\nu}$. Therefore, one of these equations, which we can take to be the $\mu = 0$

equation, is a consequence of the other three. The $\mu = 0$ equation is, in the $1 + 3$ notation,

$$\frac{d}{dx^0}\left(\frac{m_0}{\sqrt{1 - u^2}}\right) = e\mathbf{E} \cdot \mathbf{u}. \tag{7-12.8}$$

The left-hand side is the change in the mechanical energy of the particle; the right-hand side is therefore the work done on the particle per unit time. As in the Newtonian case the magnetic field does no work on the particle, since the force it exerts is always perpendicular to the velocity of the particle.

As we have already said, Eq. (7-12.7) must be considered as an approximation to a larger system of equations of motion that properly take account of the action of the particle moving in the field $F^{\mu\nu}$ back on the sources of this field. Of special importance is the case when the source consists of a single unaccelerated charge. If we map so that this charge is at rest at the origin of spatial coordinates, it produces a field that can be gotten directly from Eqs. (7-9.10) and (7-9.11). In this case these equations give

$$\mathbf{E} = \frac{1}{2}\{\mathbf{E}^+ + \mathbf{E}^-\} = \frac{e}{r^3}\mathbf{r}, \qquad \mathbf{B} = 0,$$

that is, the usual Coulomb field of a point charge. If we substitute this field into Eq. (7-12.7), we obtain the relativistic equations of motion of a particle moving in an inverse square force field; they are therefore the relativistic analogue of the equations of motion for the Newtonian Kepler problem. These equations can therefore be properly considered to be an approximation to a rigorous set of relativistically invariant equations of motion for two interacting point charges when the rest mass of one of them is much larger than that of the other.

In three-vector notation the Lagrangian (7-12.1) can be written as

$$L = -m_0\sqrt{1 - \mathbf{u}^2} + e\mathbf{A} \cdot \mathbf{u} - e\phi, \tag{7-12.9}$$

where we have taken

$$A^{\mu} = (\phi, \mathbf{A}) \tag{7-12.10}$$

so as to conform to the usual expression for the four-potential in terms of the vector potential $\mathbf{A}$ and the scalar potential ϕ. We can define the canonical momentum $\boldsymbol{\pi}$ conjugate to $\mathbf{z}$ in the usual fashion to be

$$\boldsymbol{\pi} = \frac{\partial L}{\partial \mathbf{u}} = \frac{m_0 \mathbf{u}}{\sqrt{1 - \mathbf{u}^2}} + e\mathbf{A}$$

$$= \mathbf{p} + e\mathbf{A}. \tag{7-12.11}$$

The canonical momentum thus differs from the kinetic momentum $\mathbf{p}$ by the additional term $e\mathbf{A}$. The Hamiltonian is also given by the usual equation

$$H = \boldsymbol{\pi} \cdot \mathbf{u} - L$$

$$= \frac{m_0}{\sqrt{1 - \mathbf{u}^2}} + e\phi. \tag{7-12.12}$$

Expressed in terms of π, H has the form

$$H = \sqrt{m_0{}^2 + (\boldsymbol{\pi} - e\mathbf{A})^2} + e\phi. \tag{7-12.13}$$

The canonical equations of motion corresponding to this Hamiltonian are

$$\frac{d}{dx^0} z^r = \frac{\partial H}{\partial \pi^r} = \frac{(\pi^r - eA^r)}{\sqrt{m_0{}^2 + (\boldsymbol{\pi} - e\mathbf{A})^2}} = u^r \tag{7-12.14}$$

and

$$\frac{d}{dx^0} \pi^r = -\frac{\partial H}{\partial z^r} = \frac{eA_{s,r}(\pi_s - eA_s)}{\sqrt{m_0{}^2 + (\boldsymbol{\pi} - e\mathbf{A})^2}} - e\phi_{,r}. \tag{7-12.15}$$

Using the definition of $\boldsymbol{\pi}$, these equations can then be rewritten in the familiar form

$$\frac{d}{dx^0} \mathbf{z} = \frac{\mathbf{p}}{\sqrt{m_0{}^2 + \mathbf{p}^2}}, \tag{7-12.16}$$

$$\frac{d}{dx^0} \mathbf{p} = e\left[\mathbf{u} \times (\nabla \times \mathbf{A}) - \frac{\partial A}{\partial x^0} - \nabla\phi\right], \tag{7-12.17}$$

where we have made use of the fact that $\mathbf{A}$ depends on x^0, both explicitly and implicitly, through its dependence on $z^r(x^0)$. Thus

$$\frac{d}{dx^0} A^r = \frac{\partial A_r}{\partial x^0} + A_{r,s}\dot{z}^s.$$

From the definition of $F_{\mu\nu}$ in terms of the potentials we have, with the identifications (7-12.5) and 7-12.6),

$$\mathbf{B} = \nabla \times \mathbf{A} \tag{7-12.18}$$

and

$$\mathbf{E} = -\frac{\partial}{\partial x^0} \mathbf{A} - \nabla\phi, \tag{7-12.19}$$

so that Eq. (7-12.16) can be written as

$$\frac{d}{dx^0} \mathbf{p} = e(\mathbf{u} \times \mathbf{B}) + e\mathbf{E}.$$

In the nonrelativistic limit, $p^2 \ll m^2$, so that

$$H \approx m_0 + \frac{1}{2m_0}(\boldsymbol{\pi} - e\mathbf{A})^2 + e\phi,$$

which, aside from the rest-mass term, is identical to the Newtonian expression for the Hamiltonian of a charged particle in a combined electric and magnetic field.

7-13. Gauge Invariance

From Eq. (7-12.2) we see that it is the electromagnetic field tensor $F_{\mu\nu}$ that determines the motion of a particle in this field and not the potential A_μ, which appears in the Lagrangian (7-12.1). All would be well were it not for the fact that there is not a unique potential for a given field. Thus, if $F_{\mu\nu} = (A_{\nu,\mu} - A_{\mu,\nu})$, then also $F_{\mu\nu} = (\bar{A}_{\nu,\mu} - \bar{A}_{\mu,\nu})$ where

$$\bar{A}_\mu = A_\mu + \xi_{,\mu}. \tag{7-13.1}$$

The transformation leading from A_μ to $\bar{A}_\mu$ is called a *gauge transformation.* Under a gauge transformation the fields $F_{\mu\nu}$ remain unchanged, and hence such a transformation leaves invariant the particle trajectories. This invariance is reflected in the action principle by the requirement that if

$$S(A) = \int_{\lambda_1}^{\lambda_2} \left\{ -m_0\sqrt{\dot{z}^2} - eA_\mu(z)\dot{z}^\mu \right\} d\lambda$$

is stationary for a dpt, then so is $S(\bar{A})$ for the same dpt. Now

$$\begin{aligned} S(\bar{A}) - S(A) &= -\int_{\lambda_1}^{\lambda_2} e\xi_{,\mu}\dot{z}^\mu \, d\lambda \\ &= -e\xi \Big|_{\lambda_1}^{\lambda_2}. \end{aligned}$$

Consequently $S(\bar{A})$ will be stationary against variations of a dpt that vanish at the end points λ_1 and λ_2 if $S(A)$ has this property.

7-14. Conservation Laws for Particles in Fields

Once we are given the potentials A_μ or the fields $F_{\mu\nu}$ as specified space-time functions, the equations of motion (7-12.2) will no longer admit the full inhomogeneous Lorentz group as a symmetry group. This, of course, does not represent any fundamental contradiction to the postulates of special relativity, since these equations of motion are only approximations to a rigorous set of equations that are in accord with the postulate of special relativity, that is, they admit the full group as a symmetry group.

While the inhomogeneous Lorentz group is no longer a symmetry group of the theory, it must still be a covariance group of the theory. Under a mapping of this group the fields will be transformed into new space-time functions, and a particle trajectory for the original field will be mapped into a particle trajectory of the transformed field. Thus, if $\dot{z}'^{\mu}(\lambda)$ is a mapped trajectory and $F'_{\mu\nu}(x)$ is the mapped field, we must have

$$\frac{d}{d\lambda}\left(m_0 \frac{\dot{z}'^{\mu}}{\sqrt{\dot{z}'^2}}\right) = eF'^{\mu}_{\ \nu}(z')\dot{z}'^{\nu}. \tag{7-14.1}$$

Thus the transformed trajectories satisfy exactly the same *type* of equation as do the original trajectories. Of course the actual form of the equations in terms of the particle variables will in general change under a mapping. The inhomogeneous Lorentz group can now be thought of as the covariance group of the theory (it is what is left of the original covariance group of arbitrary mappings when $g_{\mu\nu} = \eta_{\mu\nu}$) and the fields $F_{\mu\nu}$ as absolute objects in the theory. Consequently whatever symmetry group the equations of motion still possesses will be determined by the motion group of $F_{\mu\nu}$.

To find out how the fields transform under a Lorentz mapping, we could go back to the defining equations (7-9.4) and (7-9.5), and from a knowledge of the transformation properties of trajectories determine those of the fields. Alternatively we could demand that Eqs. (7-14.1) be satisfied if the same equations with unprimed quantities are satisfied. If our formalism is to be consistent, the two procedures must give the same result. Under an arbitrary Lorentz mapping

$$z^{\mu} \to z'^{\mu} = \Lambda_{\nu}^{\ \mu} z'^{\nu} + b^{\mu}.$$

Using this relation we have

$$\frac{d}{d\lambda}\left(m_0 \frac{\dot{z}'^{\mu}}{\sqrt{\dot{z}'^2}}\right) - eF'^{\mu}_{\ \nu}(z')\dot{z}'^{\nu} = \frac{d}{d\lambda}\left(m_0 \Lambda_{\rho}^{\ \mu} \frac{\dot{z}^{\rho}}{\sqrt{\dot{z}^2}}\right) - eF'^{\mu}_{\ \nu}(z')\Lambda_{\sigma}^{\ \nu}\dot{z}^{\sigma}$$

$$= \Lambda_{\rho}^{\ \mu}\left[\frac{d}{d\lambda}\left(m_0 \frac{\dot{z}^{\rho}}{\sqrt{\dot{z}^2}}\right) - \Lambda^{-1}{}_{\tau}^{\ \rho}\, F'^{\tau}_{\ \nu}(z')\Lambda_{\sigma}^{\ \nu}\dot{z}^{\sigma}\right].$$

The right-hand side of this equation will be zero for z^{ρ}, satisfying the equations of motion, with the field $F_{\mu\nu}$, provided

$$F_{\sigma}^{\ \rho}(z) = \Lambda^{-1}{}_{\tau}^{\ \rho}\Lambda_{\sigma}^{\ \nu} F'^{\tau}_{\ \nu}(z')$$

or

$$F'^{\mu\nu}(z') = \Lambda_{\rho}^{\ \mu}\Lambda_{\sigma}^{\ \nu} F^{\rho\sigma}(z). \tag{7-14.2}$$

Thus the electromagnetic field is seen to transform like a second-rank antisymmetric contratensor. Correspondingly the four-potential A^μ can be taken to transform as a contravector.

We said, and shall presently verify, that the motion group of the field $F_\nu{}^\mu$ determines the symmetry group of the equations of motion. According to Nöther's theorem any such symmetry should lead to a constant of the motion along the particle trajectory. However, in deriving these constants of the motion from the Nöther theorem, a difficulty arises because it is the potentials that appear in the Lagrangian (7-12.1). Because of gauge invariance not all the potentials that lead to a given field will have the symmetry of the field. Thus, if A_μ possesses the same symmetry as does $F_\nu{}^\mu$, then in general $\bar{A}_\mu = A_\mu + \xi_{,\mu}$ will not possess this symmetry unless $\xi_{,\mu}$ itself possesses this symmetry. This then leads to the equation: Which of the many potentials leading to a given field should one use to construct the Lagrangian? The answer, as we will see, is that one should use a potential that has symmetry properties of the field itself. That at least one such potential should exist is clear. However, it will not in general be unique. Fortunately this nonuniqueness will lead only to a nonuniqueness in the corresponding conserved quantity of an additive constant.

As an example of the above discussion let us take the case of a field that is constant in the x^1-direction so that $F_{\mu\nu,1} = 0$. This implies that

$$(A_{\nu,\mu} - A_{\mu,\nu})_{,1} = 0,$$

which will be satisfied if $A_{\mu,1} = \psi_{,\mu}$, where ψ is an arbitrary space-time function. Let us now perform a gauge transformation. The transformed potential $\bar{A}_\mu$ will then satisfy $\bar{A}_{\mu,1} - \xi_{,1\mu} = \psi_{,\mu}$. By choosing ξ such that $\xi_{,1} = -\psi$, it follows that $\bar{A}_{\mu,1} = 0$, so that $\bar{A}_\mu$ has the same symmetry as $F_{\mu\nu}$. $\bar{A}_\mu$ is not uniquely determined; we can always perform a further gauge transformation, using a ξ such that $\xi_{,1} = \text{const}$. Such a gauge transformation, however, only changes $\bar{A}$, by adding a constant to it. Since only A_1 appears in the expression for the corresponding conserved quantity, this remaining arbitrariness is unimportant. Similarly in the case of a rotational symmetry the vector potential is again determined only up to a gauge transformation with a ξ that possesses this symmetry. However, again the corresponding conserved quantity is determined up to an additive constant.

To obtain expressions for the conserved quantities corresponding to a symmetry of the field, we note that if under a Lorentz mapping

$$z^\mu \to z'^\mu = \Lambda_\nu{}^\mu z^\nu + a^\mu \quad \text{and} \quad A^\mu(x) \to A'^\mu(x') = \Lambda_\nu{}^\mu A^\nu(x),$$

then

$$S(A', z') = S(A, z).$$

For an infinitesimal mapping

$$S(A+\bar{\delta}A, z+\bar{\delta}z) = S(A,z) + \int d\lambda \left\{ \left(-m_0 \frac{\dot{z}_\mu}{\sqrt{\dot{z}^2}} - eA_\mu(z) \right) \bar{\delta}\dot{z}^\mu - e\,\delta A_\mu(z)\dot{z}^\mu \right\}$$

so that the second term on the right-hand side must be zero.* By performing an integration by parts we then obtain the condition

$$\int d\lambda \left\{ \frac{d}{d\lambda}\left(m_0 \frac{\dot{z}^\mu}{\sqrt{\dot{z}^2}} + eA_\mu(z) \right) \bar{\delta}z^\mu - eA_{\nu,\mu}\dot{z}^\nu\,\bar{\delta}z^\mu + eA_{\nu,\mu}\dot{z}^\nu\,\bar{\delta}z^\mu - e\,\delta A_\mu(z)\dot{z}^\mu \right\}$$
$$- \left\{ m_0 \frac{\dot{z}^\mu}{\sqrt{z^2}} + eA_\mu(z) \right\} \bar{\delta}z^\mu \Bigg|_{\lambda_1}^{\lambda_2} = 0,$$

where we have added and subtracted the term $-eA_{\nu,\mu}\dot{z}^\nu\,\bar{\delta}\dot{z}^\mu$ from the integrand. For a dpt this condition reduces to

$$\int e\,\bar{\delta}A_\mu(z)\dot{z}^\mu\,d\lambda + \left\{ m_0 \frac{\dot{z}_\mu}{\sqrt{\dot{z}^2}} + eA_\mu(z) \right\} \bar{\delta}z^\mu \Bigg|_{\lambda_1}^{\lambda_2} = 0, \qquad (7\text{-}14.3)$$

where as usual

$$\bar{\delta}A^\mu = \varepsilon_\nu{}^\mu A^\nu - A^\mu{}_{,\nu}\,\bar{\delta}z^\nu \quad \text{and} \quad \bar{\delta}z^\mu = \varepsilon_\nu{}^\mu z^\nu + \varepsilon^\mu.$$

Consequently, if for certain values of $\varepsilon_\nu{}^\mu$ and ε^μ, $\bar{\delta}A^\mu = 0$, then (since the end points λ_1 and λ_2 are arbitrary) we obtain the conservation law

$$\frac{d}{d\lambda}\left\{ \left(m_0 \frac{\dot{z}_\mu}{\sqrt{\dot{z}^2}} + eA_\mu(z) \right) \bar{\delta}z^\mu \right\} = 0. \qquad (7\text{-}14.4)$$

It may happen, of course, that no single potential has the full symmetry of the field one is dealing with. In that case it will be necessary to use different potentials (all related by a gauge transformation), each of which has one or more of the symmetries of the field in constructing the constants of the motion indicated in Eq. (7-14.4). Such a situation arises in the case of the uniform electric and magnetic fields.

(a) Translational Invariance

Let n^μ be a unit vector such that $A_{\mu,\nu}n^\nu = 0$. Then for $\bar{\delta}z^\mu = \varepsilon n^\mu$, we have that

$$\bar{\delta}A^\mu = -A^\mu{}_{,\nu}\varepsilon n^\nu = 0$$

* Note that $\delta A_\mu(z)$ and not $\bar{\delta}A_\mu(z)$ appears as a factor in the last term of the integrand. The former appears because, in taking the $\bar{\delta}$ variation of the interaction term, $-e\int A_\mu(z)\dot{z}^\mu\,d\lambda$, $A_\mu(z)$ varies both due to its direct variation $\bar{\delta}A_\mu(z)$ and also due to its dependence on z^μ given by $A_{\mu,\nu}\bar{\delta}z^\nu$. These two contributions add up to give the factor $\delta A_\mu(z)$.

and hence the quantity

$$\left(m_0 \frac{\dot{z}^\mu}{\sqrt{\dot{z}^2}} + eA_\mu\right)n^\mu \tag{7-14.5}$$

will be a constant of the motion along the trajectory of a particle moving in the field derived from this potential. We see that only those gauge transformations for which $\xi_{,\mu}n^\mu = \text{const}$ will preserve the symmetry of A_μ. Thus the conserved quantity is uniquely determined up to a constant.

If, in particular, n^μ is a timelike unit vector, we can always carry out a mapping such that $n^\mu = (1, 0, 0, 0)$. In this case $A_{\mu,0} = 0$, the fields are constant in time, and the conserved quantity is just

$$\mathscr{E} = \frac{m_0}{\sqrt{1 - \mathbf{u}^2}} + e\phi, \tag{7-14.6}$$

which is just the total energy of the charge.

Likewise, if n^μ is spacelike, we can always find a mapping such that $n^\mu = (0, \mathbf{n})$. The conserved quantity then can be written in the form

$$(\mathbf{p} + e\mathbf{A}) \cdot \mathbf{n} = \boldsymbol{\pi} \cdot \mathbf{n} \tag{7-14.7}$$

so that the component of the canonical momentum in the direction of $\mathbf{n}$ is constant along the trajectory.

(b) Rotational Symmetry

The discussion of rotational symmetry is analogous to that of translational symmetry. If $n^{\mu\nu}$ is an everywhere constant, antisymmetric tensor such that

$$n_\nu{}^\mu A^\nu - A^\mu{}_{,\nu} n_\sigma{}^\nu x^\sigma = 0,$$

then

$$\left(m_0 \frac{\dot{z}_\mu}{\sqrt{\dot{z}^2}} + eA_\mu\right)n_\nu{}^\mu z^\nu \tag{7-14.8}$$

is constant along the particle trajectory. If, in particular, $n_\nu{}^\mu$ is such that $n_\nu{}^\mu n^\nu = 0$, where n^ν is a timelike unit vector, we can always find a mapping such that $n_0{}^\mu = 0$ and the conserved quantity takes the form

$$\left(m_0 \frac{\dot{z}^r z^s}{\sqrt{\dot{z}^2}} + eA^r z^s\right)n_{rs}.$$

We can rewrite this conserved quantity in three-vector notation by introducing the three-vector $\mathbf{n} = \{\frac{1}{2}\varepsilon_{rst}n_{st}\}$, which defines the axis of rotational symmetry. In terms of $\mathbf{n}$ the conserved quantity is

$$\{(\mathbf{p} + e\mathbf{A}) \times \mathbf{r}\} \cdot \mathbf{n} = \{\boldsymbol{\pi} \times \mathbf{r}\} \cdot \mathbf{n}. \tag{7-14.9}$$

Thus the component of the canonical angular momentum three-vector $\mathbf{r} \times \boldsymbol{\pi}$ in the direction of the axis of symmetry is constant along the particle's trajectory.

7-15. Motion of a Charged Particle in a Constant Uniform Electric Field

As an application of the above discussion let us consider a charged particle moving in a uniform electric field. If we take the direction of this field to be in the 3-direction, then it follows from Eqs. (7-12.5) and (7-12.6) that the only nonvanishing components of $F_{\mu\nu}$ are $F_{03} = -F_{30} = E$, where E is the magnitude of the field. Now, under an arbitrary infinitesimal Poincaré mapping, it follows from Eq. (7-14.2) that

$$\bar{\delta} F_{\mu\nu} = -F_{\mu\rho}\varepsilon_{\nu}{}^{\rho} - F_{\rho\nu}\varepsilon_{\mu}{}^{\rho} - F_{\mu\nu,\rho}(\varepsilon_{\sigma}{}^{\rho}x^{\sigma} + \varepsilon^{\sigma})$$

when

$$\bar{\delta} z^{\mu} = \varepsilon_{\nu}{}^{\mu}z^{\nu} + \varepsilon^{\mu}.$$

We see that there are four translational symmetries with

$$n_1{}^{\mu} = (1, 0, 0, 0), \tag{7-15.1}$$

$$n_2{}^{\mu} = (0, 1, 0, 0), \tag{7-15.2}$$

$$n_3{}^{\mu} = (0, 0, 1, 0), \tag{7-15.3}$$

$$n_4{}^{\mu} = (0, 0, 0, 1). \tag{7-15.4}$$

In addition there are two rotational symmetries. The nonvanishing components of $n^{\mu\nu}$ are

$$n_1^{12} = -n_1^{21} = 1 \tag{7-15.5}$$

and

$$n_2^{03} = -n_2^{30} = 1. \tag{7-15.6}$$

Because of the high degree of symmetry of $F_{\mu\nu}$ no one potential will possess all its symmetries. We need in fact three different potentials to construct the six constants of the motion for this field. The potential $A_{(1)}{}^{\mu} = (-Ex^3, 0, 0, 0)$

has the symmetries (7-15.1), (7-15.2), (7-15.3), and (7-15.5). The corresponding constants of the motion are

$$\mathscr{E} = \frac{m_0}{\sqrt{1-\mathbf{u}^2}} - eEz^3, \tag{7-15.7}$$

$$p^1 = \frac{m_0 u^1}{\sqrt{1-\mathbf{u}^2}}, \qquad p^2 = \frac{m_0 u^2}{\sqrt{1-\mathbf{u}^2}}, \tag{7-15.8}$$

$$\Lambda^3 = \frac{m_0 u^1 z^2}{\sqrt{1-\mathbf{u}^2}} - \frac{m_0 u^2 z^1}{\sqrt{1-\mathbf{u}^2}}. \tag{7-15.9}$$

The potential $A_{(2)}{}^{\mu} = (0, 0, 0, -Ex^0)$ has the symmetries (7-15.2), (7-15.3), (7-15.4), and (7-15.5), and leads to the additional constant of the motion

$$\pi^3 = \frac{m_0 u^3}{\sqrt{1-\mathbf{u}^2}} - eEz^0. \tag{7-15.10}$$

Finally, the potential

$$A_{(3)}{}^{\mu} = (-\tfrac{1}{2}Ex^3, 0, 0, -\tfrac{1}{2}Ex^0)$$

has the symmetries (7-15.2), (7-15.3), (7-15.5), and (7-15.6), which give one additional constant

$$\Gamma^3 = \frac{m^0}{\sqrt{1-\mathbf{u}^2}} z^3 - \frac{m_0 u^3}{\sqrt{1-\mathbf{u}^2}} z^0 + \frac{1}{2} eE\{(z^0)^2 - (z^3)^2\}. \tag{7-15.11}$$

These constants of the motion provide us with a number of first integrals of motion and can be used to discuss the motion of a charge in a field.

As an example of the use of the above constants of the motion, consider the case when $p^2 = \Lambda^3 = 0$. From Eqs. (7-15.8) and (7-15.9) one finds in this case that $z^2 = 0$. We can use Eqs. (7-15.7), (7-15.10), and (7-15.11) to find z^3 as a function of z^0. Combining these equations we have

$$z^3 = \frac{-\mathscr{E} \pm \sqrt{\mathscr{E}^2 + 2E(\Gamma^3 + \pi^3 z^0 + \frac{1}{2}eEz^{0^2})}}{eE}. \tag{7-15.12}$$

In addition it follows from Eqs. (7-15.7) and (7-15.8) that

$$\frac{dz^1}{dz^0} = \frac{p^1}{\mathscr{E} + eEz^3}. \tag{7-15.13}$$

We can then use Eqs. (7-15.12) and (7-15.13) to find z^1 as a function of z^0 and also to find z^3 as a function of z^1. Consider the case when $\Gamma^3 = \pi^3 = 0$. Then, after some algebra, we find that

$$\frac{dz^1}{dz^3} = \frac{p^1}{eE} \frac{1}{\sqrt{z^{3^2} + (2\mathscr{E}/eE)z^3}}. \tag{7-15.14}$$

A particular solution is

$$z^3 = \frac{\mathscr{E}}{eE} \cosh \frac{eE}{p^1} z^1, \tag{7-15.15}$$

which is the equation of a catenary.

Problem 7.13. A charged particle enters a uniform electric field with zero velocity. Derive an expression for its velocity after leaving this region, in terms of the potential difference V it encounters. Compare this result with the corresponding nonrelativistic expression.

7-16. Motion of a Charged Particle in a Constant Uniform Magnetic Field

As a second application of the results of Section 7-14 let us consider the motion of a charged particle in a constant homogeneous magnetic field. Taking the field to lie in the positive 3-direction, we see from Eq. (7-12.6) that the only nonvanishing components of $F_{\mu\nu}$ are $F_{21} = -F_{12} = B$. As might be expected, the symmetries of this field are the same as for the uniform, constant electric field given by Eqs. (7-15.1) to (7-15.6). Here again it will be necessary to use several different equivalent four-potentials A^μ to construct the six corresponding constants of the motion.

The potential $A_{(1)}{}^\mu = (0, -\frac{1}{2}Bx^2, \frac{1}{2}Bx^1, 0)$ has the symmetries (7-15.1), (7-15.4), (7-15.5), and (7-15.6). The corresponding constants of the motion are

$$\mathscr{E} = \frac{m_0}{\sqrt{1 - \mathbf{u}^2}}, \tag{7-16.1}$$

$$p^3 = \frac{m_0 u^3}{\sqrt{1 - \mathbf{u}^2}}, \tag{7-16.2}$$

$$\Gamma^3 = m_0 \frac{z^3 - u^3 z^0}{\sqrt{1 - \mathbf{u}^2}}, \tag{7-16.3}$$

and

$$\Lambda^3 = m_0 \frac{u^1 z^2 - u^2 z^1}{\sqrt{1 - \mathbf{u}^2}} - \frac{1}{2} eB\{(z^1)^2 + (z^2)^2\}. \tag{7-16.4}$$

The potential $A_{(2)}{}^\mu = (0, -Bx^2, 0, 0)$ has the symmetries (7-15.1), (7-15.2), (7-15.4), and (7-15.6) and yields the additional constant

$$\pi^1 = m_0 \frac{u^1}{\sqrt{1 - \mathbf{u}^2}} - eBz^2, \tag{7-16.5}$$

while the potential $A^\mu = (0, 0, Bx^1, 0)$ has the symmetries (7-15.1), (7-15.3), (7-15.4), and (7-15.6) and leads to the final constant

$$\pi^2 = m_0 \frac{u^2}{\sqrt{1 - \mathbf{u}^2}} + eB_0 z^1. \tag{7-16.6}$$

Combining Eqs. (7-16.1) and (7-16.2), we find that

$$u^3 = \frac{p^3}{\mathscr{E}}$$

so that

$$z^3 = \left(\frac{p^3}{\mathscr{E}}\right) z^0 + z_0{}^3. \tag{7-16.7}$$

It then follows from Eq. (7-16.3) that $\Gamma^3 = \mathscr{E} z_0{}^3$. The remainder of the solution can be obtained by making use of Eqs. (7-16.5) and (7-16.6). One finds

$$z^1 = r \sin(\omega z^0 + \alpha) + z_0{}^1, \tag{7-16.8a}$$

$$z^2 = r \cos(\omega z^0 + \alpha) + z_0{}^2, \tag{7-16.8b}$$

where

$$z_0{}^2 = -\frac{\pi^1}{eB}, \qquad z_0{}^1 = \frac{\pi^2}{eB},$$

$$\Lambda^3 = \tfrac{1}{2} eB(r^2 - (z_0{}^1)^2 - (z_0{}^2)^2),$$

and

$$\omega = \frac{eB}{\mathscr{E}}. \tag{7-16.9}$$

The particle thus moves uniformly along a circular helix of radius r whose axis is in the direction of the field, and screw sense is positive. In the nonrelativistic limit $\mathscr{E} \approx m_0 c^2$,

$$\omega = \frac{eB}{m_0 c}$$

which is the cyclotron frequency of a particle moving in a uniform magnetic field.

Problem 7.14. Find the symmetries and the corresponding constants of the motion for a particle moving in the following fields:

1. Combined uniform electric and magnetic field with the two fields parallel, and
2. Combined uniform electric and magnetic fields equal in magnitude and mutually perpendicular.

Discuss the motion of the charge.

7-17. Radiation Damping

While the action-at-a-distance formulation of the electromagnetic interactions of charged particles has a number of quite satisfactory features (for example, it does not bring in the concept of the self-action of a charge on itself), it cannot be considered as a complete theory. As it stands, it fails to account for two well-established facts associated with the electromagnetic interaction. It is well known that accelerated charges radiate and as a consequence their mechanical energy decreases. This loss is produced by a radiation reaction force acting on the particle, which for slowly moving particles is proportional to the time rate of change of the acceleration of the particle. The equations of motion (7-9.3) contain no such radiation reaction term. Closely related to this difficulty is the fact that the interaction between charges is symmetric, containing both advanced and retarded interactions, whereas only retarded interactions occur in nature, at least in the macroscopic domain.

In order to overcome these difficulties Wheeler and Feynman extended the action-at-a-distance formalism to take account of the presence of an absorber during the emission of radiation. In their paper they have considered the interaction of the absorber with a radiator from several different points of view. We will discuss here only one of their approaches.

In the equations of motion (7-9.3) for the ith particle the field, $F_{\mu\nu}$ is a sum of $2(N-1)$ terms, two each for every other particle. Each particle produces a field that is the sum of an advanced and a retarded part. Thus the potential $A_{k\mu}(z_i)$ of the kth particle is given by Eq. (7-9.7), where $A^+_{k\mu}(z_i)$ is the retarded part of this potential and $A^-_{k\mu}(z_i)$ is the advanced part. Corresponding to $A^\pm_{k\mu}(z_i)$ we have the retarded field $F^+_{k\mu\nu}(z_i)$ and the advanced field $F^-_{k\mu\nu}(z_i)$ given by Eq. (7-9.9). In terms of these partial fields, Eq. (7-9.3) takes the form

$$m_{0_i}\ddot{z}_i^{\mu} = e_i \sum_{k\neq i} \{F^{+\,\mu}_{k\nu}(z_i) + F^{-\,\mu}_{k\nu}(z_i)\}\dot{z}_i^{\nu},$$

which we can rewrite as

$$\begin{aligned} m_{0_i}\ddot{z}_i^{\mu} &= e_i\left\{\sum_{k\neq i} 2F^{+\,\mu}_{k\nu}(z_i) + \sum_{k\neq i}(F^{-\,\mu}_{k\nu}(z_i) - F^{+\,\mu}_{k\nu}(z_i))\right\}\dot{z}_i^{\nu} \\ &= e_i\left\{\sum_{k\neq i} 2F^{+\,\mu}_{k\nu}(z_i) + \sum_{k}(F^{-\,\mu}_{k\nu}(z_i) - F^{+\,\mu}_{k\nu}(z_i)) \right. \\ &\qquad \left. -(F^{-\,\mu}_{i\nu}(z_i) - F^{+\,\mu}_{i\nu}(z_i))\right\}\dot{z}_i^{\nu}. \end{aligned} \tag{7-17.1}$$

In this form we can take the first term to represent the retarded effect of all the other particles on the ith particle. The last term, representing an action of the ith particle on itself, is the term that is responsible for the radiation damping term. The evaluation of this term is rather involved and will not be undertaken here. It was first computed by Dirac[17] who showed that

$$-e_i(F_{i\nu}^{-\mu}(z_i) - F_{i\nu}^{+\mu}(z_i))\dot{z}_i^{\nu} = \tfrac{2}{3}e_i^{2}(\dddot{z}_i^{\mu} + \dot{z}_i^{\mu}\ddot{z}^2). \qquad (7\text{-}17.2)$$

For slowly moving particles we can neglect the second term in this equation and replace derivatives with respect to proper time by derivatives with respect to x^0 to obtain the observed radiation reaction force $(2/3)e_i^{2}\dddot{z}_i$.

Let us now turn our attention to the second term in the force term of Eq. (7-17.1). In order to deal with this term Wheeler and Feynman make the assumption that the total system of charges one has to deal with in nature is such that any radiation emitted by charges in the system is completely absorbed by the system. If one makes the additional assumption that the system as a whole is electrically neutral, then at distances sufficiently far removed from the system the total field should be zero. This will be the case if, outside the system,

$$\sum_k (A_{k\mu}^{+} + A_{k\mu}^{-}) = 0$$

for all times. Since the sum vanishes everywhere outside the system it follows that each of the two sums must vanish separately outside the system for all times. The first sum represents, at large distances, an outgoing wave and the other represents an incoming wave. Complete destructive interference between such waves in a finite space-time domain is impossible, so we arrive at the conclusion stated above. But then it follows that the difference of the two sums also vanishes outside the system for all times:

$$\sum_k (A_{k\mu}^{+} - A_{k\mu}^{-}) = 0.$$

Now one can show that the quantity $\sum_k(A_{k\mu}^{+} + A_{k\mu}^{-})$ satisfies

$$\Box \sum_k (A_{k\mu}^{+} - A_{k\mu}^{-}) = 0$$

over the whole of space-time. Consequently, if it vanishes everywhere outside the system for all times, it must vanish throughout the whole of space-time. Thus, if the Wheeler-Feynman assumption of a completely absorbing universe is valid, the second term in the force term of Eq. (7-17.1) vanishes and we finally obtain as the equation of motion of the ith particle

$$m_{0_i}\ddot{z}_i^{\mu} = e_i \sum_{k \neq i} 2F_{k\nu}^{+\mu}(z_i)\dot{z}_i^{\nu} + \tfrac{2}{3}e_1^{2}(\dddot{z}_i^{\mu} + \dot{z}_i^{\mu}\ddot{z}^2). \qquad (7\text{-}17.3)$$

At this point we might feel that we have arrived at a satisfactory description of the known facts concerning the electromagnetic interactions of charged

[17] P. A. M. Dirac, *Proc. Roy. Soc.* (London), **A167**, 148 (1938).

particles via an action-at-a-distance formalism. Such, however, is not the case. As Wheeler and Feynman point out, by reversing the roles of the advanced and retarded fields one could also arrive at an equation of motion for the ith particle in a completely absorbing universe of the form

$$m_{0_i}\ddot{z}_i = e_i \sum_{k \neq i} 2F_{\nu}^{-\mu}(z_i)\dot{z}_i^{\nu} - \tfrac{2}{3}e_i^2(\dddot{z}_i^{\mu} + \dot{z}_i^{\mu}\ddot{z}^2). \qquad (7\text{-}17.4)$$

Now, however, it is the advanced fields of the other particles that act on the ith particle, and the radiation reaction force appears with the sign opposite to its usual one.

We can reconcile the two points of view associated with Eqs. (7-17.3) and (7-17.4) by imagining that up to the time of emission of radiation by the ith particle, the other particles were in random motion and randomly distributed throughout space. If this is the case then the retarded actions of these other particles will cancel out each other. Hence the sum, $\sum_{k \neq i} 2F_{k\nu}^{+\mu}$ will be small compared with the radiation reaction term. On the other hand, $\sum_{k \neq i} 2F_{k\nu}^{-\mu}$ will not be small compared with the radiation reaction term. Since the radiation emitted by the ith particle must be totally absorbed by the other particles, they must be put into motion in just such a way as to make this happen. Consequently one can imagine that $\sum_{k \neq i} 2F_{\nu}^{-\mu}$ contains a contribution that is twice the magnitude of the damping force over and above any contribution due to the random initial motion of the absorbing charges. Thus once again we obtain the correct expression for the radiation reaction term.

From this point of view, radiation is essentially a consequence of statistical mechanics rather than electrodynamics, a point of view first set forth by Einstein.[18] One could imagine equally well an initial state of the system in which the absorbing particles are organized to the extent that at the moment the radiator radiates, it receives an impulse that is just sufficient to reverse the sign of the radiation reaction term. Such a state of affairs is not observed in nature because the *a priori* probabilities of such an initial arrangement of absorbing particles is small compared to a random arrangement. Only with this additional interpretation does one obtain a completely satisfactory description of the radiative process within the framework of an action-at-a-distance formalism.

7-18. Radiation Reaction and Preacceleration

In our discussion of the motion of a charged particle in an electromagnetic field we have assumed that the equations of motion for such a particle are given by Eq. (7-12.2), which in turn is obtained from the original Fokker

[18] W. Ritz and A. Einstein, *Zeits. Physik.*, **10**, 323 (1909).

equations discussed in Section 7-9. Actually these equations are incomplete in that they neglect the radiation reaction force given by Eq. (7-17 2).

In addition to the derivation due to Wheeler and Feynman, this force has also been derived by Dirac[19] on the assumption that Maxwell's equations are valid down to the position of a point singularity. By subtracting out infinities at the position of the particle he finds that the self-field of the particle that acts back on the particle (a notion, of course, foreign to the action-at-a-distance formulation) is just $(1/2)(F_\nu^{+\mu} - F_\nu^{-\mu})$. His derivation therefore gives the same radiation reaction force as does the Wheeler-Feynman derivation. At the same time the expression for the radiation reaction due to Dirac and that of Wheeler-Feynman agrees in the nonrelativistic approximation with the first term in an expression for this force derived by Lorentz.[20]

In his derivation Lorentz used a charged sphere as a model for a charged particle and calculated the reactive force by considering the retarded action of one part of the sphere on another. The size of the particle appears as a parameter in an expansion for this force. The first term in this expansion is proportional to the overall acceleration of the particle and diverges as the radius of the particle approaches zero. It represents a "renormalization" of the mass of the charge. The second term is independent of the size of the particle and is proportional to the time derivative of the acceleration of the particle and represents the radiation reaction force. Higher terms in the expansion depend on higher time derivatives of the particle position but vanish in the limit of zero particle size.

If one accepts the existence of a radiation reaction force as an explanation for the energy loss in an antenna or the loss in a betatron due to syncrotron radiation then for the equations of motion of a charged particle moving in an external field $F_\nu{}^\mu$, one must take not Eq. (7-12.2) but rather the equations[21]

$$m_0 \ddot{z}^\mu = eF_\nu{}^\mu(z)\dot{z}^\nu + \tfrac{2}{3}e^2(\dddot{z}^\mu + \dot{z}^\mu \ddot{z}^2), \qquad (7\text{-}18.1)$$

where the path parameter is taken to be proper time.

Solutions to these equations were first consistently studied by Dirac[22] and later by Eliezer.[23] They have been most recently studied by Plass.[24]

[19] P. A. M. Dirac, *loc. cit.*

[20] H. A. Lorentz, *The Theory of Electrons* (1st ed., Teubner, Leipzig, 1909, pp. 49, 253) (Dover reprint, 2nd ed., 1952).

[21] From the point of view of the Wheeler-Feynman picture of radiation reaction, there is a certain inconsistency in adopting these equations of motion. The first term on the right-hand side represents the action of the other charges in the universe acting so as to produce the field $F_\nu{}^\mu$ but neglects the action of the charge e back on these other charges. On the other hand, the radiation-reaction force arises, in the Wheeler-Feynman picture, in the first instance from the effect of this charge on the other charges.

[22] Dirac, *op. cit.*

[23] C. J. Eliezer, *Rev. Mod. Phys.*, **19**, 147 (1947).

[24] G. N. Plass, *Rev. Mod. Phys.*, **33**, 37 (1961).

The fundamental difficulty involved in these equations can be most easily appreciated by considering their one-dimensional analogue. As a further simplification let us assume that only an electric field is acting and is a given function $E(\tau)$ of the proper time along the trajectory of the particle. In this case Eq. (7-18.1) can be rewritten in the form

$$\dot{v} - \frac{1}{b}\ddot{v} + \frac{1}{b}\frac{v\dot{v}^2}{1+v^2} = \frac{1}{m_0}(1+v^2)f(\tau), \qquad (7\text{-}18.2)$$

where

$$\frac{1}{b} = \frac{2}{3}\frac{e^2}{m}, \qquad f(\tau) = eE(\tau),$$

and v is the one spatial component of the four-velocity ($v = \dot{z}^1$). As usual a dot over a quantity stands for a differentiation with respect to proper time. Following Plass, we introduce a new dependent variable $w(\tau)$, defined by

$$v(\tau) = \sinh w(\tau).$$

In terms of this new variable Eq. (7-18.2) becomes

$$m\dot{w} - \frac{m}{b}\ddot{w} = f(\tau). \qquad (7\text{-}18.3)$$

The general solution to this equation can be found and is given by

$$\dot{w} = e^{b\tau}\left[\dot{w}(0) - \frac{b}{m}\int_0^\tau e^{-b\tau'}f(\tau')\,d\tau'\right].$$

For an arbitrary value of $\dot{w}(0)$, it is seen that ultimately $\dot{w}$ increases as $e^{b\tau}$, implying also that $\dot{v}$ increases without limit. These are the so-called runaway solutions. Even if $f(\tau) = 0$, such solutions exist.[25]

Dirac proposed the following solution to the problem of the runaway solutions. Without the radiative reaction term, Eq. (8-18.1) is of second differential order in the proper time. Hence a particular solution is specified by giving the position and velocity of the particle at some point on its trajectory. However, with the radiation reaction term, these equations are of third differential order. Now one must give the acceleration of the particle as well, to specify a particular trajectory. Dirac proposed that the initial acceleration must be such that in the limit as time approaches infinity, the acceleration

[25] Because of the existence of these nonphysical solutions many authors hold that Eq. (7-18.1) is valid only in a slow-motion approximation when the damping force is small compared with the force on the particle arising from external fields. According to this view, if these external forces are absent there should be no radiation reaction. For a discussion of this approach see L. D. Landau and E. M. Lifshitz, *The Classical Theory of Fields*, 2d ed. (Pergamon Press, London, 1962), Sec. 75.

must remain finite unless the force acting on the particle supplies it with the energy to produce a infinite acceleration.

If we accept Dirac's proposal then the coefficient of $e^{b\tau}$ must approach zero as $\tau \to \infty$ in order to avoid the runaway solution. Thus the initial acceleration (in our case $\dot{w}(0)$) must be chosen so that

$$m\dot{w}(0) = \int_0^\infty e^{-b\tau'} f(\tau')\, d\tau'. \tag{7-18.4}$$

Thus, unlike the situation without radiation reaction where the initial acceleration is determined by the initial force, the initial acceleration here, and hence the initial motion, is determined by the forces that act on the particle in its future. However, since $b^{-1} = 6.27 \times 10^{-24}$ sec, the initial acceleration will be effectively determined by the initial force unless the latter changes rapidly in a time interval of this size.

With the initial value of the $\dot{w}$ given by Eq. (7-18.4), $\dot{w}(\tau)$ at other times is given by

$$\begin{aligned} m\dot{w}(\tau) &= b \int_{\tau}^\infty e^{-b(\tau' - \tau)} f(\tau')\, d\tau' \\ &= b \int_0^\infty e^{-b\tau'} f(\tau + \tau')\, d\tau'. \end{aligned} \tag{7-18.5}$$

In the case that $f(\tau)$ is a delta function, that is, $f(\tau) = k\,\delta(\tau)$, $w(\tau)$ is given by

$$m\dot{w}(\tau) = \begin{cases} bke^{b\tau} & \tau < 0 \\ 0 & \tau > 0 \end{cases}$$

Thus the particle experiences the phenomenon of preacceleration. It has to start moving a short time before the delta-function pulse acts on it, in order that its acceleration at $\tau = 0$ will be such that the ensuing motion is not of the runaway type.

7-19. The Uniformly Accelerated Charge; Energy Balance

No discussion of the radiation reaction force would be complete without a mention of the case of the uniformly accelerated charge. If $f(\tau) = mk$, that is, is constant in time, then we see from Eq. (7-18.5) that $\dot{w}(\tau) = k$. In this special case the radiation reaction term in Eq. (7-18.2) is absent, and the particle moves in the same way as an uncharged particle in the presence of a constant force. Because of the absence of a radiation reaction force, the question arises (and has been answered in different ways by different writers[26]): Does a uniformly accelerated charge radiate? If it does radiate, then another question arises: Where does the radiated energy come from? It cannot come directly from the force, since no more force is needed to produce a given acceleration when the particle is charged than when it is uncharged.

Plass answered this question by examining the time rate of change of the energy of the particle. We saw in Section 7-14 that if $A_{\mu,\nu}n^\nu = 0$ for $n^\mu n_\mu = 1$, then

$$\mathscr{E} = \left(m_0 \frac{\dot{z}_\mu}{\sqrt{\dot{z}^2}} + eA_\mu\right)n^\mu$$

was a constant of the motion and could be interpreted as the energy of the particle. Let us now compute the derivative of $\mathscr{E}$ with respect to proper time, making use of the Eq. (7-18.1). One finds after some computation that

$$\frac{d}{d\tau}\{m_0(1+u^2)^{1/2} + eA_\mu n^\mu\} = -\frac{2e^2}{3}\frac{d}{d\tau}\left\{\frac{\dot{u}^\mu u_\mu}{(1+u^2)^{1/2}}\right\} + \frac{2e^2}{3}\frac{(\dot{u}^\mu u_\mu)^2 + (1-u^2)\dot{u}^2}{(1+u^2)^{1/2}}, \tag{7-19.1}$$

where u^μ is the part of the four-velocity that is normal to n^μ and is given by $u^\mu = \dot{z}^\mu - n^\mu n_\nu \dot{z}^\nu$ and $u^2 = u^\mu u_\mu$. For one-dimensional motion, with $n^\mu = (1, 0)$ and $u^\mu = (0, u)$, Eq. (7-19.1) can be written in the form

$$\frac{d}{d\tau}\left\{m_0(1+u^2)^{1/2} + e\phi(z) - \frac{2e^2}{3}u\dot{u}(1+u^2)^{-1/2}\right\} = -\frac{2}{3}\left\{e^2\dot{u}^2(1+u^2)^{-1/2}\right\}. \tag{7-19.2}$$

Since $(1+u^2)^{1/2} = (1-v^2)^{-1/2}$, the first term within the bracket on the left-hand side of this equation represents the relativistic energy, kinetic plus rest mass, of the particle. The second term is the potential energy of the particle. The third term, which arises as a consequence of the radiation reaction, has been called the *acceleration energy* by Schott.[27] It is interpreted as a reversible loss or gain of energy by radiation during the period of acceleration. The term on the right-hand side, which is always negative, is interpreted to be an irreversible loss of radiation by radiation and is the usual expression for

[26] For another discussion and also other references, see T. Fulton and F. Rohrlich, *Ann. Phys.*, **9**, 499 (1960).

[27] G. A. Schott, *Phil. Mag.*, **29**, 49 (1915).

the energy radiated by a particle.[28] If the acceleration of the particle is non-zero for only a finite period, then indeed the integral of this quantity represents the difference of $m_0(1+v^2)^{-1/2} + e\phi(z)$ at the beginning and end of the period.

In the case of a uniformly accelerated particle the fact that the integral of the right side of Eq. (7-19.2) increases monotonically with time can be taken as an indication that the particle radiates energy. According to Plass the source of this radiated energy is the acceleration energy. However, one could also argue that the energy radiated per unit proper time is the difference between the right side of Eq. (7-19.2) and the acceleration energy, in which case the uniformly accelerated particle could be said to not radiate. Clearly the question cannot be settled by arguing about definitions, which is what we are doing. In order to give a meaningful answer to our question one should take into account both the mechanism whereby the radiated energy is absorbed and the reaction of this mechanism back on the particle that is doing the radiating.

The case of a constant force that acts for a finite period of time has also been treated by Plass, with results that are more easily interpreted. If

$$f(\tau) = \begin{cases} 0 & 0 < \tau < \tau_0 \\ mk & \tau_0 < \tau < \tau_1, \\ 0 & \tau_1 < \tau \end{cases}$$

then $\dot{w}(\tau)$ as determined by Eq. (7-18.5) is given by

$$\dot{w}(\tau) = \begin{cases} k(e^{-b\tau_0} - e^{-b\tau_1})e^{b\tau} & 0 < \tau < \tau_0 \\ k(1 - e^{-b(\tau_1 - \tau)}) & \tau_0 < \tau < \tau_1, \\ 0 & \tau_1 < \tau \end{cases}$$

so not only does the particle preaccelerate but also it predecelerates. Upon integrating this expression for $\dot{w}(\tau)$ one obtains

$$w(\tau) = \begin{cases} w_0 + (k/b)(e^{-b\tau_0} - e^{-b\tau_1})(e^{b\tau} - 1) & 0 < \tau < \tau_0 \\ w_0 + k(\tau - \tau_0) - (k/b)e^{-b\tau_1}(e^{b\tau} - 1) + (k/b)(1 - e^{b\tau_0}) & \tau_0 < \tau < \tau_1 \\ w_0 + k(\tau_1 - \tau_0) - k(b)(e^{-b\tau_0} - e^{-b\tau_1}) & \tau_1 < \tau \end{cases}$$

Thus the final value of w is less by an amount $(k/b)(e^{-b\tau_0} - e^{-b\tau_1})$ than in the case of an uncharged particle (no radiation reaction force) acted upon by the same force. This could be taken as an indication that a particle acted on by such a force really does radiate.

[28] See Section 7-4.

Problem 7.15. Calculate the work done on the particle in the above discussion and compare it to the work done by the same force on an uncharged particle.

7-20. Motion of Spinning Particles in Fields

In nonrelativistic mechanics it is possible to describe the motion of particles in fields that possess electric and magnetic moments as well as net charges. As long as the inhomogeneities in the field can be neglected over dimensions of the order of magnitude of the particle, its overall motion depends only on its net charge, and its equations of motion are the slow-motion limit ($c \to \infty$) of the relativistic equations (7-12.2). If the system has a magnetic dipole moment $\boldsymbol{\mu}$ it will be related to its angular momentum $\mathbf{l}$ by the usual relation between these two quantities:

$$\boldsymbol{\mu} = \frac{e}{2mc}\,\mathbf{l}.$$

When a magnetic field is present the angular momentum vector will in general precess. Since the torque acting on it is $\boldsymbol{\mu} \times \mathbf{B}$, we have the nonrelativistic equation of motion for $\mathbf{l}$:

$$\frac{d}{dt}\mathbf{l} = \frac{e}{2mc}\,\mathbf{l} \times \mathbf{B}. \tag{7-20.1}$$

It is known that the electron and other particles possess magnetic moments. While it is not possible to picture these moments as arising from current loops, we can nevertheless assign an intrinsic angular momentum to such particles. This quantity, called its *spin* $\mathbf{s}$, is related to its magnetic moment $\boldsymbol{\mu}$ by

$$\boldsymbol{\mu} = g\,\frac{e}{2mc}\,\mathbf{s}, \tag{7-20.2}$$

where g is the gyromagnetic ratio peculiar to the particle in question. For electrons and muons $g \approx 2$, while for a proton $g_p = 5.59$ and for a neutron $g_n = -3.83$. The difference between the actual value of $\boldsymbol{\mu}$ for the particle and the value obtained by taking $g = 2$ is called the *anomalous magnetic moment*.

We would now like to develop, for a particle with spin, a relativistic set of equations of motion that reduce in the slow-motion approximation to Eq. (7-20.1). In doing so, we will assume that the effect of all field gradients on the spin of the particle can be neglected. Consequently we can take Eq. (7-12.2) to be the equation of motion that determines the trajectory of the particle. In order to describe the spin motion of the particle we introduce a spin tensor $s^{\mu\nu} = -s^{\nu\mu}$, with the property that

$$s_{\mu\nu}\dot{z}^{\nu} = 0. \tag{7-20.3}$$

In the slow-motion approximation, $\dot{z}^{\mu} \to (1, 0, 0, 0)$, so that $s^{0\mu} \to 0$. If we take $\varepsilon_{rst}s^{rs} \to s_t$, then $s^{\mu\nu}$ is seen to be a possible generalization of the spin of a particle.

We now ask for a relativistic equation of motion that reduces in the slow-motion approximation to Eq. (7-20.1), with **l** replaced by **s**. Let us choose the proper time τ to be the path parameter. Then a possible set of equations of motion for $s^{\mu\nu}$ that satisfy our requirements is

$$\frac{ds_{\mu\nu}}{d\tau} = g\left(\frac{e}{2m}\right)(s_{\nu\rho}F_{\mu}{}^{\rho} - s_{\mu\rho}F_{\nu}{}^{\rho}). \tag{7-20.4}$$

The question arises: To what extent is Eq. (7-20.4) unique? The left side is unique. Therefore, we must now ask: Given the quantities $\dot{z}^{\mu}$, $F^{\mu\nu}$, and $s^{\mu\nu}$, what antisymmetric tensor can we construct from them that is linear in $s^{\mu\nu}$? Because of the condition (7-20.3) the only possibility other than the right side of Eq. (7-20.4) is a term $g^*(e/2m)(s_{\nu\rho}F^*{}_{\mu}{}^{\rho} - s_{\mu\rho}F^*{}_{\nu}{}^{\rho})$, where $F^*_{\mu\nu} = \frac{1}{2}\varepsilon_{\mu\nu\rho\sigma}F^{\rho\sigma}$. However, this reduces in the slow-motion limit to $g^*(e/2m)(\mathbf{s} \times \mathbf{E})$. Such a term would be present in Eq. (7-20.1) if the particle, in addition to having a magnetic moment given by Eq. (7-20.2) had an electric dipole moment **d**, given by $\mathbf{d} = g^*(e/2mc)\mathbf{s}$. While nothing in principle rules out this possibility, it leads to equations of motion that are not invariant under time reversal. Further, since no particles have been observed to have an intrinsic electric dipole moment, we will assume that Eq. (7-20.4) is the correct relativistic generalization of Eq. (7-20.1).

Let us now inquire if the equations of motion (7-20.4) are consistent with the equations of motion (7-12.2) for the trajectory of our spinning particle. If we multiply Eq. (7-20.4) by $\dot{z}^{\nu}$ and make use of Eq. (7-20.3) we find that the left side gives

$$\frac{ds_{\mu\nu}}{d\tau}\dot{z}^{\nu} = -s_{\mu\nu}\ddot{z}^{\nu},$$

and making use of Eq. (7-12.2), we obtain for the right side the result

$$g\left(\frac{e}{2m}\right)(s_{\nu\rho}F_{\mu}{}^{\rho} - s_{\mu\rho}F_{\nu}{}^{\rho})\dot{z}^{\nu} = -g\left(\frac{e}{2m}\right)s_{\mu\nu}\left(\frac{m}{e}\right)\ddot{z}^{\nu}.$$

The equality of these two terms requires that we take $g = 2$.

Consistency therefore forces us to consider only particles with zero anomalous magnetic moment. It is well known that the magnetic moment of any particle that obeys a Dirac equation has no anomalous part. The anomalous contribution to the magnetic moment of the electron arises from radiative corrections to the Dirac theory, calculated from quantum field theory. Bhabha and Corben[29] have shown that if one takes account of radiation reaction, then g has the value 2 only in a first approximation. They show that $s_{\mu\nu}$ will still obey an equation of the form (7-20.4) when radiation reaction effects are included in the next order of approximation but that it need no longer have the value $g = 2$.

Instead of $s_{\mu\nu}$ it is more convenient, for purposes of interpretation, to work with the four-vector w_μ, the polarization vector, defined by

$$w_\mu = \tfrac{1}{2}\varepsilon_{\mu\nu\rho\sigma}s^{\nu\rho}p^\sigma. \tag{7-20.5}$$

Since in the slow-motion limit $p^\sigma \to (m_0, 0, 0, 0)$, it follows that in this limit, $w^\mu \to (0, m_0\mathbf{s})$. One can solve Eq. (7-20.5) for $s^{\mu\nu}$. The result is that

$$s_{\mu\nu} = \frac{1}{{m_0}^2}\,\varepsilon_{\mu\nu\rho\sigma}w^\rho p^\sigma. \tag{7-20.6}$$

It follows from the fact that $p^\mu p_\mu = {m_0}^2$ and Eq. (7-20.3) that $w^\mu w_\mu$ is a spacelike scalar. Its value will be taken to be

$$w^\mu w_\mu = -{m_0}^2{s_0}^2. \tag{7-20.7}$$

This scalar is directly related to the quantity W introduced in Section 6-5 in our discussion of the representations of the Poincaré group, just as $p^\mu p_\mu$ is related to the P defined there. We see that in the limit $c \to \infty$, ${s_0}^2 \to \mathbf{s}^2$.

The equation of motion for the polarization vector can be obtained directly from Eq. (7-20.4) by making use of Eq. (1-7.10). A short calculation leads to

$$\dot{w}_\mu = -\dot{z}_\mu(w_\nu\ddot{z}^\nu) + g\left(\frac{e}{2m}\right)(F_{\mu\nu} - \dot{z}_\mu\dot{z}^\rho F_{\rho\nu})w^\nu. \tag{7-20.8}$$

From this equation and Eq. (7-12.2) it follows that

$$\frac{d}{d\tau}(w_\mu w^\mu) = 0 \quad \text{and} \quad \frac{d}{d\tau}(w_\mu p^\mu) = 0.$$

It also follows from Eq. (7-12.2) that

$$\frac{d}{d\tau}(p^\mu p_\mu) = 0.$$

Thus we see that m_0 and s_0 are constants of the motion for a spinning particle moving in a field $F_{\mu\nu}$.

[29] H. J. Bhabha and H. C. Corben, *Proc. Roy. Soc.* (London), **A178**, 273 (1941).

With the help of Eq. (7-12.2) we can rewrite Eq. (7-20.8) in the form

$$\dot{w}_\mu = (g-2)\left(\frac{e}{2m}\right)\dot{z}_\mu \dot{z}^\rho F_\rho{}^\nu + g\left(\frac{e}{2m}\right)F_{\mu\nu}w^\nu. \tag{7-20.9}$$

When $g = 2$ the first term on the right drops out. In this special case one can show that in a pure magnetic field ($\mathbf{E} = 0$), the quantity $\mathbf{p}\cdot\mathbf{w}$ where $w^\mu = (w_0, \mathbf{w})$ and $p^\mu = (\mathscr{E}, \mathbf{p})$, is a constant of the motion. Since the magnitudes of $\mathbf{p}$ and $\mathbf{w}$ are also constant in such a field, it follows that the component of the spin along $\mathbf{p}$ does not change in time. Consequently the spin and momentum precess together in a magnetic field. However, if $g \neq 2$ we see from Eq. (7-20.9) that the magnitude of $\mathbf{w}$ is no longer a constant of the motion, and hence $\mathbf{p}$ and $\mathbf{w}$ no longer precess together. One can make use of this fact to measure $(g-2)$ for the muon.[30]

Let us now return to Eq. (7-20.8). At first sight it would seem that even if $F_{\mu\nu} = 0$, $\dot{w}_\mu$ would be nonzero. However, if $F_{\mu\nu} = 0$ and there were no other forces present, $\ddot{z}^\mu$ would vanish and hence so would $\dot{w}_\mu$. (One could, of course, assume the existence of other nonelectrical forces that cause $\ddot{z}^\nu$ to be nonzero. However, such forces might also act on w_μ directly so that Eq. (7-20.8) would no longer be valid.) On the other hand, if $F^{\mu\nu} \neq 0$, even with $g = 0$, we see that $\dot{w}_\mu \neq 0$. Thus a spinning charged particle will precess in an electromagnetic field even if it has no magnetic moment. This precession is a pure relativistic effect known as the *Thomas precession.*

In order to apply Eq. (7-20.8) to specific problems it is convenient to introduce a three-vector $\mathbf{s}$ by the equation

$$\mathbf{s} = \frac{1}{m_0}\left(\mathbf{w} - \frac{\mathbf{p}w^0}{\mathscr{E} + m_0}\right), \tag{7-20.10}$$

since the magnitude of $\mathbf{s}$, unlike $\mathbf{w}$, is constant in time. Since $w^\mu p_\mu = 0$ it follows from Eq. (7-20.10) that

$$w^0 = \mathbf{p}\cdot\mathbf{s} \quad \text{and} \quad \mathbf{w} = m_0\mathbf{s} + \frac{\mathbf{p}(\mathbf{p}\cdot\mathbf{s})}{\mathscr{E} + m_0}.$$

With the help of these relations one can work out the equations of motion for $\mathbf{s}$. One has, after a somewhat lengthy calculation,

$$\frac{d\mathbf{s}}{dx^0} = \frac{m_0}{\mathscr{E}}\frac{ge}{2m_0}(\mathbf{s}\times\mathbf{B}_0) + \frac{\mathscr{E}^2}{m_0(m_0+\mathscr{E})}\mathbf{s}\times\left(\mathbf{u}\times\frac{d\mathbf{u}}{dx^0}\right), \tag{7-20.11}$$

where

$$\mathbf{B}_0 = \mathbf{B} - \frac{\mathscr{E}^2}{m_0(m_0+\mathscr{E})}\mathbf{u}\times(\mathbf{u}\times\mathbf{B}) - \frac{\mathscr{E}}{m_0}\mathbf{u}\times\mathbf{E}. \tag{7-20.12}$$

[30] G. Charpak *et al.*, *Phys. Rev. Letters*, **6**, 128 (1961).

By making use of the equations of motion (7-12.2) one can rewrite Eq. (7-20.11) in the form

$$\frac{d\mathbf{s}}{dx^0} = \frac{m_0}{\mathscr{E}}\left[(g-2)\left(\frac{e}{2m_0}\right)(\mathbf{s}\times\mathbf{B}_0) + \left(\frac{e}{m_0}\right)\mathbf{s}\times\left(\mathbf{B} - \frac{\mathscr{E}}{m_0+\mathscr{E}}\mathbf{u}\times\mathbf{E}\right)\right]. \tag{7-20.13}$$

Problem 7.16. Apply Eq. (7-20.11) to the case of a particle moving with constant speed in a circular orbit. Take $g = 0$ and let $\boldsymbol{\omega}$ be the angular velocity of the orbital motion, that is, $d\mathbf{u}/dx^0 = \boldsymbol{\omega}\times\mathbf{u}$. In particular show that the spin precesses by an amount $-2\pi((1-\mathbf{u}^2)^{-1/2}-1)$ per revolution of the particle.

Let us now take the slow-motion limit of these equations, retaining terms up to the first order in c^{-1}. Because of the inclusion of these first-order terms the resulting equations will not possess the Galilean velocity mapping as a symmetry group. This does not disturb us, since Eq. (7-20.13) does not by itself admit the full Poincaré symmetry group (it is, of course, part of a larger set of equations that do admit this symmetry). In particular, $\mathbf{s}$ is not part of a four-vector, depending as it does on both $\mathbf{w}$ and w^0. If we perform an arbitrary velocity mapping, $\mathbf{s}$ will have to be recomputed from the transformed values of w^μ and p^μ. However, this new $\mathbf{s}$ will satisfy an equation of the form (7-20.13) with $\mathbf{B}_0$, $\mathbf{B}$, and $\mathbf{E}$ being computed from the transformed $F^{\mu\nu}$.

To take the slow-motion limit of Eq. (7-20.13) we must replace x^0 by ct, $\mathbf{u}$ by $\mathbf{u}/c$ and m_0 by m_0c^2, and similarly in Eq. (7-20.12). When these substitutions are made one finds that

$$\mathbf{B}_0 \to \mathbf{B} - \frac{1}{c}\mathbf{u}\times\mathbf{E}$$

so that Eq. (7-20.13) reduces to

$$\frac{d\mathbf{s}}{dt} = \left[(g-2)\frac{e}{2m_0c}\mathbf{s}\times\left(\mathbf{B} - \frac{1}{c}\mathbf{u}\times\mathbf{E}\right) + \frac{e}{m_0c}\mathbf{s}\times\left(\mathbf{B} - \frac{1}{2c}\mathbf{u}\times\mathbf{E}\right)\right] \tag{7-20.14}$$

In atomic physics one usually takes g for the electron to be 2. In this case only the last term in Eq. (7-20.14) survives. It gives rise to the spin-orbit coupling term in the quantum mechanical Hamiltonian of an atom of the form

$$\frac{1}{2}\frac{\hbar^2 e}{m_0{}^2c^2}(\mathbf{l}\cdot\mathbf{s})\frac{1}{r}\frac{d\phi}{dr}$$

The factor 1/2 appearing in this term is a consequence of the Thomas precession; it is absent in the anomalous term ($g - 2 \neq 0$) in Eq. (7-20.14). For a further discussion of the motion of spinning particles the reader is referred to the original paper of Bargmann, Michel, and Telegdi.[31]

7-21. Gravidynamics

In his description of the gravitational interaction between massive bodies Newton made use of the concept of an instantaneous action-at-a-distance. It is natural, therefore, to ask if it is possible to describe this interaction along similar lines within the framework of special relativity. Just such a description was in fact developed by Whitehead[32] in 1922. It has been largely ignored because by that time a successful and beautiful description of all gravitational phenomena had been given by Einstein in his general theory of relativity. Nevertheless, since the necessity of that theory has been questioned from time to time, it is worth while to see how far one can go in special relativity in describing gravitational phenomena.

The greatest difficulty that one faces in constructing a theory of gravity within a relativistic framework is the lack of experimental information concerning the gravitational interaction between massive bodies. Only a few astronomical observations such as the advance of the perihelion of Mercury cannot be accounted for adequately by the Newtonian theory.

The reason for our lack of knowledge concerning gravity is due, in the final analysis, to the weakness of the gravitational force, compared with that of other forces occurring in nature. Consider, for example, the relative strengths of the static electric and gravitational forces between two protons. The electrical force is, of course, given by e^2/r^2, while the gravitational force is given by $Gm_p{}^2/r^2$. Using accepted values for e, m_p, and G, one finds for the ratio of these two forces $Gm_p{}^2/e^2 = 8.08 \times 10^{-37}$. Trautman has emphasized this point further by calculating the first Bohr orbit for a gravitational hydrogen atom, that is, an electron-proton system held together by gravitational rather than electric forces. The usual expression for the first Bohr orbit is $a_0 = \hbar^2/m_e e^2 = 0.53 \times 10^{-8}$ cm. The first Bohr orbit, b_0, of the gravitational hydrogen atom is given by $b_0 = \hbar^2/Gm_e{}^2 m_p = 1.2 \times 10^{31}$ cm. From these considerations we see that we can only hope to observe gravitational effects with macroscopic bodies. However, because of the small velocities involved, relativistic effects are not easily observed when one has to deal with such objects. On the astronomical level, $(v/c) = 10^{-7}$ for the earth, and for

[31] V. Bargmann, L. Michel, and V. L. Telegdi, *Phys. Rev. Letters*, **2**, 435 (1959).
[32] A. N. Whitehead, *The Principle of Relativity* (Cambridge Univ. Press, Cambridge, 1922).

Mercury, the fastest planet, $(v/c) = 1.6 \times 10^{-7}$. It is therefore no wonder that observational checks on relativistic theories of gravity are so hard to come by.

If we try to pattern our treatment of the gravitational interaction after that of the electromagnetic interaction of Section 7-9, we have available to us three relatively simple interactions which we term scalar, vector, and tensor for reasons that will be obvious. We have already dealt extensively with the vector type of interaction in connection with our discussion of electrodynamics. Furthermore we will see later that a combination of scalar plus tensor interaction is sufficient to explain all known astronomical observations. We therefore consider only these two types of interactions here.

While there are a number of possible expressions that we could use for the partial Lagrangian L_i, we will choose one that leads to particle equations of motion that are in accord with the known information on the gravitational interaction obtained from observations on planetary motions. This Lagrangian has the form

$$L_i = -m_i\sqrt{g_{i\mu\nu}(z_i)\dot{z}_i^{\mu}\dot{z}_i^{\nu}} \tag{7-21.1}$$

where

$$g_{i\mu\nu}(z_i) = \eta_{\mu\nu} + G\sum_{j\neq i} g_j \int_{-\infty}^{+\infty} \delta((z_i - z_j)^2)(\alpha\eta_{\mu\nu}\eta_{\rho\sigma} + \beta\eta_{\mu\rho}\eta_{\nu\sigma}) \frac{\dot{z}_j^{\rho}\dot{z}_j^{\sigma}}{\sqrt{\dot{z}_j^{2}}}\, d\lambda_j \tag{7-21.2}$$

and where α and β are two adjustable parameters. The value of the gravitational constant G appearing in this expression is taken so that the gravitational charge g_i of a particle is numerically equal to its inertial mass m_i. The "gravitational field" $g_{i\mu\nu}$ is seen to satisfy the equation

$$\Box\, g_{i\mu\nu}(z_i) = 4\pi G(\alpha\eta_{\mu\nu}\eta_{\rho\sigma} + \beta\eta_{\mu\rho}\eta_{\nu\sigma})\Theta_i^{\rho\sigma}(z_i) \tag{7-21.3}$$

where

$$\Theta_i^{\mu\nu}(z_i) = \sum_{j\neq i} g_j \int_{-\infty}^{+\infty} \delta^4(z_i - z_j) \frac{\dot{z}_j^{\mu}\dot{z}_j^{\nu}}{\sqrt{\dot{z}_j^{2}}}\, d\lambda_j$$

is the stress-energy tensor (see Section 8-5) of all but the ith particle.

The equations of motion associated with the partial Lagrangian (7-21.1) follow from the requirement that

$$\delta S_i = \delta \int L_i\, d\lambda_i = 0.$$

When the path parameter λ_i is so chosen that

$$g_{i\mu\nu}\dot{z}_i^{\mu}\dot{z}_i^{\nu} = 1, \tag{7-21.4}$$

these equations of motion have the form

$$\ddot{z}_i^{\mu} + \begin{Bmatrix} \mu \\ \rho\sigma \end{Bmatrix}_i \dot{z}_i^{\rho} \dot{z}_i^{\sigma} = 0 \tag{7-21.5}$$

where $\{^{\mu}_{\rho\sigma}\}_i$ is a Christoffel symbol constructed from $g_{i\mu\nu}$ by treating it formally like a metric. These equations are seen to have the same form as the geodesic equations (3-2.3) and it is for this reason that L_i was taken to have the form (7-21.1). But if indeed these are the equations of motion of a particle moving in the gravitational field of a collection of other particles, we are faced with a similar situation as that encountered in Section 5-3 where we discussed the Newtonian theory of the gravitational interaction. A test body will only be able to measure the total flat affinity $\{^{\mu}_{\rho\sigma}\}$ in a space-time region and again, an attempt to decompose it into a flat affinity plus a gravitational part will not lead to a unique result. Thus if $\mathring{g}_{\mu\nu}$ is a flat metric, $\mathring{g}_{\mu\nu} + \mathring{g}_{\rho\sigma}\phi^{\rho}{}_{,\mu}\phi^{\sigma}{}_{,\nu}$ will also be a flat metric where ϕ^{ρ} is an arbitrary vector field. Only in the case that the sources of the gravitational field are bounded will it be possible to so normalize the gravitational field that such a decomposition is unique.

Let us now consider a gravitational two-body problem where the mass of one of the bodies is much larger than that of the other. We will denote the parameters of the more massive body by M, Z, and Λ and those of the other body by m, z, and λ with $m/M \ll 1$. To the extent that we can neglect the action of the lighter body on the more massive one, this latter body will move as a free particle, and hence we can always find a mapping such that its motion is described by the equations

$$Z^1 = Z^2 = Z^3 = 0, \qquad \frac{dZ^0}{d\Lambda} = 1. \tag{7-21.6}$$

For the particle of mass m, the partial field $g_{\mu\nu}$ then has the form

$$g_{\mu\nu} = \left(1 + \alpha \frac{GM}{r}\right)\eta_{\mu\nu} + \beta \frac{GM}{r} \eta_{0\mu}\eta_{0\nu}, \tag{7-21.7}$$

where $r^2 = (x^1)^2 + (z^2)^2 + (z^3)^2$. For $GM/r \ll 1$, Eq. (7-21.5) reduces to the Newtonian gravitational equations of motion in the limit $c \to \infty$, provided

$$\alpha + \beta = -2. \tag{7-21.8}$$

We will not explore the further consequences of this theory as they bear on astronomical observations at this point, but will defer the discussion to Chapter 12 where we can compare them with similar consequences of the general theory of relativity.

Problem 7.17. Show that in the limit of small velocities the partial Lagrangian (7-21.1) reduces to that for gravitationally interacting masses in the Newtonian theory, provided $\alpha + \beta = -2$.

8

Relativistic Continuum Mechanics; Microscopic Theory

Historically the field description of the interaction of charged particles preceded the action-at-a-distance description discussed in Chapter 7. The field concept grew out of the work of Faraday and reached maturity with the electrodynamics of Maxwell in 1864. In many respects the field description has advantages over the action-at-a-distance description. The introduction of the electromagnetic field allows one to formulate an initial value problem for interacting particles and, perhaps more important, to construct local conservation laws for energy and momentum. However, along with these advantages comes the great problem that a charge can now act on itself through its self-field. While this self-interaction leads to a relatively simple derivation of radiation reaction, it also leads to the well-known self-energy difficulties of the classical electron theory. We shall see that these difficulties can be only partially overcome by the renormalization procedures of Poincaré and Dirac.

In this chapter we discuss a number of properties of the electromagnetic field, both with regard to its behavior as an independent physical entity and in its interaction with matter. In particular we are concerned with those properties that are a consequence of the relativistic invariance of electrodynamics, since these properties have their analogues in most other relativistic field theories.

In addition to the Maxwell theory, which makes use of a four-vector as the basic field variable, other relativistic field theories have been developed since the early days of relativity theory, to describe atomic and elementary particle phenomena. We will discuss two such typical theories, one that makes use of a scalar field variable and has as its single field equation the Klein-Gordon equation, and another that makes use of a spinor field satisfying the Dirac equation. Finally we discuss the problem of constructing a field theoretic description of the gravitational interaction of massive bodies. By representing the gravitational field by a symmetric second-rank tensor, one can construct a theory that correctly describes all known observational data. However, associated with it are difficulties having to do with energy considerations, which can be only partially overcome within the framework of special relativity.

8-1. The Electromagnetic Field, Maxwell-Lorentz Equations

In Section 7-9 we introduced the electromagnetic tensor $F_{\mu\nu}$ and the associated vector potential A^μ as a convenient shorthand notation for the functions of the particle parameters that appear in their definition. Further, while we determined the values of these quantities at an arbitrary space-time point (for example, the Lieńard-Wiechert potential given by Eq. (7-9.7)), the only use we made of these results was to determine the force acting on a charged particle at that point. In this view of things the electromagnetic field did not exist as an independent entity to be determined by its own set of dynamical laws. The alternate, and more conventional, description of the electromagnetic interaction of charged particles is by means of the electromagnetic field that satisfies the well-known Maxwell-Lorentz equations. In this description one introduces additional degrees of dynamical freedom over and above those associated with the particles in the system; in fact, one introduces an infinity of them.

The usual description of the electromagnetic field is in terms of the electric field $\mathbf{E}$ and the magnetic induction $\mathbf{B}$, which in vacuo are assumed to satisfy the Maxwell-Lorentz equations

$$\nabla \cdot \mathbf{E} = 4\pi\rho, \tag{8-1.1a}$$

$$\nabla \times \mathbf{B} - \frac{1}{c}\frac{\partial}{\partial t}\mathbf{E} = 4\pi\mathbf{j}, \tag{8-1.1b}$$

and

$$\nabla \times \mathbf{B} = 0, \tag{8-1.2a}$$

$$\nabla \times \mathbf{E} + \frac{1}{c}\frac{\partial}{\partial t}\mathbf{B} = 0, \tag{8-1.2b}$$

where $\rho(x)$ and $\mathbf{j}(x)$ are respectively the charge and current densities that produce these fields. It follows from Eqs. (8-1.1a) and (8-1.1b) that ρ and $\mathbf{j}$ must satisfy an equation of continuity:

$$\frac{\partial}{\partial t}\rho + \nabla\cdot\mathbf{j} = 0. \tag{8-1.3}$$

From the discussion of Section 6-5 and Eqs. (6-5.15) and (6-5.16) we see that if we make the identification

$$\mathbf{F} \Rightarrow \mathbf{E} \quad \text{and} \quad \mathfrak{F} \Rightarrow -\mathbf{B},$$

Eqs. (8-1.1) are the (1 + 3) form of the four-dimensional equations

$$F^{\mu\nu}{}_{;\nu} = -4\pi j^{\mu}, \tag{8-1.4}$$

while Eqs. (8-1.2) are the (1 + 3) form of the equations

$$\tfrac{1}{2}\varepsilon^{\mu\nu\rho\sigma}F_{\nu\sigma,\rho} = 0, \tag{8-1.5}$$

where $F_{\mu\nu}$ and $F^{\mu\nu}$ are related to each other by

$$F^{\mu\nu} = \eta^{\mu\rho}\eta^{\nu\sigma}F_{\rho\sigma}.$$

The electromagnetic field tensor appearing in these equations is therefore formed from $\mathbf{E}$ and $\mathbf{B}$ according to the scheme

$$F^{\mu\nu} = \begin{Bmatrix} 0 & -E_x & -E_y & -E_z \\ E_x & 0 & -B_z & B_y \\ E_y & B_z & 0 & -B_x \\ E_z & -B_y & B_x & 0 \end{Bmatrix}, \tag{8-1.6}$$

and the current density four-vector $j^{\mu}(x)$ is given by

$$j^{\mu}(x) = \{\rho(x), \mathbf{j}(x)\}. \tag{8-1.7}$$

If we differentiate Eq. (8-1.4) with respect to x^{μ} and make use of the antisymmetry of $F^{\mu\nu}$ we see that j^{μ} must satisfy

$$j^{\mu}{}_{,\mu} = 0, \tag{8-1.8}$$

which is just the equation of continuity (8-1.3), written in four-dimensional form.

The expressions for the fields in terms of the scalar potential ϕ and the vector potential $\mathbf{A}$,

$$\mathbf{B} = \nabla\times\mathbf{A}, \qquad \mathbf{E} = -\nabla\phi - \frac{\partial}{\partial t}\mathbf{A} \tag{8-1.9}$$

can also be written in four-dimensional form by introducing the four-potential

$$A^\mu \equiv \{\phi, \mathbf{A}\} \tag{8-1.10}$$

as

$$F_{\mu\nu} = A_{\nu,\mu} - A_{\mu,\nu} \tag{8-1.11}$$

where

$$A_\mu = \eta_{\mu\nu} A^\nu.$$

Let us now examine the relation between the Maxwell fields introduced here, satisfying Eqs. (8-1.4) and (8-1.5), and the expressions given by Eqs. (7-9.4) and (7-9.5) for A_μ and $F_{\mu\nu}$. By an extension of Eq. (7-9.5) we will take, as the vector potential due to a system of point charges, the expression

$$A^\mu(x) \equiv \sum_i e_i \int_{-\infty}^{+\infty} \delta((x - z_i)^2) \dot{z}_i^{\,\mu} \, d\lambda_i, \tag{8-1.12}$$

where the sum now extends over all the particles of the system and where the field-point x^μ is assumed to have finite coordinates. Under these conditions one finds by direct calculation that

$$\eta^{\mu\nu} A_{\mu,\nu} = 0. \tag{8-1.13}$$

Further, by making use of Eq. (7-9.6), $\delta(x^2)$ can be shown to satisfy the equation

$$\Box\, \delta(x^2) = 4\pi \delta^4(x), \qquad \left(\Box \equiv \eta^{\mu\nu} \frac{\partial}{\partial_x\mu} \frac{\partial}{\partial_x\nu}\right). \tag{8-1.14}$$

It follows therefore that

$$\Box\, A_\mu(x) = 4\pi \sum_i e_i \int_{-\infty}^{+\infty} \delta^4(x - z_i) \dot{z}_{i\mu} \, d\lambda_i. \tag{8-1.15}$$

Let us now compute $F^{\mu\nu}{}_{,\nu}$ for the field associated with this four-potential. One has

$$\begin{aligned} F^{\mu\nu}{}_{,\nu} &= \eta^{\mu\rho} \eta^{\nu\sigma} (A_{\sigma,\rho} - A_{\rho,\sigma}) \\ &= -\Box A^\mu \\ &= -4\pi j^\mu(x), \end{aligned} \tag{8-1.16}$$

where

$$j^\mu(x) \equiv \sum_i e_i \int_{-\infty}^{+\infty} \delta^4(x - z_i) \dot{z}_i^{\,\mu} \, d\lambda_i. \tag{8-1.17}$$

Furthermore

$$j^{\mu}{}_{,\mu} = \sum_i e_i \int_{-\infty}^{+\infty} \frac{\partial}{\partial x^{\mu}} \delta^4(x - z_i)\dot{z}_i{}^{\mu}\, d\lambda_i$$

$$= -\sum_i e_i \int_{-\infty}^{+\infty} \frac{\partial}{\partial z_i{}^{\mu}} \delta^4(x - z_i)\dot{z}_i{}^{\mu}\, d\lambda_i$$

$$= -\sum_i e_i \int_{-\infty}^{+\infty} \frac{d}{d\lambda_i} \delta^4(x - z_i)\, d\lambda_i = 0,$$

provided the coordinates of the point x^{μ} are all finite. Equations (8-1.16) is thus seen to be identical in form to Eq. (8-1.4), with j^{μ} given by Eq. (8-1.17). The four-potential given by Eq. (8-1.12) is therefore a special solution of Maxwell's equations corresponding to a system of arbitrarily moving charges.

To complete the system of Maxwell equations (8-1.4) and (8-1.5) we must give the equations of motion for the charges that enter into the expression for j^{μ}. For a system of point charges the natural choice is Eq. (7-9.3):

$$m_{0i} \frac{d}{d\lambda_i} \left(\frac{\dot{z}_i{}^{\mu}}{\sqrt{\dot{z}_i{}^2}} \right) = e_i F_{\nu}{}^{\mu}(z_i)\dot{z}_i{}^{\nu}, \tag{8-1.18}$$

where $F_{\nu}{}^{\mu}(x)$ is a solution of Eq. (8-1.4), with $j^{\mu}(x)$ given by Eq. (8-1.17). This combined set of equations is *not* equivalent to the action-at-a-distance equations of motion for point charges, however. They differ from these equations in two important respects. First, the field that appears in Eq. (8-1.18), being a solution of Eqs. (8-1.4) and (8-1.5), need not be identical to the field associated with the potential A^{μ} given by Eq. (8-1.12); it can differ from this field by a solution of the homogeneous equations obtained by setting $j^{\mu} = 0$. Second, and more important, the field that appears in Eq. (8-1.18) for the ith particle is the *total* field produced by the system of charges and includes a contribution from the ith particle itself. These equations therefore allow a particle to act on itself by means of its self-field. Since this self-field is infinite at the position of the particle we can expect that serious difficulties will arise if we try to find solutions to these equations for a system of point charges.

Lorentz was well aware of these difficulties and considered that Eqs. (8-1.4) and (8-1.5) held only for continuous distributions of charge, that is, a charged fluid. In the Lorentz picture, all charged particles, including the so-called elementary particles (for example, electrons, protons), were pictured as finite-sized distributions of charge. Unfortunately such a model of a charged particle brings with it a whole Pandora's box of other difficulties. Since the different parts of the particle must repel each other, it is necessary to introduce other, nonelectrical, forces to hold the particle together. Furthermore the equations of motion of such a particle are extremely complicated and involve the structure of the particle in an intrinsic manner. They involve time derivatives of the position of the particle to all orders and are thus essentially nonlocal equations in the time variable. Only by going to the limit of zero size can these higher derivative terms be eliminated from the equations of motion. However, in this limit, the term that multiplies the second derivative of the position diverges and gives rise to the infinite self-energy problems of the theory.

Attempts to remove the divergence problems from electrodynamics have continued down to the present day.[1] The renormalization procedures of Tomonaga, Schwinger, Feynman, and other have been partially successful in overcoming the divergence problems in quantum electrodynamics, but no one believes that they represent a final solution. The formulation of an internally consistent, relativistic, quantum mechanical description of interacting particles still eludes us and represents one of the great unsolved problems of modern theoretical physics. At the classical level, the action-at-a-distance formulation is, in the author's view, the best that we have; its chief drawback is that, so far, no one has any idea of how to construct a quantum version of this theory.

In the discussion of the Maxwell-Lorentz theory that follows we will ignore the above-mentioned self-energy difficulties as we develop a number of consequences of these equations. However, we must be prepared to encounter problems in this approach and will discuss them as they arise.

The combined Maxwell-Lorentz equation (8-1.4) and 8-1.5) and the equations of motion (8-1.18) can be obtained from a formal variational principle. We consider a space-time region R bounded by two nonintersecting spacelike surfaces σ_1 and σ_2 and form the action

$$S = +\frac{1}{8\pi}\int_R \left\{\frac{1}{2} F_{\mu\nu}F^{\mu\nu} - (A_{\nu,\mu} - A_{\mu,\nu})F^{\mu\nu}\right\} d^4x$$
$$- \sum_i m_i \int_{\sigma_1}^{\sigma_2} \sqrt{\dot{z}_i^{\,2}}\, d\lambda_i - \sum_i e_i \int_R \int_{\sigma_1}^{\sigma_2} A_\mu(x)\, \delta^4(x - z_i)\dot{z}_i^{\,\mu}\, d\lambda_i\, d^4x. \quad (8\text{-}1.19)$$

The integrals over λ_i are to extend from its value where the ith trajectory intersects the surface σ_1 to its value where the trajectory intersects the surface σ_2. We then require that S be stationary for variations of the fields A_μ and $F_{\mu\nu}$ and the particle variables $z_i^{\,\mu}$ that vanish on σ_1 and σ_2. Variations of the $z_i^{\,\mu}$

[1] For a thorough discussion of the self-energy problem in electrodynamics and a proposed solution on the classical level, see F. Rohrlich, *Classical Charged Particles* (Addison-Wesley, Reading, Mass., 1965).

leads to Eqs. (8-1.18), variation of A_μ leads to Eq. (8-1.4), with $j^\mu(x)$ given by Eq. (8-1.17), and variation of $F_{\mu\nu}$ leads to the equations

$$F_{\mu\nu} = A_{\nu,\mu} - A_{\mu,\nu},$$

which are equivalent to Eq. (8-1.5). We emphasize the formal nature of this derivation. Since the field appearing in the second term on the right of Eq. (8-1.19) is the total field produced by the charges, this term will include contributions due to the self-action of these charges, and hence, properly speaking, the value of S will diverge for any dpt of the system. We can perhaps justify the variational derivation by the somewhat dubious procedure of requiring the varied fields to diverge at the location of the particles in the same way that the actual fields diverge there. Then, although both S and S', the varied action, diverge, their difference can be expected to be finite.

8-2. Transformation Properties of the Maxwell Field

From the form of the Maxwell equations (8-1.4) it is clear that if $F^{\mu\nu}(x)$ is a solution for a given $j^\mu(x)$, then its transform $F'^{\mu\nu}(x)$ under a Poincaré mapping formed by treating $F^{\mu\nu}(x)$ as a contratensor will be a solution of a similar set of equations, with $j^\mu(x)$ replaced by its transform $j'^\mu(x)$ formed by treating it as a contravector. Thus, if $F^{\mu\nu}(x)$ and $j^\mu(x)$ satisfy these equations then so will

$$F'^{\mu\nu}(x') = \Lambda_\rho{}^\mu \Lambda_\sigma{}^\nu F^{\rho\sigma}(x) \tag{8-2.1}$$

and

$$j'^\mu(x') = \Lambda_\rho{}^\mu j^\rho(x), \tag{8-2.2}$$

where $\Lambda_\rho{}^\mu$ is a Lorentz matrix and where

$$x'^\mu = \Lambda_\rho{}^\mu x^\rho + b^\mu.$$

The fact that j^μ is required to transform like a contravector is consistent with the form of the transformation law for j^μ which follows from Eq. (8-1.17) as a consequence of the transformation laws for the particle variables appearing therein.

From the transformation law (8-2.1) for the field tensor $F^{\mu\nu}$, one can obtain the transformation laws for the fields $\mathbf{E}$ and $\mathbf{B}$ for a special Lorentz mapping characterized by the velocity $\mathbf{v}$ with the help of Eqs. (6-5.24a, b):

$$\mathbf{E}' = \gamma\left[\mathbf{E} - (\mathbf{v} \times \mathbf{B}) + \frac{\mathbf{v}}{\mathbf{v}^2}(\mathbf{v} \cdot \mathbf{E})\left(\frac{1}{\gamma} - 1\right)\right], \tag{8-2.3}$$

$$\mathbf{B}' = \gamma\left[\mathbf{B} + (\mathbf{v} \times \mathbf{E}) + \frac{\mathbf{v}}{\mathbf{v}^2}(\mathbf{v} \cdot \mathbf{B})\left(\frac{1}{\gamma} - 1\right)\right]. \tag{8-2.4}$$

For the special case when $\mathbf{v} = (v, 0, 0)$ these equations reduce to

$$E'_x = E_x, \quad E'_y = \gamma[E_y + vB_z], \quad E'_z = \gamma[E_z - vB_y] \tag{8-2.5}$$

and

$$B'_x = B_x, \quad B'_y = \gamma[B_y - vE_z], \quad B'_z = \gamma[B_z + vE_y]. \tag{8-2.6}$$

We see from these transformation laws that a field that was originally purely electric, that is, $\mathbf{B} = 0$, acquires a magnetic part under a Lorentz mapping.

In discussing electromagnetic fields it is sometimes convenient to make use of the two algebraically independent scalars that one can construct from the electromagnetic field tensor $F_{\mu\nu}$. These scalars are the eigenvalues of $F_{\mu\nu}$ and hence are obtained from the determinantal equation

$$\det(F_{\mu\nu} - \lambda\eta_{\mu\nu}) = 0.$$

Expressed in terms of $\mathbf{E}$ and $\mathbf{B}$ this equation becomes

$$\lambda^4 + (\mathbf{B}^2 - \mathbf{E}^2)\lambda^2 - (\mathbf{B}\cdot\mathbf{E})^2 = 0,$$

so $\mathbf{B}^2 - \mathbf{E}^2$ and $\mathbf{B}\cdot\mathbf{E}$ are scalars under a Poincaré mapping. This also follows from the fact that these quantities can be expressed in terms of the field tensor $F^{\mu\nu}$ as

$$\mathbf{B}^2 - \mathbf{E}^2 = \tfrac{1}{2}F_{\mu\nu}F^{\mu\nu} \quad \text{and} \quad \mathbf{B}\cdot\mathbf{E} = \tfrac{1}{6}\varepsilon^{\mu\nu\rho\sigma}F_{\mu\nu}F_{\rho\sigma},$$

and the quantities on the right clearly transform as scalars.

Problem 8.1. Show that if the two scalars $\mathbf{E}\cdot\mathbf{B}$ and $\mathbf{B}^2 - \mathbf{E}^2$ are both zero, there exists a Lorentz mapping such that the transformed fields satisfy

$$E'_x = B'_x = 0, \quad E'_y = B'_z, \quad E'_z = -B'_y.$$

Problem 8.2. Show that if both $\mathbf{B}\cdot\mathbf{E}$ and $\mathbf{B}^2 - \mathbf{E}^2$ are zero, there exist Lorentz mappings for which the transformed field strengths can be made arbitrarily small or arbitrarily large. Give a physical interpretation to these mappings.

Problem 8.3. Show that, except when both $\mathbf{B}\cdot\mathbf{E}$ and $\mathbf{B}^2 - \mathbf{E}^2$ are zero, there exists velocity mappings such that the transformed fields are parallel to each other (that is, $\mathbf{E}' \times \mathbf{B}' = 0$) and that the velocity which characterizes this mapping is given by

$$\frac{\mathbf{v}}{1 + \mathbf{v}^2} = \frac{\mathbf{E}\times\mathbf{B}}{\mathbf{E}^2 + \mathbf{B}^2}.$$

Problem 8.4. If $\mathbf{E} \cdot \mathbf{B} = 0$, show that there exists Lorentz mappings such that $\mathbf{E}' = 0$ if $\mathbf{B}^2 - \mathbf{E}^2 > 0$ and $\mathbf{B}' = 0$ if $\mathbf{B}^2 - \mathbf{E}^2 < 0$.

As an example of use of the transformation formulas (8-2.3) and (8-2.4), let us determine the field of a uniformly moving charge from a knowledge of the field when the particle is at rest. For a charge q at rest at the origin of the spatial coordinates, this field is just the usual Coulomb field

$$\mathbf{E} = q\frac{\mathbf{x}}{|\mathbf{x}|^3},$$

$$\mathbf{B} = 0.$$

Let us now perform a velocity mapping of the form (6-5.23). Under such a mapping the trajectory of the charge will be converted into the trajectory of a charge moving with velocity v in the positive x^1-direction. The transformed fields are given by

$$\mathbf{E}'(\mathbf{x}') = q\left\{\frac{x^1}{|\mathbf{x}|^3}, \gamma\frac{x^2}{|\mathbf{x}|^3}, \gamma\frac{x^3}{|\mathbf{x}|^3}\right\} \tag{8-2.7a}$$

and

$$\mathbf{B}'(x') = q\left\{0, -\gamma v\frac{x^3}{|\mathbf{x}|^3}, \gamma v\frac{x^2}{|\mathbf{x}|}\right\}. \tag{8-2.7b}$$

We are not through, however, since we must substitute for x^μ on the right in terms of x'^μ, to obtain explicit expressions for $\mathbf{E}$ and $\mathbf{B}$ in terms of the coordinates. For the mapping considered,

$$x'^0 = \gamma(x^0 + vx^1), \quad x'^1 = \gamma(x^1 + vx^0), \quad x'^2 = x^2, \quad x'^3 = x^3,$$

so we have

$$\mathbf{E}'(x) = q\gamma\left\{\frac{x^1 - vx^0}{R^3}, \frac{x^2}{R^3}, \frac{x^3}{R^3}\right\} \tag{8-2.8a}$$

and

$$\mathbf{B}'(x) = \mathbf{v} \times \mathbf{E}'(x), \tag{8-2.8b}$$

where

$$R \equiv \sqrt{\frac{(x^1 - vx^0)^2}{1 - v^2} + (x^2)^2 + (x^3)^2}.$$

(Note that we have dropped primes on coordinates after substituting for x^ν in terms of x'^μ on the right of Eqs. (8-2.8a, b); we are allowed to do this because x'^μ is an arbitrary point of manifold.) Finally we can re-express the fields in terms of the instantaneous position vector $\mathbf{r}_0$ between the charge and the observation point x^μ (see Fig. 7.4). Since the trajectory of the charge is given by $x^1 = vx^0$, $x^2 = x^3 = 0$, this position vector has components $\mathbf{r}_0 = (x^1 - vx^0, x^2, x^3)$ so that

$$\mathbf{E}'(x) = \gamma q \frac{\mathbf{r}_0}{\{\mathbf{r}_0{}^2 + ((\mathbf{r}_0 \cdot \mathbf{v})^2/(1 - \mathbf{v}^2))\}^{3/2}} \tag{8-2.9a}$$

$$\mathbf{B}'(x) = \mathbf{v} \times \mathbf{E}'(x). \tag{8-2.9b}$$

These expressions for $\mathbf{E}'(x)$ and $\mathbf{B}'(x)$ are seen to agree with the expressions for $\mathbf{E}^+$ and $\mathbf{B}^+$ given by Eqs. (7-9.14). (They also agree with the corresponding expressions for $\mathbf{E}^-$ and $\mathbf{B}^-$, since for a uniformly moving charge, $\mathbf{E}^- = \mathbf{E}^+$ and $\mathbf{B}^- = \mathbf{B}^+$.) We see that $\mathbf{E}'$ is still a radial field and that the magnetic induction is perpendicular to both $\mathbf{r}$ and $\mathbf{v}$. However, the equipotential surfaces are no longer spheres but are rather Heaviside ellipsoids. By introducing the angle θ between $\mathbf{r}$ and $\mathbf{v}$ we have

$$\mathbf{E}'(x) = \frac{q}{r^3} \frac{(1 - v^2)}{(1 - v^2 \sin^2 \theta)^{3/2}} \mathbf{r}$$

so that, as $v \to 1$ for $\theta = \pi/2$, $\mathbf{E}$ is increased by a factor $1/\sqrt{1 - v^2}$ over its original value, while for $\theta = 0$ it is decreased by a factor $(1 - v^2)$. For very high velocities the electric field is therefore highly concentrated in directions at right angle to the direction of motion. An observer thus sees an electromagnetic field composed of nearly equal transverse electric and magnetic field perpendicular to each other. Such a field is essentially that of a pulse of plane-polarized radiation propagating in the x^1-direction.

Problem 8.5. By means of an appropriate Lorentz mapping find the magnetic induction due to an infinitely long straight current, using the knowledge that the electric field is due to an infinitely long, straight charge distribution.

8-3. Solutions of Maxwell's Equations

The potential A^μ given by Eq. (8-1.12) is a solution of the Maxwell equations (8-1.4). We saw in Section 7-9 how it led to the Lieńard-Wiechert potential as a result of an arbitrarily moving charge. Not only is A^μ a solution but so also are the retarded and advanced potentials $A^{+\mu}$ and $A^{-\mu}$ respectively, given by

$$A^{\pm\mu}(x) = \sum_i e_i \int_{-\infty}^{+\infty} \delta^\pm((x - z_i)^2)\dot{z}_i^\mu \, d\lambda_i, \tag{8-3.1}$$

where the function $\delta^{\pm}(x)$ are defined as

$$\delta^{\pm}(x^2) \equiv \frac{1}{|\mathbf{x}|}\{\delta(|\mathbf{x}| \mp x^0)\} \tag{8-3.2}$$

and satisfy

$$\Box\, \delta^{\pm}(x^2) = 4\pi\delta^4(x). \tag{8-3.3}$$

Furthermore we can make use of the properties of the $\delta^{\pm}$ functions to construct the well-known retarded and advanced solutions of the Maxwell equations for arbitrary charge and current distributions. These solutions are given by

$$A^{\pm\mu}(x) = \int \delta^{\pm}((x - x')^2) j^{\mu}(x')\, d^4x'. \tag{8-3.4}$$

Like A^{μ} given by Eq. (8-1.12), these potentials satisfy

$$A^{\pm\mu}{}_{,\mu} = 0 \tag{8-3.5}$$

and thus lead to fields $F^{\pm}_{\mu\nu}$ that satisfy the Maxwell equations. By making use of the expressions (8-3.2) for $\delta^{\pm}(x)$ and performing the integration in Eq. (8-3.4) over x'^0, the potentials $A^{\pm\mu}$ can be made to assume their more familiar form:

$$A^{\pm\mu}(x) = \int \frac{j^{\mu}(x^0 \pm |\mathbf{x} - \mathbf{x}'|, \mathbf{x}')}{|\mathbf{x} - \mathbf{x}'|}\, d^3\mathbf{x}'. \tag{8-3.6}$$

Usually one finds these solution first and then obtains the Lieńard-Wiechert potentials (7-9.7) by taking $j^{\mu}(x)$ to be the current due to an arbitrarily moving point charge. Either way we see that the Lieńard-Wiechert potential yields a solution to Maxwell's equations, with a source term given by

$$j^{\mu}(x) = e\int_{-\infty}^{+\infty} \delta^4(x - z)\dot{z}^{\mu}\, d\lambda,$$

where $z^{\mu}(\lambda)$ is an arbitrary timelike world-line.

Once having constructed the retarded and advanced solutions given by Eq. (8-3.6), we would like to know to what extent they are unique. To answer this equation we make use of a uniqueness theorem for the scalar wave equation

$$\Box\, \psi(x) = 0. \tag{8-3.7}$$

Let us impose the following conditions on $\psi(x)$:

1. The function ψ is bounded,

$$|\psi| < M_0$$

for all space-time points, where M_0 is a positive constant.

2. The function ψ, together with its first derivatives, should go to zero at spatial infinity at least as fast as $1/r$. More explicitly, if r is the geodesic distance from some fixed point in the surface $t = \text{const}$, if $\partial_n \psi$ is the change in ψ normal to σ, and $\nabla\psi$ is the change in ψ in the surface, then

$$\lim_{r\to\infty} r|\psi| < M \tag{8-3.8a}$$

and

$$\lim_{r\to\infty} r|\nabla\psi| < M_1; \; \lim_{r\to\infty} r|\partial_n\psi| < M_1, \tag{8-3.8b}$$

where M and M_1 are positive constants.

3. The function ψ satisfies the outward (inward) radiation condition

$$\lim_{r\to\infty}\left\{\frac{\partial(r\psi)}{\partial r} \pm \frac{\partial(r\psi)}{\partial x^0}\right\} = 0 \tag{8-3.9}$$

for all $t \pm (r/c)$ in an arbitrary finite interval. The plus sign is used to eliminate incoming waves and the minus sign is used to eliminate outgoing waves.

One can then show[2] that the only solution of Eq. (8-3.7) which satisfies the conditions 1, 2, and 3 is identically zero.

The above-stated uniqueness theorem can now be applied to the advanced and retarded potentials given by Eq. (8-3.6). We have seen that each component A^μ separately satisfies an inhomogeneous scalar wave equation of the form (8-1.15). Furthermore, for large $r = |\mathbf{x}|$,

$$A^{\pm\mu}(x) \simeq \frac{1}{r}\int j^\mu\left(x^0 \mp r + \frac{\mathbf{x}\cdot\mathbf{x}'}{r}, \mathbf{x}'\right) d^3x;$$

these $A^{\pm\mu}(x)$ are seen to satisfy the boundary conditions 1, 2, and 3. Hence it follows from the uniqueness theorem that these are the unique solutions of Eq. (8-1.15) satisfying these boundary conditions.

While the Lieńard-Wiechert potentials lead to the only solutions of Maxwell's equations satisfying the boundary conditions 1, 2, and 3 above and are also the only solutions that have analogues in the action-at-a-distance formalism, there are other important solutions of these equations. Of particular importance are the *plane-wave* solutions. Let us look for a solution of the empty-space equations (Eq. 8-1.4 and (8-1.5) with $j^\mu(x) = 0$) of the form

$$F^{\mu\nu}(x) = F_0^{\mu\nu}\exp[-ik_\rho x^\rho + i\alpha]. \tag{8-3.10}$$

It follows from Eq. (8-1.4) that $F_0^{\mu\nu}$ and k_μ must be related by

$$F_0^{\mu\nu}k_\nu = 0, \tag{8-3.11}$$

[2] For a derivation, see V. Fock, *The Theory of Space, Time, and Gravitation* (Pergamon Press, Oxford, 1964), § 92.

while Eq. (8-1.5) leads to the relation

$$\varepsilon^{\mu\nu\rho\sigma}F_{0\nu\rho}k_{\sigma} = 0. \tag{8-3.12}$$

Expressed in terms of the fields **E** and **B** these relations become

$$\mathbf{E}_0 \cdot \mathbf{k} = 0, \qquad \mathbf{B}_0 \times \mathbf{k} - \mathbf{E}_0\omega = 0$$

and

$$\mathbf{B}_0 \cdot \mathbf{k} = 0, \qquad \mathbf{E}_0 \times \mathbf{k} - \mathbf{B}_0\omega = 0,$$

where

$$k^{\mu} = (\omega, \mathbf{k}).$$

The quantity ω is the frequency associated with the plane wave and **k** is its wave-number vector: we will show presently that together they constitute the components of a four-vector.

Since Eqs. (8-3.11) and (8-3.12) are linear and homogeneous in both $F_0^{\mu\nu}$ and k_{μ}, they will have only nontrivial solutions, provided these quantities satisfy a number of conditions. We will show that nontrivial solutions exist when

$$F_{0\mu\nu}^{\mu\nu} = \mathbf{B}_0{}^2 - \mathbf{E}_0{}^2 = 0, \tag{8-3.13}$$

$$\tfrac{1}{6}\varepsilon^{\mu\nu\rho\sigma}F_{0\mu\nu}F_{0\rho\sigma} = \mathbf{B}_0 \cdot \mathbf{E}_0 = 0, \tag{8-3.14}$$

and

$$k^{\mu}k_{\mu} = \omega^2 - \mathbf{k}^2 = 0. \tag{8-3.15}$$

The plane-wave fields **E** and **B** are thus seen to be transverse to the propagation vector **k** perpendicular to each other and equal in magnitude. One can alternately characterize a plane wave as an electromagnetic field having these properties.

In order to demonstrate that Eqs. (8-3.13), (8-3.14), and (8-3.15) are satisfied by all nontrivial solutions of Eqs. (8-3.11) and (8-3.12) we will need to know how $F_0^{\mu\nu}$ and k_{μ} transform under a Poincaré mapping. To obtain this information let us calculate the transform of the plane-wave field (8-3.10) under the mapping

$$x^{\mu} \to x'^{\mu} = \Lambda_{\nu}{}^{\mu}x^{\nu} + a^{\mu}.$$

With the help of the transformation law (8-2.1) we find

$$F'^{\mu\nu}(x) = F_0^{\mu\nu} \exp[-ik'_{\rho}x^{\rho} + i\alpha'],$$

where

$$F'^{\mu\nu}_0 = \Lambda_{\rho}{}^{\mu}\Lambda_{\sigma}{}^{\nu}F_0^{\rho\sigma}, \tag{8-3.16}$$

$$k'_{\mu} = \Lambda_{\mu}^{-1\nu}k_{\nu}, \tag{8-3.17}$$

and

$$\alpha' = \alpha - k_{\rho}a^{\rho}.$$

It follows, therefore, that $F_0^{\mu\nu}$ transforms like a contravector under a Poincaré mapping whereas k_μ transforms like a covector. We will occasionally refer to k_μ as the propagation four-vector associated with a plane wave.

Since k_μ is a four-vector it follows that it can be characterized in an invariant manner according as to whether it is a timelike, spacelike, or null vector. To investigate the nature of the solutions of Eqs. (8-3.11) and (8-3.12) we will therefore assume that k_μ is one or another of these types and try to determine the corresponding possible $F_0^{\mu\nu}$. Let us assume first that k_μ is timelike. Then we can always find a Lorentz mapping such that only the zero component of the transform of k_μ is nonzero, that is, such that $k'_\mu = (k'_0, 0, 0, 0)$. It then follows immediately from Eqs. (8-3.11) and (8-3.12) that $F'^{\mu\nu}_0 = 0$ when k'_μ has this form. Since $F_0^{\mu\nu}$ transforms according to Eq. (8-3.16) it follows therefore, that $F_0^{\mu\nu} = 0$ for all timelike propagation vectors k_μ. In a like manner we can show that $F_0^{\mu\nu} = 0$ whenever k_μ is spacelike. In this case one can transform k_μ so that $k'_\mu = (0, k_1, 0, 0)$, and again it is an easy matter to show that the corresponding $F'^{\mu\nu}_0$ must be zero. We can therefore conclude that Eqs. (8-3.11) and (8-3.12) have only trivial solutions when k_μ is timelike or spacelike.

Let us now consider the only other possibility, namely, that k_μ is null. Then, by means of a suitable Lorentz mapping, it can be transformed so that $k'^\mu = (k_0, k_x, 0, 0)$, where $k_x = \pm k_0$. In this case Eqs. (8-3.11) and (8-3.12) are seen to have nontrivial solutions. We find that such solutions must satisfy

$$E_x = B_x = 0, \quad \text{and} \quad E_y = \pm B_z, \quad E_z = \mp B_y,$$

depending on whether $k_x = \pm k_0$. We see that for these solutions, Eqs. (8-3.13) and (8-3.14) are satisfied. It follows, therefore, that Eqs. (8-3.11) and (8-3.12) will always possess nontrivial solutions when k_μ is a null vector, since the transform of a solution is also a solution, and that these solutions will all satisfy Eqs. (8-3.13) and (8-3.14), which is what we set out to prove. It also follows from this analysis that for a given k_μ satisfying Eq. (8-3.15), there are two linearly independent solutions to Eqs. (8-3.11) and (8-3.12) (in the above special case, B_y and B_z can be assigned arbitrary values) corresponding to the two states of circular polarization of a plane wave.

It is worth while to point out here that the technique employed in the above discussion is another example of the general procedure outlined in Section 3-4 for checking the validity of invariant relations in a theory with a covariance group. One looks for a transformation that reduces the quantities appearing in the relation to a form for which the proof of the relation is particularly simple. Then, if the relation can be shown to preserve its validity under arbitrary transformations of the covariance group, we can conclude that it will still be valid when the quantities appearing in it do not have any special form. Thus, in the above discussion, it was a simple matter to show that $F_0^{\mu\nu} = 0$ when k_μ had the special form $(k_0, 0, 0, 0)$. Then, since the vanishing of all components of a tensor ensures the vanishing of its transformed components under any Poincaré mapping, we were able to conclude that $F_0^{\mu\nu} = 0$ for any timelike vector.

Problem 8.6. Find the symmetries of a plane wave and show that they constitute a five-parameter group.

Let us now make use of the fact that k_μ transforms like a covector under Lorentz mappings to give a simple derivation of the Doppler effect for an electromagnetic wave. From Eq. (6-5.22) it follows that for a velocity mapping characterized by $\mathbf{v}$,

$$\omega' = \gamma(\omega + \mathbf{v} \cdot \mathbf{k}) \tag{8-3.18a}$$

and

$$\mathbf{k}' = \mathbf{k} + \frac{\mathbf{v}(\mathbf{v} \cdot \mathbf{k})}{\mathbf{v}^2}(\gamma - 1) + \gamma \mathbf{v} \omega. \tag{8-3.18b}$$

Let us consider a wave source moving along a trajectory $x^1 = vx^0$. As it approaches the origin $(x^0 < 0)$ the waves reaching an observer stationed there will appear to have a frequency ω_+ and hence a propagation vector $\mathbf{k}_+ = (\omega_+, 0, 0)$, while as it recedes, the waves reaching the origin will appear to have a frequency ω_- and a propagation vector $\mathbf{k}_- = (-\omega_-, 0, 0)$. If we now effect a velocity mapping characterized by $\mathbf{v} = (-v, 0, 0)$ the source will be brought to rest and the observed frequency $\omega' \equiv \omega_0$, its rest frequency, will be related to $\omega_\pm$ by Eq. (8-3.18a). Hence, for a source approaching an observer,

$$\omega_+ = \omega_0 \sqrt{\frac{1+v}{1-v}}, \tag{8-3.19a}$$

while if the source is receding from the observer,

$$\omega_- = \omega_0 \sqrt{\frac{1-v}{1+v}}. \tag{8-3.19b}$$

Finally we observe that the transformation law (8-3.18) also allows for the possibility of a transverse Doppler effect when $\mathbf{v} \cdot \mathbf{k} = 0$ because of the presence of the factor γ in Eq. (8-3.18a). In this case $\omega = \omega_0 \sqrt{1 - \mathbf{v}^2}$, which is a second-order effect in the velocity v. This transverse Doppler effect was not actually observed until 1938 by Ives and Stillwell[3] due to the difficulty of obtaining an

[3] H. E. Ives and C. R. Stillwell, *J. Opt. Soc. Am.*, **28**, 215 (1938).

exactly transverse wave; any deviation from pure transversality would produce a longitudinal Doppler shift proportional to the first power of v, which would completely mask the second-order transverse shift.

Problem 8.7. Use the transformation formulae (8-3.18a, b) to derive the relativistic expression for the aberration of starlight.

Because of the linearity of the Maxwell equations we can construct solutions to these equations by taking a linear superposition of the plane-wave solutions (8-3.10). In particular we can construct wave-packet solutions of the form

$$F_{\mu\nu}(x) = \int f_{\mu\nu}(k) \exp[-ik_\alpha x^\alpha]\delta(k^2)\, d^4k, \tag{8-3.20}$$

where $f_{\mu\nu}(k)$ is nonzero only in a small domain of k space centered around $\bar{k}^\mu$. The δ function appears here to ensure that only propagation vectors k^μ satisfying Eq. (8-3.15) contribute to the integral. For $\bar{k}^0$ positive we can perform the integration over k^0 by making use of the identity

$$\delta(k^2) = \frac{1}{2|\mathbf{k}|}\{\delta(k^0 - |\mathbf{k}|) + \delta(k^0 + |\mathbf{k}|)\}$$

to obtain the result

$$F_{\mu\nu}(x) = \int f'_{\mu\nu}(\mathbf{k}) \exp[-i(\mathbf{k}\cdot\mathbf{x} - \omega(\mathbf{k})x^0)]\, d^3k, \tag{8-3.21}$$

where $\omega(\mathbf{k}) = |\mathbf{k}|$ and $f'_{\mu\nu}(\mathbf{k}) = f_{\mu\nu}(\omega(\mathbf{k}), \mathbf{k})/|\mathbf{k}|$. The functions $f'_{\mu\nu}(\mathbf{k})$ are now such that they vanish outside a small region S around $\bar{\mathbf{k}}$. If the phase of the exponential oscillates rapidly in this region the contributions to the integral from different parts of S interfere destructively and $F_{\mu\nu}$ will be negligible. It will be appreciable only when the phase remains practically constant in S, that is, in a region of stationary phase. In such a region

$$\nabla_{\mathbf{k}}(\mathbf{k}\cdot\mathbf{x} - \omega(\mathbf{k})x^0) = \mathbf{x} - \nabla_{\mathbf{k}}\omega(\mathbf{k})x^0 \approx 0. \tag{8-3.22}$$

Consequently $F_{\mu\nu}$ will be significantly different from zero only for values of x^μ that satisfy this condition. The center of the packet thus moves along a straight path with a *group velocity*

$$\mathbf{v}_g = \nabla_k\omega(k) = \frac{\mathbf{k}}{|\mathbf{k}|}, \tag{8-3.23}$$

whose magnitude is unity. Its trajectory is therefore a null geodesic in space-time. Since a light signal corresponds to an electromagnetic wave packet, we see that these objects propagate along null geodesics.

8-4. Continuity Equations and Conservation Laws

Before we examine the continuity equations of electrodynamics that arise as a consequence of Poincaré invariance it will be instructive to consider first the current continuity equation (8-1.3 or 8). Let us integrate Eq. (8-1.8) over a space-time region bounded by two nonintersecting spacelike surfaces σ_1 and σ_2. We have, using Gauss' theorem,

$$\int_R j^\mu{}_{,\mu}\, d^4x = \oint_B j^\mu\, dS_\mu = 0,$$

where B is the boundary of R. If j^μ is localized in a finite region of space, the only contributions to the surface integral come from the two surfaces σ_1 and σ_2. Consequently

$$\int_{\sigma_2} j^\mu\, dS_\mu - \int_{\sigma_1} j^\mu\, dS_\mu = 0. \tag{8-4.1}$$

Since the two surfaces σ_1 and σ_2 are arbitrary spacelike surfaces it follows that

$$Q(\sigma) \equiv \int_\sigma j^\mu\, dS_\mu \tag{8-4.2}$$

is independent of the surface used in evaluating the integral. For $j^\mu(x)$ given by Eq. (8-1.17), one has

$$Q = \sum_i e_i. \tag{8-4.3}$$

The easiest way to derive this result is to choose σ in Eq. (8-4.2) to be the surface $x^0 = \text{const}$, and to take the parameters λ_i appearing in the expression for j^μ to x^0. Then

$$Q = \int \rho\, dV$$

and

$$\rho = \sum_i e_i\, \delta^3(x - z_i(x^0)),$$

giving the above results.

Let us now consider the transformation properties of Q under a Poincaré mapping. Since $j^\mu\, dS_\mu$ is a scalar,

$$Q'(\sigma') = \int_{\sigma'} j'^\mu(x)\, dS_\mu$$

is equal to $Q(\sigma)$ given by Eq. (8-4.1) if σ' is the image of σ under the mapping. It also follows from Eq. (8-4.1) that

$$Q'(\sigma) = \int_\sigma j'^\mu(x)\, dS_\mu = Q'(\sigma) = Q(\sigma). \tag{8-4.4}$$

Thus the value of the charge of a system obtained by integrating j^μ over a given surface σ is equal to the value obtained by integrating the transform of j^μ over the *same* surface.

Let us now consider the conservation laws associated with the empty-space Maxwell equations. These equations follow, as we have seen, from an action principle, with an action

$$S_F = \int \mathfrak{L}(x)\, d^4x = \frac{1}{8\pi}\int \left\{\frac{1}{2} F_{\mu\nu}F^{\mu\nu} - (A_{\nu,\mu} - A_{\mu,\nu})F^{\mu\nu}\right\} d^4x. \tag{8-4.5}$$

From the form of the integrand we see that $\mathfrak{L}(x)$ is a scalar under Poincaré mappings. Therefore we can apply the results of Section 4-5 to obtain a number of continuity equations. For space-time translations we obtain, using Eq. (4-5.16) together with the fact that, for translations, $\bar{\delta} A_\mu = -A_{\mu,\nu}\varepsilon^\nu$,

$$\begin{aligned} t^\mu_{em\nu} &= -\mathfrak{L}\,\delta_\nu{}^\mu - \frac{\partial \mathfrak{L}}{\partial A_{\rho,\mu}}\, w_{\rho\nu} \\ &= \frac{1}{16\pi} F_{\rho\sigma}F^{\rho\sigma}\,\delta_\nu{}^\mu + \frac{1}{4\pi} F^{\rho\mu}A_{\rho,\nu} \end{aligned} \tag{8-4.6}$$

as conserved quantities; that is,

$$t^\mu_{em\nu,\mu} = 0 \tag{8-4.7}$$

for dpt.

As it stands, $t^\mu_{em\nu}$ is not gauge invariant because of the explicit appearance of A_μ in Eq. (8-4.6), nor is $t^{\mu\nu}_{em}$ symmetrical in μ and ν. While this second deficiency is not serious, it is necessary to remedy the first one. We can do so by adding a curl to $t^\mu_{em\nu}$, that is, a term $W_\nu^{[\mu\rho]}{}_{,\rho}$ where $W_\nu^{[\mu\rho]} = -W_\nu^{[\rho\mu]}$. Then if $t^\mu_{em\nu}$ satisfies Eq. (8-4.7),

$$\bar{t}^\mu_{em\nu} \equiv t^\mu_{em\nu} + W_\nu^{[\mu\rho]}{}_{,\rho}$$

also satisfies this continuity equation. If we take

$$W_\nu^{[\mu\rho]} = -\frac{1}{4\pi} F^{\rho\mu}A_\nu$$

then

$$\bar{t}^\mu_{em\nu} = \frac{1}{16\pi} F_{\rho\sigma}F^{\rho\sigma}\,\delta_\nu{}^\mu - \frac{1}{4\pi} F^{\rho\mu}F_{\rho\nu} \tag{8-4.8}$$

for fields that satisfy the empty-space Maxwell equations. In the future we will drop the bar over $t^{\mu}_{em\nu}$ and use this quantity to represent the right side of Eq. (8-4.8). We note that

$$t^{\mu\nu}_{em} = \frac{1}{16\pi} F_{\rho\sigma} F^{\rho\sigma} \eta^{\mu\nu} - \frac{1}{4\pi} F^{\rho\mu} F^{\sigma\nu} \eta_{\rho\sigma}$$

is symmetrical in μ and ν and that $t^{\mu\nu}_{em}$ is traceless, that is, $t^{\mu}_{em\mu} = 0$.

Expressed in terms of **E** and **B**, the components of $t^{\mu\nu}_{em}$ are given by

$$t^{00}_{em} \equiv W = \frac{1}{8\pi} (\mathbf{E}^2 + \mathbf{B}^2), \tag{8-4.9a}$$

$$t^{r0}_{em} \equiv S^r = \frac{1}{4\pi} \varepsilon^{rst} E_s B_t , \tag{8-4.9b}$$

$$t^{rs}_{em} = \frac{1}{8\pi} (\mathbf{B}^2 + \mathbf{E}^2)\delta^{rs} - \frac{1}{4\pi} (E_r E_s + B_r B_s). \tag{8-4.9c}$$

We see that W is just the usual expression for the energy density of an electromagnetic field, $\mathbf{S} = (1/4\pi)\mathbf{E} \times \mathbf{B}$ is the Poynting vector and $-t^{rs}_{em}$ is the Maxwell stress tensor. As a consequence of the continuity equation (8-4.7) we have

$$\frac{\partial}{\partial x^0} W + \nabla \cdot \mathbf{S} = 0,$$

which allows us to interpret S as an energy flux. Also,

$$\frac{\partial}{\partial x^0} S^r + t^{rs}_{em,s} = 0.$$

Now t^{rs}_{em} is the rth component of the force per unit area transmitted across a surface whose normal points in the x^s-direction and hence is equal to the rth component of the momentum flux across this surface. Therefore S is also the momentum density of the field. For these reasons $t^{\mu\nu}_{em}$ is referred to as the *stress-energy tensor* of the electromagnetic field.

Problem 8.8. Show that except when $\mathbf{B} \cdot \mathbf{E} = \mathbf{B}^2 - \mathbf{E}^2 = 0$, $t^{\mu\nu}_{em}$ can be diagonalized by a suitable Lorentz mapping.

Problem 8.9. A plane electromagnetic wave is incident on a plane reflecting surface with an angle of incidence θ. If W is the incident energy density of the wave and $W' = RW$ is the energy density of the reflected wave, where

R is the reflection coefficient of the surface, show that the normal pressure is given by

$$p_n = W(1 + R)\cos^2\theta$$

and the tangential force per unit area is given by

$$p_t = W(1 - R)\sin\theta\cos\theta.$$

In the special case that $F^{\mu\nu}$ is the field of an arbitrarily moving point charge given by Eq. (7-9.9), one finds, with the same notation,

$$\begin{aligned}4\pi t_{em}^{\pm\mu\nu} = {} & -\frac{e^2}{\rho^4}\left[\ddot{z}_\rho\ddot{z}^\rho + \frac{1}{\rho^2}(\ddot{z}_\rho r^\rho)^2\right]r^\mu r^\nu \\ & + 2\frac{e^2}{\rho^6}\ddot{z}_\rho r^\rho r^\mu r^\nu + \frac{e^2}{\rho^4}(\ddot{z}^\mu r^\nu + \ddot{z}^\nu r^\mu) - \frac{e^2}{\rho^5}\ddot{z}_\rho r^\rho(\dot{z}^\mu r^\nu + \dot{z}^\nu r^\mu) \\ & + \frac{e^2}{\rho^5}(\dot{z}^\mu r^\nu + \dot{z}^\nu r^\mu) - \frac{e^2}{\rho^6}r^\mu r^\nu - \frac{e^2}{\rho^4}\eta^{\mu\nu}.\end{aligned} \tag{8-4.10}$$

The right side is evaluated at the past (future) position of the particle determined by $\theta_+ = 0$ ($\theta_- = 0$) for the retarded (advanced) solution. We see that for large values of r^μ and ρ, the terms in the three lines of the right side of Eq. (8-4.10) go as ρ^{-2}, ρ^{-3}, and ρ^{-4}, respectively. The terms that go as ρ^{-2} arise from terms in $F_{\mu\nu}$ that go as ρ^{-1}, and can be interpreted as the radiation part of the field. The terms that go as ρ^{-4} come from Coulomb-type terms in $F_{\mu\nu}$ that fall off as ρ^{-2}.

So far we have dealt with the continuity equations, and the quantities appearing therein, that are associated with the space-time translational symmetry of Maxwell's equations. There are also continuity equations that follow from the Lorentz symmetry of the theory. One can obtain these equations and the associated conserved quantities in a manner analogous to that used to construct the stress-energy tensor above; there is, however, an easier way. Let us construct the quantity

$$t_{em}^{\mu\rho\sigma} = t_{em}^{\mu\rho}x^\sigma - t_{em}^{\mu\sigma}x^\rho. \tag{8-4.11}$$

As a consequence of Eq. (8-4.7) and the symmetry of $t^{\mu\nu}$ it follows directly that

$$t_{em,\mu}^{\mu\rho\sigma} = 0. \tag{8-4.12}$$

From its structure we see that $t_{em}^{\mu\rho\sigma}$ transforms like a tensor under a Lorentz mapping. It is the generalized angular-momentum tensor of the electromagnetic field and is equal to the tensor one would have obtained by a direct application of Eq. (4-5.16) modulo a curl. In particular,

$$t_{em}^{0rs} = S^r x^s - S^s x^r,$$

which, with the interpretation of **S** as a momentum density, is analogous to the spatial part of the angular momentum tensor for a particle.

We can discuss integrals of $t_{em}^{\mu\nu}$ and $t_{em}^{\mu\nu\sigma}$ over spacelike surfaces in a manner analogous to that for integrals of j^μ, with only a slight modification. For simplicity we will concentrate our discussion on the stress-energy tensor. Let us integrate Eq. (8-4.7) over the region R bounded by σ_1 and σ_2, two non-intersecting spacelike surfaces, and apply Gauss' theorem. If $t_{em}^{\mu\nu}$ is spatially localized, or falls off faster than $1/r^{-2}$, then there is no contribution from the "sides" of R and we have

$$P^\mu(\sigma_1) = P^\mu(\sigma_2), \tag{8-4.13}$$

where

$$P^\mu(\sigma) = \int_\sigma t_{em}^{\mu\nu}\, dS_\nu . \tag{8-4.14}$$

Let us now introduce on our manifold an everywhere constant timelike vector λ_μ and form $\int_\sigma \lambda_\mu t_{em}^{\mu\nu}\, dS_\nu$. We can now use the same argument as for the current vector to conclude that under a Poincaré mapping,

$$\int_{\sigma'} \lambda'_\mu t'^{\mu\nu}_{em}\, dS_\nu = \int_\sigma \lambda_\mu t_{em}^{\mu\nu}\, dS_\nu$$

and hence, from Eq. (8-4.13), that

$$\int_\sigma \lambda'_\mu t'^{\mu\nu}_{em}\, dS_\nu = \int_\sigma \lambda_\mu t_{em}^{\mu\nu}\, dS_\nu .$$

Then, since $\lambda'_\mu = \Lambda^{-1}{}_\mu{}^\nu \lambda_\nu$ and λ_μ is constant over the manifold, it follows that

$$P'^\mu(\sigma) = \int_\sigma t'^{\mu\nu}_{em}\, dS_\nu = \Lambda_\nu{}^\mu P^\nu(\sigma). \tag{8-4.15}$$

Thus $P^\mu(\sigma)$ can be interpreted as the momentum contravector of a spatially localized radiation pulse.

Problem 8.10. Show that $P^\mu P_\mu = 0$ for a plane wave.

These results do not hold if we are dealing with the field of an arbitrarily moving point charge, for two reasons. Since $F_{\mu\nu}$ no longer satisfies the empty-space Maxwell equations it is no longer true that $t_{em,\nu}^{\mu\nu} = 0$ everywhere. Also, $t_{em}^{\mu\nu}$ no longer falls off faster than $1/r^2$ for such a field, and hence the contributions from the sides of R no longer vanish. If we compute $t_{em,\nu}^{\mu\nu}$ we find that

$$\begin{aligned} t_{em,\nu}^{\mu\nu} &= \frac{1}{8\pi} F_{\rho\sigma,\nu} F^{\rho\sigma} \eta^{\mu\nu} - \frac{1}{4\pi} F^{\rho\mu}{}_{,\nu} F^{\sigma\nu} \eta_{\rho\sigma} - \frac{1}{4\pi} F^{\rho\mu} F^{\sigma\nu}{}_{,\nu} \eta_{\rho\sigma} \\ &= -F^{\mu\nu} j_\nu , \end{aligned} \tag{8-4.16}$$

since the first two terms on the right vanish as a consequence of Eq. (8-1.5). Therefore, even when there are sources of the field, $t^{\mu\nu}_{em,\nu} = 0$ in regions where j^μ vanishes.

Schild[4] has shown how to remedy the second difficulty mentioned above, and in doing so, has given a definition for the energy and momentum radiated by a moving charge. Let Q and Q' be two points on the trajectory of this charge (see Fig. 8.1). We now construct two future light cones $\sum$ and $\sum'$ with

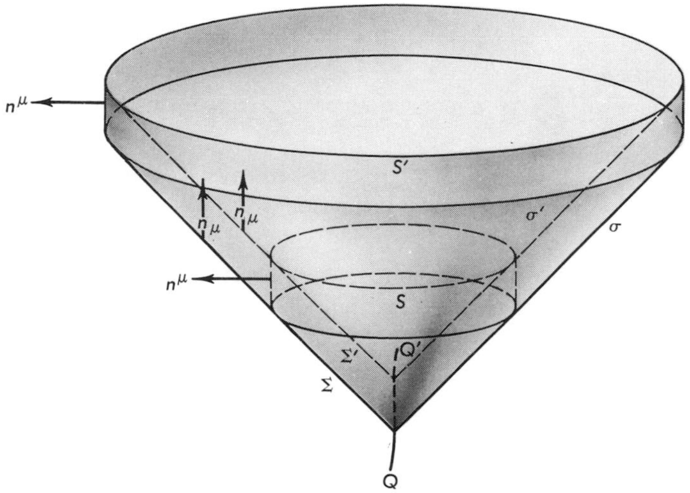

Fig. 8.1.

vertices at Q and Q', respectively. In addition we construct two hypersurfaces S and S' lying between these null cones and denote the portions of these cones lying between S and S' by σ and σ'. Following Schild, we now integrate $t^{\mu\nu}_{em,\nu}$ over the space-time region bounded by S, S', σ, and σ'. Since $t^{\mu\nu}_{em,\nu} = 0$ in this region, we have, with the help of Gauss' theorem,

$$\int_{S'} t^{\mu\nu}_{em}\, dS_\nu - \int_S t^{\mu\nu}_{em}\, dS_\nu + \int_{\sigma'} t^{\mu\nu}_{em}\, dS_\nu - \int_\sigma t^{\mu\nu}_{em}\, dS_\nu = 0,$$

where the elements dS_ν have been so chosen that $dS_\nu n^\nu > 0$ for the normals n^μ shown in Fig. 8.1. Now dS_μ on the null cone σ is a null vector proportional to $R_\mu = \eta_{\mu\nu}(x^\mu - z^\mu(\lambda))$[5]:

$$dS_\mu = R_\mu\, d\omega.$$

[4] A. Schild, *J. Math. Analy. and Appl.*, **1**, 127 (1960).

[5] See J. L. Synge, *Relativity, the Special Theory*, Appendix D (North Holland Publishing, Amsterdam, 1956).

Using the expression (8-4.10) for $t_{em}^{\mu\nu}$ we see that only the terms in the last line contribute to $t_{em}^{\mu\nu}\, dS_\nu$ on σ and σ', so that on these surfaces,

$$t_{em}^{\mu\nu}\, dS_\nu = \frac{1}{8\pi}\frac{e^2}{\rho^4} R^\mu\, d\omega.$$

One can then show that the integrals over σ and σ' differ from each other by terms of the order $\bar{\rho}^{-1}$, where $\bar{\rho}$ is the minimum value of ρ for the points on S and S'. It therefore follows that

$$P^\mu = \lim_{\bar{\rho}\to\infty} \int_S t_{em}^{\mu\nu}\, dS_\nu$$

is independent of S. As a consequence P^μ transforms, under a Lorentz mapping, according to

$$P'^\mu = \lim_{\bar{\rho}\to\infty} \int_S t'^{\mu\nu}_{em}\, dS_\nu = \Lambda_\nu{}^\mu P^\mu,$$

that is, like a contravector. One finds, then, by straightforward calculation,[6] that if the arc QQ' is of infinitesimal length $d\tau$,

$$dP^\mu = -\tfrac{2}{3}e^2 \ddot{z}_\rho \ddot{z}^\rho \dot{z}^\mu\, d\tau, \tag{8-4.17}$$

while for finite intervals,

$$P^\mu = -\tfrac{2}{3}e^2 \int_Q^{Q'} \ddot{z}_\rho \ddot{z}^\rho \dot{z}^\mu\, d\tau. \tag{8-4.18}$$

We see that P^μ measures the energy and momentum that flows through S and that P^0 is the relativistic generalization for the energy radiated by an accelerated charge. In the field picture it is this energy loss that results in the radiation reaction term appearing in Eq. (7-18.1). In fact $dP^\mu/d\tau$ is just the negative of the last term appearing in this equation.

As a final application of the continuity equation (8-4.7) we will outline the Dirac derivation of the equations of motion of a radiating charge.[7] Dirac considers the total field present, $F_{\text{tot}}^{\mu\nu}$. It consists of two parts: $F_{\text{in}}^{\mu\nu}$, the incoming electromagnetic field incident on the charge, plus $F^{+\mu\nu}$, the retarded field of the charge given by Eq. (7-9.9). Thus

$$F_{\text{tot}}^{\mu\nu} = F_{\text{in}}^{\mu\nu} + F^{+\mu\nu}. \tag{8-4.19}$$

Alternatively

$$F_{\text{tot}}^{\mu\nu} = F_{\text{out}}^{\mu\nu} + F^{-\mu\nu},$$

[6] J. L. Synge, *loc. cit.*, Appendix B.

[7] P. A. M. Dirac, *Proc. Roy. Soc.* (London), **A167**, 148 (1938).

where $F^{\mu\nu}_{\text{out}}$ is the outgoing field from the particle. The radiation field $F^{\mu\nu}_{\text{rad}}$ of the electron is then defined by Dirac to be

$$\begin{aligned} F^{\mu\nu}_{\text{rad}} &= F^{\mu\nu}_{\text{out}} - F^{\mu\nu}_{\text{in}} \\ &= F^{+\mu\nu} - F^{-\mu\nu}. \end{aligned}$$

On the trajectory of the particle, $F^{\mu\nu}_{\text{rad}}$ is nonsingular and is given by

$$F^{\mu\nu}_{\text{rad}} = \frac{4e}{3}(\dddot{z}^{\mu}\dot{z}^{\nu} - \dddot{z}^{\nu}\dot{z}^{\mu}). \tag{8-4.20}$$

Usually the radiation field is taken to be $F^{+\mu\nu}$ at large distances from the charge and at correspondingly large times after an acceleration takes place. However, $F^{-\mu\nu}$, the advanced field, will be zero in this region of space-time so that $F^{\mu\nu}_{\text{rad}} = F^{+\mu\nu}$ there. By introducing $F^{\mu\nu}_{\text{rad}}$, Dirac succeeded in giving a definition of the radiation field that is valid everywhere, including the neighborhood of the charge. In the usual definition the radiation field is inextricably mixed up with the Coulomb field near the particle.

To obtain equations of motion for the charge, Dirac proceeded to integrate $t^{\mu\nu}_{em,\nu}$ over a small tube surrounding its trajectory. For purposes of calculation he used a tube such that its intersection with the spacelike plane whose normal is parallel to $\dot{z}^{\mu}$ at a point on the trajectory is a sphere of radius ε (Fig. 8.2).

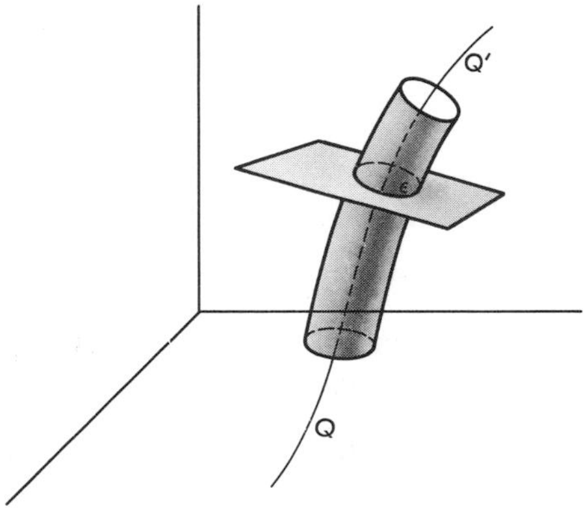

Fig. 8.2.

The results obtained are in fact independent of any assumed shape of the tube, since $t^{\mu\nu}_{em,\nu}$ is nonzero only on the trajectory of the particle. By making use of Gauss' theorem one can now convert $\int t^{\mu\nu}_{em,\nu}\, d^4x$ into an integral $\oint t^{\mu\nu}_{em}\, dS_{\nu}$ over the surface of the tube. When $t^{\mu\nu}_{em}$ is evaluated, using $F^{\mu\nu}_{\text{tot}}$ for the field appearing in its definition, the contribution to this latter integral from the sides of the tube can be evaluated and has the value

$$\int_{\text{sides}} t_{em}^{\mu\nu}\, dS_\nu = \int_{Q}^{Q'} [\tfrac{1}{2}e^2\varepsilon^{-1}\ddot{z}^\mu - e\dot{z}^\nu f_\nu{}^\mu]\, ds, \tag{8-4.21}$$

where

$$f^{\mu\nu} = F_{\text{tot}}^{\mu\nu} - \tfrac{1}{2}(F^{+\mu\nu} + F^{-\mu\nu}). \tag{8-4.22}$$

Since, now, there is no other flow out from the sides, the contribution to the surface integral from the sides must be balanced by just what comes in and leaves at the two ends. It follows from Eq. (8-4.16) that the integrand of this surface integral must be a perfect differential, that is,

$$\tfrac{1}{2}e^2\varepsilon^{-1}\ddot{z}^\mu - e\dot{z}^\nu f_\nu{}^\mu = \dot{B}^\mu, \tag{8-4.23}$$

where B^μ is as yet undetermined. As matters stand we can obtain no further information about it from the continuity equations, but we are forced to make an assumption about its form.

If we multiply Eq. (8-4.23) by $\dot{z}_\mu$ and make use of the fact that $\dot{z}^\mu\dot{z}_\mu = 1$ and $\ddot{z}^\mu\dot{z}_\mu = 0$, we see that $\dot{B}^\mu$ must satisfy $\dot{z}_\mu\dot{B}^\mu = 0$. The simplest vector that satisfies this condition is $B^\mu = k\dot{z}^\mu$, where k is a constant. There are, of course, other possibilities such as

$$B^\mu = k'[\ddot{z}^4\dot{z}^\mu + 4(\ddot{z}^\nu\dddot{z}_\nu)\ddot{z}^\mu],$$

k' being another constant. Dirac chose for B^μ the simplest expression, namely, $k\dot{z}^\mu$. With B^μ of this form we see that k must be of the form

$$k = \tfrac{1}{2}e^2\varepsilon^{-1} - m, \tag{8-4.24}$$

where m is another constant, independent of ε, in order that Eq. (8-4.23) have a definite form when $\varepsilon \to 0$. We then obtain

$$m\ddot{z}^\mu = e\dot{z}^\nu f_\nu{}^\mu$$

as the equations of motion of the charge. Making use of Eqs. (8-4.20) and (8-4.22) we see that these equations have the form

$$m\ddot{z}^\mu - \tfrac{2}{3}e^2\dddot{z}^\mu - \tfrac{2}{3}e^2(\ddot{z}^\nu\ddot{z}_\nu)\dot{z}^\mu = e\dot{z}^\nu F_{\text{in}\ \nu}^\mu \tag{8-4.25}$$

in agreement with Eq. (7-18.1).

We can now interpret Eq. (8-4.24). The constant m is seen to be the observed mass of the particle, and $\tfrac{1}{2}e^2\varepsilon^{-1}$ is the "Coulomb self-energy" of the particle. The constant k can be thought of as being the nonelectrical mass of the particle or, as it is sometimes called, its mechanical mass. It, like the Coulomb self-energy, diverges in the limit $\varepsilon \to 0$. The observed mass is the difference between these two divergent quantities. The method used here to obtain meaningful equations of motion is a classical example of the renormalization procedures employed in quantum electrodynamics to obtain meaningful results. It is far from satisfactory but is the best we have at the present time.

8-5. The Minkowski Tensor

Because of the appearance of the term $-F^{\mu\nu}j_\nu$ on the right, Eq. (8-4.16) does not have the form of a continuity equation. Nevertheless, if the general relation between continuity equations and invariance properties is valid, we should be able to reformulate these equations as continuity equations. Minkowski[8] was the first to accomplish this task. For the purpose we introduce the Minkowski, or kinetic stress-energy, tensor $\Theta^{\mu\nu}$, defined by

$$\Theta^{\mu\nu}(x) = \sum_i \int m_i\, \delta^4(x - z_i(\lambda_i)) \frac{\dot{z}_i^\mu \dot{z}_i^\nu}{\sqrt{\dot{z}_i^2}}\, d\lambda_i, \tag{8-5.1}$$

where the sum is over the charges of the system. Let us now compute $\Theta^{\mu\nu}{}_{,\nu}$. We have

$$\begin{aligned}\Theta^{\mu\nu}{}_{,\nu} &= \sum_i m_i \int \frac{\partial}{\partial x^\nu} \delta^4(x - z_i) \frac{\dot{z}_i^\mu \dot{z}_i^\nu}{\sqrt{\dot{z}_i^2}}\, d\lambda_i \\ &= -\sum_i m_i \int \frac{\partial}{\partial z_i^\nu} \delta^4(x - z_i) \frac{\dot{z}_i^\mu \dot{z}_i^\nu}{\sqrt{\dot{z}_i^2}}\, d\lambda_i \\ &= \sum_i m_i \int \frac{d}{d\lambda_i} \delta^4(x - z_i) \frac{\dot{z}_i^\mu}{\sqrt{\dot{z}_i^2}}\, d\lambda_i,\end{aligned}$$

which, upon integration by parts, gives

$$\Theta^{\mu\nu}{}_{,\nu} = -\sum_i \int \delta^4(x - z_i) \frac{d}{d\lambda_i}\left(\frac{\dot{z}_i^\mu}{\sqrt{\dot{z}_i^2}}\right) d\lambda_i.$$

Let us now make use of the equations of motion (8-1.18) to replace the factor $d(\dot{z}_i^\mu/\sqrt{\dot{z}_i^2})/d\lambda_i$ by $e_i F_\nu{}^\mu(z_i)\dot{z}_i^\nu$:

$$\Theta^{\mu\nu}{}_{,\nu} = -\sum_i \int \delta^4(x - z_i) e_i F_\nu{}^\mu(z_i) \dot{z}_i^\nu.$$

[8] H. Minkowski, *Nachr. Ges. Wiss. Göttingen*, 53 (1908).

Because of the δ function appearing in the integrand we can replace $F_\nu{}^\mu(z_i)$ by $F_\nu{}^\mu(x)$ and take $F_\nu{}^\mu(x)$ outside the integrand without changing the value of $\Theta^{\mu\nu}{}_{,\nu}$. Therefore

$$\Theta^{\mu\nu}{}_{,\nu} = -F_\nu{}^\mu(x) \int \delta^4(x - z_i) e_i \dot{z}_i{}^\nu \, d\lambda_i$$

$$= -F_\nu{}^\mu(x) j^\nu(x),$$

with $j^\mu(x)$ given by Eq. (8-1.17). Thus we see that Eq. (8-4.16) can be written as

$$T^{\mu\nu}{}_{,\nu} = 0, \tag{8-5.2}$$

where $T^{\mu\nu}$, the total stress-energy tensor, is given by

$$T^{\mu\nu} = t^{\mu\nu}_{em} + \Theta^{\mu\nu}. \tag{8-5.3}$$

Problem 8.11. Show that Eq. (8-5.2), with $T^{\mu\nu}$ given by Eq. (8-5.3), follows directly from the action (8-1.17) with the help of Eq. (4-5.16).

Equation (8-5.2) can be used to derive a relativistic virial theorem for a system of charged particles. Let us compute the time average of this equation when $\mu = 1, 2, 3$. In taking this average we will assume that the motion of the particles is such that their positions, velocities, etc., are bounded and that likewise the fields that they produce are also bounded. Then the time average of the term $T^{r0}{}_{,0}$ will be zero for such motions since

$$\overline{T}^{r0}{}_{,0} \equiv \lim_{X^0 \to \infty} \frac{1}{X^0} \int_0^{X^0} T^{r0}{}_{,0} \, dx^0$$

$$= \lim_{X^0 \to \infty} \frac{T^{r0}(X^0) - T^{r0}(0)}{X^0},$$

and by assumption $\lim_{X^0 \to \infty} T^{r0}(X^0)$ is finite. Thus we have $\overline{T}^{rs}{}_{,s} = 0$.

Now multiply this equation by $\delta_{ru} x^u$ and integrate over all space. If we make the further assumption that the field produced by the charges is asymptotically Coulombic, or that the system is enclosed in a box with reflecting walls, we can integrate by parts in this integral and obtain the result

$$\int \delta_{ru} x^u \overline{T}^{rs}{}_{,s} \, d^3x = -\int \delta_{ru} x^u{}_{,s} \overline{T}^{rs} \, d^3x$$

$$= -\int \delta_{rs} \overline{T}^{rs} \, d^3x = 0.$$

Therefore

$$\int \eta_{\mu\nu} \overline{T}^{\mu\nu} \, d^3x = \int \overline{T}^{00} \, d^3x = \mathscr{E}, \tag{8-5.4}$$

where $\mathscr{E}$ is the total energy of the system. Now $\eta_{\mu\nu} t^{\mu\nu}_{em} = 0$ and

$$\eta_{\mu\nu}\theta^{\mu\nu} = \sum_i m_i \sqrt{1 - u_i^2}\, \delta^3(x - z_i)$$

when we take the path parameters in the expression (8-5.1) for $\Theta^{\mu\nu}$ to be z_i^0. Consequently Eq. (8-5.4) reduces to

$$\mathscr{E} = \sum_i m_i \overline{\sqrt{1 - u_i^2}} \tag{8-5.5}$$

and is the relativistic expression of the virial theorem for a closed system of charged particles, or in fact for any closed system of particles interacting via a field whose stress-energy tensor has zero trace.

In the limit $c \to \infty$, that is, in the Newtonian limit, Eq. (8-5.5) reduces to

$$\mathscr{E} - \sum_i m_i c^2 = \sum_i \tfrac{1}{2} m_i \overline{u_i^2} \tag{8-5.6}$$

so that in this limit the time-averaged kinetic energy of the particles is equal to the total energy minus their rest-mass energy. But this is just the virial theorem for a closed system of particles interacting Coulombically and is to be expected, since in the Newtonian limit the advanced, retarded electromagnetic interaction between charged particles reduces to an instantaneous Coulomb interaction.

8-6. Classical Electron Theory

After the development of the Maxwell-Lorentz electrodynamics a considerable amount of effort was expended to develop a classical model of the electron.[9] Lorentz[10] suggested that the mass and momentum of an electron could be completely electromagnetic in origin, the electromagnetic field itself arising from the charge distribution in the electron. However, one encounters difficulties in such a picture from the start, because, according to Ernshaw's theorem, it is impossible to have a stationary nonneutral charge distribution held together by purely electric forces. For stability it appears necessary to introduce nonelectrical forces to hold the electron together, and these forces will in general contribute to the mass and momentum of the electron through the stress-energy tensor associated with them.

The simplest example of an electron model is that of a charge e distributed uniformly over the surface of a sphere of radius r_0, as measured by an

[9] For a review of these attempts, see W. Pauli, *Theory of Relativity*, Part V (Pergamon Press, New York, 1958).

[10] H. A. Lorentz, *Das Relativitätsprinzip* (3 Haarlemer Vorträge, Leipzig, 1914).

observer at rest with respect to the sphere.[11] If the sphere is at rest at the origin of spatial coordinates, then the field produced is given by

$$\mathbf{E}_0 = \frac{e}{|\mathbf{x}|^3}\,\mathbf{x} \qquad \text{for } |\mathbf{x}| > r_0$$
$$= 0 \qquad \text{for } |\mathbf{x}| < r_0$$

and

$$\mathbf{B}_0 = 0,$$

where the subscript zero indicates the field of the charge at rest. We can now make use of Eqs. (8-4.9) to obtain the nonvanishing components of the electromagnetic stress tensor. They are

$$t^{00}_{em0} = \frac{e^2}{8\pi}\frac{1}{|\mathbf{x}|^4}, \qquad |\mathbf{x}| > r_0 \tag{8-6.1}$$

and

$$t^{rs}_{em0} = \frac{1}{4\pi}\left\{\frac{e^2}{2|\mathbf{x}|^4}\,\delta^{rs} - \frac{e^2}{|\mathbf{x}|^6}\,x^r x^s\right\}. \tag{8-6.2}$$

If we integrate t^{00}_{em0} over all space, as an expression for the total energy contained in the electric field we obtain

$$\mathscr{E}_0 = \int t^{00}_{em0}\,dV = \frac{e^2}{2r_0}. \tag{8-6.3}$$

By taking $\mathscr{E}_0$ to be the observed mass of an electron (multiplied by c^2 in cgs units) one obtains the value $r_0 = 1.41 \times 10^{-13}$ cm. The length r_0 is referred to as the classical radius of the electron. The electric force per unit area acting on the charge is obtained using t^{rs}_{em0}:

$$f_0{}^r = -t^{rs}_{m0e}\,\frac{x^s}{|\mathbf{x}|}\bigg|_{|\mathbf{x}|=r_0} = \frac{1}{8\pi}\frac{e^2}{|\mathbf{x}|^5}\,x^r\bigg|_{|\mathbf{x}|=r_0} \tag{8-6.4}$$

This surface force, of magnitude 0.23×10^{32} dyne-cm^{-2}, is directed outward and normal to the surface of the sphere, and unless balanced by some non-electric force acting between the charges on the surface would cause them to fly apart.[12] We see that both $\mathscr{E}_0$ and $f_0{}^r$ diverge as r_0 approaches zero.

Suppose now that we wish to calculate the energy and momentum of a uniformly moving electron. We cannot make use of the results of Section

[11] The consequences of such a model were first worked out by M. Abraham, *Ann. Phys.* (*Leipzig*), **10** (1903).

[12] M. Abraham, *Phys. Z.*, **5**, 576 (1904).

8-4 to obtain these quantities. In particular, Eq. (8-4.15) will no longer be valid even though the field falls off sufficiently rapidly, since $t_{em,\nu}^{\mu\nu}$ does not equal zero everywhere. An explicit calculation also shows this to be the case. Thus, by making use of the transformation law for the stress-energy tensor one has

$$\begin{aligned} P'^{\mu} &= \int t'^{\mu 0}_{em}(x)\, dV \\ &= \Lambda_{\rho}{}^{\mu}\Lambda_{\sigma}{}^{0} \int t^{\rho\sigma}_{em0}(\Lambda_{\beta}^{-1\alpha}x^{\beta})\, dV. \end{aligned} \tag{8-6.5}$$

For a velocity mapping (6–5.23), which maps the trajectory of the center of charge onto the curve $x^{\mu} = \gamma(\tau, v\tau, 0, 0)$, one, therefore, obtains

$$\begin{aligned} \mathscr{E}' &= \gamma^2 \int t^{00}_{em0}(-v\gamma x^0 + \gamma x^1, x^2, x^3)\, dV \\ &\quad + v^2\gamma^2 \int t^{11}_{em0}(-v\gamma x^0 + \gamma x^1, x^2, x^3)\, dV \\ &= \gamma \int t^{00}_{em0}(\mathbf{x})\, dV + v^2\gamma \int t^{11}_{em0}(\mathbf{x})\, dV \end{aligned} \tag{8-6.6}$$

where a change of variable has been made to obtain the last result. Since $\int t^{11}_{em0}\, dV = \int \frac{1}{3}\, t^{00}_{em0}\, dV$ for a spherically symmetric field, it follows finally that

$$\mathscr{E}' = \gamma(1 + \tfrac{1}{3}v^2)\mathscr{E}_0. \tag{8-6.7}$$

A similar calculation leads to the result that

$$P'^1 = \tfrac{4}{3}\gamma v \mathscr{E}_0, \qquad P'^2 = P'^3 = 0. \tag{8-6.8}$$

On the other hand, a direct application of Eq. (8-4.15) leads to the result

$$\mathscr{E}' = \gamma\mathscr{E}_0, \quad P'^1 = v\gamma\mathscr{E}_0, \quad P'^2 = P'^3 = 0, \tag{8-6.9}$$

which is seen not to agree with the expressions (8-6.7) and (8-6.8) for $\mathscr{E}'$ and P'^1.

There have been a number of attempts to overcome this difficulty, namely, that $P^{\mu}(x^0) \neq \Lambda_{\nu}{}^{\mu}P^{\mu}(x^0)$. Poincaré[13] suggested that the energy and momentum of the electron should not be computed using the electromagnetic stress-energy tensor alone but rather by using the total stress-energy tensor $T^{\mu\nu}$. If $T^{\mu\nu}$ satisfies a continuity equation, Eq. (8-4.15) will hold and we will have the

[13] H. Poincaré, *Rend. Pal.*, **21**, 129 (1906).

desired transformation law for $P^{\mu}(\sigma)$. Poincaré's suggestion also overcomes the difficulty that $\mathscr{E}_0$ given by Eq. (8-6.3) diverges. In effect, the nonelectrical contribution to $\mathscr{E}_0$ must differ from $-e^2/2r_0$ by an amount just equal to the observed mass of the electron, that is,

$$\mathscr{E}_{0\,\mathrm{mech}} = -\frac{e^2}{2r_0} + m.$$

The sum of $\mathscr{E}_{0\,\mathrm{mech}}$ and $\mathscr{E}_{0\,\mathrm{elec}}$ is then just m. We see that $\mathscr{E}_{0\,\mathrm{mech}}$ is analogous to the quantity $-k$ introduced by Dirac in deriving the equations of motion of a radiating charge.

An alternate proposal was put forth by Born and Infeld[14] to deal with the self-energy of a classical electron. They suggested that in regions very near to charges, the linear electrodynamics of Maxwell-Lorentz breaks down and should be replaced by a nonlinear theory. Of course, once one admits the possibility of nonlinear equations for the field tensor $F^{\mu\nu}$, there is no longer an obvious and unique choice for the field equations as there is in the linear case. However, to a large extent the results are independent of the precise form of these equations, provided, of course, that in the limit of weak fields these equations reduce to the usual Maxwell-Lorentz equations. The fundamental feature of Born-Infeld electrodynamics is a stress-energy tensor whose divergence vanishes everywhere, even in the presence of point source, and such that $\mathscr{E}_0$ computed with this tensor is finite for such a source. In many respects, the Born-Infeld electrodynamics represents the best classical solution to the self-energy problem. The main drawback to this theory is the inability (to date) to construct a quantized version without destroying the nonlinear features.

Finally we should mention the suggestion of Rohrlich and others, namely, that in computing the energy of a moving charge one should not integrate over the surface $x^0 = \text{const}$ but rather over the surface whose normal is everywhere parallel to the velocity of the charge. Since this surface is the image of the $x^0 = \text{const}$ surface, under a Lorentz mapping that transforms the trajectory of the charge from the line $\mathbf{x} = 0$ to the line $\mathbf{x} = \mathbf{v}x^0$, it follows that $P^{\mu}(\sigma) = P'^{\mu}(\sigma')$, where σ is the surface $x^0 = \text{const.}$ and σ' is its image. While Rohrlich's procedure is unobjectionable if one accepts his method for computing the self-energy of a moving charge, it does not solve the self-energy problem for a point charge but rather brushes it under the rug, so to speak. Depending on the kind of questions asked and the uses that one puts the Rohrlich definition to, this may nevertheless be an acceptable and satisfactory method of dealing with the self-energy problem.

[14] M. Born and L. Infeld, *Proc. Roy. Soc.* (London), **A144**, 425 (1934).

8-7. Structure of the Electromagnetic Equations and the Initial Value Problem

When written in terms of the potentials in a 1 + 3 notation, Maxwell's equations take the form

$$-\nabla \cdot (\dot{\mathbf{A}} + \nabla \phi) = \rho \tag{8-7.1}$$

and

$$\nabla \times (\nabla \times \mathbf{A}) + \ddot{\mathbf{A}} + \nabla \dot{\phi} = \mathbf{j}, \tag{8-7.2}$$

where dots over a quantity denote derivatives with respect to x^0. We see immediately that the highest order x^0 derivatives, $\ddot{A}$ and $\dot{\phi}$, all appear in the second set of three equations. The equations (8-7.1) and (8-7.2) are therefore not of the Cauchy-Kowalewski type. This feature is, of course, due to their gauge invariance. This invariance is also responsible for the Bianchi identity:

$$F^{\mu\nu}{}_{,\mu\nu} \equiv 0. \tag{8-7.3}$$

To facilitate the discussion of the initial value problem for the Maxwell field, it is convenient to introduce the electric field $\mathbf{E} = -\dot{\mathbf{A}} - \nabla\phi$ explicitly. With its help the Maxwell equations can be written in the form

$$\nabla \cdot \mathbf{E} = \rho, \tag{8-7.4}$$

$$\dot{\mathbf{E}} = \nabla \times (\nabla \times \mathbf{A}) - \mathbf{j}, \tag{8-7.5a}$$

$$\dot{\mathbf{A}} = -\mathbf{E} - \nabla\phi. \tag{8-7.5b}$$

As written, Eqs. (8-7.5) appear to be in Cauchy-Kowalewski form, since the right-hand sides do not involve any x^0 differentiated terms. Given $\mathbf{A}$, $\mathbf{E}$, and ϕ on an initial data surface, that is, a spacelike surface $x^0 = \text{const}$, with $\mathbf{E}$ satisfying Eq. (8-7.4), one can compute $\dot{\mathbf{E}}$ and $\dot{\mathbf{A}}$ and so determine $\mathbf{E}$ and $\mathbf{A}$ (and hence $\mathbf{B} = \nabla \times \mathbf{A}$) on a surface parallel to and infinitesimally separated from the initial data surface. However, these equations do not allow one to compute $\ddot{\mathbf{A}}$ and higher derivatives on this surface from these initial values, so that we cannot determine $\mathbf{E}$ and $\mathbf{A}$ on surfaces finitely separated from it. To do so we need to know first and higher derivatives of ϕ, which are indeterminate.

The way out of this difficulty is to impose a gauge condition by requiring that a single, nongauge covariant relation between $\mathbf{A}$ and $\mathbf{E}$ be satisfied; for example the Coulomb condition

$$\nabla \cdot \mathbf{A} = 0. \tag{8-7.6}$$

This condition together with the constraint Eq. (8-7.4) now allows us to eliminate two of the six quantities A_r and E_r from consideration as dynamical variables. It also serves to determine ϕ (modulo boundary conditions), since in general the gauge condition will remain satisfied on later $x^0 = \text{const}$ surfaces only if ϕ is chosen appropriately. In the case of the Coulomb gauge we find, by taking the divergence of Eq. (8-7.5b) and making use of Eq. (8-7.4), that ϕ must satisfy

$$\nabla^2\phi + \rho = 0. \tag{8-7.7}$$

With the ϕ so determined, Eqs. (8-7.5) allow one to compute the necessary x^0 derivatives of the four undetermined components of **A** and **E** in terms of the initial value data needed for finding the values of **A** and **E** on other $x^0 = \text{const}$ surfaces. From these considerations we see that the electromagnetic field describes the kpt of a system with two degrees of freedom per space point, corresponding to the two possible states of polarization of electromagnetic plane waves.

Problem 8.12. Make use of the Helmholtz decomposition of a vector $\mathbf{K} = \mathbf{K}^{tr} + \mathbf{K}^{lo}$, where

$$\nabla \cdot \mathbf{K}^{tr} \equiv 0 \quad \text{and} \quad \nabla \times \mathbf{K}^{lo} \equiv 0,$$

to rewrite Eqs. (8-7.4) and (8-7.5) in terms of $\mathbf{A}^{tr}$, $\mathbf{A}^{lo}$, $\mathbf{E}^{tr}$, and $\mathbf{E}^{lo}$. By making use of Eq. (8-7.6), show that $\mathbf{A}^{tr}$ and $\mathbf{E}^{tr}$ are the components of **A** and **E** which are determined by Eqs. (8-7.5) from the initial value data.

8-8. The Klein-Gordon Equation

While electrodynamics is the most important classical example of a theory that admits the Poincaré group as a symmetry group, other such theories play an important role in quantum field theory. Although a discussion of quantum processes lies outside the scope of this book, it is desirable to discuss some of the classical properties of these laws that carry over in the quantum theory. In this section we examine the Klein-Gordon equation, and in the next section, the Dirac equation.

Perhaps the simplest Poincaré invariant theory is one in which the kpt are characterized by a scalar or pseudoscalar (scalar density) field $\phi(x)$. The dynamical law is taken to be

$$\Box\phi + \alpha^2\phi = 0, \qquad (\Box \equiv \eta^{\mu\nu}\partial_\mu\,\partial_\nu). \tag{8-8.1}$$

Since this theory is the starting point of the quantum field theory of mesons, ϕ is sometimes called the *meson field.*[15]

[15] Equation (8-8.1) is known as the *Klein-Gordon equation*. When $\alpha = 0$, it reduces to the scalar wave equation.

Problem 8.13. Construct a wave packet from plane-wave solutions of Eq. (8-8.1) and derive an expression for the group velocity of this packet.

Equation (8-8.1) can be derived from a variational principle with an action given by[16]

$$S_F = \int \mathfrak{L}(x)\, d^4x = \tfrac{1}{2} \int \{\eta^{\mu\nu}\phi_{,\mu}\phi_{,\nu} - \alpha^2\phi^2\}\, d^4x. \tag{8-8.2}$$

By making use of Eq. (4-5.16), we can obtain an expression for the stress-energy tensor for the scalar field. Since, for a space-time translation $\bar{\delta}\phi = -\phi_{,\mu}\varepsilon^\mu$ we have

$$\begin{aligned} t^{\rho}_{sc\ \nu} &= -\mathfrak{L}\delta_\nu{}^\mu + \left(\frac{\partial \mathfrak{L}}{\partial \phi_{,\mu}}\right)\phi_{,\nu} \\ &= \{\eta^{\mu\rho}\delta_\nu{}^\sigma - \tfrac{1}{2}\eta^{\rho\sigma}\delta_\nu{}^\mu\}\phi_{,\rho}\phi_{,\sigma} + \tfrac{1}{2}\alpha^2\phi^2\, \delta_\nu{}^\mu. \end{aligned} \tag{8-8.3}$$

We see in particular that the energy density

$$t^{0}_{sc\ 0} = \tfrac{1}{2}\{\phi_{,0}\phi_{,0} + \alpha^2\phi^2\}$$

is positive definite. The requirement of positive-definiteness of the energy density can thus be used to fix the sign of the action S_F, which would otherwise be arbitrary.

The scalar meson field can be made to interact with "mesonically" charged matter in much the same way that the electromagnetic field interacts with electrically charged matter. The total action for the system of particles plus field can be written in a form analogous to that for the system of electrically charged particles plus field, given by Eq. (8-1.19):

$$S = S_F - \sum_i m_i \int_{\sigma_1}^{\sigma_2} \sqrt{\dot{z}_i{}^2}\, d\lambda_i - \sum_i g_i \int_{\sigma_1}^{\sigma_2} \int_R \phi(x)\, \delta_4(x - z_i)\sqrt{\dot{z}_i{}^2}\, d\lambda_i\, d^4x_i, \tag{8-8.4}$$

where g_i is the mesonic charge, m_i is the mass of the ith particle, and S_F is given by Eq. (8-8.2). Variation of S with respect to ϕ then leads to the modified field equations

$$(\Box + \alpha^2)\phi(x) = -\rho(x), \tag{8-8.5}$$

where $\rho(x)$, the source density, is given by

$$\rho(x) = \sum_i g_i \int_{\sigma_1}^{\sigma_2} \delta(x - z_i)\sqrt{\dot{z}_i{}^2}\, d\lambda_i. \tag{8-8.6}$$

[16] See, for example, S. S. Schweber, *An Introduction to Relativistic Quantum Field Theory* (Row, Peterson, Evanston, Illinois, 1961), Chap. 3.

Likewise, variation with respect to z_i leads to the ith particle equations of motion

$$-g_i\phi_{,\mu}(z_i) + \frac{d}{d\tau_i}\{(m_i + g_i\phi(z_i))\dot{z}_{i\mu}\} = 0 \tag{8-8.7}$$

when the path parameters are taken to be the particle proper times.

Having obtained a field theoretic description of the mesonic interaction between particles, we now ask if it is possible to construct from it an action-at-a-distance description of this interaction analogous to that for the electronic interaction discussed in Section 7-9. The most obvious procedure for obtaining such direct-particle equations of motion would be to solve Eq. (8-8.5) for $\phi(x)$ as a function of $\rho(x)$ and substitute this solution into Eq. (8-8.7). However, in doing so we would be including self-interactions of the particles, and these interactions in general diverge. To overcome this difficulty we must work with partial fields. The field to be substituted into the equation of motion for the ith particle is required to satisfy the equation

$$(\Box + \alpha^2)\phi_i(x) = -\rho_i(x), \tag{8-8.8}$$

where

$$\rho_i(x) = \sum_{j\neq i} g_j \int_{\sigma_1}^{\sigma_2} \delta_4(x - z_j)\sqrt{\dot{z}_j{}^2}\, d\lambda_j\,. \tag{8-8.9}$$

The ith partial field $\phi_i(x)$ is thus seen to be independent of the coordinates of the ith particle.

Solutions of Eq. (8-8.8) can be expressed in terms of Green's functions for the Klein-Gordon operator $(\Box + \alpha^2)$. Such a Green's function $\Delta(x, x')$ satisfies

$$(\Box + \alpha^2)\,\Delta(x, x') = \delta_4(x - x'). \tag{8-8.10}$$

There are. of course, many different Green's functions that satisfy this equation, corresponding to different boundary conditions, but for the moment we will not decide between them. However, because of the translational invariance of Eq. (8-8.10), they must all have the property that $\Delta(x, x') = \Delta(x - x')$. Independently of which Green's function we choose, we can construct a particular solution to Eq. (8-8.10) of the form

$$\phi_i(x) = -\int d^4x'\,\Delta(x - x')\rho_i(x'). \tag{8-8.11}$$

If we now substitute this solution into Eq. (8-8.7) we obtain the action-at-a-distance equations of motion for the ith particle.

The ith direct-particle equations of motion can be obtained from an action principle with a partial action S_i given by

$$\begin{aligned} S_i &= -m_i \int_{\sigma_1}^{\sigma_2} \sqrt{\dot{z}_i^2}\, d\lambda_i - g_i \int_{\sigma_1}^{\sigma_2} \phi_i(z_i)\sqrt{\dot{z}_i^2}\, d\lambda_i \\ &= -m_i \int_{\sigma_1}^{\sigma_2} \sqrt{\dot{z}_i^2}\, d\lambda_i + g_i \sum_{j \neq i} g_j \int_{\sigma_1}^{\sigma_2} \int_{(\sigma_1)}^{(\sigma_2)} \Delta(z_i - z_j)\sqrt{\dot{z}_i^2}\sqrt{\dot{z}_j^2}\, d\lambda_j\, d\lambda_i . \end{aligned} \tag{8-8.12}$$

The integration over λ_j in the second term above extends far enough beyond the surfaces σ_1 and σ_2 to include the total effect of the jth particle on the part of the trajectory of the ith particle lying between these two surfaces. We have indicated this by enclosing the limits of integration of λ_j in parentheses (see Section 7-4).

We cannot add the partial actions (8-8.12) to obtain a single action for the totality of direct-particle equations of motion for all the particles. The only exception is if $\Delta(x)$ is symmetric in its arguments, that is, if $\Delta(x) = \Delta(-x)$. In this case the total action has the form of the general action given in Eq. (7-4.1):

$$S = -\sum_i m_i \int_{\sigma_1}^{\sigma_2} \sqrt{\dot{z}_i^2}\, d\lambda_i + \frac{1}{2} \sum_{i \neq j}\sum g_i g_j \iint_{(\sigma_1)}^{(\sigma_2)} \Delta(z_i - z_j)\sqrt{\dot{z}_i^2}\sqrt{\dot{z}_j^2}\, d\lambda_i\, d\lambda_j . \tag{8-8.13}$$

Having obtained this single comprehensive action we can then employ the methods of Section 7-5 to derive conservation laws for the system of mesonically charged particles.

The requirement that the Green's function $\Delta(x)$ be a symmetric function of its arguments greatly reduces the number of solutions of Eq. (8-8.10). We can obtain a unique solution if we impose what amounts to a causality condition on $\Delta(x)$. We will require that a given particle can interact with another particle only if the latter is within the light cone of the former. From the form of the action (8-8.13), we see that this will be the case, provided $\Delta(x) = 0$ for $x^2 = x_\mu x^\mu < 0$. An explicit expression for the particular Green's function satisfying the above two conditions, denoted by $\Delta_P(x)$ in the literature, is given by [17]

$$\Delta_P(x) = \frac{1}{4\pi}\,\delta(x^2) - \begin{cases} \dfrac{1}{8\pi}\dfrac{\alpha}{\sqrt{x^2}}\, J_1(\alpha\sqrt{x^2}) & x^2 > 0 \\ 0 & x^2 < 0 \end{cases},$$

[17] Cf. J. Schwinger, *Phys. Rev.*, **75**, 651 (1949).

where J_1 is the cylindrical Bessel function of order 1. We see that in the limit $\alpha = 0$, Δ_P reduces to the Green's function appearing in the expression (8-1.12) for the electromagnetic four-potential $A^\mu(x)$. This particular form of the four-potential corresponds to taking a sum of half-advanced, half-retarded solutions of Maxwell's equations. Therefore Δ_P is the "half-advanced, half-retarded" Green's function for the Klein-Gordon equation. It differs, however, from the Green's function for the Maxwell equations in one important respect, namely, while the latter is nonzero only on the light cone, the former is nonzero both on and inside the light cone. The integrals appearing in the interaction term of the action (8–8.13) must therefore extend from minus to plus infinity.

To understand better the nature of the scalar mesic interaction let us consider the motion of a particle with mesic charge g_1, mass m_1, moving in the field of another particle, and with mesic charge g_2, held fixed at the origin of coordinates. The field of this fixed source particle is gotten by solving Eq. (8-8.8) with

$$\rho_1(x) = g_2\,\delta^3(\mathbf{x}).$$

Since ρ_1 is independent of x^0, ϕ_1 will also have this property, and hence Eq. (8-8.8) will reduce to

$$(-\nabla^2 + \alpha^2)\phi_1(\mathbf{x}) = -g_2\,\delta^3(\mathbf{x}). \tag{8-8.14}$$

This equation has, as a solution,

$$\phi_1(\mathbf{x}) = -\frac{g_2}{4\pi}\frac{e^{-\alpha r}}{r}, \tag{8-8.15}$$

as can be verified by direct substitution into Eq. (8-8.14) and integration over all space.

With ϕ_1 independent of x^0 we see that the equations of motion (8-8.7) have a first integral

$$\{m_1 + g_1\phi_1(\mathbf{z}_1)\}\dot{z}_1{}^0 = \text{const}.$$

This is just the energy integral for the particle, and the constant can be taken to be its total energy. We can express the energy integral in terms of the three-velocity $\mathbf{u}_1$ of the particle as

$$\mathscr{E} = \frac{m_1 + g_1\phi_1(\mathbf{z}_1)}{\sqrt{1 - \mathbf{u}_1{}^2}}.$$

In the nonrelativistic limit this result reduces to

$$\mathscr{E} = m_1 c^2 + \frac{1}{2}m_1\mathbf{u}_1{}^2 - \frac{g_1 g_2}{4\pi}\frac{e^{-\alpha r_1}}{r_1}. \tag{8-8.16}$$

We see from Eq. (8-8.16) that the interaction potential is short range, with a range $\sim 1/\alpha$. This potential is the famous Yukawa potential, introduced by him in his meson theory of nuclear forces. In the limit $\alpha = 0$ it reduces to the ordinary Coulomb potential. We also see that the potential is attractive if g_1 and g_2 have the same sign, and is repulsive if their signs are different. The situation is thus the opposite of that in electrodynamics where like charges repel and unlike charges attract. It is instructive to see how this difference arises in the two cases. It is due, in the final analysis, to the relative sign between the free-field and free-particle parts of the total action for the field-plus-particle system. The signs of these two terms are fixed in turn by the requirement that the free-field and free-particle energies that result from these terms are positive definite. In both electromagnetic (8-1.19) and mesic (8-8.4) actions, the signs of the free-field and free-particle contributions have been chosen so as to satisfy this requirement. The sign of the interaction term, on the other hand, is arbitrary in both cases. However, if one traces through the steps leading to Eq. (8-8.16), for instance, one sees that a change in sign of the interaction term will not affect the sign of the interaction energy.

We will now show that it is possible to represent a charged scalar meson by a complex field ϕ satisfying, in the absence of electromagnetic fields, the Klein-Gordon equation. The action leading to Eq. (8-8.1) and its complex conjugate is

$$S = \tfrac{1}{2} \int d^4x \{\eta^{\mu\nu}\phi^*{}_{,\mu}\phi_{,\nu} - \alpha^2\phi^*\phi\}. \tag{8-8.17}$$

This action is invariant under the internal transformation

$$\phi \to \phi' = e^{i\lambda}\phi, \tag{8-8.18a}$$

$$\phi^* \to \phi'^* = e^{-i\lambda}\phi^*, \tag{8-8.18b}$$

where λ is an arbitrary constant. For an infinitesimal transformation we have

$$\bar{\delta}\phi = i\varepsilon\phi \quad \text{and} \quad \bar{\delta}\phi^* = -i\varepsilon\phi^*.$$

It therefore follows from Eq. (4-5.16) that

$$j^\mu = \frac{i}{2}\eta^{\mu\nu}(\phi^*{}_{,\nu}\phi - \phi^*\phi_{,\nu}) \tag{8-8.19}$$

is conserved, that is, $j^\mu{}_{,\mu} = 0$. The vector j^μ therefore satisfies the same continuity equation as does the electric current.

We can now couple the scalar field to the electromagnetic field by letting the constant λ appearing in the transformation law (8-8.18) be an arbitrary space-time function (actually a constant e times an arbitrary function) and apply the results of Section 2-2. The analogue of Eq. (2-2.1) in the present case is thus

$$\bar{\delta}\phi = ie\varepsilon(x)\phi(x) \quad \text{and} \quad \bar{\delta}\phi^* = -ie\varepsilon(x)\phi^*(x).$$

Since we are dealing with an Abelian group we need only a single vector field A_μ, which we will identify with the electromagnetic potential, to form an affinity and the covariant derivative of ϕ and ϕ^*:

$$\phi_{;\mu} = \phi_{,\mu} - ieA_\mu\phi \quad \text{and} \quad \phi^*{}_{,\mu} = \phi^*{}_{;\mu} + ieA_\mu\phi^*,$$

with

$$\bar{\delta}A_\mu = \varepsilon(x)_{,\mu}.$$

We can now form an invariant action by replacing ordinary derivatives in Eq. (8-8.17) by covariant derivatives. The total action for the system of scalar field plus electromagnetic field is thus

$$S = \frac{1}{2}\int \{\eta^{\mu\nu}(\phi^*{}_{,\mu} + ieA_\mu\phi^*)(\phi_{,\nu} - ieA_\nu\phi) - \alpha^2\phi^*\phi\}$$

$$+ \frac{1}{8\pi}\int d^4x\left\{\frac{1}{2}F^{\mu\nu}F_{\mu\nu} - (A_{\nu,\mu} - A_{\mu,\nu})F^{\mu\nu}\right\} d^4x. \qquad (8\text{-}8.20)$$

The equations obtained by varying A_μ are

$$F^{\mu\nu}{}_{,\nu} = -4\pi j^\mu, \qquad (8\text{-}8.21)$$

with

$$j^\mu = -\tfrac{1}{2}ie\eta^{\mu\nu}(\phi^*{}_{;\nu}\phi - \phi^*\phi_{;\nu}) \qquad (8\text{-}8.22)$$

which is seen to reduce to the expression (8-8.19) (multiplied by $-e$) when $A_\mu = 0$. We see that the Maxwell equation (8-8.21) can be interpreted as the equation for the Riemann tensor (in this case, $F^{\mu\nu}$) associated with the internal affinity.

Problem 8.13. Show that $j^\mu{}_{,\mu} = 0$, with j^μ given by Eq. (8-8.22) as a consequence of the equations of motion for ϕ and ϕ^*.

8-9. The Dirac Equation

As a second example of a Poincaré invariant theory we will consider the Dirac equation. This equation was developed by Dirac[18] in an attempt to obtain a relativistic wave equation for the electron, which, unlike the Klein-Gordon equation, was first order in the derivative with respect to time. The

[18] P. A. M. Dirac, *Proc. Roy. Soc.* (London), **A117**, 610 (1928).

brilliant success of Dirac's theory is well known: It predicted correctly the observed spectrum of the hydrogen atom, it contained an explanation of the intrinsic spin of the electron, and it ultimately led to the discovery of the positron. It would again take us beyond the scope of this work to develop these consequences of the theory; we shall content ourselves with a discussion of the formal properties of the Dirac equation.

Dirac proposed, as the wave equation to be associated with a free electron,

$$i\hbar\gamma^{\mu}\psi_{,\mu} - m\psi = 0, \tag{8-9.1}$$

where the γ^{μ} are a set of four, everywhere constant, noncommuting operators. They are restricted by the requirement that any solution of Eq. (8-9.1) must also be a solution of the Klein-Gordon equation (8-8.1). To determine the form of these restrictions we multiply Eq. (8-9.1) by the operator $i\gamma^{\nu}\,\partial\nu + m$ to obtain

$$\hbar^2\gamma^{\mu}\gamma^{\nu}\psi_{,\mu\nu} + m^2\psi = 0.$$

If we symmetrize the coefficient of $\psi_{,\mu\nu}$ in this equation we obtain the Klein-Gordon equation, provided the γ satisfy

$$\gamma^{\mu}\gamma^{\nu} + \gamma^{\nu}\gamma^{\mu} = 2\eta^{\mu\nu}. \tag{8-9.2}$$

The simplest representation for a set of four matrices satisfying these conditions is in terms of 4×4 matrices.[19] For our purposes it is not necessary to have an explicit representation for the γ, the anticommutation relations (8-9.2) sufficing.

In order that the Dirac equation admit the Poincaré group as a symmetry group it is necessary (1) that the ψ form the basis of a representation of the Poincaré group, and (2) that solutions of Eq. (8-9.1) be transformed into solutions under the action of this group. While condition (1) can be satisfied by treating ψ as a scalar under the Poincaré group, condition (2) will not in general be satisfied. In an attempt to satisfy both conditions let us assume that under an infinitesimal Poincaré mapping for which

$$x^{\mu} \to x'^{\mu} = x^{\mu} + \varepsilon_{\nu}{}^{\mu}x^{\nu} + \varepsilon^{\mu},$$

the Dirac field transforms in such a way that

$$\bar{\delta}\psi = i\varepsilon_{\mu\nu}S^{\mu\nu}\psi - \psi_{,\mu}(\varepsilon_{\nu}{}^{\mu}x^{\nu} + \varepsilon^{\mu}). \tag{8-9.3}$$

[19] See, for example, A. Messiah, *Quantum Mechanics* (Wiley, New York, 1962), Chap. XX, for a discussion of these matrices and other properties of the Dirac equation.

Since Eq. (8-9.1) is linear, condition (2) above will be satisfied, provided $\bar{\delta}\psi$ satisfies this equation when ψ satisfies it. The unknowns, $S_{\mu\nu}$, are therefore determined by the equation

$$(i\gamma^\mu S^{\rho\sigma}\psi_{,\mu} - iS^{\rho\sigma}\gamma^\mu\psi_{,\mu} - \eta^{\mu\rho}\gamma^\sigma\psi_{,\mu})\varepsilon_{\rho\sigma} = 0,$$

where we have made explicit use of the fact that ψ satisfies Eq. (8-9.1). Since both $\varepsilon_{\rho\sigma}$ and $\psi_{,\nu}$ are arbitrary at any one space-time point, it follows that $S^{\rho\sigma}$ must satisfy the equation

$$S^{\rho\sigma}\gamma^\mu - \gamma^\mu S^{\rho\sigma} = \frac{i}{2}(\eta^{\mu\sigma}\gamma^\rho - \eta^{\mu\rho}\gamma^\sigma). \tag{8-9.4}$$

This equation can be satisfied if we take

$$S^{\rho\sigma} = \frac{i}{4}(\gamma^\rho\gamma^\sigma - \gamma^\sigma\gamma^\rho). \tag{8-9.5}$$

Having found a form for $S^{\rho\sigma}$ such that condition (2) above is satisfied, we must still check that the commutator of two mappings, $\bar{\delta}_2(\bar{\delta}_1\psi) - \bar{\delta}_1(\bar{\delta}_2\psi)$, is again of the form $\bar{\delta}_3\psi$ when $S^{\rho\sigma}$ is given by Eq. (8-9.4), and when $\varepsilon^\mu_{3\nu}$ is related to $\varepsilon^\mu_{1\nu}$ and $\varepsilon^\mu_{2\nu}$ as for a Lorentz mapping; that is,

$$\varepsilon^\mu_{3\nu} = \varepsilon^\mu_{1\rho}\varepsilon^\rho_{2\nu} - \varepsilon^\mu_{2\rho}\varepsilon^\rho_{1\nu}.$$

A somewhat lengthy calculation shows this to be the case. Objects that transform according to Eq. (8-9.3), with $S^{\rho\sigma}$ given by Eq. (8-9.4) are called *spinors*. They form the basis of a representation of the Poincaré group corresponding to a value of S of 1/2 as mentioned at the end of Section 6-5.

Having found the form of $S^{\rho\sigma}$ appearing in Eq. (8-9.3) we can follow the procedure used in Section 6-5 to obtain the finite forms of the Lorentz matrices, given by Eqs. (6-5.18) and (6-5.20), to find the finite forms of the transformation law of ψ. For the Poincaré mapping $x^\mu \to x^\mu = \Lambda_\nu{}^\mu x^\nu + a^\mu$ we have

$$\psi'(x') = \Omega(\Lambda)\psi(x), \tag{8-9.6}$$

where

$$\Omega(\Lambda) = e^{iS^{\rho\sigma}\Lambda_{\rho\sigma}}. \tag{8-9.7}$$

Corresponding to the special Lorentz mapping $V(\mathbf{v})$ given by Eq. (6-5.20) and (6-5.21), one finds, with the help of the anticommutation relations (8-9.2),

$$\Omega(\mathbf{v}) = \frac{\gamma}{\sqrt{2(1+\gamma)}}\left\{1 + \frac{1}{\gamma} - (\boldsymbol{\alpha}\cdot\mathbf{v})\right\}, \qquad \gamma = \frac{1}{\sqrt{1-\mathbf{v}^2}}, \tag{8-9.8}$$

where $\boldsymbol{\alpha} = (2/i)(S^{10}, S^{20}, S^{30})$. For a spatial notation $R(\theta, u)$ given by Eq. (6-5.18) we have

$$\Omega(\theta, \mathbf{u}) = \cos\tfrac{1}{2}\theta + i(\boldsymbol{\sigma}\cdot\mathbf{u})\sin\tfrac{1}{2}\theta \tag{8-9.9}$$

where $\boldsymbol{\sigma} = 2(S^{23}, S^{31}, S^{12})$. We see from this result that $\Omega(2\pi, \mathbf{u}) = -1$ as a consequence of the half-angle appearing in (8-9.9). The spinor ψ is thus the basis of a *double-valued* representation of the three-dimensional rotation group, since $\Omega(\theta, \mathbf{u}) \neq \Omega(\theta + 2\pi, \mathbf{u})$.

The Dirac equation can be derived from a variational principle with an action

$$S = \int d^4x \bar{\psi}(i\hbar\gamma^\mu \psi_{,\mu} - m\psi), \tag{8-9.10}$$

where $\bar{\psi} = \psi^\dagger \gamma^0$, $\psi^\dagger$ being the Hermitian conjugate of ψ. By varying ψ and $\bar{\psi}$ independently in this action one obtains for ψ Eq. (8-9.1), and for $\bar{\psi}$, its Hermitian conjugate,

$$-i\bar{\psi}_{,\mu}\gamma^\mu - m\bar{\psi} = 0.$$

Then, since S is invariant under the internal transformation,

$$\psi \to \psi' = e^{i\lambda}\psi, \qquad \bar{\psi} \to \bar{\psi}' = e^{-i\lambda}\bar{\psi},$$

we can parallel the discussion of the preceding section about coupling the complex scalar field to the electromagnetic field, to obtain the coupling of the Dirac field to the electromagnetic field. To do so we merely replace the ordinary derivative in the action (8-9.10) by the covariant derivative formed with A_μ, to obtain the new action

$$S = \int \bar{\psi}[i\hbar\gamma^\mu(\psi_{,\mu} - ieA_\mu\psi) - m\psi].$$

If we combine this action with that for the free electromagnetic field S_{em}, we again obtain Eq. (8-8.21), now with

$$j^\mu = -ie\bar{\psi}\gamma^\mu\psi.$$

Problem 8.14. Obtain an expression for the stress-energy tensor for the Dirac field and show that $t_0{}^0$ is positive-definite.

8-10. The Gravitational Field in Special Relativity

Attempts to formulate a field theory of gravity within the framework of special relativity go back to the early days of relativity theory.[20-22] However, these and subsequent attempts were soon overshadowed by the Einstein

[20] A. Einstein, *Ann. Phys. Lpz.*, **38**, 355 and 443 (1912).
[21] M. Abraham, *Phys. Z.*, **13**, 1, 4, and 793 (1912).
[22] G. Nordström, *Phys. Z.*, **13**, 1126 (1912); *Ann. Phys. Lpz.*, **40**, 856 (1913); **42**, 533 (1913); **43**, 1101 (1914).

theory of general relativity. Although most physicists will agree that his latter theory affords the most satisfying description of gravitational phenomena, it is nevertheless useful to review here the problems attendant on a special relativistic gravitational field theory, if for no other reason than to lend support to the general theory of relativity. We will see, in fact, that at least one attempt to formulate a consistent special relativistic theory of the gravitational field leads one to the field equations of general relativity.

If we want to describe the gravitational interaction of massive bodies by means of a field, we most naturally look for the simplest description to start with that reduces to the Newtonian theory described in Section 5-3 in a suitable nonrelativistic limit. In the Newtonian theory the gravitational field can be constructed from a single scalar field that satisfies the inhomogeneous Laplacian equation (5-3.6). This suggests that we should try to represent the gravitational field in special relativity also by a scalar field $\phi(x)$ and replace the inhomogeneous Laplacian equation by the inhomogeneous wave equation

$$\Box\phi(x) = -4\pi G\rho(x),$$

where now

$$\rho(x) = \sum_i m_i \int \delta^4(x - z_i)\sqrt{\dot{z}_i^{\,2}}\, d\lambda_i .$$

Such a theory is seen to be equivalent to the meson field theory discussed in Section 8-8, with $\alpha = 0$ and the mesic charge replaced by m_i. We can therefore use the action (8-8.4) with these substitutions to obtain the field-plus-particle equations of motion of this theory. It follows, then, that the interaction potential between particles is Coulombic in the nonrelativistic limit and that like charges attract each other.

In all respects this scalar theory seems to be an excellent candidate for a field description of gravitational phenomena in special relativity. However, it does not appear to agree with certain currently accepted observational facts. In particular the scalar theory predicts a perihelion retrogression of planetary orbits rather than the observed precession (see Chapter 12 for a discussion of the pertinent observational data). Furthermore it does not predict a bending of light in a gravitational field.

After a scalar field the next simplest tensorial object (in the sense of fewest number of independent components) is a vector field. However, a vector theory of gravity would be equivalent in all respect to electrodynamics and hence would predict, among other things, that massive bodies should repel each other. Furthermore, while a vector theory predicts a perihelion advance for the planetary orbits, the amount is only one-sixth of that observed. We can rule out the possibility of using an antisymmetric second-rank tensor to describe the gravitational field because there is no way to couple it to matter other than by introducing a vector potential, as in electrodynamics. We come, therefore, to consider a symmetric second-rank tensor as the next simplest object to associate with the gravitational field. A special relativistic tensor description of gravity was first proposed by Birkhoff[23] but suffered from the fact that it could not be obtained from a variational principle. Later, Belinfante[24] constructed a tensor theory that was free of this defect. As we will see, a tensor theory sufficies to explain all observed gravitational phenomena. Nevertheless there are attendant difficulties that make it somewhat less than satisfactory from a theoretical point of view.

The first problem that confronts us in constructing a tensor theory of gravity is to find suitable free-field equations of motion. Even if we limit ourselves to linear, second-order differential equations there are, unlike in the scalar case, several different such equations for a second-rank symmetric tensor. This arbitrariness can be reduced somewhat if we take into account the fact that such a tensor is reducible under the Lorentz group to a traceless tensor plus a scalar field. Thus, we can decompose the gravitational field tensor $h_{\mu\nu}$ according to

$$h_{\mu\nu} = \psi_{\mu\nu} + \tfrac{1}{4}\eta_{\mu\nu}\phi \tag{8-10.1}$$

where $\eta^{\mu\nu}\psi_{\mu\nu} = 0$. Since we are only considering here linear theories of the gravitational field we can treat separately the traceless part and the trace. The latter is, however, just a scalar and hence must satisfy, in the free-field case, the scalar wave equation in order that the gravitational interaction resulting from such a field be long range and reduce to Coulombic form in the nonrelativistic limit.

The most general equations for the traceless part of the gravitational field that we can write down which are linear, of second differential order, Lorentz invariant, and Coulombic in the nonrelativistic limit are of the form

$$\alpha\Box\psi_{\mu\nu} + \beta\psi_{\mu}{}^{\rho}{}_{,\nu\rho} = 0.$$

However, as they stand these equations suffer from a serious defect; namely, the energy density associated with a $\psi_{\mu\nu}$ satisfying these equations will not, in general, be positive definitive. Fierz[25] has discussed the problem of the positive-definiteness of the energy density for various types of fields and has shown that only if $\psi_{\mu\nu}$ satisfies the two equations

$$\Box\psi_{\mu\nu} = 0 \tag{8-10.2}$$

and

$$\psi_{\mu}{}^{\rho}{}_{,\rho} = 0 \tag{8-10.3}$$

[23] G. D. Birkhoff, *Proc. Nat. Ac.*, **30**, 54 (1944); A. Barajas, G. D. Birkhoff, C. Graf, and M. Sandoval-Valarta, *Phys. Rev.*, **66**, 138 (1944).

[24] F. J. Belinfante, *Phys. Rev.*, **89**, 914 (1953).

[25] M. Fierz, *Helv. Phys. Acta*, **12**, 3 (1939).

will the energy density be positive-definite. It also turns out that only in this case will the spin associated with $\psi_{\mu\nu}$ have a definite value, namely, 2.

The first of the above equations for $\psi_{\mu\nu}$ can be considered to be a field equation and can be derived from a variational principle with an action $\mathfrak{L}_\psi$ given by

$$\mathfrak{L}_\psi = \tfrac{1}{2}\eta^{\rho\sigma}\psi_{\mu\nu,\rho}\psi^{\mu\nu}{}_{,\sigma}. \tag{8-10.4}$$

For such an action the results of Section 4-5 lead to a stress-energy tensor of the form

$$t_\psi^{\mu\nu} = (\eta^{\mu\rho}\eta^{\nu\sigma} - \tfrac{1}{2}\eta^{\mu\nu}\eta^{\rho\sigma})\psi_{\alpha\beta,\rho}\psi^{\alpha\beta}{}_{,\sigma}. \tag{8-10.5}$$

The second equation for $\psi_{\mu\nu}$ on the other hand does not follow from a variational principle and must be treated as a supplementary condition on $\psi_{\mu\nu}$.

Problem 8.15. By considering the various types of plane-wave solutions of Eq. (8-10.2) satisfying the supplementary condition (8-10.3) show that t_ψ^{00} is positive-definite in all cases.

Let us now consider how the field $\psi_{\mu\nu}$ can be coupled to matter or to some other field. In the presence of sources we can expect that Eq. (8-10.2) will get replaced by an equation of the form

$$\Box\psi_{\mu\nu} = j_{\mu\nu} \tag{8-10.6}$$

where $j_{\mu\nu}$ is the "current" associated with these sources. If we believe that all forms of energy can act as sources of the gravitational field, it follows that $j_{\mu\nu}$ must be related in some way to the stress-energy tensor associated with these sources. (We will consider shortly the possibility that the gravitational field can act as its own source because of its energy content.)

At this point we now encounter a fundamental difficulty in our theory. If $\psi_{\mu\nu}$ is required to satisfy the supplementary condition (8-10.3), it follows from Eq. (8-10.6) that $j_{\mu\nu}$ must satisfy a continuity equation

$$j^{\mu\nu}{}_{,\mu} = 0. \tag{8-10.7}$$

On the face of it the situation here is very much like that in electrodynamics where the vector potential A_μ can be made to satisfy

$$\Box A_\mu = j_\mu$$

by imposing the Lorentz condition $A^\mu{}_{,\mu} = 0$. The current j_μ also satisfies a continuity equation $j^\mu{}_{,\mu} = 0$. However, there is an essential difference in the two cases. The continuity equation for the electric current is consistent with the equations of motion for the sources that contribute to this current; the continuity equation (8-10.7) is not. Thus if we take $j_{\mu\nu}$ to be the Minkowski tensor (8-5.1), we see that Eq. (8-10.7) implies that $\ddot{z}_i{}^\mu = 0$ which will certainly not be the case for gravitationally interacting particles.

A number of authors have suggested alternative approaches to constructing a tensor theory of the gravitational field within the framework of special relativity when sources are present. Belinfante proposed that one should drop the requirement of positive definiteness of t_ψ^{00}. If we do so, however, we must contend with the arbitrariness in the field equations for $\psi_{\mu\nu}$. A possible way out of this difficulty is to require that t_ψ^{00} be positive-definitive only for *free* gravitational fields, that is, for fields that are not associated with sources. In this case $j_{\mu\nu}$ need no longer be conserved by itself; only the total stress-energy tensor of the system, gravitational field-plus sources, will be conserved as a consequence of the overall Poincaré invariance of the theory.

If the sources of the gravitational field are point masses, we must add to the Lagrangian of the gravitational field a term which is a function of $\psi_{\mu\nu}$, ϕ, and $z_i{}^\mu$ where the latter quantities are the coordinates of the ith particle. (We will see later that it is necessary to include the trace part of the gravitational field in the theory in order to get agreement with the observed motions of massive bodies in a gravitational field.) If one requires that the resulting particle equations of motion contain no higher than quadratic terms in the particle velocities, the total action for the system will be of the form

$$S = \int_R \frac{1}{2} \{\eta^{\rho\sigma}\psi_{\mu\nu,\rho}\psi^{\mu\nu}{}_{,\sigma} + \eta^{\rho\sigma}\phi_{,\rho}\phi_{,\sigma}\}\, d^4 x - \sum_i m_{0i} \int_R \int_{\sigma_1}^{\sigma_2} \sqrt{g_{\mu\nu}(x)\dot{z}_i{}^\mu \dot{z}_i{}^\nu}\, \delta^4(x - z_i)\, d\lambda_i\, d^4x \qquad (8\text{-}10.8)$$

where

$$g_{\mu\nu}(x) \equiv \eta_{\mu\nu} + \alpha\eta_{\mu\nu}\phi + \beta\psi_{\mu\nu} \qquad (8\text{-}10.9)$$

and α and β are constants to be determined. Note that we have assumed here that the gravitational charges of the sources are numerically equal to their inertial masses in writing down this expression for S. We could have used a form for S in which this was not the case by taking α and β to depend on these particle parameters.

By varying S with respect to $z_i{}^\mu$ we obtain the particle equations of motion

$$\ddot{z}_i{}^\mu + \begin{Bmatrix} \mu \\ \rho\sigma \end{Bmatrix}_i \dot{z}_i{}^\rho \dot{z}_i{}^\sigma = 0 \qquad (8\text{-}10.10)$$

when the path parameters are chosen so that

$$g_{\mu\nu}(z_i)\dot{z}_i{}^\mu \dot{z}_i{}^\nu = 1 \qquad (8\text{-}10.11)$$

and where $\begin{Bmatrix} \mu \\ \rho\sigma \end{Bmatrix}_i$ is a Christoffel symbol formed from $g_{\mu\nu}(z_i)$. The equations of motion for $\psi_{\mu\nu}$ then follow by varying S with respect to $\psi_{\mu\nu}$ and have the form

$$\Box\psi_{\mu\nu} = -\frac{\beta}{2}\sum_i m_{0i}\int_{\sigma_1}^{\sigma_2} \dot{z}_{i\mu}\dot{z}_{i\nu}\delta^4(x - z_i)\, d\lambda_i \tag{8-10.12}$$

while the equations of motion obtained by varying S with respect to φ have the form

$$\Box\phi = -\frac{\alpha}{2}\sum_i m_{0i}\int_{\sigma_1}^{\sigma_2} \eta_{\mu\nu}\dot{z}_i{}^\mu\dot{z}_i{}^\nu\delta^4(x - z_i)\, d\lambda_i . \tag{8-10.13}$$

We see incidentally that if we had required that $\psi_{\mu\nu}$ satisfy the supplementary condition (8-10.3), $\ddot{z}_i{}^\mu$ would have had to be zero in contradiction to the equations of motion (8-10.10). We see also that the flat-space metric $\eta_{\mu\nu}$ no longer appears in these latter equations by itself but only through the appearance of $g_{\mu\nu}$. The situation here is therefore similar to that encountered in our discussion of the Newtonian equations of motion of a particle in a gravitational field in Section 5-3 and all of the comments made there apply, with slight modifications, to the present case. In particular these equations of motion again only allow us to determine the total field $g_{\mu\nu}$ by observations of the motion of particles when gravitational fields are present.[26] This of course would not have been the case if the ratio of gravitational charge to inertial mass was not the same for all particles.

We can now obtain direct-particle equations of motion from Eqs. (8-10.10), (8-10.12), and (8-10.13) by following a procedure similar to that employed in Section 8-8 in the case of a scalar field. If we use the half-advanced, half-retarded solutions of Eqs. (8-10.12) and (8-10.13), we see that the ith partial field $g_{i\mu\nu}$ has the form

$$g_{i\mu\nu}(z_i) = \eta_{\mu\nu} - \frac{1}{8\pi}\sum_{j\neq i} m_{0i}\int_{\sigma_1}^{\sigma_2} \delta((z_i - z_j)^2)(\alpha\eta_{\mu\nu}\eta_{\rho\sigma} + \beta\eta_{\mu\rho}\eta_{\nu\sigma})\dot{z}_j{}^\rho\dot{z}_j{}^\sigma\, d\lambda_j \tag{8-10.14}$$

If we compare the resulting direct-particle equation of motion obtained by substituting this solution into Eq. (8-10.10) with those of Section 7-21, we see that they will reduce to the correct Newtonian limit for gravitationally interacting particles, provided we take $\alpha + \beta = 16\pi G$. Furthermore, as we will see in Chapter 12, these equations correctly predict the observed advances of the planetary perihelia. The only difficulty with this scheme is that when we take into account the effect of gravitational radiation on a system of gravitating bodies, the possibility arises that they may radiate away negative energy, since t_{gr}^{00} need not in general be positive. Havas has investigated this problem

[26] See also in this connection W. Thirring, *Ann. of Phys.*, **16**, 96 (1961).

and concludes that two unequal masses moving in elliptic orbits about each other will tend to gain energy by such a process. However, owing to the smallness of the gravitational coupling constant, such a gain would have virtually no effect on actual physical systems. Thus, while the planets would tend to slow down and spiral outward from the sun, such an effect would be unobservable even over geological times. In the atomic systems, electric forces completely overwhelm gravitational forces so that such systems would always tend to lose energy because of radiation. Nevertheless it is disturbing to have a theory that allows for the possibility of a system capable, in principle, of gaining energy by a radiative process. Furthermore the possibility of such an occurrence leads to serious difficulties when one takes quantum theory into account.

A more fundamental approach to the problem of constructing a consistent theory of the gravitational field has been suggested by Gupta[27] and Kraichnan.[28] These authors require that the source of the gravitational field should be the total stress-energy tensor, *including* a contribution from the gravitational field itself. In developing the consequences of this requirement, Kraichnan made use of the following result due to Eddington:[29] Let z_a constitute the components of the dynamical objects of a theory and $g_{0\mu\nu}$ be the flat-space metric. The z_a may represent field variables, particle variables, or a combination of both. Also, the flat-space metric $g_{0\mu\nu}$ need not take on the Minkowski values. The dynamical equations of motion for the z_a are assumed to follow from a variational principle, with an action of the form

$$S = \int \mathfrak{L}(z_a, g_{0\mu\nu})\, d^4x,$$

where the Lagrangian $\mathfrak{L}$ transforms as a scalar density of weight $+1$ under an arbitrary coordinate mapping. (The symmetry group of the theory is, of course, still the Poincaré group, since $g_{0\mu\nu}$ is an absolute object.) Consider now an infinitesimal mapping that reduces to the identity mapping on the boundary of and outside some finite space-time region R, and let us integrate Nöther's identity (4-5.10) over this region. We obtain thereby

$$\int (\bar{\delta} t^{\mu})_{,\mu}\, d^4x = \int \frac{\delta \mathfrak{L}}{\delta y_A}\, \bar{\delta} y_A\, d^4x, \tag{8-10.15}$$

where $y_A = \{z_a, g_{0\mu\nu}\}$. Now the integral on the left can be converted into a surface integral over the boundary of R with the help of Gauss' theorem, and since all $\bar{\delta}$ variations vanish on this boundary, the integral itself must vanish.

[27] S. N. Gupta, *Phys. Rev.*, **96**, 1683 (1954).

[28] R. H. Kraichnan, *Phys. Rev.*, **98**, 1118 (1955).

[29] A. S. Eddington, *Mathematical Theory of Relativity* (Cambridge Univ. Press, Cambridge, 1937), 2d. ed., pp. 140–141.

When the equations of motion for the z_a are satisfied, that is, when $\delta\mathfrak{L}/\delta z_a = 0$, it follows that

$$\int \mathfrak{T}^{\mu\nu}\bar{\delta}g_{0\mu\nu}\, d^4x = 0, \tag{8-10.16}$$

where

$$\mathfrak{T}^{\mu\nu} \equiv \frac{\delta\mathfrak{L}}{\delta g_{0\mu\nu}}. \tag{8-10.17}$$

For the mappings here considered, $\bar{\delta}g_{0\mu\nu} = -g_{0\mu\rho}\xi^{\rho}{}_{,\nu} - g_{0\rho\nu}\xi^{0}{}_{,\mu} - g_{0\mu\nu,\rho}\xi^{\rho}$, so that Eq. (8-10.16) can, by partial integration, be put in the form

$$2\int \mathfrak{T}^{\mu\nu}{}_{;\nu}g_{0\mu\rho}\xi^{\rho}\, d^4x = 0.$$

Then, since ξ^{ρ} is arbitrary inside R, it follows finally that

$$\mathfrak{T}^{\mu\nu}{}_{;\nu} = 0, \tag{8-10.18}$$

which is the result obtained by Eddington.

When $g_{0\mu\nu}$ takes on its Minkowski values the co-derivative appearing in Eq. (8-10.18) reduces to an ordinary derivative. In this case $\mathfrak{T}^{\mu\nu}$ is seen to satisfy the same kind of continuity equation as the usual stress-energy tensor $t^{\mu\nu}$ constructed with the help of Eq. (4-5.16). In fact one can show in this case that these two objects differ from each other, aside from a numerical factor, by at most a curl. Thus, when the dynamical equations for z_a are satisfied, the Nöther identity (4-5.10) becomes

$$\begin{aligned}(\bar{\delta}t^{\mu})_{,\mu} &= \mathfrak{T}^{\mu\nu}\bar{\delta}g_{0\mu\nu} \\ &= -2(\mathfrak{T}^{\mu\nu}g_{0\mu\rho}\xi^{\rho})_{,\nu} + 2\mathfrak{T}^{\mu\nu}{}_{;\nu}g_{0\mu\rho}\xi^{\rho},\end{aligned} \tag{8-10.19}$$

where the second line is obtained by using the expression for $\bar{\delta}g_{0\mu\nu}$ and differentiating by parts. When $g_{0\mu\nu}$ takes on its Minkowski values and ξ^{ρ} is taken to be an arbitrary small constant, Eq. (8-10.19) reduces, with the help of Eq. (8-10.18), to

$$t^{\mu}_{\nu,\mu} = -2\mathfrak{T}^{\mu}_{\nu,\mu}.$$

It follows, therefore, that $t_{\nu}{}^{\mu}$ and $-2\mathfrak{T}_{\nu}{}^{\mu}$ differ from each other by at most a curl, that is,

$$t_{\nu}{}^{\mu} = -2\mathfrak{T}_{\nu}{}^{\mu} + W_{\nu}^{[\rho\mu]}{}_{,\rho},$$

where $W_{\nu}^{[\rho\mu]} = -W_{\nu}^{[\mu\rho]}$. Since the stress-energy tensor of a theory is determined only up to a curl we see that we can use either $t_{\nu}{}^{\mu}$ or $\mathfrak{T}_{\nu}{}^{\mu}$ to characterize the stress-energy tensor of the system described by the theory. It is the latter quantity that plays the role of the source of the gravitational field in the Kraichnan approach.

The field equations for the gravitational field $h_{\mu\nu}$ are assumed by Kraichnan to have the form

$$\mathfrak{D}^{\mu\nu}(h) - \lambda\mathfrak{T}^{\mu\nu} = 0, \tag{8-10.20}$$

where $\mathfrak{D}^{\mu\nu}$ is a linear differential operator that, because of Eq. (8-10.18), must satisfy

$$[\mathfrak{D}^{\mu\nu}(h)]_{;\nu} = 0. \tag{8-10.21}$$

Such an operator is given by

$$\mathfrak{D}^{\mu\nu}(h) = \sqrt{-g_0}\{\Box h^{\mu\nu} - h^{\mu\rho}{}_{;\rho}{}^{\nu} - h^{\rho\nu}{}_{;\rho}{}^{\mu} + g_0^{\mu\nu} h^{\rho\sigma}{}_{;\rho\sigma}\}$$

where here $\Box$ is the generally covariant generalization of the Dalembert operator. Following Kraichnan, we now require that Eq. (8-10.20) be derivable from a variational principle with an action $S = \int \mathfrak{L}(z_a, h_{\mu\nu}, g_{0\mu\nu})\, d^4x$ such that $\delta\mathfrak{L}/\delta g_{0\mu\nu} = \mathfrak{T}^{\mu\nu}$. If such an action exists, then

$$\begin{aligned}\frac{\delta\mathfrak{L}}{\delta h_{\mu\nu}} &= \mathfrak{D}^{\mu\nu}(h) - \lambda\mathfrak{T}^{\mu\nu} \\ &= \mathfrak{D}^{\mu\nu}(h) - \lambda\frac{\delta\mathfrak{L}}{\delta g_{0\mu\nu}}.\end{aligned} \tag{8-10.22}$$

Now set $\mathfrak{L} = \mathfrak{L}_1 + \mathfrak{L}_2$ where

$$\mathfrak{L}_2 = -\left(\frac{2}{\lambda}\right)\sqrt{-g_0}\, h_{\mu\nu} R^{\mu\nu}(g_0),$$

$R^{\mu\nu}$ being the Ricci tensor formed from $g_{0\mu\nu}$. Since $R^{\mu\nu}(g_0)$ vanishes for a flat-space metric, it follows that $\delta\mathfrak{L}_2/\delta h_{\mu\nu} = 0$ for dpt. On the other hand, one can show easily that

$$\lambda\left(\frac{\delta\mathfrak{L}_2}{\delta g_{0\mu\nu}}\right) = \mathfrak{D}^{\mu\nu}(h),$$

so Eq. (8-10.22) is equivalent to the following equation for $\mathfrak{L}_1$:

$$\frac{\delta\mathfrak{L}_1}{\delta h_{\mu\nu}} = -\lambda\frac{\delta\mathfrak{L}_1}{\delta g_{0\mu\nu}}.$$

It follows, therefore, that $\mathfrak{L}_1$ must depend on $h_{\mu\nu}$ and $g_{0\mu\nu}$ in the combination $g_{0\mu\nu} + \lambda h_{\mu\nu}$. We can then take $g_{\mu\nu} \equiv g_{0\mu\nu} + \lambda h_{\mu\nu}$ to be a new gravitational field variable in place of $h_{\mu\nu}$. Since $\delta\mathfrak{L}_2/\delta g_{\mu\nu} = \delta\mathfrak{L}_2/\delta h_{\mu\nu} = 0$, it follows that the field equations for $g_{\mu\nu}$ follow from an action principle, with $S = \mathfrak{R}(g_{\mu\nu}, z_a)\, d^4x$, where $\mathfrak{R}$ transforms like a scalar density under arbitrary space-time mappings. In the absence of matter the simplest possibility for $\mathfrak{R}$ is

$\sqrt{-g}\, g_{\mu\nu} R^{\mu\nu}(g)$, which leads to the field equations of general relativity for $g_{\mu\nu}$. Furthermore, since $g_{0\mu\nu}$ does not appear in these equations, nor does $g_{\mu\nu}$ appear in the equations for $g_{0\mu\nu}$, we can dispense entirely with this latter quantity in our description of the gravitational field. But then we no longer have any absolute objects in our theory, and the MMG becomes a symmetry group of the system described by this theory. This is indeed a surprising conclusion, but it seems inescapable if the gravitational field can act as its own source. We will not now pursue further the problem of constructing a theory to describe the gravitational field, but we will return to it when we discuss the general theory of relativity.

9

Relativistic Continuum Mechanics; Macroscopic Theory

Ideally, a macroscopic description of a continuum system such as a fluid, an elastic solid, or a dielectric proceeds from some underlying microscopic theory of the atomic structure of the system. By forming suitable averages one obtains relations between macroscopically measurable quantities such as pressure, density, electric displacement, and so forth. In general these relations will involve parameters characteristic of the system such as elastic constants or electric susceptibilities. They may also be incomplete, needing for instance, an equation of state for their completion. While the former relations, for example, the equations of hydrodynamics, will be independent of the particular system involved, the parameters or equations of state must be deduced from a detailed analysis of the microstructure of the system, usually by the methods of statistical mechanics.

When one comes to construct a relativistic macroscopic theory of a continuum system, one can expect modifications of the corresponding nonrelativistic theory at two levels. The macroscopic relations between measurable quantities will be modified, usually as a consequence of the equivalence of mass and energy. Thus a transport of energy will contribute to the net momentum of the system. In general, modifications of this kind can be obtained with relative ease. Given a set of nonrelativistic relations, one can usually generalize them so that they are Poincaré invariant and reduce to the nonrelativistic relations in the limit $c \to \infty$. The second type of modification is in a way more fundamental. It involves changes in the parameters and equations of state due to the relativistic behavior of the microscopic constituents of the system. To deal with these latter modifications one would need a complete relativistic quantum theory and statistical mechanics, neither of which is available today. It is, however, reasonable to believe that relativistic corrections to the parameters of a system or its equation of state can be neglected except under extreme circumstances, for example, the interior of a white dwarf star. We will therefore ignore this second type of modification in our discussion of the relativistic description of relativistic systems.

9-1. Matter in Bulk

On the microscopic level an elastic solid or a fluid can be considered as consisting of atoms and molecules interacting primarily via the electric and magnetic fields they produce. To the extent that we can neglect quantum effects we can characterize these particles by means of their trajectories. The ith particle is thus imagined to follow a trajectory in space-time given by

$$x^\mu = z_i^\mu(\lambda_i)$$

where λ_i is a monotonically increasing parameter along the trajectory.

Let us then consider the following two objects: the particle flux density $\sigma^\mu(x)$ defined by

$$\sigma^\mu(x) = \sum_i m_i \int_{-\infty}^{+\infty} \delta^4(x - z_i)\dot{z}_i^\mu \, d\lambda_i \tag{9-1.1}$$

and the kinetic stress-energy density $\Theta^{\mu\nu}(x)$ discussed in Section 8-5

$$\Theta^{\mu\nu}(x) = \sum_i m_i \int_{-\infty}^{+\infty} \delta^4(x - z_i) \frac{\dot{z}_i^\mu \dot{z}_i^\nu}{\sqrt{\dot{z}_i^2}} \, d\lambda_i \tag{9-1.2}$$

where m_i is the rest-mass of the ith particle. From its definition it follows that σ^μ satisfies

$$\begin{aligned} \sigma^\mu{}_{,\mu}(x) &= \sum_i m_i \int_{-\infty}^{+\infty} \frac{d}{d\lambda_i} \delta^4(x - z_i) \, d\lambda_i \\ &= 0. \end{aligned} \tag{9-1.3}$$

The particle flux density is thus seen to satisfy a continuity equation as it must if the number of particles in the system is to be conserved.

A similar calculation leads to the result that

$$\Theta^{\mu\nu}{}_{,\nu} = \sum_i \int_{-\infty}^{+\infty} \delta^4(x - z_i) m_i \frac{d}{d\lambda_i} \left(\frac{\dot{z}_i^\mu}{\sqrt{z_i^2}} \right) d\lambda_i .$$

If the equation of motions of the ith particle are of the form

$$m_i \frac{d}{d\lambda_i}\left(\frac{\dot{z}_i^{\mu}}{\sqrt{\dot{z}_i^{2}}}\right) = f_i^{\mu}$$

then

$$\Theta^{\mu\nu}{}_{,\nu} = f^{\mu} \tag{9-1.4}$$

where

$$f^{\mu}(x) = \sum_i \int_{-\infty}^{+\infty} \delta^4(x - z_i) f_i^{\mu}\, d\lambda_i . \tag{9-1.5}$$

We have already discussed the interpretation of $\Theta^{\mu\nu}$ in Section 8-5. Eq. (9-1.4) gives its change due to the forces f_i^{μ} acting on the particles.

The transition from the above microdescription of a system of particles to a macrodescription is accomplished by forming suitable averages of σ^{μ} and $\Theta^{\mu\nu}$. In order to form these averages in an invariant manner, however, we need first to introduce the average velocity field $u^{\mu}(x)$ at the point x^{μ}. Its definition in terms of the microscopic parameters of the system will be given shortly. For the moment we merely assume the existence of such a velocity field with the property that

$$u^{\mu} u_{\mu} = 1. \tag{9-1.6}$$

Let us now construct a small, flat, spacelike hypersurface at x^{μ} whose normal is equal to u^{μ} there and whose volume is ΔV_0. We then define the average $\bar{\phi}(x)$ of any microscopic quantity $\phi(x)$ such as $\sigma^{\mu}(x)$ or $\theta^{\mu\nu}(x)$ as the integral over this surface divided by ΔV_0:

$$\bar{\phi}(x) \equiv \frac{1}{\Delta V_0} \int_{\Delta V_0} \phi(x)\, dV \tag{9-1.7}$$

where $dV = u^{\mu}\, dS_{\mu}$ with dS_{μ} an element of surface "area."

In constructing the various averages we shall need, it is convenient to express the velocity of a particle whose trajectory passes through the hypersurface at x^{μ} as

$$\dot{z}_i^{\mu} = u^{\mu}(x) + \varepsilon_i^{\mu} \tag{9-1.8}$$

when the path parameter λ_i is chosen so that $\dot{z}_i^{\mu} u_{\mu} = 1$.[1] Then, since $u^{\mu} u_{\mu} = 1$, it follows that

$$\varepsilon_i^{\mu} u_{\mu} = 0. \tag{9-1.9}$$

[1] It is necessary to specify the path parameter for which Eq. (9-1.8) holds since u^{μ} is parameter independent while $\dot{z}_i^{\mu}$ is not.

To interpret ε_i^{μ} let us map so that $u^{\mu} = (1, \mathbf{0})$. Then the three-velocity $\dot{z}_i^r/\dot{z}_i^0$ of the ith particle is equal to ε_i^r. The velocity ε_i^{μ} is thus related to the random thermal motion of the ith particle that is added to the average velocity field to give the total velocity of this particle.

Let us now consider the average value of $\sigma^{\mu}(x)$. Using the expression (9-1.1) for this quantity together with Eq. (9-1.8) we have, since $u^{\mu}(x)$ is constant over the region of integration,

$$\bar{\sigma}(x) = \rho(x)u^{\mu}(x) + v^{\mu}(x) \tag{9-1.10}$$

where

$$\rho(x) = \frac{1}{\Delta V_0} \int_{\Delta V_0} \sum_i m_i \int_{-\infty}^{+\infty} \delta^4(x - z_i)\, d\lambda_i\, dV$$

and

$$v^{\mu}(x) = \frac{1}{\Delta V_0} \int_{\Delta V_0} \sum_i m_i \int_{-\infty}^{+\infty} \delta^4(x - z_i)\varepsilon_i^{\mu}\, d\lambda_i\, dV.$$

With the choice of path parameters for which the decomposition (9–1.8) holds, the integrals over the delta functions have the value one if the ith trajectory passes through the volume ΔV_0 and zero otherwise. Therefore,

$$\rho(x) = \frac{1}{\Delta V_0} \sum_i m_i \tag{9-1.11}$$

and

$$v^{\mu}(x) = \frac{1}{\Delta V_0} \sum_i m_i \varepsilon_i^{\mu} \tag{9-1.12}$$

where the sums now extend over all the particles in ΔV_0. The quantity $\rho(x)$ is the *proper density* of the system. If all of the particles comprising the system have the same rest mass, then $\rho(x)$ is equal to rest mass multiplied by the number of particles per unit proper volume—that is, per unit volume as measured by an observer at rest in the system. We see from Eq. (9-1.11) that $\rho(x)$ transforms like a scalar under a Poincaré mapping. Likewise $v^{\mu}(x)$ is seen to transform like a contravector.

So far we have not given a precise definition of the average velocity field $u^{\mu}(x)$. We do so now by requiring that it be such that

$$v^{\mu}(x) = 0. \tag{9-1.13}$$

This requirement is consistent with the requirement that the internal motions of the particles comprising the system are random. Thus, if we map so that $u^{\mu} = (1, \mathbf{0})$ at a point, it follows from Eq. (9-1.9) that $\varepsilon_i^0 = 0$ and condition (9-1.13) reduces to

$$\sum_i m_i \boldsymbol{\varepsilon}_i = 0.$$

Therefore, when Eq. (9-1.13) is satisfied, Eq. (9-1.10) reduces to

$$\bar{\sigma}^{\mu}(x) = \rho(x)u^{\mu}(x). \tag{9-1.14}$$

Furthermore, since differentiation and the averaging process employed here commute, it follows from Eq. (9-1.3) that

$$\bar{\sigma}^{\mu}(x)_{,\mu} = 0. \tag{9-1.15}$$

Let us now consider the average of the kinetic stress-energy tensor $\Theta^{\mu\nu}$ given by Eq. (9-1.2). After performing the integrations that appear in the expression for this average we find that

$$\overline{\Theta}^{\mu\nu}(x) = \frac{1}{\Delta V_0} \sum_i m_i \frac{(u^{\mu} + \varepsilon_i^{\ \mu})(u^{\nu} + \varepsilon_i^{\ \nu})}{\sqrt{(u + \varepsilon_i)^2}} \tag{9-1.16}$$

To interpret the various components of this tensor let us map so that $u^{\mu} = (1, \mathbf{0})$ at a point in the system. Then at this point the components of $\overline{\Theta}^{\mu\nu}$ have the form

$$\overline{\Theta}^{00} = \frac{1}{\Delta V_0} \sum_i m_i \frac{1}{\sqrt{1 - \boldsymbol{\varepsilon}_i^{\ 2}}} \tag{9-1.17a}$$

$$\overline{\Theta}^{r0} = \overline{\Theta}^{0r} = \frac{1}{\Delta V_0} \sum_i m_i \frac{\varepsilon_i^{\ r}}{\sqrt{1 - \boldsymbol{\varepsilon}_i^{\ 2}}} \tag{9-1.17b}$$

and

$$\overline{\Theta}^{rs} = \frac{1}{\Delta V_0} \sum_i m_i \frac{\varepsilon_i^{\ r} \varepsilon_i^{\ s}}{\sqrt{1 - \boldsymbol{\varepsilon}_i^{\ 2}}} \tag{9-1.17c}$$

The values of the averages appearing in the above expressions will depend on the assumptions one makes about the distribution of the velocities of the particles. If there is no heat flux and no shear in the system, then we will assume that

$$\overline{\Theta}^{00} = e \tag{9-1.18a}$$

$$\overline{\Theta}^{0r} = 0 \tag{9-1.18b}$$

and

$$\overline{\Theta}^{rs} = p\delta^{rs} \tag{9-1.18c}$$

where

$$e = \frac{1}{\Delta V_0} \sum_i \frac{m_i}{\sqrt{1 - \boldsymbol{\varepsilon}_i^{\ 2}}}, \tag{9-1.19}$$

$$p = \frac{1}{3} \frac{1}{\Delta V_0} \sum_i m_i \frac{\boldsymbol{\varepsilon}_i^{\ 2}}{\sqrt{1 - \boldsymbol{\varepsilon}_i^{\ 2}}}. \tag{9-1.20}$$

We see that if the thermal velocities of the particles are small compared to unity, that is, if $\varepsilon_1{}^2 \ll 1$, e is just the sum of the rest-mass energy plus the thermal kinetic energy of the particles per unit proper volume. Likewise, the expression for p reduces to the usual kinetic theory expression for the pressure of an ideal gas. The assumptions we have made here concerning the distribution of velocities is therefore equivalent to the usual assumptions of kinetic theory concerning this distribution for an ideal gas. Finally, we note that both e and p transform like scalars under a Poincaré mapping.

If there is a net heat flux in the system then in general $\overline{\Theta}^{0r} \neq 0$ as we should expect since the $0, r$ component of the stress-energy tensor represents an energy flux. Also, if $\varepsilon_i{}^2 \ll 1$, then

$$\overline{\Theta}^{0r} \simeq \frac{1}{2}\frac{1}{\Delta V_0}\sum_i m_i \varepsilon_i{}^2 \varepsilon_i{}^r,$$

which in classical theory is just the quantity that represents the heat flux in a system. Likewise, if the system can support a shear stress or is viscous, $\overline{\Theta}^{rs}$ will no longer in general be diagonal.

Independent of what assumptions we make concerning the distribution of velocities in a system it follows from Eq. (9-1.4) that

$$\overline{\Theta}^{\mu\nu}{}_{,\nu} = \bar{f}^{\mu}. \tag{9-1.21}$$

From this point on in our discussion we will no longer have need of the microscopic quantities $\sigma^\mu(x)$, etc., so we can dispense with bars over these quantities to indicate their averages. In general, the forces acting on the particles of a system will be composed of two parts: forces originating from outside the system, for example, external electric fields; and internal or particle-particle forces. Consequently f^μ will consist of two parts:

$$f^\mu = f^\mu_{\text{ext}} + f^\mu_{\text{int}}$$

We now make the assumption that f^μ_{int} can be written as

$$f^\mu_{\text{int}} = -t^{\mu\nu}{}_{,\nu},$$

$$t^{\mu\nu} = t^{\nu\mu}.$$

On the microscopic level electric forces between particles have this form (see Eq. 8-5.2) and since macroscopic forces such as elastic forces are, in the final analysis, due to electric forces, this assumption is reasonable. With this assumption concerning $t^{\mu\nu}$ we can then rewrite Eq. (9-1.21) in the form

$$T^{\mu\nu}{}_{,\nu} = f^\mu_{\text{ext}} \tag{9-1.22}$$

where

$$T^{\mu\nu} = \Theta^{\mu\nu} + t^{\mu\nu} \tag{9-1.23}$$

is the total stress-energy tensor for the system and is seen to be symmetric in μ and ν. Eq. (9-1.22) together with the continuity equation (9-1.15) constitute the basic relativistic equations governing the motion of matter in bulk.

9-2. The Perfect Fluid, Relativistic Thermodynamics

In classical hydrodynamics a perfect fluid is defined to be a continuum system that is incapable of supporting shear stresses and for which Pascal's law hold good. We can use these criteria to construct the stress-energy tensor for a perfect fluid in special relativity. Let us consider an element of the fluid and map so that the velocity vector of this element has components $(1, \mathbf{0})$; that is, so that it is at rest. Then T^{0r}, the energy flux of the fluid, must be zero if there is no heat conduction while T^{00} must just be equal to the total internal energy density e of the fluid. Furthermore, $T^{rs}\, dS_s$ is the rth component of the force exerted on the surface element dS_r assumed to be at rest with respect to the element of fluid considered. Since according to Pascal's law this force is isotropic and normal to the surface on which it acts, it follows that $T^{rs} = p\delta^{rs}$, where p is the pressure measured by an observer moving with the fluid. Consequently, for an element of fluid for which $u^\mu = (1, \mathbf{0})$ the total stress-energy tensor $T^{\mu\nu}$ has the form

$$T^{\mu\nu} = \begin{pmatrix} e & 0 & 0 & 0 \\ 0 & p & 0 & 0 \\ 0 & 0 & p & 0 \\ 0 & 0 & 0 & p \end{pmatrix} \tag{9-2.1}$$

We note that the expression (9-2.1) is just the form assumed by $\Theta^{\mu\nu}$ as given by Eqs. (9-1.18) for an ideal gas. For such a system $t^{\mu\nu}$ is zero. Even when $t^{\mu\nu}$ is nonzero, as in a liquid, however, the system can still behave like a perfect fluid, in which case $T^{\mu\nu}$ would again have the form given by Eq. (9-2.1). In the general case, of course, e and p will no longer be given by the simple expressions (9-1.19), (9-1.20) that are valid for the ideal gas.

To find an expression for $T^{\mu\nu}$ for a moving element of the fluid one can either map so that u^μ takes on its general form and transform $T^{\mu\nu}$ as given by Eq. (9-2.1) accordingly or else look for an expression that reduces to this form when $u^\mu = (1, \mathbf{0})$. In either case one finds that

$$T^{\mu\nu} = hu^\mu u^\nu - p\eta^{\mu\nu} \tag{9-2.2}$$

where $h = e + p$ is the enthalpy density of the fluid. Both e and p appearing in this equation are taken to be the values of these quantities as measured by an observer moving with the fluid, that is, these quantities are assumed to transform like scalars under Poincaré mappings. Such an assumption is consistent with the transformation properties of e and p given by Eqs. (9-1.19) and (9-1.20) that follow from these equations.

Let us now examine the equations of motion (9-1.22) for a perfect fluid when $f^{\mu}_{\text{ext}} = 0$. These equations, together with the continuity equation (9-1.15) and an equation of state of the form $e = e(p, \rho)$ serve to determine the quantities u^μ, e, p, and ρ. If we multiply Eq. (9-1.22) by u_μ and make use of the fact that $u^\mu u_\mu = 1$, and hence $u^\mu{}_{,\nu} u_\mu = 0$, we obtain

$$(hu^\mu)_{,\mu} - p_{,\mu} u^\mu = 0. \tag{9-2.3}$$

Let us now make use of the equation of continuity (9-1.15) to rewrite this equation, after dividing through by ρ, in the form

$$u^\mu \{(h/\rho)_{,\mu} - p_{,\mu}/\rho\} = 0. \tag{9-2.4}$$

It then follows from the thermodynamic relation

$$d\left(\frac{h}{\rho}\right) - \frac{dp}{\rho} = Td\left(\frac{\sigma}{\rho}\right), \tag{9-2.5}$$

where T is the temperature and σ is the entropy density of a fluid element as measured by an observer at rest with respect to it, that

$$u^\mu \left(\frac{\sigma}{\rho}\right)_{,\mu} = 0. \tag{9-2.6}$$

Thus the specific entropy σ/ρ is constant along the trajectory of each fluid element. The motion of the fluid, therefore, is adiabatic as we should expect since we have explicitly assumed that there is no heat flow or internal friction, that is, viscosity, in obtaining the expression (9-2.2) for $T^{\mu\nu}$ (see also the discussion following Eq. (9-1.20).) We note finally that by making use of the continuity equation (9-1.15) we can rewrite Eq. (9-2.6) in the form

$$(\sigma u^\mu)_{,\mu} = 0, \tag{9-2.7}$$

that is, as a continuity equation for the entropy flux σu^μ.

Before we continue our discussion of Eq. (9-1.22) for the perfect fluid it is necessary to comment on the assumed transformation properties of the various thermodynamic quantities appearing in the expression (9-2.2) for $T^{\mu\nu}$ and the use of the second law of thermodynamics as expressed in Eq. (9-2.5). As we said, all of the thermodynamic quantities employed, e, p, T, and σ are defined by measurements made by an observer moving with the fluid element under consideration. It might appear, therefore, that we have violated the spirit, if not the principle, of relativity physics by singling out a special class of observers in our discussion. There is no such violation, however, since this special class of observers can be characterized in a completely invariant manner. Further, if we keep in mind the fact that all of the quantities appearing in the description of a physical system represent the outcomes of measurements made on the system, the fact that e, p, T, and σ are to be measured in the rest frame of a fluid element does not represent any deficiency in our description since these are well-defined measurements. (We have, of course, made the tacit assumption that the fluid is in a state of local thermodynamic equilibrium; otherwise it would be meaningless to introduce thermodynamic quantities in the first place.) Nevertheless, various authors[2] have proposed transformation laws for the thermodynamic quantities that differ from those employed in our treatment. Thus Planck and Einstein find that the temperature T' of a fluid element moving with respect to an observer is given by

$$T' = T\sqrt{1 - \mathbf{U}^2} \tag{9-2.8}$$

where T is the rest temperature of the element and $\mathbf{U}$ is its velocity relative to the observer.

There are two problems associated with a law of transformation such as given in Eq. (9-2.8). There is, of course, the purely formal problem that there are no geometrical objects which, by themselves, transform in this manner under a Lorentz mapping. More important, however, is the problem of what we mean by T'; that is, how are we to measure such a quantity. While the measurement of the rest temperature is well defined and, in fact, can be performed in a number of ways all giving the same result, the outcome of a measurement of T' will in general depend on how the measurement is made. Thus the mere insertion of a thermometer into the fluid will certainly not give the value T' predicted by Eq. (9-2.8). There may well be a way of measuring T' that verifies this equation but the equation by itself does not tell us how this measurement is to be made. The situation here is materially different from that associated, for example, with the transformation laws (8-2.3) and (8-2.4) for the electromagnetic field. There both the original and the transformed fields are measured in the same way by all observers and hence we can directly verify these transformation laws by means of measurements made on the field. In fact, it is just because the electromagnetic field is measured in the same way by all observers that we can describe it by means of a geometrical object. We could, of course, take the position that T' is to be measured by first measuring T and $\mathbf{U}$ and then computing it using Eq. (9-2.8). But then this equation would be merely a definition of a derived quantity and since it is the rest temperature that appears in all of our thermodynamic equations we could dispose with it entirely.

[2] M. Planck, *S.B. preuss. Akad. Wiss.*, 542 (1907); *Ann. Phys. Lpz.*, **76**, 1 (1908); A. Einstein, *Jahrb. Radioakt.*, **4**, 443 (1907). Cf. also R. C. Tolman, *Relativity, Thermodynamics and Relativity* (Clarendon Press, Oxford, 1934).

There is one way in which we could hope to give meaning to a transformation law like (9-2.8). Suppose we form a temperature four vector $T^\mu \equiv Tu^\mu$. In the rest frame of the fluid, $T^\mu = (T, \mathbf{0})$ so that in any other frame we have

$$T'^0 = \frac{T^0}{\sqrt{1 - \mathbf{U}^2}} \tag{9-2.9}$$

which, aside from the fact that $\sqrt{1 - \mathbf{U}^2}$ appears in the denominator rather than in the numerator, is similar to Eq. (9-2.8). It might be then that we could prescribe a standard technique for measuring T^0 valid for all observers, in which case we could verify Eq. (9-2.9). We gain nothing by this procedure, however, since again it is the rest temperature that appears in all of our equations. The essential point of our treatment is that we are making the assumption that the classical laws of thermodynamics are valid in the rest frame of a fluid, as evidence by our use of the classical thermodynamic relation (9-2.5). It could well be, of course, that this assumption is invalid in special relativity and that the thermodynamic temperature is really the timelike component of a four vector that does not reduce simply to $(T, \mathbf{0})$ in the rest frame of the fluid. But then it would not obey a simple transformation law such as (9-2.8) or (9-2.9). Since there is no experimental evidence at present that bears on the question and since, at least in the case of the ideal gas, our discussion of statistical mechanics in Section 7-8 indicates that the internal state of a fluid can be characterized by a scalar temperature, it is not unreasonable to apply the results of thermodynamics in special relativity.

To complete our discussion of the equations of motion (9-1.22) for a perfect fluid, let us make use of Eq. (9-2.3) in these equations to obtain the equations

$$hu^\mu{}_{,\nu}u^\nu + (u^\mu u^\nu - \eta^{\mu\nu})p_{,\nu} = 0. \tag{9-2.10}$$

These equations are the relativistic analogue of the nonrelativistic Euler equations of motion for a perfect fluid and reduce to these equations in the limit $c \to \infty$. In this limit $u^\mu \to (1, \mathbf{U}/c)$ where $\mathbf{U}$ is the velocity of the fluid element. Also in this limit the largest contribution to h comes from the rest-energy density ρc^2 included in the internal energy e. Therefore, the nonrelativistic limit of Eq. (9-2.10) with $\mu = 1, 2, 3$ is

$$\rho \frac{\partial}{\partial t} U_r + \rho U_{r,s} U_s - p_{,r} = 0$$

which are just the Euler equations for a perfect fluid. (The $\mu = 0$ equation is not an independent equation but is a consequence of the $\mu = 1, 2, 3$ equations.) Likewise the nonrelativistic limit of Eq. (9-2.3) is

$$\frac{\partial}{\partial t} \rho + (\rho U_r)_{,r} = 0,$$

which is just the classical continuity equation expressing mass conservation.

In arriving at the above nonrelativistic limits of Eqs. (9-2.3) and (9-2.10) we assumed both that **U** was small compared to c and that the rest-energy density ρc^2 was the dominant term in h. Let us now consider the case where this latter assumption is no longer valid. Such a situation arises in the propagation of sound through a medium with a relativistic equation of state. In discussing such propagation we will make the usual linear assumption that ρ, e, and p differ from their equilibrium values by small amounts ρ', e', and p', respectively. With these assumptions and the assumption that **U** is small compared to c, Eq. (9-2.3) reduces to

$$\frac{\partial}{\partial t} e' + h_0 \nabla \cdot \mathbf{U} = 0$$

while Eq. (9-2.10) reduces to

$$\frac{1}{c^2} h_0 \frac{\partial}{\partial t} \mathbf{U} + \nabla p = 0$$

where h_0 is the equilibrium value of h. If we differentiate the first of these equations with respect to t and take the divergence of the second, we obtain

$$\frac{1}{c^2} \frac{\partial^2}{\partial t^2} e' - \nabla^2 p = 0.$$

Then if e is considered to be a function of p and the entropy density σ, we have

$$e' \simeq \left.\frac{\partial e}{\partial p}\right|_\sigma p'$$

since the propagation of sound is an adiabatic process, Therefore, the velocity of sound in a relativistic fluid is given by

$$v_{\text{sound}} = \frac{c}{\sqrt{(\partial e/\partial p)|_\sigma}}, \tag{9-2.11}$$

In Section 7-7 we saw that in the extreme relativistic limit $p = e/3$ for all material systems so that $c/\sqrt{3}$ is the limiting velocity of sound in any substance. In the nonrelativistic limit $e \simeq \rho c^2$ and Eq. (9-2.11) reduces to the usual Newtonian expression for the velocity of sound in a fluid.

9-3. Kinematics of Velocity Fields

To discuss the various kinds of velocity fields that can exist in a continuous substance it is convenient to introduce the Lagrangian coordinates a^i, $i = 1, 2, 3$, of an element of the substance. The space time coordinates of this element will then be functions of the a^i and the proper time τ measured along the element;

$$x^\mu = x^\mu(\tau, a^i)$$

are the points lying along a streamline. Let us further define the operator δ by

$$\delta \equiv \delta a^i \frac{\partial}{\partial a^i}. \tag{9-3.1}$$

When applied to any function of the Lagrangian coordinates, δ gives the change in this function normal to the streamlines. Thus

$$\delta x^\mu = \delta \frac{\partial x^\mu}{\partial a^i} \tag{9-3.2}$$

are the displacements between streamlines with Lagrangian coordinates a^i and $a^i + \delta a^i$. The velocity u^μ of an element is given by

$$u^\mu = \frac{\partial x^\mu}{\partial \tau} \tag{9-3.3}$$

with

$$u^\mu u_\mu = 1.$$

It follows then that

$$\frac{\partial}{\partial \tau} \delta x^\mu = \delta a^i \frac{\partial}{\partial a^i} \frac{\partial x^\mu}{\partial \tau} = \delta a^i \frac{\partial}{\partial a^i} u^\mu$$

$$= \delta a^i \frac{\partial u^\mu}{\partial x^\nu} \frac{\partial x^\nu}{\partial a^i} = \delta x^\nu u^\mu{}_{,\nu}. \tag{9-3.4}$$

Likewise, the acceleration a^μ is given by

$$a^\mu = \frac{\partial u^\mu}{\partial \tau} = u^\mu{}_{,\nu} u^\nu. \tag{9-3.5}$$

We now define the "metric" $h_{\mu\nu}$ of the space normal to a streamline by

$$h_{\mu\nu} = \eta_{\mu\nu} - u_\mu u_\nu. \tag{9-3.6}$$

It satisfies

$$h_{\mu\nu} u^\nu = 0 \tag{9-3.7}$$

and

$$h_\nu{}^\mu h_\rho{}^\nu = h_\rho{}^\mu \tag{9-3.8}$$

where

$$h_\nu{}^\mu = \delta_\nu{}^\mu - u^\mu u_\nu. \tag{9-3.9}$$

The projection $\bar{\delta}x^\mu$ of δx^μ onto this space is then defined by

$$\bar{\delta}x^\mu \equiv h_\nu{}^\mu \delta x^\nu \tag{9-2.10}$$

and is the position vector of element $a^i + \delta a^i$ with respect to the element a^i. In general, $\partial(\bar{\delta}x^\mu)/\partial\tau$ will not be orthogonal to u_μ. However, its projection

$$\bar{\delta}v^\mu \equiv h_\nu{}^\mu \frac{\partial}{\partial\tau} \bar{\delta}x^\mu \tag{9-3.11}$$

has this property. In fact

$$\begin{aligned} \delta v^\mu &= h_\nu{}^\mu h_\rho{}^\nu \frac{\partial}{\partial\tau} \delta x^\rho + h_\nu{}^\mu \frac{\partial}{\partial\tau} h_\rho{}^\nu \delta x^\rho \\ &= h_\rho{}^\mu \delta x^\nu u^\rho{}_{,\nu} - a^\mu v_\rho \, \delta x^\rho \\ &= \bar{\delta}x^\nu u^\mu{}_{,\nu} \end{aligned} \tag{9-3.12}$$

where we have made use of Eq. (9-3.4) in going from the first to the second line and Eq. (9-3.5) in going from the second to the third line of this derivation. It then follows that

$$\bar{\delta}v^\mu u_\mu = 0.$$

The vector $\bar{\delta}v^\mu$ is thus the velocity of the element $a^i + \delta a^i$ relative to a^i.

Let us now define the following objects:

$$\omega_{\mu\nu} \equiv u_{[\mu,\nu]} + u_{[\mu}a_{\nu]} = h_\mu{}^\rho h_\nu^\sigma u_{[\rho,\sigma]} \tag{9-3.13}$$

$$\theta \equiv u^\mu{}_{,\mu} \tag{9-3.14}$$

and

$$\begin{aligned} \sigma_{\mu\nu} &\equiv u_{(\mu,\nu)} + u_{(\mu}a_{\nu)} - \tfrac{1}{3}\,\theta h_{\mu\nu} \\ &= h_\mu{}^\rho h_\nu{}^\sigma u_{(\rho,\sigma)} - \tfrac{1}{3}\,\theta h_{\mu\nu}. \end{aligned} \tag{9-3.15}$$

It is a simple matter to show that these quantities satisfy the following relations:

$$\omega_{\mu\nu} = -\omega_{\nu\mu} \quad \text{and} \quad \omega_{\mu\nu}u^\nu = 0 \tag{9-3.16}$$

and

$$\sigma_{\mu\nu} = \sigma_{\nu\mu}, \quad \sigma_{\mu\nu}\eta^{\mu\nu} = 0 \quad \text{and} \quad \sigma_{\mu\nu}u^\nu = 0. \tag{9-3.17}$$

With the help of the above defined quantities we can write the velocity field $\bar{\delta}v^\mu$ in the neighborhood of the element a^i as

$$\bar{\delta}v^\mu = \bar{\delta}x^\nu(\omega_\nu{}^\mu + \sigma_\nu{}^\mu + \tfrac{1}{3}\,\theta\delta_\nu{}^\mu). \tag{9-3.18}$$

$\bar{\delta}v^\mu$ is thus seen to be a superposition of three parts due, respectively, to rotation, shear, and expansion of the fluid.

To justify our above characterization of the three parts of $\bar{\delta}v^\mu$ let us calculate the quantity $1/\delta s\ \partial(\delta s)/\partial\tau$ where

$$\delta s^2 = \eta_{\mu\nu}\,\bar{\delta}x^\mu\,\bar{\delta}x^\nu$$

is the distance between the elements a^i and $a^i + \delta a^i$. We find, with the help of Eq. (9-3.18)

$$\frac{1}{\delta s}\frac{\partial\delta s}{\partial\tau} = \frac{1}{(\delta s)^2}\,\eta_{\mu\nu}\bar{\delta}x^\mu\,\frac{\partial\bar{\delta}x^\nu}{\partial\tau} = \frac{1}{(\delta s)^2}\,\eta_{\mu\nu}\,\bar{\delta}x^\mu\,\bar{\delta}v^\nu$$

$$= \sigma_{\mu\nu}\,\frac{\bar{\delta}x^\mu}{\delta s}\,\frac{\bar{\delta}x^\nu}{\delta s} + \tfrac{1}{3}\,\theta. \tag{9-3.19}$$

Since $\sigma_{\mu\nu}\eta^{\mu\nu} = 0$, the first term on the right gives the relative change in shape of an element while the second gives the relative change in its volume. We see that $\omega_{\mu\nu}$ does not enter into this expression for change of local distance and hence must represent a rotation.

In the early days of relativity much thought was given to the problem of defining a rigid body within the framework of this theory. Born[3] defined a rigid body to be one for which

$$\frac{\partial\delta s}{\partial t} = 0 \tag{9-3.20}$$

for all motions of the body. However, Herglotz[4] and Nöther[5] proved that any body that obeyed this criterian had only three degrees of freedom instead of the usual six for a rigid body in classical mechanics. One can understand the reason for this result; as long as a body moves uniformly, corresponding to its three translational degrees of freedom, there are no violations of the principles of relativity. If one tried to rotate a rigid bar of sufficient length, however, its ends would have to travel faster than c if it were to accord with the Born definition. Laue[6] argued that any continuum system must have, in effect, an infinite number of degrees of freedom. The gist of his argument was that, since the maximum velocity of sound is $c/\sqrt{3}$, an impulse given to the body simultaneously at n different points must, at least initially, produce a motion with n degrees of freedom. Consequently, an ideal rigid body in the classical sense, that is, a body of finite extension with only six degrees of freedom—cannot exist in nature.

While continuum systems can execute motions other than those permitted by the Born condition (9-3.20) they can also, under certain conditions,

[3] M. Born, *Phys. Z.*, **11**, 233 (1910).
[4] G. Herglotz, *Ann. Phys. Lpz.*, **31**, 393 (1910).
[5] E. Nöther, *Ann. Phys. Lpz.*, **31**, 919 (1910).
[6] M. v. Laue, *Phys. Z.*, **12**, 85 (1911).

execute motions compatible with these conditions. Such motions are sometimes called *rigid body motions.* It follows from Eq. (9-3.19) that for such motions

$$\sigma_{\mu\nu} = 0 \quad \text{and} \quad \theta = 0;$$

that is, they are shear free and expansion free.

Problem 9.1. Show that a fluid motion described by the one-parameter family of streamlines

$$(x^1)^2 - (x^0)^2 = \frac{1}{a^2}, \quad 0 < a < \infty$$

satisfies the Born condition for rigid body motion.

So far we have neglected transport processes in our discussion of relativistic fluid dynamics. The reason is that one does not have today a clear-cut, unambiguous way of describing such phenomena in a relativistic theory. To date there have been two proposals for such a description, one due to Landau and Lifshitz[7] and another due to Eckart.[8] In the Landau-Lifshitz treatment the expression for the stress-energy tensor (9-2.2) is maintained but the expression (9-1.14) for the particle flux density is modified by adding to it a term ν^μ with $\nu^\mu u_\mu = 0$. On the other hand, Eckart maintains the expression for the particle flux density but adds to the expression for $T^{\mu\nu}$ a term $2q^{(\mu}u^{\nu)}$ with $q^\mu u_\mu = 0$ when there is heat flow in the system. In the Eckart treatment q^μ is directly related to the heat flux, while in the Landau-Lifshitz treatment it is only indirectly related to this quantity. Unfortunately, the two treatments do not agree except in the limit when $q^\mu/(e + p) \ll 1$.

Of the two treatments, Eckart's appears to be more satisfactory, particularly if one maintains the interpretation of σ^μ as a particle flux density. What is perhaps not so obvious in his treatment is that $T^{\mu\nu}$ should be modified only to the extent of adding the term $2q^{(\mu}u^{\nu)}$. In essence, the difference between these two treatments resides in the definition of the rest frame of a fluid element. In the Landau-Lifshitz treatment the rest frame is defined by the condition that

$$\sum_i m_i \frac{\varepsilon_i}{\sqrt{1 - \varepsilon_i^2}} = 0$$

for the particles in the fluid element while in the Eckart treatment it is defined by Eq. (9-1.13), that is,

$$\sum_i m_i \varepsilon_i = 0.$$

[7] L. D. Landau and E. M. Lifshitz, *Fluid Mechanics* (Addison-Wesley, Reading, Mass., 1962), Chap. XV.

[8] Carl Eckart, *Phys. Rev.*, **58**, 919 (1940).

While either definition of the rest frame is acceptable, it is not clear that one should accept the expressions for σ^μ and $T^{\mu\nu}$ in these frames as proposed by the respective authors. For the case of momentum transport both authors appear to agree that such processes can be accounted for by adding to $T^{\mu\nu}$ a term $\tau^{\mu\nu}$ where $\tau^{\mu\nu} = \tau^{\nu\mu}$ and $\tau^{\mu\nu}u_\nu = 0$, but even here there is room for doubt. A final solution to the problem of describing transport processes within the framework of special relativity can only come when we have a satisfactory statistical mechanics for relativistic systems.

Problem 9.2. The relativistic analogue of the Navier-Stokes equations can be obtained by adding to $T^{\mu\nu}$ a term proportional to $\sigma^{\mu\nu}$. (For compressible flow one would also add a term proportional to $\theta h^{\mu\nu}$.) Show that, in the classical limit, the equations

$$(T^{\mu\nu} + \eta\sigma^{\mu\nu})_{,\nu} = 0,$$

with η a viscosity coefficient, reduce to the Navier-Stokes equations.

9-4. Electrodynamics in Matter

Maxwell's equations in a material medium are obtained from the Maxwell-Lorentz equations (8-1.1 and 8-1.2) by forming averages of the microscopic fields appearing therein. In doing so one introduces a polarization vector **P** and a magnetization vector **M** to describe the effects of the dipoles induced in the medium. These in turn give rise to a charge density

$$\rho_M = -\nabla \cdot \mathbf{P} \tag{9-4.1}$$

and a current

$$\mathbf{j}_M = \nabla \times \mathbf{M} + \frac{\partial \mathbf{P}}{\partial x^0} \tag{9-4.2}$$

that acts as sources of additional fields. By defining new fields

$$\mathbf{D} = \mathbf{E} + \mathbf{P} \tag{9-4.3}$$

and

$$\mathbf{H} = \mathbf{B} - \mathbf{M} \tag{9-4.4}$$

one can write Maxwell's equations for the macroscopic fields as

$$\nabla \cdot \mathbf{D} = 4\pi\rho, \quad \nabla \times \mathbf{H} = 4\pi\mathbf{j} + \frac{\partial}{\partial x^0}\mathbf{E}, \tag{9-4.5}$$

and

$$\nabla \cdot \mathbf{B} = 0 \quad \nabla \times \mathbf{E} + \frac{\partial}{\partial x^0}\mathbf{B} = 0, \tag{9-4.6}$$

where ρ and $\mathbf{j}$ are sums of free source terms plus ρ_M and $\mathbf{j}_M$, respectively.

In the case of sufficiently weak fields **D** and **H** are related to **E** and **B** by the constitutive equations

$$\mathbf{D} = \varepsilon \mathbf{E} \quad \text{and} \quad \mathbf{H} = \frac{1}{\mu}\mathbf{B} \tag{9-4.7}$$

where ε is the dielectric constant and μ the permeability of the medium. If in addition the medium is conducting, there will be a conduction current j_c given by Ohm's law

$$\mathbf{j}_c = \sigma \mathbf{E} \tag{9-4.8}$$

where σ is the conductivity of the material.

We can now express this body of formulae in four-dimensional language in much the same way that we obtained Eqs. (8-1.4 and 8-1.5) from Eqs. (8-1.1 and 8-1.2). To this end we introduce the fields $H^{\mu\nu}$ and $M^{\mu\nu}$ by

$$H^{\mu\nu} = \begin{pmatrix} 0 & -D_x & -D_y & -D_z \\ D_x & 0 & -H_z & H_y \\ D_y & H_z & 0 & -H_x \\ D_z & -H_y & H_x & 0 \end{pmatrix} \tag{9-4.9}$$

and

$$M^{\mu\nu} = \begin{pmatrix} 0 & -P_x & -P_y & -P_z \\ P_x & 0 & M_z & -M_y \\ P_y & M_z & 0 & -M_x \\ P_z & -M_y & M_x & 0 \end{pmatrix} \tag{9-4.10}$$

and the magnetization current

$$j_M{}^{\mu} = (\rho_M, \mathbf{j}_M). \tag{9-4.11}$$

In terms of these quantities Eqs. (9-4.1 and 9-4.2) combine to give

$$-j_M{}^{\mu} = M^{\mu}{}_{,\nu}. \tag{9-4.12}$$

Eqs. (9-4.3 and 9-4.4) become

$$H^{\mu\nu} = F^{\mu\nu} - 4\pi M^{\nu\nu} \tag{9-4.13}$$

and the Maxwell equations (9-4.5 and 9-4.6) take the form

$$H^{\mu\nu}{}_{,\nu} = -4\pi j^{\mu} \tag{9-4.14}$$

and

$$\varepsilon^{\mu\nu\rho\sigma} F_{\mu\nu,\rho} = 0. \tag{9-4.15}$$

When we come to write the constitutive equations, we find a certain arbitrariness. Following Minkowski[9] we shall require that they reduce to Eqs. (9-4.7) when the medium is at rest. The simplest possibility is to take

$$H_{\mu\nu}u^{\nu} = \varepsilon F_{\mu\nu}u^{\nu} \tag{9-4.16a}$$

and

$$\varepsilon^{\mu\nu\rho\sigma}H_{\mu\nu}u^{\rho} = \frac{1}{\mu}\,\varepsilon^{\ \nu\mu\rho\sigma}F_{\mu\nu}u_{\rho} \tag{9-4.16b}$$

where u^{μ} is the four velocity of the medium assumed to be moving as a whole. These relations are assumed to be valid for a homogeneous, isotropic medium. In the general case one would have to introduce a four-component object $\chi^{\mu\nu\rho\sigma}$ and write

$$H^{\mu\nu} = \chi^{\mu\nu\rho\sigma}F_{\rho\sigma}. \tag{9-4.17}$$

The conduction current $j_c{}^{\mu}$ can be defined to be

$$j_c{}^{\mu} = j^{\mu} - j_{\nu}u^{\nu}u_{\mu} \tag{9-4.18}$$

and is related to $F_{\mu\nu}$ by

$$j_c = -\sigma F_{\mu\nu}u^{\nu} \tag{9-4.19}$$

which reduces to Eq. (9–4.8) when $u^{\mu} = (1, \mathbf{0})$.

Unfortunately, few observable conclusions can be drawn from the above formalism due to the difficulty of obtaining sufficiently high velocities for material media. One interesting conclusion can be drawn from the transformation formulae for **P** and **M**. By making use of Eqs. (6-5.24a and b) we find, for a Lorentz mapping $V(\mathbf{v})$,

$$\mathbf{P}' = \gamma\left[\mathbf{P} + (\mathbf{v}\times\mathbf{M}) + \frac{\mathbf{v}}{\mathbf{v}^2}(\mathbf{v}\cdot\mathbf{P})\left(\frac{1}{\gamma} - 1\right)\right] \tag{9-4.20}$$

and

$$\mathbf{M}' = \gamma\left[\mathbf{M} - (\mathbf{v}\times\mathbf{P}) + \frac{\mathbf{v}}{\mathbf{v}^2}(\mathbf{v}\cdot\mathbf{M})\left(\frac{1}{\gamma} - 1\right)\right] \tag{9-4.21}$$

The second of these equations tells us that a polarized moving body will appear to be magnetized. This conclusion is not too surprising since moving charge distributions produce currents. What is perhaps surprising, however, is that Eq. (9-4.20) predicts that a magnetized moving body will appear to be electrically polarized. This effect can be traced back to the transformation law for j^{μ}. In particular, under a Lorentz mapping $V(\mathbf{v})$

$$\rho' = \gamma(\rho + \mathbf{j}\cdot\mathbf{v})$$

so that with $\rho = 0$ one still has a nonvanishing ρ'. Because of this result one

[9] H. Minkowski, *Gött. Nachr.*, **53** (1908); *Math. Ann.* **68**, 472 (1910).

would expect that a moving slab of magnetized material should appear to have an effective emf developed in a direction normal to **M** and **v**. Such an effect is, in fact, observed with a Faraday disk.

The problem of constructing a stress-energy tensor for the macroscopic electromagnetic field has occupied the attention of many people. Since the macroscopic field equations cannot be derived from a variational principle, the results of Section 4-5 can not be used. There have been several different proposals for the stress-energy tensor that leads to different predictions. Unfortunately, the differences are too small to allow one to choose between the different expressions[10] at the present time by experimental means. The most natural choice for this tensor is the one made by Minkowski.[11] He took

$$t^{\mu\nu} = \frac{1}{1\partial\pi} F_{\rho\sigma} H^{\rho\sigma} \eta^{\mu\nu} - \frac{1}{4\pi} F^{\rho\mu} H^{\sigma\nu} \eta_{\rho\sigma}. \tag{9-4.22}$$

With this choice

$$t^{00} = \frac{1}{4\pi} (\mathbf{E} \cdot \mathbf{D} + \mathbf{H} \cdot \mathbf{B})$$

$$t^{0r} = \frac{1}{4\pi} \varepsilon^{rst} E_s H_t$$

$$t^{r0} = \frac{1}{4\pi} \varepsilon^{rst} D_s B_t$$

and

$$t^{rs} = \frac{1}{8\pi} (\mathbf{E} \cdot \mathbf{D} + \mathbf{H} \cdot \mathbf{B}) \delta^{rs} - \frac{1}{4\pi} (E_r D_s + H_r B_s)$$

The Minkowski tensor is thus no longer symmetric. This led Abraham[12] and others to propose a symmetric stress-energy tensor. Actually, there is no reason to require that $t^{\mu\nu}$ should be symmetric since it is only part of the total stress-energy tensor of the system which also contains contributions from the matter in it. It is only the total stress-energy tensor that need be symmetric to insure the conservation of angular momentum of the system as a whole. We note, finally, that the Minkowski tensor can be shown to satisfy

$$t^{\mu\nu}{}_{,\nu} = -F^{\mu\nu} j_\nu$$

For matter at rest,

$$-F^{0\nu} j_\nu = \mathbf{E} \cdot \mathbf{j}_c$$

which is just the Joule heat developed per unit volume and time by the current $\mathbf{j}_c$.

[10] For a detailed discussion of the problem of constructing a stress-energy tensor for macroscopic electrodynamics see C. Møller *The Theory of Relativity* (Oxford Univ. Press, London, 1952), § 75.

[11] H. Minkowski, *loc. cit.*

[12] M. Abraham, *Rend. Pal.* **28** (1909); *Ann. der Phys.* **44,** 537 (1914).

PART III

DYNAMICAL SPACE-TIME THEORIES

10

Foundations of General Relativity

In 1915 Einstein proposed a far-reaching modification of the space-time description of physical systems of special relativity. In that description, as in the older Newtonian description, the geometry of space-time was fixed once and for all; it was not affected by the presence of other physical systems existing in space and time. In our terminology, the geometry of space-time appeared as an absolute element in all theories whose covariance group was the manifold mapping group. Einstein's proposal consisted in removing the geometry of space-time from the realm of the absolute; in his new description of nature it became a dynamical element on an equal footing with the other dynamical elements of physical systems. This world-picture of Einstein is known today as the *general theory of relativity.*

The geometry of space-time in the general theory was assumed by Einstein to be characterized by a Riemannian metric, as in special relativity. Being a dynamical element, it was to be determined by a set of dynamical laws in much the same way that the electromagnetic field is determined by the Maxwell-Lorentz equations. Because of this dynamical property of the metric, Einstein was then able to put forth a second fundamental proposal. He assumed that in the general theory, the gravitational field was not to be described by a separate geometrical object but was to be described *by the metric tensor.* By this second assumption Einstein succeeded actually in eliminating geometry from the space-time description of physical systems by letting the gravitational field take over all its functions. While it is still convenient to use geometrical terminology in discussing the general theory, nowhere is it necessary to use the gravitational field tensor $g_{\mu\nu}$ explicitly as a metric (for example, in an expression for the distance between two neighboring space-time points).

Einstein was led to search for a more general world-picture than that afforded by the special theory, because of his dissatisfaction with certain aspects of this theory. In particular he was disturbed by the inability of either Newtonian mechanics or special relativity to explain the universal constancy of the ratio of inertial to gravitational mass of material bodies. In addition he objected to the existence of certain types of absolute motions, namely, accelerated motions, in these two theories. On the basis of arguments put forth by Mach, he felt that all motion should be relative, not just uniform motion.

In the search that ultimately led to the general theory, Einstein was guided by three general principles. The first was Mach's principle, the second was the principle of equivalence, and the third was the principle of general invariance. In this chapter we examine each of these principles and show how they led Einstein to the general theory. We then introduce the field equations proposed by him for the gravitational field $g_{\mu\nu}$ and discuss their general structure.

10-1. Mach's Principle

The general philosophical dictum set forth by Mach, which today has become known as Mach's principle, held that only relative motion between bodies existed. It denied, therefore, one of the essential tenets of the Newtonian world-picture. Thus Mach wrote[1]:

> "For me only relative motions exist .... When a body rotates relatively to the fixed stars, centrifugal forces are produced; when it rotates relatively to some different body not relative to the fixed stars, no centrifugal forces are produced. I have no objection to calling the first rotation as long as it be remembered that nothing is meant except relative motion with respect to the fixed stars."

Mach's point of view was a direct outgrowth of the idealist metaphysics of Bishop Berkeley, for whom also the notion of absolute motion was unacceptable. At one point in his writings Berkeley commented[2]:

> "Let us imagine two globes, and that besides them nothing else material exists, then the motion in a circle of these two globes round their common center cannot be imagined. But suppose that the heaven of the fixed

[1] E. Mach, *The Science of Mechanics*, 5th English ed. (LaSalle, Ill., 1942), Chap. 1.

[2] G. Berkeley, *The Principles of Human Knowledge* (A. Brown & Sons, London, 1937).

stars was suddenly created and we shall be in a position to imagine the motion of the globes by their relative position to the different parts of the heaven."

It is not known to us whether Berkeley considered the possibility of observing absolute motion by dynamical rather than visual means, as suggested by Newton in his discussion of the two rotating globes. Mach took this possibility into account by ascribing all inertial effects to the mutual interaction of matter. For Mach, the tension in a rope connecting the two rotating globes of Newton is due to the interaction of these globes with the rest of the matter in the universe. As a consequence, all that one could say when a tension was observed to exist in the rope was that the globes were rotating relative to that matter.

On the surface, the Machian point of view seems reasonable. One could imagine inertial effects arising as a consequence of advanced interactions in much the same way that Wheeler and Feynman imagined the radiation reaction force to arise. In fact, if one took the Machian point of view seriously, it would require a complete action at a distance description of the interaction of material bodies. Thus we could describe the interaction of bodies by means of a theory whose dynamical laws are given by Eq. (8-8.7) and without any difficulty set m_{0i}, the inertial mass of the ith body, equal to zero. There would still be an inertial type term, that is, a term involving the acceleration of the body, proportional to the field ϕ_i due to the other bodies present. We could then imagine that the main part of this field, due to distant matter, is homogeneous and isotropic so that the force term on the left side of Eq. (8-8.7) would arise from the presence of nearby matter. In fact, Dicke[3] has discussed just such a theory.

There are a number of difficulties attendant on the interpretation of Mach's principle as implying an action-at-a-distance theory of the inertial properties of matter. As we will see, the purely scalar theory with equations of motion (8-8.7) is not in accord with observations of the advance of the perihelion of the planet Mercury. To obtain agreement one must use a combination of the scalar and tensor theories, the latter having equations of motion of the form (8-10.10). However, in the latter theory, we see that the inertial forces are velocity dependent. In fact, Mach's ideas do not really contribute to an understanding of why there appears to be such a fundamental distinction between uniform and accelerated motion in nature. While suggesting that inertial effects are due to the interaction of matter with the distant stars, it does not explain why this interaction is velocity independent and hence why the various states of uniform motion relative to these stars are indistinguishable while various

[3] R. H. Dicke, "The Many Faces of Mach," in *Gravitation and Relativity*, Hong-Yee Chiu and W. F. Hoffmann, eds. (Benjamin, New York, 1964).

types of accelerative motion relative to them are distinguishable from each other and from uniform relative motion. In effect, Mach only replaced Newton's absolute space by the distant stars but learned nothing new thereby.

Recently Hughes[4] has carried out an extensive series of measurements with the purpose of testing Mach's hypothesis, based on a suggestion of Cocconi and Salpeter.[5] They suggested that if inertial effects are due to the distribution of matter in the universe, there should be a small anisotropy in these effects due to the anisotropic distribution of mass in our galaxy relative to the solar system. If one assumes, as did Cocconi and Salpeter, that the contribution Δm to the inertial mass of a body, due to a mass ΔM located a distance r away from this body, is proportional to $\Delta M/r^{\nu}$, then the total contribution m_0 due to matter outside our galaxy would be given by

$$m_0 = \kappa \int_0^R 4\pi r^{2-\nu} \rho \, dr = \kappa \frac{4\pi\rho R^{3-\nu}}{3-\nu}.$$

Here R is the radius of the universe, ρ the average density of matter (assumed to be distributed homogeneously and isotropically), and κ a proportionality constant. The range of values of the exponent ν can be restricted by the following considerations: If $\nu < 0$ then the more distant a given amount of mass, the greater will be its contribution to m, which is unreasonable. On the other hand, if $\nu > 1$, then nearby matter (for example, the sun) would dominate in determining m, which is contrary to observations on planetary motion. Therefore Cocconi and Salpeter assume that $0 < \nu < 1$.

The contribution Δm to m due to the mass M_0 in the galaxy, assumed concentrated at a distance R_0 from m, would be

$$\Delta m = \kappa \frac{M_0}{(R_0)^{\nu}}.$$

The ratio of these two contributions to m is thus

$$\frac{\Delta m}{m_0} = \frac{M_0(3-\nu)}{(R_0)^{\nu} 4\pi\rho R^{3-\nu}}.$$

Using the presently accepted values

$$M_0 = 3 \times 10^{44} \text{ g},$$
$$R_0 = 2.5 \times 10^{22} \text{ cm},$$
$$R = 3 \times 10^{27} \text{ cm},$$
$$\rho = 10^{-29} \text{gm/cm}^3,$$

[4] V. W. Hughes, "Mach's Principle and Experiments on Mass Anisotropy," in *Gravitation and Relativity*, Hong-Yee Chiu and W. F. Hoffmann, eds. (Benjamin, New York, 1964).

[5] G. Cocconi and E. E. Salpeter, *Nuovo Cimento*, **10**, 646 (1958).

one finds for $v = 0$:

$$\frac{\Delta m}{m} = 3 \times 10^{-10};$$

while for $v = 1$,

$$\frac{\Delta m}{m} = 2 \times 10^{-5}.$$

Hughes has tested the Cocconi-Salpeter hypothesis by a series of standard nuclear magnetic-resonance experiments, using a Li^7 nucleus in the ground state. The nuclear spin of this ground state is 3/2, so that in a magnetic field the Li^7 nucleus will have four energy levels. These would be equally spaced if there were no mass anisotropy, and a single nuclear resonance line should be observed. If there is a mass anisotropy the spacing will no longer be uniform and one should observe a triplet nuclear-resonance line if the structure is resolved or a single broadened line if the structure is unresolved. Hughes observed the Li^7 resonance over a 12-hr period. A single line of width 1.2 cps was observed, which changed by less than 0.2 cps over this period. On the basis of a simple shell-model calculation, Hughes obtained as a limit $\Delta m/m_0 < 10^{-22}$, which seems to rule out any mass anisotropy. Dicke, however, has argued that Mach's principle should apply not only to inertial mass but to fields as well, so that the propagation of photons would be determined by the distribution of matter in the universe. Likewise, nuclear forces should also exhibit an anisotropy and hence one should expect a null result from the Hughes experiment. The interpretation of Hughes' experiment, at least as it bears on Mach's principle, must therefore be considered as an open question.

Whether one accepts Mach's point of view or not, it is important to recognize the role it played in guiding Einstein to the general theory. Thus he remarks[6]: "But in the second place the theory of relativity makes it appear probable that Mach was on the right road in his thought that inertia depends upon a mutual action of matter." He was led then to argue that if inertial effects are a manifestation of the geometry of space and time, it follows from Mach's principle that this geometry must be determined, or at least influenced, by the distribution of matter in the universe. As a consequence, geometry cannot be considered to be an absolute element as it is in these theories but must be treated as a dynamical element on an equal footing with all other dynamical elements. After he developed the general theory, Einstein came to realize that it did not satisfy Mach's principle in the strict sense. Nevertheless the removal of geometry from the realm of the absolute was essential for the

[6] A. Einstein, *The Meaning of Relativity* (Princeton Univ. Press, Princeton, New Jersey, 1955), p. 100.

development of the general theory and represents a fundamental change in one's notions of the nature of space and time. As for Mach's principle itself, several writers[7,8] have suggested that it provides boundary conditions for the general theory but does not bear directly on the field equations of that theory.

10-2. The Principle of Equivalence

Mach's principle suggested to Einstein that the geometry of space-time should be a dynamical element in all physical theories. As we shall see in this section his own principle of equivalence suggested that geometry and gravitation were one and the same thing. The assumed identity of these hitherto unrelated concepts represented the second major step taken by Einstein in arriving at his general theory. The principle of equivalence itself was an outgrowth of the recognition, by Einstein, of the importance of an experimental fact that had, up to then, been considered to be an interesting accident of nature. This accident was the apparent universal constancy of the ratio of inertial mass to gravitational charge for material bodies.

The constancy of the ratio of inertial mass to gravitational mass, which we will denote here by Q, had been known, of course, to science since the time of Galileo, who observed that, in the earth's gravitational field, all bodies fall with the same acceleration. Since the same is not true for the motion of charged bodies in an electric field, for instance, Galileo's observation bespoke of a very special property of the gravitational field. Yet in both the Newtonian and special relativistic gravitational theories the constancy of Q had to be assumed as an additional hypothesis rather than being a consequence of these theories. Since the time of Galileo the constancy of Q has been tested with ever-increasing accuracy. In the *Principia*, Newton described a pendulum experiment that fixed the constancy of Q to one part in a thousand. Later, Bessel,[9] also using a pendulum method, improved the accuracy to one part in 60,000. The next significant improvement in accuracy came with the observations of Eötvös[10] and Eötvös, Pekar, and Fekete.[11] The ultimate accuracy of these latter workers established the constancy of Q to one part in 10^8. Later, Southerns,[12] working with radioactive uranium oxide, fixed Q to one

[7] J. A. Wheeler, "Mach's Principle as a Boundary Condition for Einstein's Equations," in *Gravitation and Relativity*, Hong-Yee Chiu and W. F. Hoffmann, eds. (Benjamin, New York, 1964).

[8] F. Gürsey, *Ann. Physik*, **24**, 211 (1963).

[9] F. W. Bessel, *Abhandl. Berl. Akad.*, **30**, 41 (1830).

[10] R. Eötvös, *Math. natur. Berichte Ungarn*, **8**, 65 (1891); *Ann. Phys. Chemie*, **59**, 354 (1896).

[11] R. Eötvös, D. Pekar, and E. Fekete, *Ann. Physik*, **68**, 11 (1922).

[12] L. Southerns, *Proc. Roy. Soc.* (London), **A84**, 325 (1910).

part in 2×10^5. While not so accurate as the Eötvös measurements, Southerns' experiment was sufficient to establish the constancy of Q for the mass equivalent of the binding energy of nuclei. Zeeman[13] was able later to increase this accuracy to one part in 2×10^7. More recently, Dicke,[14] working with aluminum and gold, has been able to extend the accuracy of the Eötvös result to one part in 10^{11}. On the basis of the available data, Wapstra and Nijgh[15] were able to argue for the constancy of Q for the various constituents of matter. Thus the Q ratio for a proton plus electron is shown by them to be equal to Q for a neutron to about one part in 10^7, although neither of these values can be measured directly. To conclude this discussion, then, we can take it as an extremely well-established experimental fact that the Q ratio for matter of all kinds is a universal constant.

Problem 10.1. How would you most easily establish the equality of Q for an electron plus proton and a neutron? the constancy of Q for the orbital binding energy of electrons?

If Q has the same value for all material bodies, they should all behave in the same way in an arbitrary, but given, gravitational field. In all of our discussions of particle motion in a gravitational field we saw that the equations of motion were of the form

$$\frac{d^2 z^\mu}{d\lambda^2} + \Gamma^\mu_{\rho\sigma} \frac{dz^\rho}{d\lambda} \frac{dz^\sigma}{d\lambda} = 0 \tag{10-2.1}$$

for a suitable choice of the path parameter λ. Consequently, to the extent that we can neglect their action back on the sources of the gravitational field, observations on the motion of otherwise free particles will all serve to determine the same affinity in a given space-time region.

Einstein went on to generalize the above result to all physical systems, in what we call here the *principle of equivalence*. This principle asserts that to the extent we can neglect their action back on the sources of the gravitational field, measurements made on any physical system will serve to determine the same affinity in a given space-time region. Thus weak light signals should propagate along rays determined by the same affinity as appears in the particle equations of motion (10-2.1). It is, of course, not obvious that the principle of equivalence is universally valid, although what little data we have seems to support it. We should perhaps mention the possibility, discussed by Trautman,[16] that the actual physical affinity is asymmetrical, in which case there would exist physical

[13] P. Zeeman, *Proc. Roy. Amsterdam Acad.*, **20:1**, 542 (1918).
[14] R. H. Dicke, *Sci. Am.*, **205(6)**, 84 (1961).
[15] A. H. Wapstra and G. J. Nijgh, *Physica*, **21**, 796 (1955).
[16] A. Trautman, to be published in Uspekhi, USSR.

systems that could be used to determine its antisymmetric part. (Of course, if no such systems exist in nature, there is no need to consider asymmetric affinities.) Since measurements made on particle motions and light rays can serve to determine only the symmetric part of the affinity, the principle of equivalence would have to be generalized in this case in an obvious manner.

As we have stated it, the principle of equivalence strongly suggests that we can do away with the notion of a flat space-time affinity in our space-time descriptions. Only in some approximate sense would it be meaningful to consider the affinity as being flat in a region where the gravitational field is weak. It also suggests that we should treat the affinity appearing in the equations of motion of a physical system as a dynamical, rather than an absolute, object since it is a manifestation of the gravitational field. We say "suggests" since the principle of equivalence is only applicable in those cases where one can neglect the action of a physical system back on the sources of whatever gravitational field may be present. Thus, if we were to use a sufficiently massive body to measure the affinity in some space-time region, it would interact strongly with these sources and hence the affinity determined by observations on the motion of this particle would be different from that determined by using a less massive body. It might be that by using a number of different systems we could separate out, in a unique manner, a flat affinity from among the various affinities determined by observations made on these systems. Also the possibility exists, as we discussed in Section 4-3, that by means of observations made on systems sufficiently far removed from the sources of the gravitational field we can separate out uniquely a flat affinity from whatever affinity is observed to exist in any given space-time region. In the next section we will rule out these possibilities by means of the principle of general invariance.

In arguing for the principle of equivalence, Einstein made use of his famous falling-elevator experiment. He imagined an elevator that was freely falling in a gravitational field. To an observer inside the elevator, a material body that was also freely falling would appear to be in a state of uniform motion. Likewise, in an accelerated elevator in a gravity-free region of space, a material body would appear to behave as though it were falling in a uniform gravitational field. Also, light rays would appear to travel along curved trajectories in this latter elevator. It was, in fact, just this argument that led Einstein to conclude that the path of a light ray should be affected by a gravitational field. On the basis of this "gedanken" experiment Einstein then concluded that it should be impossible, on the basis of purely local measurements, to distinguish between uniform motion of an observer in a gravitational field and accelerated motion in gravity-free space. He was thus led to the notion of the relativity of all motion, in conformity with the Machian point of view. It was for this reason that he named his new world-picture the "general theory of relativity."

A number of authors have taken the local equivalence of gravitational and accelerative effects to be fundamental to the development of general relativity and have called it the principle of equivalence. Actually it constitutes a separate assumption over and above the principle of equivalence as we have stated it. In effect it postulates that only the affinity determined by the gravitational field appear in the dynamical laws for these systems, to the extent that these laws are local laws. This restatement follows from the fact that space-time measurements made in an accelerated reference frame in flat space can, as we saw in Section 6-16, be described by performing a mapping such that the flat-space affinity no longer has its Minkowski values. Alternatively, one can always find a mapping—that is, a reference frame—such that any affinity can be made to take on Minkowski values and its derivatives made to vanish at a point of the space-time manifold. We see, therefore, that the assumed local equivalence of accelerative and gravitational effects amounts to requiring that in a sufficiently small region of space-time, the laws of special relativity are valid.

Because it is an additional requirement or assumption, we shall, in what follows, refer to the local equivalence of gravitational and inertial effects as the principle of *minimal coupling of the gravitational field* to matter, or as the principle of minimal coupling, for short. The chief difficulty in applying this principle lies in the use of the word "local" to mean a sufficiently small region of space-time. Thus we can imagine a dimensionless particle that is characterized by a mass m and, in addition, a spin tensor $\sigma^{\mu\nu}$, for example, an electron. Its trajectories might then be determined by the geodesic equation, to which would be added a term proportional to $R_{\mu\nu\iota\kappa}\sigma^{\mu\nu}\sigma^{\iota\kappa}$. Such a dynamical law would violate the principle of minimal coupling as we have stated it. However, one could also argue that a particle such as the one we have considered is in reality not a local structure and so does not violate the principle. But, of course, this objection is in reality a quibble; if such a particle did exist in nature, one could never find a region of space-time that was sufficiently small to cause such a spin-dependent term to vanish.

It is important to recognize the differences between the principles of equivalence and minimal coupling as we have stated them. The first is a nonlocal statement and is crucial to the general theory of relativity. On the other hand, the principle of minimal coupling may or may not have universal validity. Its violation would in no way affect the validity of the general theory of relativity, unless one insisted on taking it to be a principle of this theory. Let us also emphasize that accelerative effects are not the same thing as geometrical effects, although they have been taken to be so by a number of authors. The fact that a free particle in flat space moves along a geodesic of the flat metric is a geometrical fact. That it appears to move along a curved trajectory when viewed from an accelerated reference frame does not alter this fact. Geometrical effects are intrinsic to the physical system considered while accelerative effects are extrinsic to it.

10-3. The Principle of General Invariance

We come now to the third principle that led Einstein to the general theory, the principle of general invariance, which is usually referred to as the principle of general covariance. There is still a good deal of confusion concerning just what content Einstein implied by this principle, due in part to his own writing on the subject. He took for the principle the requirement that the form of the laws of physics as determined with the help of any one reference frame must be the same as those determined with the help of any other reference frame. Such a requirement is equivalent to demanding that the manifold mapping group be a covariance group of all physical theories. However, since every space-time theory can be made generally covariant by introducing absolute elements into the description of the kpt of the theory, Kretschmann[17] argued that the requirement of general covariance made no assertion about the content of these laws (that is, about the equivalence classes of dpt), and Einstein[18] concurred with this view.

In spite of Einstein's acceptance of the Kretschmann position, it is clear that he intended more by the principle of general covariance than is contained in the requirement that the MMG be a covariance group of all physical theories, since he was seeking a generalization of the principle of special relativity. Einstein referred many times to the principle of general invariance as the principle of general relativity and also referred to his new world-picture as the general theory of relativity. But the principle of special relativity is a symmetry requirement on physical systems and not a covariance requirement. It is therefore reasonable to assume that Einstein also viewed the principle of general invariance as a symmetry requirement. In any event we will take the principle to mean that the MMG is a symmetry group of all physical systems.

Before we discuss the consequences of accepting the principle of general invariance as a fundamental principle of nature let us discuss why we should accept it in the first place. There are four reasons for doing so. First, what little experimental evidence we have tends to support it. Thus, as we shall see presently, the universal constancy of the ratio of inertial mass to gravitational charge is a direct consequence of this principle. (Since the principle of equivalence is a generalization based on this result, the latter cannot be said to be a consequence of the former.)

[17] E. Kretschmann, *Ann. Physik* (*Leipzig*), **53**, 575 (1917).
[18] A. Einstein, *Ann. Physik* (*Leipzig*), **55**, 241 (1918).

The second reason for accepting the principle of general invariance is that it leads to new physics. While it is possible to describe all known gravitational phenomena within the framework of special relativity one learns very little more by doing so. Most of the new effects predicted by such a theory are much too small to be experimentally verifiable by present-day techniques. On the other hand, the principle of general invariance forces on us theories that predict essentially new types of phenomena having no counterparts in the flat, linear world of special relativity. It must always be considered as a mark in favor of a hypothesis that it results in such predictions.

The third argument for the principle is that it rules out completely the possibility of all absolute objects from our space-time descriptions. In a world with absolute objects, parts of this world, the absolute objects, influence the behavior of the remainder without, however, being influenced in turn. Such a world therefore lacks reciprocity. It seems reasonable, however, and more in keeping with all our other experience that there is operative in nature a generalized law of action and reaction. While the principle of equivalence suggests that the space-time affinity should be a dynamical object, it does not rule out completely the possibility of the existence of a flat affinity. Furthermore, it says nothing concerning the existence of other absolute objects such as the Newtonian planes of absolute simultaneity. The principle of general invariance on the other hand does eliminate all such objects from our space-time descriptions. Thus the principle of equivalence can be considered to be a consequence of the principle of general invariance. In fact, if we accept this principle, we can dispense both with Mach's principle and the principle of equivalence as foundations of the general theory of relativity.

There have been a number of attempts to introduce absolute objects into the general theory of relativity, primarily for the purpose of defining a stress-energy tensor for the gravitational field.[19–21] Of course, if such objects do indeed exist in nature the appelation "general relativity" would no longer be applicable. We should perhaps emphasize that we are discussing here universal absolute objects, which must appear in the description of every dpt of our space-time description. It is quite possible that, for a subclass of dpt, one or more dynamical objects satisfy the criteria of Section 4-3 and so play the role of absolute objects for these dpt. It is also possible that one or more dynamical objects of a subclass of dpt all satisfy the same boundary conditions. Such boundary conditions would involve objects that, for these dpt, behave as asymptotic absolute objects. Thus we saw in our discussion of Newtonian gravitational theory that if we consider systems of bounded sources, we could introduce a

[19] N. Rosen, *Phys. Rev.* **57**, 147, 150 (1940).

[20] C. Møller, *Nat. Fys. Skr. Dan. Selsk.* **1**, No. 10 (1961).

[21] J. Plebański, in *Proc. of Conference on Gravitation in Warsaw and Jabonna*, L. Infeld, ed. (Gauthiers-Villars, Paris 1964).

globally flat affinity. Likewise, Fock[22] has argued in effect that, for a large class of gravitational fields, it is possible to introduce a flat metric on the space-time manifold.

The existence of such special subclasses of dpt as those discussed above does not, of course, constitute a violation of the principle of general invariance as we have formulated it. Only the existence of universal absolute objects would do so. But then the existence of such objects must be experimentally verifiable. In particular, they must appear in an essential way in the equations of motion for some dynamical system. Otherwise they can be eliminated from our space-time description in the same way that the congruence of curves associated with the absolute space of Newton could be eliminated from the description of the Newtonian systems. As long as we do not take account of gravitational phenomena such objects must be introduced in an essential way into our space-time descriptions. Once we include such phenomena, however, they can be dispensed with since the gravitational field can take over all of their functions. Moreover, at least in one treatment of the gravitational field using absolute objects—namely, Kraichnan's—they were seen to be superfluous. We might also point out that once we admit absolute objects then there is no longer any justification for keeping the Einstein equations for the gravitational field.

The final argument that we shall advance is in some respects the weakest argument we can give for any hypothesis and yet it has been one of the most powerful guides in the development of physics. Above all else, the principle of general invariance is simple and beautiful. Certainly its formulation by Einstein will stand for all time as one of the supreme achievements of the human intellect. Of course, the criteria of simplicity and beauty are subjective. Nevertheless, there are few physicists who would, I believe, deny the simplicity and beauty of the general theory that flows from this principle.

While the above arguments lend support to the principle of general invariance they do not force us to accept it; as with all physical hypotheses, nature is the final arbiter. Nevertheless they are, in my opinion, sufficient to justify a study of the general theory and the expenditure of a considerable amount of continuing effort to develop the consequences of this theory.

Let us now ask: What are the simplest generally invariant theories that we can construct? First, such theories must contain fields since at least one field is needed in order to form the basis of a faithful realization of the MMG. Furthermore, linear homogeneous objects yield the simplest realizations of this group. Finally, we might require that a single irreducible field should be used. If we accept these restrictions, then the simplest object we can use to construct a nontrivial theory must be a symmetric, second-rank tensor.

[22] V. Fock, *The Theory of Space, Time, and Gravitation*, trans. from the Russian by N. Kemmer (Macmillan, New York, 1964).

It is clear that the simplest object must contain at least five kinematically independent components since we can always, because of the assumed invariance properties of the theory, impose four coordinate conditions on the dpt of the theory analogous to the gauge condition of electrodynamics. We might try, therefore, to use an antisymmetric second-rank tensor density as our basic field variable. However, it is impossible to construct an invariant theory with such an object that restricts it sufficiently. Since we do not have at our disposal an affinity or a metric, whatever equations we require this object to satisfy must involve only ordinary derivatives. If we make use of the results of Sections 1-6 and 1-7, we see that there are only two possibilities. We can take our field variable to be a cotensor $F_{\mu\nu}$ and require that

$$F_{[\mu\nu,\rho]} = 0.$$

Or we can take it to be a contratensor density $\mathfrak{F}^{\mu\nu}$ and require it to satisfy

$$\mathfrak{F}^{\mu\nu}{}_{,\nu} = 0.$$

In either case we see that we have only four equations to determine the six components $F_{\mu\nu}$ or $\mathfrak{F}^{\mu\nu}$. We are led, therefore, to consider the next simplest object, namely, a symmetric second-rank tensor. If we take this object to be a second-rank cotensor, we will see that we can construct a determinate set of field equations for such an object.

There are, in fact, three possible generally covariant, second-order differential equations that a symmetric second-rank cotensor $g_{\mu\nu}$ can satisfy. We could require that it satisfy $R_{\mu\nu\rho\sigma} = 0$, where $R_{\mu\nu\rho\sigma}$ is the Riemann-Christoffel formed using $g_{\mu\nu}$ as a metric. Such a theory is not generally invariant, however, since it makes $g_{\mu\nu}$ an absolute object. We could require that the Weyl tensor formed from $g_{\mu\nu}$ vanish. But in this case, $(-g)^{-1/4}g_{\mu\nu}$ can always be mapped so as to take on Minkowski values everywhere and hence is an absolute object. Nordström, in 1912, did develop a scalar theory of the gravitational field that was equivalent to one[23] in which the Weyl tensor $C_{\mu\nu\rho\sigma} = 0$ and where $g = \det g_{\mu\nu}$ was determined by the equation

$$R = \kappa g_{\mu\nu} T^{\mu\nu}$$

where $T^{\mu\nu}$ was the stress-energy tensor associated with whatever other fields or matter were present and where R was the curvature scalar formed from $g_{\mu\nu}$. The symmetry group of the theory was, however, just the Poincaré group. Furthermore it did not predict a bending of light in a gravitational field ($g_{\mu\nu}T^{\mu\nu} = 0$ for the electromagnetic field) and it gave the wrong sign for the

[23] G. Nordström, *Phys. Z.*, **13**, 1126 (1912). The generally covariant form of the Nordström theory was given by A. Einstein and A. D. Fokker, *Ann. Physik (Leipzig)*, **44**, 321 (1914).

advance of the planetary perihelia. The only other possible second-order differential equations that do not result in $g_{\mu\nu}$ being an absolute object can then be shown to be of the form

$$R_{\mu\nu} + \alpha R g_{\mu\nu} + \Lambda g_{\mu\nu} = \kappa T_{\mu\nu} \tag{10-3.1}$$

where $R_{\mu\nu}$ is the Ricci tensor formed from $g_{\mu\nu}$ and α, Λ, and κ are arbitrary constants. If we require further that these equations follow from a variational principle, then, as we shall show in the next section, α must have the value $-1/2$.

These equations, with $\alpha = -1/2$ are in fact just the equations adopted by Einstein for the metric tensor of space-time. However, from the point of view of the principle of general invariance we need not interpret $g_{\mu\nu}$ as a metric nor $R_{\mu\nu}$ as a Ricci tensor. Equations (10-3.1) do not rest on such an interpretation; one can show that they are the only dynamical equations of second differential order for a symmetric tensor $g_{\mu\nu}$ that are in accord with the principle of general invariance as we have interpreted it. While it is convenient to continue to make use of geometrical terms (for example, we will continue to call $R_{\mu\nu}$ the Ricci tensor and R the curvature scalar), nothing of what we will do in what follows depends on such an interpretation. As in all physical theories we will look for consequences of Eqs. (10-3.1) that will lead us to associate $g_{\mu\nu}$ with some observable element of the physical world. In doing so, we will see that this element is the gravitational field.

It is interesting to contrast the situation here with that in electrodynamics. In the latter theory the basic field variable is the vector potential A_μ. If we require only that it satisfy a second-order differential equation, which is both Poincaré invariant and gauge invariant, there are a continuous infinity of possible field equations. We saw in Section 8-2 that we could form the two first-order scalars $F^{\mu\nu}F_{\mu\nu}$ and $\varepsilon^{\mu\nu\rho\sigma}F_{\mu\nu}F_{\rho\sigma}$ from A_μ where $F_{\mu\nu} = A_{\nu,\mu} - A_{\mu,\nu}$. Therefore any function of these two scalars that is taken as a Lagrangian in an action principle will lead to second-order differential equations for A_μ. Only if we impose the additional requirement of linearity do we obtain essentially unique equations for A_μ. In the present case the single additional requirement over and above the requirement of general invariance that the equations for $g_{\mu\nu}$ be of second differential order is sufficient to lead to essentially unique field equations for this object.

We see from the above discussion that the principle of general invariance as we have interpreted it here is, in fact, a very strong symmetry principle. Regardless of what other objects we introduce the above discussion shows that we must always include $g_{\mu\nu}$ or some other more complicated object such as an affinity in the list of variables that describe the kpt of any generally invariant theory. Without such an object we would be unable to construct an invariant theory. Furthermore, if we require that the equations of motion for this tensor be of second differential order and follow from a variational principle, we are led to an essentially unique expression for these equations. These consequences are in marked contrast to the situation in special relativity. Poincaré invariance does not force us to include the gravitational field in the description of a physical system; and when we do include this field, the dynamical equations for it are far from unique unless we impose additional conditions of linearity and positive definiteness of the gravitational energy.

The power of the principle of general invariance led Einstein and others to try to generalize it in the hope of finding a single *unified theory* of nature. Einstein never liked the necessity of including a right-hand side in his field equations—that is, the term $\kappa T^{\mu\nu}$ in Eq. (10-3.1)—and considered it to be only a provisional element in the theory to be discarded in a more complete theory. The aim of the unified theories was to postulate a single invariance group that would lead to essentially unique equations for the simplest object for which such equations could be formulated. In most cases one tried to enlarge the MMG on one or another grounds so that the basic object of the theory would have enough components to describe both the gravitational and electromagnetic field. In particular the five-dimensional theories of Kaluza and Klein had some success in this regard.[24] At one point Einstein considered a theory in which the basic object was an asymmetric second-rank tensor. Since such a tensor is not reducible under the MMG, however, the resulting theory cannot be said to be unified nor are the field equations for such an object in any sense unique. With the discovery of other fields in nature attempts at a unified theory have, to a large extent, been abandoned and as many different objects as necessary are being used to describe the kpt of the theories of these fields. But the idea of a unified theory still is attractive from the point of view of simplicity and beauty and it is therefore all the more a pity that to date it has not been more fruitful. We will not discuss such theories here but rather concentrate our attention on those theories that follow from the principle of general invariance.

10-4. The Einstein Field Equations

Once Einstein had become convinced of the identity of geometrical and gravitational effects, he set out to discover the form of the equations that would determine the field $g_{\mu\nu}$. The guiding principle employed by him in his search for these equations was the principle of general invariance. At one point of the search he even believed that he had proved that the equations for $g_{\mu\nu}$ could not be generally invariant. However, he soon returned to the principle

[24] See, for example, W. Pauli, *Relativity Theory* (Pergamon Press, New York, 1958) for a brief description of these and other so-called unified theories.

and succeeded, with its help, in arriving at his now famous equations for the gravitational field.[25]

In arriving at these equations Einstein was guided by the additional requirement that, in an appropriate limit, his theory should reduce to the Newtonian gravitational theory. If we look upon that theory as a field theory, then (as we saw in Section 5-3) the gravitational potential ϕ is determined by the Poisson equation (5-3.6):

$$\nabla^2\phi = 4\pi G\rho.$$

This suggested to Einstein that, in addition to being generally invariant, the field equations for $g_{\mu\nu}$ should (1) be of second differential order in $g_{\mu\nu}$, and (2) should involve the total stress-energy tensor $T_{\mu\nu}$ linearly. The only equations that meet all these requirements must be of the form (10-3.1). Einstein was unsure of the values to assign to α and Λ. He originally took $\alpha = -(1/2)$ so that $T_{\mu\nu}$ would satisfy a covariant continuity equation of the form

$$g^{\rho\sigma}T_{\mu\rho;\sigma} = 0. \qquad (10\text{-}4.1)$$

With $\alpha = -(1/2)$, $R_{\mu\nu} - (1/2)g_{\mu\nu}R \equiv G_{\mu\nu}$ satisfies the contracted Bianchi identity (3-4.9) and hence Eq. (10-4.1) is a consequence of Eq. (10-3.1). In addition, Einstein originally took the so-called cosmological constant Λ to have zero value. Thus the field equations for $g_{\mu\nu}$ proposed by Einstein had the form

$$R_{\mu\nu} - \tfrac{1}{2}g_{\mu\nu}R = \kappa T_{\mu\nu}. \qquad (10\text{-}4.2)$$

Later Einstein[26] suggested, on the basis of considerations having to do with the structure of matter, that α should have the value of $-(1/4)$. However, he soon rejected this possibility and returned to the equations (10-4.2).

In the same year (1915) that Einstein proposed his field equations, Hilbert[27] also formulated them independently, using as a starting point the requirement that they should be derivable from a variational principle. It is interesting to note the attitude toward variational principles in those days. In his 1921 article on the theory of relativity in the *Mathematic Encyclopedia*,[28] Pauli wrote: " His presentation, though, would not seem to be acceptable to physicists for . . . the existence of a variational principle is introduced as an axiom." Today most physicists would be not only willing to accept as axiomatic the existence of a variational principle but would be also loath to accept any dynamical equations that were not derivable from such a principle. In fact, then, if we require that the equations for the determination of $g_{\mu\nu}$ follow from

[25] A. Einstein, *S. B. preuss. Akad. Wiss.*, 778, 799, and 844 (1915); *Ann. Phys. Lpz.*, **49**, 769 (1916); English translation available in *The Principle of Relativity*, Dover reprint.

[26] A. Einstein, *S. B. preuss. Akad.*, 349, (1919).

[27] D. Hilbert, Nachr. Ges. Wiss. Gottingen, 395 (1915).

[28] W. Pauli, *op cit.*

a variational principle, coupled with the requirements that they be generally invariant and of second differential order, we are led almost uniquely to the Eqs. (10-4.2).

Let us consider first the equations for $g_{\mu\nu}$ in the absence of matter and fields. In this case the most general action that will lead to equations of motion of the desired type is

$$S_G = -\frac{1}{2\kappa}\int\sqrt{-g}\{R - 2\Lambda\}\, d^4x, \tag{10-4.3}$$

where κ and Λ are constants to be determined and where R is given by

$$R = g^{\mu\nu}[\Gamma^{\rho}_{\mu\rho,\nu} - \Gamma^{\rho}_{\mu\nu,\rho} + \Gamma^{\rho}_{\mu\sigma}\Gamma^{\sigma}_{\rho\nu} - \Gamma^{\rho}_{\mu\nu}\Gamma^{\sigma}_{\rho\sigma}]. \tag{10-4.4}$$

In deriving the Euler-Lagrange equations associated with S_G, it is customary to regard the components $\Gamma^{\rho}_{\mu\nu}$ of the affine connection appearing therein as being equal to the Christoffel symbols

$$\left\{{\rho \atop \mu\nu}\right\} \equiv \tfrac{1}{2}g^{\rho\sigma}(g_{\mu\sigma,\nu} + g_{\sigma\nu,\mu} - g_{\mu\nu,\sigma})$$

and hence as known functions of the $g_{\mu\nu}$. However, one can also treat both the $g_{\mu\nu}$ and the $\Gamma^{\rho}_{\mu\nu}$ as independent variants and obtain thereby equations that determine both objects.[29] Such a procedure is known as the *Palatini formalism.* This procedure is analogous to the one employed in deriving the electromagnetic field equations from a variational principle where both the field $F^{\mu\nu}$ and the potential A_μ are variants of an action principle.

In taking the variation of S_G we will need the following results:

$$\delta\sqrt{-g} = \frac{\partial\sqrt{-g}}{\partial g_{\mu\nu}}\,\delta g_{\mu\nu} = \frac{1}{2}\sqrt{-g}\,\frac{1}{g}\frac{\partial g}{\partial g_{\mu\nu}}\,\delta g_{\mu\nu}$$

$$= \frac{1}{2}\sqrt{-g}\; g^{\mu\nu}\,\delta g_{\mu\nu} \tag{10-4.5}$$

and

$$\delta g^{\rho\sigma} = -g^{\rho\mu}g^{\sigma\nu}\delta g_{\mu\nu}, \tag{10-4.6}$$

this last result following from the fact that $g^{\rho\sigma}g_{\sigma\lambda} = \delta_\lambda{}^\rho$. Making use of these relations one has

$$\delta S_G = \frac{1}{2\kappa}\int\sqrt{-g}\{R^{\mu\nu} - \frac{1}{2}g^{\mu\nu}R + \Lambda g^{\mu\nu})\delta g_{\mu\nu} + g^{\mu\nu}\delta R_{\mu\nu}\}\, d^4x. \tag{10-4.7}$$

[29] A. Einstein, *S. B. preuss. Akad. Wiss.*, 414 (1925).

Since $R_{\mu\nu}$ involves only the $\Gamma^\rho_{\mu\nu}$, we have as the condition that $\delta S_G = 0$ for arbitrary variations of the $g_{\mu\nu}$ that vanish on the boundary of the region of integration, the equations

$$R^{\mu\nu} - \tfrac{1}{2}g^{\mu\nu}R + \Lambda g^{\mu\nu} = 0. \tag{10-4.8}$$

Thus the requirement that the equations for $g_{\mu\nu}$ follow from a variational principle fixes the value of the constant α in Eq. (10-4.1). The coefficient Λ of the so-called cosmological term, however, is undetermined.

Let us now look at the last term of the integrand in Eq. (10-4.7). Using the product rule for covariant differentiation, one finds that it can be written as

$$\sqrt{-g}\, g^{\mu\nu}\, \delta R_{\mu\nu} = \{\sqrt{-g}(g^{\mu\rho}\delta\Gamma^\nu_{\mu\nu} - g^{\mu\nu}\delta\Gamma^\rho_{\mu\nu})\}_{,\rho} + \{(\sqrt{-g}\, g^{\mu\nu})_{;\rho} - (\sqrt{-g}\, g^{\mu\sigma})_{;\sigma}\delta_\rho{}^\nu\}\delta\Gamma^\rho_{\mu\nu} \tag{10-4.9}$$

on the assumption that $\Gamma^\rho_{\mu\nu} = \Gamma^\rho_{\nu\mu}$. The first term on the right of Eq. (10-4.9), being a complete divergence, will, when integrated over in forming δS_G, lead to a surface integral with the help of Gauss' theorem. For variations of the $\delta\Gamma^\rho_{\mu\nu}$ that vanish on this surface this term itself will vanish and hence not contribute to the equations for $\Gamma^\rho_{\mu\nu}$. It then follows from Eq. (10-4.9) that the Euler-Lagrange equations obtained by varying $\Gamma^\rho_{\mu\nu}$ in S_G are

$$(\sqrt{-g}\, g^{\mu\nu})_{;\rho} - (\sqrt{-g}\, g^{\mu\sigma})_{;\sigma}\delta_\rho{}^\nu = 0. \tag{10-4.10}$$

These equations in turn imply that $g^{\mu\nu}{}_{;\rho} = 0$ and hence that $\Gamma^\rho_{\mu\nu} = \{{}^{\,\rho}_{\mu\nu}\}$.

Problem 10.2. Show that Eq. (10-4.10) implies $g^{\mu\nu}{}_{;\rho} = 0$.

Let us consider the interaction of matter or fields, or both, with the gravitational field. In keeping with our variational approach we will assume that the equations of motion for these objects also follow from a variational principle with an action S_M. We then require that the total action leading to the equations for $g_{\mu\nu}$ be the sum of S_G and S_M. However, if we treat the $\Gamma^\rho_{\mu\nu}$ as independent variants again, we should require that they do not appear in the expression for S_M, since otherwise the Eqs. (10-4.10) would contain additional terms and we would no longer have the simple relation $\Gamma^\rho_{\mu\nu} = \{{}^{\,\rho}_{\mu\nu}\}$. This requirement for the form of S_M is also related to our principle of minimal coupling; if S_M involves the $\Gamma^\rho_{\mu\nu}$, then the equations of motion for matter and fields will in general involve derivatives of the $\Gamma^\rho_{\mu\nu}$ at a point. If the action S_M involves only the $g_{\mu\nu}$, then the variation of S_M resulting from a variation of $g_{\mu\nu}$ must be of the form

$$\delta S_M \equiv -\tfrac{1}{2}\int \sqrt{-g}\, T^{\mu\nu}\delta g_{\mu\nu}\, d^4x, \tag{10-4.11}$$

where the quantity $T^{\mu\nu}$ is defined through this equation. It then follows that the equations of motion for $g_{\mu\nu}$ resulting from varying the total action $S_G + S_M$ are

$$R^{\mu\nu} - \tfrac{1}{2}g^{\mu\nu}R + \Lambda g^{\mu\nu} = \kappa T^{\mu\nu}. \tag{10-4.12}$$

Problem 10.3. Derive expressions for $T^{\mu\nu}$ for the following systems:

(a) A system of neutral particles of rest-mass m_0 with

$$S_M = -m_0 \int d^4x \sum_i \int d\lambda_i\, \delta^4(x - z_i)\sqrt{g_{\mu\nu}(x)\dot{z}_i^{\ \mu}\dot{z}_i^{\ \nu}}. \tag{10-4.13}$$

(b) The electromagnetic field with

$$S_M = -\frac{1}{8\pi}\int d^4x\sqrt{-g}\, g^{\mu\rho}g^{\nu\sigma}\left\{\frac{1}{2}F_{\mu\nu}F_{\rho\sigma} - F_{\mu\nu}(A_{\sigma,\rho} - A_{\rho,\sigma})\right\}. \tag{10-4.14}$$

(c) A scalar field ϕ with

$$S_M = \tfrac{1}{2}\int d^4x\sqrt{-g}\{g^{\mu\nu}\phi_{,\mu}\phi_{,\nu} - m^2\phi^2\}. \tag{10-4.15}$$

In his original formulation of the field equations, Einstein did not include the cosmological term $\Lambda g^{\mu\nu}$. Later, however, he was led to consider such a term based partly on cosmological considerations and partly on the basis of arguments having to do with Mach's principle. He pointed out that in the case $T^{\mu\nu} = 0$, the Eqs. (10-4.12), with $\Lambda = 0$, possess the solution $g_{\mu\nu} = \eta_{\mu\nu}$. A small body moving in such a field would possess all the inertia it is observed to have, even though from the Machian point of view there is no other matter present to produce this inertia. Einstein emphasized[30] that the existence of such solutions is directly related to the boundary conditions employed in connection with the field equations.

Any system of local differential equations will in general admit a large number of physically inequivalent solutions. To decide which of these solutions are actually realized in nature, it is customary to supplement the equations with boundary conditions. In the gravitational case one might require, for instance, that asymptotically $R_{\mu\nu\rho\sigma} = 0$ as one approached spatial infinity. But these are just the boundary conditions that lead to the "anti-Machian" solution $g_{\mu\nu} = \eta_{\mu\nu}$ in the case $T^{\mu\nu} = 0$. Of course it is possible that one might have to give different boundary conditions for different $T^{\mu\nu}$, so that in the case of the empty space one could require $g_{\mu\nu} = 0$ asymptotically, thereby ruling out the solution $g_{\mu\nu} = \eta_{\mu\nu}$. However, Einstein rejected this formulation of the Mach principle as being ad hoc and artificial. Instead he suggested that one

[30] A. Einstein, *S. B. preuss. Akad. Wiss.*, 142 (1917).

should not give boundary conditions to supplement the field equations but rather should require that all solutions lead to geometries that are spatially closed. In this case the overall topology of space would be that of a cylinder or some other such closed structure rather than that of a plane.

Einstein went on to apply his ideas to the case of a universe uniformly and isotropically filled with matter. Believing at that time that this matter was statically distributed (this was before the discovery of the galactic red shifts), he was forced to include the cosmological term in order to find a static, closed solution to the field equations for $g_{\mu\nu}$. The inclusion of the cosmological term also ruled out the flat-space solution, even in the case of an infinite space. However, it was soon discovered that the matter in the universe was not at rest but rather the galaxies appeared to be receding from each other. At the same time, nonstatic bounded solutions of the field equations, without the cosmological term corresponding to an expanding distribution of matter, were found. Also, solutions of the equations with the cosmological term and with $T^{\mu\nu} = 0$ were found by de Sitter.[31] These discoveries led Einstein to reject the cosmological term, and for the most part it is not included in modern discussions of the field equations. We will, therefore, when referring to the Einstein equations, mean the Eqs. (10-4.12) with $\Lambda = 0$.

The question of boundary conditions is still to a large extent an open one. As we mentioned in Section 10-1, Wheeler and Gursey independently have again suggested that the boundary conditions employed should depend upon $T^{\mu\nu}$ so as to get a theory that is more in accord with Mach's principle. However, these considerations seem to rule out the Schwarzschild solution of the field equations for a point mass, as does the requirement of spatially closed solutions. But the Schwarzschild solution is the only solution, to date, of the Einstein field equations that has any experimental support. If we are willing to give up Mach's principle, and there seems to be little reason at present to maintain it, then perhaps asymptotically flat boundary conditions could be assumed.

10-5. Stationary and Static Gravitational Fields

In a number of the discussions to follow we will have to deal with a special subclass of kinematically possible gravitational fields, the so-called *stationary* fields. A gravitational field $g_{\mu\nu}$ is called "stationary" if it possesses an everywhere timelike Killing vector τ^μ. Such a vector satisfies Eq. (4-4.3), which can be rewritten as

$$\tau_{\mu;\nu} + \tau_{\nu;\mu}, \qquad \tau_\mu = g_{\mu\nu}\tau^\nu, \tag{10-5.1}$$

[31] W. de Sitter, *Proc. Acad. Sci. Amst.*, **19**, 1217 (1917); **20**, 229 (1917).

together with the condition

$$\tau^2 = \tau_\mu \tau^\mu > 0. \tag{10-5.2}$$

If such a vector exists, one can always find a mapping such that in a finite region of space-time, $\tau^\mu = \{1, \mathbf{0}\}$. In this case Eqs. (10-5.1, 2) reduce to

$$g_{\mu\nu,0} = 0 \quad \text{and} \quad g_{00} > 0.$$

Given a timelike vector field τ^μ (not necessarily a Killing field) defined on the space-time manifold, it is possible to construct a three-parameter family of curves whose tangents are equal to τ^μ. If the points on such a curve are characterized by

$$x^\mu = \xi^\mu(\lambda, y^i), \tag{10-5.3}$$

where λ is a parameter along the curve and the y^i, $i = 1, 2, 3$, are the parameters that characterize a given curve, then

$$\frac{\partial \xi^\mu}{\partial \lambda} = \tau^\mu(\xi(\lambda, y^i)) \tag{10-5.4}$$

are the defining equations of these curves. When $\tau^\mu = \{1, \mathbf{0}\}$ these curves are seen to be just the curves $x^r = \text{const}$ and the parameter λ is equal to x^0. In this case the parameters y^i of a particular curve can be taken to be equal numerically to the "spatial" coordinates x^r of a point lying on that curve. Thus a timelike vector field furnishes us with a method of defining a space V_3 similar to the absolute space of Newtonian mechanics. A point of this space consists of an equivalence class of points of the space-time manifold, two points being considered as equivalent for the purpose of constructing this space if they lie on the same curve (10-5.3). The parameters y^i can then be taken as the coordinates of a "point" in this space. In effect we have projected the four-dimensional space-time manifold onto a three-dimensional spatial manifold. Under this projection a trajectory in space-time gets projected onto a curve in V_3. This curve will be referred to as the *orbit* associated with the trajectory.

We can decompose an arbitrary displacement dx^μ at a point of space-time into a component parallel to τ^μ and a component normal to τ^μ, according to

$$dx^\mu = d\alpha\tau^\mu + d\beta^\mu, \tag{10-5.5}$$

where $\tau_\mu \, d\beta^\mu = 0$. It follows that

$$d\alpha = \frac{\tau_\mu \, dx^\mu}{\tau^2} \tag{10-5.6}$$

and

$$d\beta^\mu = dx^\mu - \frac{\tau^\mu \tau_\nu \, dx^\nu}{\tau^2}$$

$$= \mathfrak{P}_\nu{}^\mu \, dx^\nu, \tag{10-5.7}$$

where

$$\mathfrak{P}_\nu{}^\mu \equiv \delta_\nu{}^\mu - \frac{\tau^\mu \tau^\nu}{\tau^2}. \tag{10-5.8}$$

The tensor $\mathfrak{P}_\nu{}^\mu$ is called the *projection operator* onto the hyperplane normal to τ^μ. We see in particular that $\mathfrak{P}_\nu{}^\mu \tau^\nu = 0$.

Let us now define the "length" dl of the projection of a displacement dx^μ in the plane normal to τ^μ. From Eq. (10-5.7) we have

$$dl^2 \equiv -g_{\mu\nu} \mathfrak{P}_\rho{}^\mu \mathfrak{P}_\sigma{}^\nu \, dx^\rho \, dx^\sigma$$

$$= h_{\mu\nu} \, dx^\mu \, dx^\nu, \tag{10-5.9}$$

where

$$h_{\mu\nu} = -\mathfrak{P}_\mu{}^\rho \mathfrak{P}_\nu{}^\sigma g_{\rho\sigma} = -g_{\mu\nu} + \frac{\tau_\mu \tau_\nu}{\tau^2}. \tag{10-5.10}$$

The tensor $h_{\mu\nu}$ is thus the projection of $g_{\mu\nu}$ onto the plane normal to τ^μ and plays the role of a metric in this plane. When $\tau^\mu = \{1, \mathbf{0}\}$ we see that $h_{\mu\nu}$ has the form

$$h_{00} = h_{0r} = 0,$$

$$h_{rs} = -g_{rs} + \frac{g_{0r} g_{0s}}{g_{00}}. \tag{10-5.11}$$

In a like manner we can define the "temporal" difference dt of the component of the displacement dx^μ parallel to τ^μ. From Eq. (10-5.6) we have

$$dt^2 \equiv g_{\mu\nu} (\tau^\mu \, d\alpha)(\tau^\nu \, d\alpha)$$

$$= p_{\mu\nu} \, dx^\mu \, dx^\nu, \tag{10-5.12}$$

where

$$p_{\mu\nu} = \frac{\tau_\mu \tau_\nu}{\tau^2}. \tag{10-5.13}$$

When $\tau^\mu = \{1, \mathbf{0}\}$ we have

$$dt^2 = \frac{g_{0\mu} g_{0\nu}}{g_{00}} \, dx^\mu \, dx^\nu.$$

If τ^μ is not a Killing vector, dl^2 will in general vary along the curves (10-5.3) and hence there will not be a unique distance between the points of the three-dimensional space V_3 associated with these curves. If τ^μ is a Killing vector however, we can define a metric on this space by making use of the expression (10-5.9) for dl^2.

A displacement of the form $dx^\mu = \xi^\mu{}_{,i}\, d\lambda^i$ is orthogonal to τ^μ and, using Eq. (10-5.9), has a length dl^2, given by

$$\begin{aligned} dl^2 &= h_{\mu\nu}\xi^\mu{}_{,i}\xi^\nu{}_{,j}\, d\lambda^i\, d\lambda^i \\ &= \gamma_{ij}\, d\lambda^i\, d\lambda^j. \end{aligned} \tag{10-5.14}$$

The "spatial metric" $\gamma_{ij} = h_{\mu\nu}\xi^\mu{}_{,i}\xi^\nu{}_{,j}$ must, of course, be independent of the parameter λ if it is to be a unique metric on the projective space V_3. To show that this is in fact the case, let us map so that $\tau^\mu = \{1, \mathbf{0}\}$. Then

$$\gamma_{ij} = \delta_i{}^r\delta_j{}^s h_{rs},$$

where h_{rs} is given by Eq. (10-5.11) and $\delta_i{}^r = 1$ if $r = i$, and zero otherwise. But h_{rs} is independent of x^0 and so γ_{ij} is independent of λ, which coincides with x^0 in this case. Since γ_{ij} and λ are clearly scalars under space-time mappings, γ_{ij} must be independent of λ in general.

While we can define a hyperplane normal to τ^μ at each point of the manifold where τ^μ is defined, we cannot in general form a family of hypersurfaces with the τ^μ as normals. When we can, we say that τ^μ is *hypersurface orthogonal*. If these surfaces are characterized by $\phi(x) = \phi_0$, τ^μ must be of the form (see Section 3-6)

$$\tau_\mu = \zeta(x)\phi_{,\mu} \tag{10-5.15}$$

that is, τ_μ must be proportional to the gradient of a scalar. To see that τ^μ is not in general hypersurface orthogonal, let us map so that $\tau^\mu = \{1, \mathbf{0}\}$. Then $\tau_\mu = g_{\mu 0}$ so that Eq. (10-5.15) reduces to

$$g_{\mu 0} = \zeta\phi_{,\mu},$$

which is in general not the case.

A special subclass of stationary gravitational fields are those for which the associated timelike Killing vector field τ^μ is hypersurface orthogonal. From Eq. (10-5.15) it follows that for these fields,

$$(\tau_\mu/\zeta)_{,\nu} - (\tau_\nu/\zeta)_{,\mu} = 0.$$

Therefore the projection of $\tau_{\mu,\nu} - \tau_{\nu,\mu}$ onto the plane normal to τ^μ must vanish in general. Thus, in addition to satisfying Eqs. (10-5.1) and (10-5.2), τ^μ must also satisfy

$$\mathfrak{P}_\rho{}^\mu\phi_\sigma{}^\nu(\tau_{\mu,\nu} - \tau_{\nu,\mu}) = 0. \tag{10-5.16}$$

Gravitational fields that possess a hypersurface orthogonal timelike Killing vector are called *static*.

Problem 10.4. Show that one can always find a mapping such that in a simply-connected region of space-time, $g_{0r} = 0$ for a static field.

When a timelike vector field is defined on the space-time manifold, it is meaningful to project the space-time manifold onto a three-dimensional "space." If in addition the vector field is hypersurface orthogonal we can also project onto a one-dimensional "time" manifold. The points of time are again equivalence classes of points of the space-time manifold; in this case the points lying on a given hypersurface. We can characterize the hypersurfaces by the value of the parameter λ along any one of the curves associated with the vector field. The parameters on the other curves can then be adjusted so that, at the point of intersection of any one of the curves with a given hypersurface, the parameters all have the same common value. Furthermore, such a parameterization is consistent with the original parameterization for which Eq. (10-5.4) is valid. (To see this, map so that $\tau^\mu = \{1, \mathbf{0}\}$). The hypersurfaces are then characterized by the equations $x^0 = \text{const}$. However, in this case Eq. (10-5.4) is satisfied with $\lambda = x^0$.

10-6. Interaction of Physical Systems with the Gravitational Field

In the problems immediately following Eq. (10-4.12) we gave a number of expressions for S_M for different kinds of systems in interaction with the gravitational field $g_{\mu\nu}$. They were obtained from their counterparts in the special theory by the device of replacing everywhere the Minkowski matrix by the gravitational field. Our justification for doing so rests on the principle of minimal coupling. According to this principle the dynamical laws, and consequently the actions leading to these laws, should be reducible at a space-time point to their special relativistic form by means of a suitable mapping. If we did not invoke this principle but only the principle of general invariance, then there would be nothing to rule out, for instance, a term proportional to $F^{\mu\nu}F^{\rho\sigma}R_{\mu\nu\rho\sigma}$ in the expression for S_M for the electromagnetic field. In this section we will discuss some of the consequences and difficulties of coupling various systems to the gravitational field when we make use of the principles of general invariance and minimal coupling.

(a) Motion of Particles in a Gravitational Field

The motion of an otherwise free particle in a gravitational field will be governed by the geodesic equation

$$\ddot{z}^\mu + \begin{Bmatrix} \mu \\ \rho\sigma \end{Bmatrix} \dot{z}^\rho \dot{z}^\sigma = 0 \tag{10-6.1}$$

with the supplementary condition

$$g_{\mu\nu}(z)\dot{z}^\mu \dot{z}^\nu = 1. \tag{10-6.2}$$

The gravitational field appearing in these equations is the *total* gravitational field, which is influenced by the presence of the particle both directly through the contribution of its own field and through its effect on whatever other sources of gravitational fields are present. Because of the intrinsic nonlinearity of the Einstein equations, these two contributions are inextricably intertwined in the total field so that there is no simple way that we can eliminate self-interactions as we did in the case of the electromagnetic field. Only for the limiting case in which we can neglect the influence of the particle on the other sources can we subtract off this self-interaction and take Eq. (10-6.1) and (10-6.2) to describe the motion of a free particle in a gravitational field.

If the gravitational field $g_{\mu\nu}$ possesses a symmetry, that is, if there exists a Killing vector k^μ such that

$$k_{\mu;\nu} + k_{\nu;\mu} = 0, \tag{10-6.3}$$

where $k_\mu = g_{\mu\nu}k^\nu$, then it is easy to show that $\dot{z}^\mu k_\mu$ is a constant of the motion along the trajectory of the particle. Thus

$$\begin{aligned} \frac{d}{d\tau}(\dot{z}^\mu k^\mu) &= \ddot{z}^\mu k_\mu + \dot{z}^\mu k_{\mu,\nu}\dot{z}^\nu \\ &= 0 \end{aligned}$$

as a consequence of Eqs. (10-6.1) and (10-6.3). If $g_{\mu\nu}$ is stationary then the quantity $\mathscr{E} \equiv m\tau_\mu \dot{z}^\mu$, where τ_μ is the timelike Killing vector associated with this field, can be taken to represent the energy of the particle in analogy with the definition of this quantity in special relativity.

We can re-express $\mathscr{E} = \tau_\mu \dot{z}^\mu$, in a way that brings out its analogy, with the expression (7-14.6) for the energy of a charged particle moving in a constant electromagnetic field. To this end we introduce the "normal" velocity of the particle as the change in the displacement normal to τ^μ relative to its displacement parallel to τ^μ. By using Eqs. (10-5.5), (10-5.6), and (10-5.7) we find, for the components u^μ of this normal velocity,

$$u^\mu \equiv \frac{\mathfrak{P}_\nu{}^\mu \, dz^\nu}{\tau_\rho \, dz^\rho/\tau^2} \tag{10-6.4}$$

where dz^μ is a displacement along the trajectory of the particle.

When $\tau^\rho = \{1, \mathbf{0}\}$ the expression for u^μ reduces to

$$\begin{aligned} u^\mu &= \frac{1}{\sqrt{h}(dz^0 + g_r \, dz^r)}\{-g_s \, dz^s, dz^r\} \\ &= \{-\mathbf{g}\cdot\mathbf{u}, \mathbf{u}\}, \end{aligned} \tag{10-6.5}$$

where

$$h = g_{00} \quad \text{and} \quad g_r = \frac{g_{0r}}{g_{00}}.$$

It then follows from Eq. (10-6.4) that

$$\dot{z}^\mu = \frac{1}{\sqrt{1-\mathbf{u}^2}}(u^\mu + \hat{\tau}^\mu), \qquad \hat{\tau}^\mu = \frac{\tau^\mu}{\tau^2}, \tag{10-6.6}$$

where

$$\begin{aligned}\mathbf{u}^2 &= h_{\mu\nu}u^\mu u^\nu \\ &= \left(\frac{dl}{dt}\right)^2.\end{aligned}$$

For the case when $\tau^\mu = \{1, \mathbf{0}\}$,

$$\dot{z}^\mu = \frac{1}{\sqrt{1-\mathbf{u}^2}}\left\{\frac{1}{\sqrt{h}} - \mathbf{g}\cdot\mathbf{u}, \mathbf{u}\right\}. \tag{10-6.7}$$

By comparing this relation with its special relativistic counterpart (6-8.4), we see that **u** is the analogue of the particle three-velocity. Since the above results depend only on the fact that τ^μ is timelike, we can always define a normal velocity u^r in an arbitrary gravitational field relative to a timelike vector by means of Eq. (10-6.4). Of course, $\tau_\mu \dot{z}^\mu$ will no longer be a constant of the motion along a geodesic in this case.

To obtain now the desired expression for $\mathscr{E}$, we make use of the fact that $\tau_\mu u^\mu = 0$. Therefore

$$\mathscr{E} = m\tau_\mu \dot{z}^\mu = m\frac{\sqrt{g_{\mu\nu}\tau^\mu\tau^\nu}}{\sqrt{1-\mathbf{u}^2}}, \tag{10-6.8}$$

which reduces when $\tau^\mu = \{1, \mathbf{0}\}$ to

$$\mathscr{E} = \frac{m\sqrt{h}}{\sqrt{1-\mathbf{u}^2}}. \tag{10-6.9}$$

In the next section we will see that in the limit of a weak gravitational field,

$$g_{00} \simeq 1 + 2\phi$$

where ϕ is the Newtonian potential for this field. Hence, in this limit,

$$\mathscr{E} \simeq \frac{m}{\sqrt{1-u^2}} + \frac{m\phi}{\sqrt{1-u^2}}, \tag{10-6.10}$$

which is the gravitational analogue of Eq. (7-14.6) in the limit of weak fields. Finally, in the low-velocity limit, that is, the limit $c \to \infty$ with m, u^2 and ϕ replaced by mc^2, u^2/c^2, and ϕ/c^2, we obtain

$$\mathscr{E} \simeq mc^2 + \tfrac{1}{2}mu^2 + m\phi,$$

which, aside from the rest-energy term mc^2, is the usual Newtonian expression for the energy of a particle moving in a gravitational field.

We might ask finally if it is possible to construct a theory of the gravitational interaction which is analogous to the action-at-a-distance formulation of the electromagnetic interaction discussed in Section 7-9 and which has the MMG as a symmetry group. It would appear that the answer to this question is in the negative.[32] In order for the MMG to be a symmetry group it must be a covariance group, so that the elements that describe the kpt of the theory must form the basis of a *faithful* realization of this group. However, as we pointed out in Section 1-8, fields appear to be necessary ingredients of such a basis, and fields are not allowed in an action-at-a-distance formalism. Furthermore, if a theory involving only particle trajectories described a system whose symmetry group was the MMG, it would always be possible to find a mapping such that these trajectories were parallel straight lines. (Of course, if we restrict ourselves to Poincaré invariance, such a formalism for the gravitational interaction is possible, as discussed in Section 7-21).

(b) The Interaction of the Electromagnetic and Gravitational Fields

The dynamical laws for an electromagnetic field in interaction with a gravitational field can be obtained by varying $F_{\mu\nu}$ and A_μ in the action (10-4.14). Upon varying $F_{\mu\nu}$ we have the usual relation between the field and the corresponding potential:

$$F_{\mu\nu} = A_{\nu,\mu} - A_{\mu,\nu}, \tag{10-6.11}$$

while varying A_μ gives

$$(\sqrt{-g}F^{\mu\nu})_{,\nu} = 0. \tag{10-6.12}$$

If in addition there is a source term $-\int A_\mu(x) j^\mu(x)\, d^4x$ in the action, Eq. (10-6.12) becomes

$$(\sqrt{-g}F^{\mu\nu})_{,\nu} = -4\pi j^\mu, \tag{10-6.13}$$

from which it follows immediately that

$$j^\mu{}_{,\mu} = 0. \tag{10-6.14}$$

[32] For an attempt at an action-at-a-distance theory that is invariant under the MMG, see F. Hoyle and J. V. Narlikar, *Proc. Roy. Soc.* (London), **A282**, 190 (1964). This theory is, however, not free of fields and so cannot be said to be a pure action-at-a-distance theory. See also the comments of S. Deser and F. A. E. Pirani, *Proc. Roy. Soc.* (London), **A288**, 133 (1965).

This is a covariant equation, since j^μ is a contravector of weight $+1$, as follows either from Eq. (10-6.13) or the fact that $A_\mu j^\mu \, d^4x$ must be a scalar. If, in particular,

$$j^\mu(x) = \sum_i e_i \int d\lambda_i \, \delta^4(x - z_i)\dot{z}_i{}^\mu,$$

Eq. (10-6.1) gets modified to

$$\ddot{z}^\mu + \begin{Bmatrix} \mu \\ \rho\sigma \end{Bmatrix} \dot{z}^\rho \dot{z}^\sigma = eF_\nu{}^\mu(z)\dot{z}^\nu. \tag{10-6.15}$$

The problem of characterizing a light signal is considerably more difficult in general relativity than in special relativity, for several reasons. For one thing, all such signals contribute to the total stress-energy tensor which is the source of the gravitational field. Since Maxwell's equations in turn involve the gravitational field, there will be a complicated dynamical interaction between these two systems. Even if one could neglect the effect of the signal on the gravitational field, one could not in general find simple plane-wave solutions to the Maxwell equations in a given gravitational field for use in constructing a wave packet to represent such a signal. Only if the dimensions of the signal are small compared with the dimensions of regions over which the gravitational field can be considered constant is such a construction possible. In that case one can map so that in such a region, $g_{\mu\nu}$ takes on Minkowski values and its derivatives vanish. Then, since the Maxwell equations satisfy the principle of minimal coupling, they reduce to their usual flat-space form and so possess local plane-wave solutions. One can therefore proceed to construct a wave packet from these solutions as was done in Section 8-3, where such a packet was seen to travel along a null geodesic. Hence the same will be true in the present case, at least when the packet remains in a region where $g_{\mu\nu}$ is approximately constant. However, since such regions can be made to overlap, one can argue that a light signal of sufficiently small dimensions will propagate along a null geodesic throughout space-time. Its trajectory will therefore be determined by the equation

$$\ddot{z}^\mu + \begin{Bmatrix} \mu \\ \rho\sigma \end{Bmatrix} \dot{z}^\rho \dot{z}^\rho = 0 \tag{10-6.16}$$

for a suitable choice of the path parameter λ.

If a gravitational field is stationary we can reformulate the condition that the trajectory of a light signal will be a null geodesic in a way that is analogous to Fermat's principle of least time in classical optics. Consider the integral

$$S = \int_{\lambda_1}^{\lambda_2} g_{\mu\nu}\dot{z}^\mu \dot{z}^\nu \, d\lambda$$

evaluated along two neighboring trajectories $x^\mu = z^\mu(\lambda)$ and $x^\mu = z'^\mu(\lambda) = z^\mu(\lambda) + \delta z^\mu(\lambda)$. The difference in the values of this integral, $\delta S = S' - S$, along these two trajectories is given by

$$\delta S = \int_{\lambda_1}^{\lambda_2} \left\{ g_{\mu\nu,\rho} \dot{z}^\mu \dot{z}^\nu - \frac{d}{d\lambda}(g_{\mu\rho}\dot{z}^\mu) \right\} \delta z^\mu \, d\lambda + g_{\mu\nu}\dot{z}^\mu \delta z^\nu \Big|_{\lambda_1}^{\lambda_2} \tag{10-6.17}$$

after an integration by parts. If now $z^\mu(\lambda)$ is a null geodesic and $z'^\mu(\lambda)$ is a null trajectory (but not necessarily a geodesic), the value of S along both trajectories will be zero, so that $\delta S = 0$, and by Eq. (10-6.16) the integral on the right side of Eq. (10-6.17) vanishes. Therefore, for such trajectories,

$$g_{\mu\nu}\dot{z}^\mu \, \delta z^\nu \Big|_{\lambda_1}^{\lambda_2} = 0. \tag{10-6.18}$$

If now δz^μ is taken parallel to the timelike Killing vector τ^μ of the gravitational field,

$$\delta z^\mu = \delta \upsilon \tau^\mu,$$

and if we make use of the fact that $\tau_\mu \dot{z}^\mu$ is a constant along a null geodesic, we have

$$\delta \upsilon \Big|_{\lambda_1}^{\lambda_2} = \delta \int_{\lambda_1}^{\lambda_2} d\upsilon = 0 \tag{10-6.19}$$

along neighboring null trajectories, one of which is a null geodesic. When $\tau^\mu = \{1, \mathbf{0}\}$, $\upsilon = z^0$ and Eq. (10-6.19) becomes

$$\delta \int_{\lambda_1}^{\lambda_2} dz^0 = 0. \tag{10-6.20}$$

The integral of dz^0 along a path can be interpreted as the total time taken for a signal to go from λ_1 to λ_2 on the path. Equation (10-6.20) then tells us that of all null trajectories, a light signal takes the least time (or the most time) to travel from one point of space to another; this is Fermat's principle. Then, since along a light ray, $g_{\mu\nu} dz^\mu \, dz^\nu = 0$, Eq. (10-6.20) can be rewritten as

$$\delta \int_{\lambda_1}^{\lambda_2} \left\{ -g_{r\text{o}} \, dz^r + \frac{\sqrt{(g_{r\text{o}} g_{s\text{o}} - g_{rs}) dz^r \, dz^s}}{\sqrt{h}} \right\} = \delta \int_{\lambda_1}^{\lambda_2} \left\{ -g_{r\text{o}} \, dz^r + \frac{dl}{\sqrt{h}} \right\} = 0 \tag{10-6.21}$$

when $g_{\mu\nu}$ is static, $g_{r\text{o}} = 0$ and Eq. (10-6.21) reduces to

$$\delta \int_{\lambda_1}^{\lambda_2} \frac{dl}{\sqrt{h}} = 0. \tag{10-6.22}$$

In a flat-space-time, $\sqrt{h} = 1$ and Eq. (10-6.22) reduces to the condition that the projection of a light ray in space is a straight line. In an arbitrary static field, when $\tau^\mu = \{1, \mathbf{0}\}$, $\sqrt{h}$ acts like a variable index of refraction. It is this property of the gravitational field that leads to a bending of light near the edge of the sun (see Section 12-2).

(c) The Dirac Equation in a Gravitational Field

In Section 8-9 we discussed the Dirac equation within the framework of special relativity and showed that the Dirac wave function transformed like the basis of a spinor representation of the Poincaré group. However, there are no finite dimensional spinor type, that is, double-valued, representations of the MMG of general relativity. It is possible in general relativity to write an equation that in many respects resembles the Dirac equation but which nevertheless differs fundamentally from the Dirac equation in special relativity. To understand the nature of this difference it will be instructive to discuss the invariance of the Dirac equation in special relativity by an alternative method to that employed in Section 8-9.

Instead of considering the γ^μ operators appearing in Eq. (8-9.1) as fixed quantities, which do not transform under a Poincaré mapping, we can consider them as variables on the same footing with ψ, to be determined by additional laws. Under a Poincaré mapping, ψ is taken to transform like a scalar and γ^μ like a contravector. Then, if ψ and γ^μ satisfy the Dirac equation, their transforms under a Poincaré mapping will also satisfy it. In order to complete the theory we now require that γ^μ satisfy the dynamical laws.

$$\gamma^\mu \gamma^\nu + \gamma^\nu \gamma^\mu = 2\eta^{\mu\nu} \quad \text{and} \quad \gamma^\mu{}_{,\nu} = 0 \tag{10-6.23}$$

so as to obtain a theory that is equivalent to the original Dirac equation (8-9.1). We see that this equation is also Poincaré covariant. However, as it stands, γ^μ is clearly an absolute object whose only symmetries are those space-time translations and Lorentz mappings that leave it invariant. Since the mappings that leave invariant a given vector are only a subgroup of the full Lorentz group, the symmetry group of the theory we have constructed is not the Poincaré group. To remedy this difficulty we make use of the fact that the covariance group of both Eqs. (8-9.1) and (10-6.23) is in fact larger than the Poincaré group. Let us assume that ψ and γ^μ transform under an internal transformation according to

$$\bar{\delta}\psi = i\varepsilon^A S_A \psi, \tag{10-6.24a}$$

$$\bar{\delta}\psi^+ = -i\varepsilon^A \psi^+ S_A, \tag{10-6.24b}$$

and

$$\bar{\delta}\gamma^\mu = i\varepsilon^A\{S_A\gamma^\mu - \gamma^\mu S_A\} \tag{10-6.25}$$

where S_A are hermitian operators to be determined and ε^A are the infinitesimal parameters of this internal group. Then one can show directly that this group is a covariance group of our theory. Therefore the symmetry group of the system being described is determined by

$$\bar{\delta}\gamma^\mu = i\varepsilon^A\{S_A\gamma^\mu - \gamma^\mu S_A\} + \varepsilon_\nu{}^\mu\gamma^\nu = 0, \tag{10-6.26}$$

where the $\varepsilon_\nu{}^\mu$ are parameters of the Poincaré group. This equation can be satisfied for arbitrary $\varepsilon_\nu{}^\mu$ if we take ε^A to correspond to $\varepsilon_{\mu\nu} = \eta_{\mu\rho}\varepsilon_\nu{}^\rho$, and S_A to be $(i/4)(\gamma^\mu\gamma^\nu - \gamma^\nu\gamma^\mu)$. Thus, by a rather roundabout procedure, we are able to recover the result of Section 8-9 that the Poincaré group is a symmetry of the system described by the Dirac equation. It does show how the intrinsic spin of the electron arising from its symmetry under the internal group gets coupled to its orbital angular momentum, which is associated with space-time Poincaré symmetry, and how it is that it is the sum of the two which is conserved during the electron's motion. Furthermore it is this procedure that must be employed if we want to construct a Dirac equation in general relativity. However, in doing so, we will destroy the relation between intrinsic spin and orbital angular momentum alluded to above.

To write a Dirac type of equation within the framework of general relativity we can proceed by generalizing Eq. (10-6.23) to

$$\gamma^\mu\gamma^\nu + \gamma^\nu\gamma^\mu = 2g^{\mu\nu} \tag{10-6.27}$$

and drop the requirement that $\gamma^\mu{}_{,\nu} = 0$. We can maintain the internal transformation group with infinitesimals given by Eqs. (10-6.24) and (10-6.25), but must take into account that the generators $S_A = (i/4)(\gamma^\mu\gamma^\nu - \gamma^\nu\gamma^\mu)$ are no longer constants. To do so, we can employ a procedure similar to that used in Section 2-2, where we dealt with a situation in which the parameters of the internal group were not constants. We define a covariant derivative of ψ by

$$\psi_{;\mu} = \psi_{,\mu} + \Gamma_\mu\psi$$

and require that it transform like ψ under an internal transformation so that, from Eq. (10-6.24),

$$\bar{\delta}\psi_{;\mu} = i\varepsilon^A S_A\psi_{;\mu}.$$

But

$$\bar{\delta}\psi_{;\mu} = i\varepsilon^A S_{A,\mu}\psi + i\varepsilon^A S_A\psi_{,\mu} + \bar{\delta}\Gamma_\mu\psi + i\varepsilon^A\Gamma_\mu S_A\psi.$$

Equating these two expressions for ψ leads to an expression for $\bar{\delta}\Gamma\mu$:

$$\bar{\delta}\Gamma_\mu = i\varepsilon^A\{S_A\Gamma_\mu - \Gamma_\mu S_A - S_{A,\mu}\} \tag{10-6.28}$$

The quantities Γ_μ were first introduced by Fock and Ivanenko[33] and are known as *Fock-Ivanenko coefficients*. They are given by

$$\Gamma_\mu = \frac{1}{8}\left[\gamma^\nu\gamma_{\mu,\nu} - \gamma_{\mu,\nu}\gamma^\nu - \left\{ {\rho \atop \mu\nu} \right\}(\gamma^\nu\gamma_\rho - \gamma_\rho\gamma^\nu)\right] \tag{10-6.29}$$

and satisfy Eq. (10-6.28) by virtue of the transformation properties (10-6.25) of the γ^μ. With the help of the Fock-Ivanenko coefficients we can now write a Dirac type of equation that has both the MMG and the internal group with infinitesimals (10-6.24) and (10-6.25) as symmetry groups:

$$i\gamma^\mu(\psi_{,\mu} + \Gamma_\mu\psi) + m\psi = 0. \tag{10-6.30}$$

However, we emphasize again that Eq. (10-6.30) together with Eq. (10-6.27) is only formally the general relativistic analogue of the Dirac equation of special relativity, since there is no longer a relation between the internal group and the group of space-time mappings.

10-7. Newtonian Limit of the Einstein Equations

Having accepted the Einstein equations we must now try to associate $g_{\mu\nu}$ with some object found in nature. We will be led to associate $g_{\mu\nu}$ with the gravitational field by considering the Newtonian limit of these equations and Eqs. (10-6.1). It is to be expected that this limit will in fact involve two approximations. We must take $g_{\mu\nu}$ as differing from being flat by a small amount, and we must take the velocities of all particles to be much smaller than the speed of light. Only if this double approximation is made will we be able to obtain the Newtonian gravitational theory from the Einstein theory.

Let us begin our construction of the Newtonian limit by considering the motion of a small mass moving in a given field $g_{\mu\nu}$. If we can neglect the effect of this mass on the sources of $g_{\mu\nu}$ and on $g_{\mu\nu}$ itself, then we can use Eq. (10-6.1) as the dynamical laws governing the motion of this mass. However, instead of using the proper time, we will take $t = z^0/c$ as a path parameter. The small velocity limit of the geodesic equations is then formed by letting $c \to \infty$. In addition we will make the weak field approximation. In this approximation we assume that there exists a mapping such that $g_{\mu\nu} = \eta_{\mu\nu} + h_{\mu\nu}$ where $\det h_{\mu\nu} \ll 1$. With these assumptions the geodesic equations take the form

$$\frac{d^2 z^r}{dt^2} = -c^2\left\{ {r \atop 00} \right\}. \tag{10-7.1}$$

[33] V. Fock and D. Ivanenko, *Compt. Rend.*, **188**, 1470 (1929); and V. Fock, *Z. Physik*, **57**, 261 (1929).

Since $\{{}^{\,r}_{00}\} = \frac{1}{2}g^{rs}(2g_{0s,0} - g_{00,s}) + \frac{1}{2}g^{r0}g_{00,0}$ for the case of static fields we have $\{{}^{\,r}_{00}\} \simeq \frac{1}{2}\gamma_{00,r}$ to first order in the $h_{\mu\nu}$. As a consequence, Eq. (10-7.1) can be written in the form

$$\frac{d^2\mathbf{z}}{dt} \simeq -\nabla\phi, \tag{10-7.2}$$

where we have set

$$\phi = \frac{c^2}{2} h_{00} \tag{10-7.3}$$

Equation (10-7.2) is indeed the equation of motion in Newtonian theory of a particle moving in a gravitational field derived from the potential ϕ.

To complete the discussion of the Newtonian limit we must now show that ϕ, given by Eq. (10-7.3), satisfies the Poisson equation (5-3.6) in this limit. Since the Newtonian theory assumes that only massive bodies are sources of the gravitational field, it suffices to take for $T_{\mu\nu}$ in equations (10-4.2) the Minkowski tensor of Section 8-5. In the limit of small velocities the only non-vanishing component of this tensor is

$$\Theta^{00}(x) = c^2 \sum_i m_{0i}\delta^3(\mathbf{x} - \mathbf{z}_i) = c^2\rho(\mathbf{x}). \tag{10-7.4}$$

Now let us multiply Eq. (10-4.2) by $g_{\mu\nu}$ to obtain the result

$$R = -\kappa g_{\mu\nu}T^{\mu\nu} \simeq -\kappa\Theta^{00} \tag{10-7.5}$$

As a consequence, the only component of Eq. (10-4.2) that involves the sources can be written in the form

$$R^{00} = \tfrac{1}{2}\kappa c^2\rho(\mathbf{x}). \tag{10-7.6}$$

In computing R^{00} we retain only terms of first order in the deviation of $g_{\mu\nu}$ from its flat value, and we also neglect all derivatives with respect to x^0. As a consequence

$$R^{00} \simeq -\left\{ {r \atop 00} \right\}_{,r} = -\frac{1}{c^2}\nabla^2\phi \tag{10-7.7}$$

so that Eq. (10-7.6) becomes

$$\nabla^2\phi = -\tfrac{1}{2}\kappa c^4\rho(\mathbf{x}). \tag{10-7.8}$$

Upon comparison with Eq. (5-3.6) we see that the Einstein equations reduce to the Newtonian limit if we take

$$\kappa = -\frac{8\pi G}{c^4}. \tag{10-7.9}$$

Thus, for a point mass, $\rho(\mathbf{x}) = m\,\delta^3(\mathbf{x})$,

$$\phi = \frac{\kappa c^4}{8\pi}\frac{m}{r},$$

and hence

$$g_{00} \simeq 1 + \frac{\kappa c^2 m}{4\pi r} = 1 - \frac{2Gm}{c^2 r} \tag{10-7.10}$$

Thus, on the basis of the above analysis we are led to associate $g_{\mu\nu}$ with the gravitational field, just as we associate $F^{\mu\nu}$ with the electromagnetic field. Additional justification for this interpretation of $g_{\mu\nu}$ must await further development of the consequences of the gravitational field equations.

10-8. Structure of the Einstein Equations

The symmetry group of the Einstein equations is a gauge group. An element of the group is defined by the four arbitrary functions that define the manifold mapping to which the element corresponds. It therefore follows from the discussion of Section 4-6 that the ten equations (10-4.2) satisfy four differential identities, the Bianchi identities. To obtain the form of these identities we make use of Eq. (4-6.4). In the present case the index A corresponds to the index pair $(\mu\nu)$ and the index i corresponds to the index μ. Thus the $g_{\mu\nu}$ play the role of the y_A and $\sqrt{-g}\{R^{\mu\nu} - (1/2)g^{\mu\nu}R\} \equiv \sqrt{-g}G^{\mu\nu}$ corresponds to $\mathfrak{L}^A$. Corresponding to an infinitesimal mapping, $\bar{\delta} g_{\mu\nu}$ is given by

$$\bar{\delta} g_{\mu\nu} = -g_{\mu\rho}\xi^\rho{}_{,\nu} - g_{\rho\nu}\xi^\rho{}_{,\mu} - g_{\mu\nu,\rho}\xi^\rho,$$

which we write in the form

$$\bar{\delta} g_{\mu\nu} = d^\sigma_{\mu\nu\rho}\xi^\rho{}_{,\sigma} + c_{\mu\nu\rho}\xi^\rho,$$

where

$$d^\sigma_{\mu\nu\rho} = -g_{\mu\rho}\delta_\nu{}^\sigma - g_{\rho\nu}\,\delta_\nu{}^\sigma \quad \text{and} \quad c_{\mu\nu\rho} = -g_{\mu\nu,\rho}\,.$$

Substituting these results into Eq. (4-6.4) we obtain the identities

$$(\sqrt{-g}\;g_{\mu\rho}G^{\mu\nu})_{,\nu} + (\sqrt{-g}\;g_{\rho\nu}G^{\mu\nu})_{,\mu} - \sqrt{-g}\;g_{\mu\nu,\rho}G^{\mu\nu} \equiv 0. \tag{10-8.1}$$

By regrouping terms one can show that these identities are equivalent to the contracted Bianchi identities (see Eq. 3-4.9)

$$G^{\mu\nu}{}_{;\nu} \equiv 0. \tag{10-8.2}$$

As a consequence of the existence of the Bianchi identities, we can expect that the coefficient of the highest x^0 derivatives of $g_{\mu\nu}$ in the Einstein equations will be singular, and indeed such is the case. By examining the form of the Einstein tensor $G^{\mu\nu}$, one can ascertain the following facts:

1. The highest order x^0 derivative factors occurring in $G^{\mu\nu}$ are $g_{rs,00}$ and $g_{0\mu,0}$.
2. The highest order x^0 derivative factors occurring in $G^{0\mu}$ are $g_{rs,0}$.
3. The G^{rs} depend linearly on $g_{rs,00}$ and $g_{0\mu,0}$.

While these facts follow directly from the expressions for $G^{\mu\nu}$, it is not easy to obtain them therefrom because of the complexity of this tensor. For the purpose, it is easier to make use of an elegant result of Dirac.[34] He showed that one can construct a Lagrangian density $\mathfrak{L}'$, which leads to the usual Einstein equations and has the property that it contains no x^0 derivatives of $g_{0\mu}$. It follows immediately that the Euler-Lagrangian equations obtained by varying $g_{0\mu}$ in the action $S' = \int \mathfrak{L}'\, d^4x$, of the form[35]

$$\frac{\delta \mathfrak{L}'}{\delta g_{0\mu}} = \frac{\partial \mathfrak{L}'}{\partial g_{0\mu}} - \left(\frac{\partial \mathfrak{L}'}{\partial g_{0\mu,r}}\right)_{,r} + \left(\frac{\partial \mathfrak{L}'}{\partial g_{0\mu,rs}}\right)_{,rs} = 0,$$

do not contain $g_{0\mu,0}$ and $g_{rs,00}$ as factors and that only first order x^0 derivatives of g_{rs} occur. If we examine the expression for $\delta\mathfrak{L}'/\delta g_{rs}$ we see that the highest x^0 derivatives are contained in the terms

$$\left(\frac{\partial \mathfrak{L}'}{\partial g_{rs,0}}\right)_{,0} \quad \text{and} \quad \left(\frac{\partial \mathfrak{L}'}{\partial g_{rs,\mu 0}}\right)_{,\mu 0},$$

Since $\mathfrak{L}'$ has the further property that it is quadratic in first derivatives of $g_{\mu\nu}$ and linear in their second derivatives (it differs from $\sqrt{-g}R$ by a complete divergence) these terms depend of $g_{rs,00}$ and $g_{0\mu,0}$ linearly.

10-9. The Initial Value Problem for the Einstein Equations

The initial value problem for the Einstein equations consists of three parts. One must find a suitable initial value surface on which to specify initial value data. One must then find suitable initial value data consistent with the Einstein equations, and finally one must integrate these equations off the initial value surface.

Because of the hyperbolic nature of the Einstein equations the initial value surface must be spacelike.[36] Ordinarily such a surface can be easily constructed on a manifold from a knowledge of the geometry on the manifold.

[34] P. A. M. Dirac, *Proc. Roy. Soc.* (London), **A246**, 333 (1958).

[35] Although $\mathfrak{L}'$ does not involve x^0 derivatives of order higher than the first, it does involve second-order x^r derivatives and mixed second-order x^r, x^0 derivatives.

[36] In this section it will be convenient to consider the gravitational field $g_{\mu\nu}$ as a metric defined on the space-time. However, all our conclusions will, in fact, be independent of such an interpretation for $g_{\mu\nu}$.

However, in the present case, the geometry itself is a dynamical component of the theory; it is to be determined from the initial value data. It would appear that we have to know the solution of the Einstein equations in order to set up the conditions that lead to this solution. Fortunately this is only an apparent difficulty, as we will now demonstrate.

If our initial value surface is to be spacelike, then the distance between any two distinct points as measured along an arbitrary curve connecting these two points and lying wholly within the surface must be finite and negative. For this to be the case it suffices to require that

$$ds^2 = g_{\mu\nu}\, dx^\mu\, dx^2 < 0 \tag{10-9.1}$$

for any two neighboring points x^μ and $x^\mu + dx^\mu$ on the surface. Our requirement that the initial data surface be spacelike imposes certain restrictions on the initial value data; they must be such that the inequality of Eq. (10-9.1) is satisfied.

Starting with the bare space-time manifold, we can in some finite region take the equation of the initial surface to be $x^0 = \text{const}$. On this surface Eq. (10-9.1) reduces to

$$g_{rs}\, dx^r\, dx^s < 0 \tag{10-9.2}$$

for arbitrary dx^r. Necessary and sufficient conditions for the satisfaction of this inequality are the inequalities following from the theory of algebraic forms. They are

$$g_{11} < 0, \qquad \begin{vmatrix} g_{11} & g_{12} \\ g_{21} & g_{22} \end{vmatrix} > 0, \qquad \begin{vmatrix} g_{11} & g_{12} & g_{13} \\ g_{21} & g_{22} & g_{23} \\ g_{21} & g_{22} & g_{23} \end{vmatrix} < 0. \tag{10-9.3}$$

Consequently, in imposing initial conditions on a surface $x^0 = \text{const}$, we must be sure that the g_{rs} satisfy these inequalities.

Problem 10.5. What are the conditions on $g_{\mu\nu}$ when the initial surface is defined by the equation $f(x^\mu) = \text{const}$ in order that it be spacelike?

In addition to requiring that the inequality of Eq. (10-9.1) be satisfied by neighboring points on the initial data surface, we must also require that the normals to this surface be timelike. If the equation of this surface is $f(x^\mu) = \text{const.}$, then the components of the normal at a point are $n_\mu = f_{,\mu}$, and hence we must require that

$$g^{\mu\nu} n_\mu n_\nu > 0. \tag{10-9.4}$$

For the case when the surface is $x^0 = \text{const}$, $n_\mu = \delta_\mu{}^0$, and this condition reduces to

$$g^{00} > 0. \tag{10-9.5}$$

Now

$$g^{\mu\nu} = \frac{1}{g}\frac{\partial g}{\partial g_{\mu\nu}},$$

so that, in particular,

$$g^{00} = \frac{1}{g}\frac{\partial g}{\partial g_{00}}$$

Then, since

$$\frac{\partial g}{\partial g_{00}} = \begin{vmatrix} g_{11} & g_{12} & g_{13} \\ g_{21} & g_{22} & g_{23} \\ g_{31} & g_{32} & g_{33} \end{vmatrix}$$

it follows from Eq. (10-9.5) that $g_{\mu\nu}$ must satisfy, in addition to the inequalities of Eq. (10-9.3), the further inequality

$$g < 0. \tag{10-9.6}$$

This inequality is consistent with the requirement that the signature of $g_{\mu\nu}$ be -2. As a consequence we can always find a mapping such that $g_{\mu\nu} = \eta_{\mu\nu}$ at a point so that $g = -1$ at this point. If we now perform the inverse mapping by a succession of infinitesimal mappings, we can see that the transformed g will always negative-definite. For it to be positive it would have to pass through the value 0, which implies that the Jacobian of the mapping must vanish for this value. However, such mappings are forbidden. When $g_{\mu\nu}$ satisfies the inequalities (10-9.3) and (10-9.5) one sometimes calls x^0 a "timelike coordinate" and x^1, x^2, x^3 "spacelike coordinates."

In discussing the remainder of the initial value problem it is convenient to make use of a canonical formulation of the gravitational field equations, due to Arnowitt, Deser, and Misner.[37] One first introduces the notation

$$N \equiv (g^{00})^{-1/2}, \qquad N_r \equiv g_{0r} \tag{10-9.7}$$

and defines quantities π_{rs} by

$$\pi_{rs} \equiv \sqrt{-g}\left[\begin{Bmatrix} 0 \\ rs \end{Bmatrix} - g_{rs}\begin{Bmatrix} 0 \\ uv \end{Bmatrix} e^{uv}\right], \tag{10-9.8}$$

where e^{uv} is the reciprocal matrix to g_{rs} ($g_{rs}e^{su} = \delta_r{}^u$). It is related to $g^{\mu\nu}$ by

$$e^{rs} = g^{rs} - \frac{g^{0r}g^{0s}}{g^{00}}. \tag{10-9.9}$$

[37] R. Arnowitt, S. Deser, and C. W. Misner, "The Dynamics of General Relativity," in *Gravitation: An Introduction to Current Research*, L. Witten, ed. (Wiley, New York, 1962), pp. 227–265.

One also has the relations

$$g^{0s} = -g^{00}e^{rs}g_{0s} \tag{10-9.10}$$

and

$$g^{00} = \frac{e}{g}, \tag{10-9.11}$$

where $e = \det g_{rs}$. One can show that the π_{rs} defined above are directly related to the second fundamental form on the surface $x^0 = \text{const}$.

Problem 10.6. Derive Eqs. (10-8.9), (10-8.10), and (10-8.11) from the fact that $g_{\mu\nu}g^{\nu\rho} = \delta_\mu{}^\rho$.

Arnowitt, Deser, and Misner[38] have shown that a set of equations equivalent to the Einstein field equations can be obtained from a variational principle in which g_{rs}, $\pi^{rs} = e^{ru}e^{sv}\pi_{uv}$, N, and N_r are independently varied. The Lagrangian density appearing in the action was shown by these authors to have the form

$$\mathfrak{L}' = -g_{rs}\pi^{rs}{}_{,0} - NR^0 - N_r R^r, \tag{10-9.12}$$

where

$$R^0 \equiv -\sqrt{-e}\{{}^3R - e^{-1}(\tfrac{1}{2}\pi^2 - \pi^{rs}\pi_{rs})\} \tag{10-9.13}$$

and

$$R^r \equiv -2\pi^{rs}{}_{|s}. \tag{10-9.14}$$

Here ${}^3R$ is the three-dimensional curvature scalar formed from the spatial metric g_{rs} and its inverse e^{rs}, a | indicates the covariant derivative with respect to this metric, using the three-dimensional Christoffel symbols ${}^3\{{}^{u}_{rs}\} \equiv \frac{1}{2}e^{uv}(g_{rv,s} + g_{sv,r} - g_{rs,v})$; spatial indices are raised and lowered using g_{rs} and e^{rs} and $\pi = e^{rs}\pi_{rs}$. We note that the only x^0 derivatives appearing in $\mathfrak{L}'$ occur in the first term. Varying the action formed from $\mathfrak{L}'$ with respect to N and N_r yields the constraint equations

$$R^0 = 0 \quad \text{and} \quad R^r = 0, \tag{10-9.15}$$

while variation with respect to g_{rs} and π^{rs} yields equations of the form

$$g_{rs,0} = -\frac{\delta}{\delta g_{rs}}\{NR^0 + N_r R^r\} \tag{10-9.16}$$

and

$$\pi^{rs}{}_{,0} = \frac{\delta}{\delta \pi^{rs}}\{NR^0 + N_r R^r\}. \tag{10-9.17}$$

[38] R. Arnowitt, S. Deser, and C. W. Misner, *Phys. Rev.*, **117**, 1595 (1960).

Equations (10-9.16) and (10-9.17) are of the Cauchy-Kowalewski type, since they are solved explicitly for $g_{rs,0}$ and $\pi^{rs}{}_{,0}$. Given g_{rs}, π^{rs}, N, and N_r on an initial data surface subject to the restrictions imposed by the constraint Eqs. (10-9.15), one can compute $g_{rs,0}$ and $\pi^{rs}{}_{,0}$ and so determine g_{rs} and π^{rs} on a surface infinitesimally removed from the initial data surface.[39] However, like the corresponding case of the Maxwell equations, one cannot use Eqs. (10-9.16) and (10-9.17) to determine these quantities on finitely removed space-like surfaces because of the undetermined quantities N and N_r appearing therein. To determine N and N_r we can follow the procedure employed in the Maxwell case, where we made use of a gauge condition to fix ϕ, the analogue of N and N_r. We impose coordinate conditions by requiring that g_{rs} and π^{rs} satisfy four noncovariant relations. For instance, we might require that

$$g_{rs} = 0, \quad r \neq s \qquad \text{and} \qquad \pi = 0,$$

the latter condition being the minimal surface condition of Dirac.[40] The requirement that these conditions hold at all space-time points then leads to a set of conditions on N and N_r analogous to the Eq. (8-7.7) for ϕ, which results from the imposition of the Coulomb gauge condition.[41]

The gauge conditions together with the constraint Eqs. (10-9.15) in principle allow one to express eight of the components of g_{rs} and π^{rs} in terms of the other four. These remaining four components describe the independent degrees of freedom of the gravitational field $g_{\mu\nu}$, two per space point, just as in the case of the Maxwell field, while the other eight components can be considered as redundant variables in the theory. Given their values on an initial $x^0 = \text{const}$ surface one can then in principle proceed to integrate Eqs. (10-9.16) and (10-9.17) to obtain a unique solution of the Einstein field equations. However, one must be prepared to encounter singularities in this solution in the future (or past) of the initial value surface. These singularities can be either coordinate singularities, which can be removed by an appropriate mapping, or intrinsic singularities, which cannot be so removed. The Schwarzschild singularity (see Section 11-4) is of the former type. Also, Raychaudhuri[42] and Komar[43] have shown, in general, that any Gaussian normal coordinate system (a coordinate system for which $g_{\mu 0} = \delta_{\mu 0}$) will, in the course of time, develop coordinate singularities. Intrinsic singularities can also develop in the course of time and solutions of the Einstein equations are known which,

[39] For an alternate approach to the initial value problem, see J. A. Wheeler, "Geometrodynamics and the Issue of the Final State," in *Relativity, Groups and Topology* (Gordon and Breach, New York, 1964), Sec. 4.

[40] P. A. M. Dirac, *Phys. Rev.*, **114**, 924 (1959).

[41] For a review of the various methods used to fix N and N_r, see J. L. Anderson, *Rev. Mod. Phys.*, **36**, 929 (1964).

[42] A. Komar, *Phys. Rev.* **104**, 544 (1956).

[43] A. Raychaudhuri, *Phys. Rev.* **98**, 1123 (1955).

starting from nonsingular initial data, develop such singularities off the initial value surface. The Kruskal form of the Schwarzschild field is an example of such a solution.

A considerable amount of effort over the past years has gone into trying to find a set of independent components such that the redundant components could be expressed in closed form. The main source of difficulty in this enterprise has been the complexity of the constraint equations (10-9.15). Foures-Bruhat[44] has studied extensively the problem of satisfying these equations. She has shown that they have an infinity of solutions in a bounded region of space-time. However, no one to date has succeeded in constructing explicit solutions of the type needed to eliminate the redundant variables from the theory. This is awkward, since without such solutions we cannot construct a self-contained set of equations of the Cauchy-Kowalewski type for the gravitational field. Such equations in turn appear to be necessary for the construction of a quantized version of the Einstein field equations.

10-10. The Linearized Einstein Equations

In our discussion of the Newtonian limit of the Einstein equations, we made two approximations; we assumed that the velocities of the sources of the gravitational field were small compared with the velocity of light, and that the gravitational field produced by these sources was weak—that is, $R_{\mu\nu\rho\sigma} \simeq 0$. In 1916 Einstein[45] developed the consequences that follow when only the second of these approximations is made. Because of the smallness of the gravitational coupling constant κ, the resulting linearized equations should be applicable to a large class of gravitational phenomena. They have the additional advantage of providing further insight into the content of the full, rigorous theory. Thus it is not completely unreasonable to expect that some of the features of solutions of the linearized solutions will be reflected in solutions of the full theory. Nevertheless one must keep in mind that the full theory is highly nonlinear, and it is well known from experiences in hydrodynamics that solutions of linearized equations may bear little or no relation to solutions of the rigorous equations. In particular, solutions of the linearized equations exist that do not approximate rigorous solutions in any sense. Furthermore it is just the nonlinearity of the Einstein equations that make them interesting from the point of view of new physics. Therefore the linearized equations we are about to discuss should not in any sense be considered as substitutes for the full nonlinear Einstein equations.

[44] Y. Foures-Bruhat, "The Cauchy Problem" in *Gravitation: An Introduction to Current Research*, L. Witten, ed. (Wiley, New York, 1962), pp. 130–168.

[45] A. Einstein, *Sitzber. Preuss. Akad. Wiss. Berlin*, 688 (1916).

If the gravitational field is weak, it should be possible to find a mapping such that

$$g_{\mu\nu} = \eta_{\mu\nu} + \varepsilon h_{\mu\nu}, \qquad (10\text{-}10.1)$$

where ε is a small parameter such that $\varepsilon h_{\mu\nu} \ll 1$. We shall assume that such is indeed the case, at least in some finite region of space-time. Having mapped $g_{\mu\nu}$ to this form we are still free to perform additional mapping that preserves this form. Arbitrary Poincaré mappings

$$x^\mu - x'^\mu = \Lambda_\nu{}^\mu x^\nu + b^\mu$$

clearly have this property, since by definition $\Lambda^{\mu\nu}$ is such that

$$\eta_{\mu\nu}\Lambda_\rho{}^\mu\Lambda_\sigma{}^\nu = \eta_{\rho\sigma}.$$

Note that under these mappings, $h_{\mu\nu}$ transforms like a cotensor. In addition to arbitrary Poincaré mappings we can perform mappings that differ from the identity mapping by first-order terms in ε:

$$x^\mu \to x'^\mu = x^\mu + \varepsilon\xi^\mu, \qquad (10\text{-}10.2)$$

where the ξ^μ are four arbitrary, bounded functions. Under such a mapping

$$\begin{aligned} g_{\mu\nu} \to g'_{\mu\nu} &= \frac{\partial x^\rho}{\partial x'^\mu}\frac{\partial x^\sigma}{\partial x'^\nu} g_{\rho\sigma} \\ &= (\delta_\mu{}^\rho - \varepsilon\xi^\rho{}_{,\mu})(\delta_\nu{}^\sigma - \varepsilon\xi^\sigma{}_{,\nu})(\eta_{\rho\sigma} + \varepsilon h_{\mu\nu}) \\ &= \eta_{\mu\nu} + \varepsilon\{h_{\mu\nu} - \xi_{\mu,\nu} - \xi_{\nu,\mu}\} + 0(\varepsilon^2), \end{aligned} \qquad (10\text{-}10.3)$$

where

$$\xi_\mu = \eta_{\mu\nu}\xi^\nu.$$

We see that $g'_{\mu\nu} = \eta_{\mu\nu} + \varepsilon h'_{\mu\nu}$ is of the form indicated in Eq. (10-10.1) and that

$$h'_{\mu\nu} = h_{\mu\nu} - \xi_{\mu,\nu} - \xi_{\nu,\mu}. \qquad (10\text{-}10.4)$$

We will shortly make use of this latter mapping freedom to simplify the linearized Einstein equations.

Let us now compute the Einstein tensor $G_{\mu\nu} = R_{\mu\nu} - \frac{1}{2}g_{\mu\nu}R$ to first order in ε. For this purpose we will need an expression for $g^{\mu\nu}$, correct to this order. A simple computation shows that

$$g^{\mu\nu} \simeq \eta^{\mu\nu} - \varepsilon\eta^{\mu\rho}\eta^{\nu\sigma}h_{\rho\sigma}. \qquad (10\text{-}10.5)$$

In computing $R_{\mu\nu}$ and R we need not consider products of Christoffel symbols since

$$\left\{ \begin{matrix} \mu \\ \rho\sigma \end{matrix} \right\} \simeq \frac{\varepsilon}{2} \eta^{\mu\nu}(h_{\rho\nu,\sigma} + h_{\sigma\nu,\rho} - h_{\rho\sigma,\nu})$$

has no zero-order term. Thus

$$R_{\mu\nu} \simeq \frac{\varepsilon}{2} \{h_{,\mu\nu} + \eta^{\rho\sigma}(h_{\mu\nu,\rho\sigma} - h_{\mu\rho,\nu\sigma} - h_{\nu\rho,\mu\sigma})\},$$

where

$$h \equiv \eta^{\mu\nu} h_{\mu\nu}$$

and

$$G_{\mu\nu} \simeq \frac{\varepsilon}{2} \{h_{,\mu\nu} + \eta^{\rho\sigma}(h_{\mu\nu,\rho\sigma} - h_{\mu\rho,\nu\sigma} - h_{\nu\rho,\mu\sigma}) - \eta_{\mu\nu}\eta^{\rho\sigma}(h_{,\rho\sigma} - \eta^{\iota\kappa} h_{\rho\iota,\sigma\kappa})\}. \qquad (10\text{-}10.6)$$

A simplification in the linearized field equations can be achieved by introducing the new variables

$$\gamma_{\mu\nu} \equiv h_{\mu\nu} - \tfrac{1}{2}\eta_{\mu\nu} h. \qquad (10\text{-}10.7)$$

Equation (10-10.7) can be solved for $h_{\mu\nu}$ in terms of $\gamma_{\mu\nu}$ to give

$$h_{\mu\nu} = \gamma_{\mu\nu} - \tfrac{1}{2}\eta_{\mu\nu}\gamma.$$

Under a Poincaré mapping, $\gamma_{\mu\nu}$ (like $h_{\mu\nu}$) transforms like a cotensor, while under a mapping described by Eq. (10-10.2),

$$\gamma_{\mu\nu} \to \gamma'_{\mu\nu} = \gamma_{\mu\nu} - \xi_{\mu,\nu} - \xi_{\nu,\mu} + \eta_{\mu\nu}\eta^{\rho\sigma}\xi_{\rho,\sigma} \qquad (10\text{-}10.8)$$

When written in terms of $\gamma_{\mu\nu}$, the tensor $G_{\mu\nu}$ takes the form

$$G_{\mu\nu} \simeq \frac{\varepsilon}{2} (\eta^{\rho\sigma}\gamma_{\mu\nu,\rho\sigma} - \tau_{\mu,\nu} - \tau_{\nu,\mu} + \eta_{\mu\nu}\eta^{\rho\sigma}\tau_{\rho,\sigma}), \qquad (10\text{-}10.9)$$

where

$$\tau_{\mu} = \eta^{\rho\sigma}\gamma_{\mu\rho,\sigma}.$$

At this point one can simplify still further the above approximate expression for $G_{\mu\nu}$ by mapping so that $\tau'_{\mu} = 0$. Under a mapping described by Eq. (10-10.2),

$$\tau_{\mu} \to \tau'_{\mu} = \tau_{\mu} - \eta^{\rho\sigma}\xi_{\mu,\rho\sigma}, \qquad (10\text{-}10.10)$$

so that a set of ξ_{μ} can always be found corresponding to a given τ_{μ} that will make τ'_μ vanish. As a consequence the Einstein equations (10-4.2) in linear approximation take the form

$$\frac{\varepsilon}{2}\eta^{\rho\sigma}\gamma_{\mu\nu,\rho\sigma} = \kappa T_{\mu\nu}, \tag{10-10.11}$$

with the supplementary condition

$$\eta^{\rho\sigma}\gamma_{\mu\rho,\sigma} = 0. \tag{10-10.12}$$

In seeking solutions to the linearized equations (10-10.11) we must make sure that they simultaneously satisfy the supplementary condition (10-10.12). If we differentiate Eqs. (10-10.11) with respect to x^ν, we see that a necessary condition for the satisfaction of Eqs. (10-10.12) is that

$$\eta^{\nu\rho}T_{\mu\nu,\rho} = 0. \tag{10-10.13}$$

However, this condition cannot be satisfied rigorously, since $T_{\mu\nu}$ must satisfy Eq. (10-4.1). This is, of course, the same kind of difficulty that we encountered in our discussion of the gravitational field in special relativity. The rigorous vanishing of the four divergence of $T^{\mu\nu}$ for otherwise free particles would again imply that they were not affected by each other's gravitational fields. However, in the present case the resolution of this difficulty is relatively simple. It comes from a realization that Eq. (10-10.13) need be satisfied only to lowest order in ε. As we will see later, deviations from straight-line motion due to gravitational interaction correspond to higher-order effects in ε. Therefore, solutions to the linearized equations need satisfy only the supplementary conditions (10-10.12) to the lowest order in ε, and this in general they will do.

11

Solutions of the Einstein Equations

The most obvious feature of the Einstein field equations for the gravitational field $g_{\mu\nu}$ is their elephantine nonlinearity. Compared to them the Navier-Stokes equations appear to be virtually linear equations. Nevertheless we possess today an enormous number of inequivalent solutions to the Einstein equations, while the solutions to the Navier-Stokes equations are few and far between. In fact, we have an embarrassment of riches; the rate at which solutions are being generated far exceeds our ability to interpret and analyze them.

Historically, the first method used to obtain exact solutions of the Einstein equations was to assume that one could find a mapping under which $g_{\mu\nu}$ assumed some particularly simple form (for example, $g_{0r} = 0$ or $g_{\mu\nu}$ independent of one or more coordinates, etc.). It was this method that was used to find the first non-trivial solution of the Einstein equations, the famous Schwarzschild solution. Later, more refined techniques were employed in constructing solutions. One assumed that $g_{\mu\nu}$ possessed certain symmetries, or one equivalently made assumptions about the form of the Killing vectors associated with the metric. Thus the assumption of axial symmetry led to the solutions found by Weyl and Levi-Civita. With the development of the Petrov and other algebraic classification schemes for the Weyl and Riemann-Christoffel tensors, whole classes of solutions were generated by requiring that $g_{\mu\nu}$ lead to a particular type of one of these tensors. In particular the requirement that the Riemann-Christoffel tensor be of Petrov type N has led to the class of solutions known as the *plane-fronted waves*. Finally, one can try to expand a solution of the Einstein equations in a Taylor series off of some initial value hypersurface, given consistent initial value data, and then to sum the series. This method has not proved too successful in finding solutions to date.

Having found a solution of the Einstein equations, one's troubles are only at a beginning. It is first necessary to make sure that it is not simply the transform of an already known solution. While there are methods that enable one to decide if two solutions belong to the same equivalence class, they are usually tedious to employ. Furthermore the large number of already extant solutions makes this task increasingly more difficult. Rather than try to solve this equivalence problem, one might try to characterize a solution in geometrical terms[1] and then show that it is unique, or at least belongs to a small class of solutions, the members of which are all known. We will see that the flat-space solution is the unique stationary solution of the Einstein equations that is everywhere regular while the Schwarzschild solution is the only spherically symmetric solution. As might be imagined, there do not exist any general uniqueness theorems for the Einstein equations as there do in the case of the Maxwell equations.

Recently, topological considerations have begun to play an important role in the characterization of gravitational fields. It is not sufficient merely to give a solution of the Einstein equations; one must also specify the topology of the manifold on which it resides. The recent discovery of the topology of the Schwarzschild field has shed wholly new light on the structure of this field and in particular on the so-called Schwarzschild singularity. As topological and global considerations are developed one can expect that one will be supplied with additional methods for constructing and characterizing solutions of the Einstein field equations in terms of topological invariants.

We begin this chapter with a discussion of solutions of the linearized Einstein field equations. Such solutions serve as useful guides in helping interpret exact solutions of these equations and are also adequate for the description of large classes of gravitational fields, owing to the smallness of the gravitational constant. The main part of this chapter is taken up with a discussion of solutions to the empty-space ($T_{\mu\nu} = 0$) equations and concludes with a brief discussion of solutions with sources ($T_{\mu\nu} \neq 0$).

11-1. Solutions of the Linearized Einstein Equations

The general solution to the linearized Einstein equations (10-10.11) can be written as a sum of the particular solution

$$\varepsilon\gamma_{\mu\nu}(x) = \frac{\kappa}{2\pi}\int d^4x'\delta((x-x')^2)T_{\mu\nu}(x') \tag{11-1.1}$$

[1] For a thorough review of the various methods of characterizing solutions by geometrical means and applications to many different types of solutions, see J. Ehlers and W. Kundt, "Exact Solutions of the Gravitational Field Equations" in *Gravitation: An Introduction to Current Research*, L. Witten, ed. (Wiley, New York, 1962), Chap. 2.

plus an arbitrary solution of the corresponding homogeneous equation. Alternately we could consider only the retarded or the advanced solution by replacing $\delta((x-x')^2)$ in Eq. (11-1.1) by $\delta^+((x-x')^2)$ or $\delta^-((x-x')^2)$, respectively, without changing the validity of the above statement. In discussing these solutions we will consider several special cases.

(a) Stationary Mass Distributions

For a point particle at rest at the origin of spatial coordinates, the only nonvanishing component of $T_{\mu\nu}$ is

$$T_{00}(x) = m\,\delta^3(\mathbf{x}),$$

so that

$$\varepsilon\gamma_{00} = \frac{\kappa}{2\pi}\frac{m}{r}, \qquad r = |\mathbf{x}|.$$

Therefore, using the expression (10-7.9) for κ and units in which the velocity of light is c, we have

$$g_{00} \simeq 1 - \frac{2Gm}{rc^2},$$

$$g_{0r} \simeq 0, \tag{11-1.2}$$

and

$$g_{rs} \simeq -1 - \frac{2Gm}{rc^2}.$$

We see that the linear approximation is valid only if $2GM/rc^2 \ll 1$. At the surface of the earth, $2Gm/rc^2 \sim 10^{-9}$.

For a general stationary mass distribution characterized by $T_{00}(x)$ with all other components of $T_{\mu\nu}$ zero, we have

$$\varepsilon\gamma_{00} = \frac{\kappa}{2\pi}\int d^3x'\,\frac{T_{00}(x)}{|\mathbf{x}-\mathbf{x}'|}. \tag{11-1.3}$$

To examine the nature of γ_{00} for large x, let us expand $1/|\mathbf{x}-\mathbf{x}'|$ as

$$\frac{1}{|\mathbf{x}-\mathbf{x}'|} = \frac{1}{r} - x'_u\left(\frac{1}{r}\right)_{,u} + \frac{1}{2}x'_u x'_v\left(\frac{1}{r}\right)_{,uv} + \cdots. \tag{11-1.4}$$

Then

$$\varepsilon\gamma_{00}(x) = \frac{\kappa}{2\pi}\left\{\frac{M}{r} - D_u\left(\frac{1}{r}\right)_{,u} + \frac{1}{6}Q_{uv}\left(\frac{1}{r}\right)_{,uv} - \cdots\right\}. \tag{11-1.5}$$

In the first term M is the total mass of system:

$$M = \int T_{00}(x)\, d^3x.$$

The second term involves the "dipole" moment

$$D_u = \int x_u T_{00}(x)\, d^3x.$$

By a suitable translation of axes D_u can be transformed to zero owing to the fact that $T_{00} > 0$ (no negative mass). The next term in the expansion (11-1.5) is the quadrupole term. Here

$$Q_{uv} \equiv \int (3x_u x_v - r^2\, \delta_{uv}) T_{00}\, d^3x$$

is the quadrupole moment tensor.

We shall see later that the pole term M/r in the expression (11-1.5) is the first term in an expansion of a rigorous solution of the Einstein equations, the Schwarzschild solution. At present it is not known if the quadrupole and higher moment terms correspond to exact solutions. Such terms correspond to fields produced by nonspherically symmetric mass distributions, for example, an oblate sun or earth.

Problem 11.1. Show that, by means of scale changes in the space and time coordinates, $g_{\mu\nu}$ inside a thin spherical shell of mass M and radius R can be made to take on Minkowski values, $g_{\mu\nu} = \eta_{\mu\nu}$.

In addition to the above stationary solutions there are also interesting solutions for the case $T_{0r} \neq 0$. If the sources are again stationary Eq. (11-1.1) reduces to

$$\varepsilon\gamma_{0r}(x) = \frac{\kappa}{2\pi} \int \frac{T_{0r}(\mathbf{x}')d^3x'}{|\mathbf{x} - \mathbf{x}'|}. \tag{11-1.6}$$

To examine the asymptotic form of γ_{0r} we again make use of the expansion (11-1.4) to obtain

$$\varepsilon\gamma_{0r} = \frac{\kappa}{2\pi}\left\{\frac{P_r}{r} + S_{rs}\left(\frac{1}{r}\right)_{,s} + \cdots\right\}. \tag{11-1.7}$$

where

$$P_r = \int T_{0r}(x)\, d^3x, \qquad S_{rs} = \int x_r T_{0s}\, d^3x, \cdots.$$

The solution thus obtained must, of course, satisfy the gauge condition (10-10.12). This condition will be satisfied, provided $\gamma_{0r,r} = 0$, which in turn will be satisfied if $P_r = 0$ and S_{rs} is of the form

$$S_{rs} = \tfrac{1}{2}\alpha_{rs} + \beta\,\delta_{rs}$$

where $\alpha_{rs} = -\alpha_{sr}$ and β are arbitrary constants. We can then eliminate the β by means of the infinitesimal mapping

$$x^0 - x'^0 = x^0 + \frac{\kappa}{2\pi}\frac{\beta}{r},$$

$$x^r \to x'^r = x^r,$$

so that finally

$$\varepsilon\gamma_{0r} = \frac{\kappa}{2\pi}\,\alpha_{rs}\left(\frac{1}{r}\right)_{,s} + \cdots. \tag{11-1.8}$$

Since $T^{\mu\sigma}x^\nu - T^{\nu\sigma}x^\mu$ can be taken to be the angular-momentum tensor density of the sources of the gravitational field,

$$M^{rs} \equiv \int \{T^{0r}x^s - T^{0s}x^r\}\,d^3x = \alpha_{rs} \tag{11-1.9}$$

represents the total angular momentum of these sources. Consequently we can interpret the field $\varepsilon\gamma_{0r}$ given by Eq. (11-1.7) as being that of a uniformly rotating mass distribution with angular momentum α_{rs}. Thirring and Lense[2] made use of this result to calculate the effect of the sun's rotation on the planetary orbits, but concluded that such effects are too small to be observed.

(b) Gravitational Waves

Let us now examine solutions to the homogeneous linearized Einstein equations, that is, Eqs. (10-10.11), with $T_{\mu\nu} = 0$. To simplify the discussion we shall assume that $\gamma_{\mu\nu}$ depends only on the variables x^0 and x^1. For a wave propagating in the positive x^1 direction the general solution is

$$\gamma_{\mu\nu}(x^0, x^1) = f_{\mu\nu}(x^0 - x^1),$$

where $f_{\mu\nu}$ is an arbitrary function of its arguments. However, satisfaction of the gauge condition (10-10.12) requires that

$$\gamma_{\mu 0,0} - \gamma_{\mu 1,1} = -\dot{f}_{\mu 0} - \dot{f}_{\mu 1} = 0,$$

where a dot denotes differentiation with respect to $x^1 - x^0$. Einstein[3] showed

[2] H. Thirring and J. Lense, *Phys. Z.*, **19**, 156 (1918).
[3] A. Einstein, *S. B. preuss. Akad. Wiss.*, 154 (1918).

that one can always find an infinitesimal mapping that preserves the gauge condition (10-10.12) for which $\gamma'_{1\mu} = \gamma'_{0\mu} = 0$ and $\gamma'_{22} + \gamma'_{33} = 0$. Thus there are two types of waves present: those with

$$\gamma_{22} = -\gamma_{33} \neq 0$$

and those for which

$$\gamma_{23} \neq 0.$$

Thus a linearized gravitational wave propagating in the x' direction has two independent states of polarization, in agreement with our discussion of the initial value problem of Section 10-9 where we concluded quite in general that the gravitational field had two degrees of freedom per space point.

Problem 11.2. Show that the above wave solution gives rise to a Riemann-Christoffel tensor that is of type II in the Petrov classification scheme.

(c) Time-varying Sources

A particular solution of the linearized Einstein equations for time-varying sources is given by Eq. (11-1.1). For sources that are spatially confined, this solution will satisfy the gauge condition (10-10.12), provided $T_{\mu\nu}$ satisfies

$$T^{\mu\nu}{}_{,\nu} = 0, \qquad (T^{\mu\nu} = \eta^{\mu\rho}\eta^{\nu\sigma}T_{\rho\sigma}). \tag{11-1.10}$$

In discussing the asymptotic form of this solution we will need certain relations between the various moments of $T^{\mu\nu}$. Let us multiply Eq. (11-1.10) by x^s and integrate over all space. We obtain thereby

$$\frac{\partial}{\partial x^0}\int T^{\mu 0}x^s\, dV = -\int T^{\mu r}{}_{,r}x^s\, dV$$

$$= -\int (T^{\mu r}x^s)_{,r}\, dV + \int T^{\mu s}\, dV.$$

For $T^{\mu\nu}$ spatially confined, as required by our solution (11-1.1), the first term on the right of the second line is seen to be zero by an application of Gauss' theorem. Therefore

$$\int T^{\mu s}\, dV = \frac{\partial}{\partial x^0}\int T^{\mu 0}x^s\, dV. \tag{11-1.11}$$

Likewise one can show that

$$\int (T^{r0}x^s + T^{s0}x^r)\, dV = \frac{\partial}{\partial x^0}\int T^{00}x^r x^s\, dV, \tag{11-1.12}$$

so finally we have

$$\int T^{rs}\,dV = \frac{1}{2}\frac{\partial^2}{\partial x^{02}}\int T^{00}x^r x^s\,dV. \qquad (11\text{-}1.13)$$

If we now make use of the relation

$$\delta(x^2) = \frac{1}{2|\mathbf{x}|}\{\delta(x^0 - |\mathbf{x}|) + \delta(x^0 + |\mathbf{x}|)\}$$

and again expand $|\mathbf{x} - \mathbf{x}'|^{-1}$ in powers in $|\mathbf{x}|$, as in Eq. (11-1.4), we find

$$\varepsilon\gamma_{rs} = \frac{\kappa}{4\pi}\frac{I''_{rs}}{r} + \cdots, \qquad (11\text{-}1.14\text{a})$$

$$\varepsilon\gamma_{r0} = \frac{\kappa}{2\pi}\left\{-\frac{I'_r}{r} - \frac{1}{2}\frac{n^s I''_{rs}}{r^2} + \cdots\right\}, \qquad n^r = \frac{x^r}{r}, \qquad (11\text{-}1.14\text{b})$$

$$\varepsilon\gamma_{00} = \frac{\kappa}{2\pi}\left\{\frac{I}{r} + \frac{n_r I'_r}{r^2} + \frac{1}{2}\frac{n^r n^s I''_{rs}}{r^3} + \cdots\right\}, \qquad (11\text{-}1.14\text{c})$$

where primes denote differentiation with respect to x^0 and where the dots indicate terms involving third and higher moments of the matter density. The moments appearing in these expressions are given by

$$I = \int \rho(x, x^0 \pm r)\,dV,$$

$$I^r = \int \rho(x, x^0 \pm r)x^r\,dV,$$

and

$$I^{rs} = \int \rho(x, x^0 \pm r)x^r x^s\,dV,$$

The $\pm$ signs appearing in these moments indicate that the value of a moment at x^0 is determined by the advanced or retarded values of the matter density ρ or a linear combination of them. In arriving at these results we have made explicit use of the relations (11-1.10), (11-1.11), and (11-1.12). Finally, the requirement that the $\gamma_{\mu\nu}$ satisfy the gauge conditions (10-10.12) leads to the conditions

$$I' = 0 \quad \text{and} \quad I'_r = 0 \qquad (11\text{-}1.15)$$

to this order of approximation. We will return to these solutions later when we come to discuss the problem of gravitational radiation.

11-2. Uniqueness of the Flat-Space Solution

The simplest solution of the empty-space Einstein equations (10-4.2 with $T^{\mu\nu} = 0$) is the flat-space solution

$$g_{\mu\nu} = \eta_{\mu\nu}. \tag{11-2.1}$$

By itself this solution would be rather uninteresting were it not for the fact that it is the only stationary solution that is asymptotically flat and everywhere regular on the whole space-time manifold $-\infty \leqq x^\mu \leqq +\infty$. In what follows we shall outline a proof of this statement first given by A. Lichnerowicz.[4]

Let us assume that $g_{\mu\nu}$ is stationary and map so that it is independent of x^0. For such a metric the components of the Ricci tensor $R_{\mu\nu}$ have the form

$$R_{00} = -\frac{1}{\xi}\nabla^2\xi + \frac{\xi^2}{2}H^2 \tag{11-2.2a}$$

$$R_{r0} = \frac{1}{2\xi^2}(\xi^3 H_r{}^s)_{|s} \tag{11-2.2b}$$

$$R_{rs} = {}^3R_{rs} - \frac{1}{\xi}(\xi_{,r})_{|s} - \frac{\xi^2}{2}H_r{}^u H_{su} \tag{11-2.2c}$$

where $\xi^2 = g_{00} > 0$, $H_{rs} = \phi_{s,r} - \phi_{r,s}$ with $\phi_r = g_{r0}/g_{00}$ and $H^2 = \frac{1}{2}H_{rs}H^{rs}$. In addition, ${}^3R_{rs}$ is the three-dimensional Ricci tensor formed from the metric $-h_{rs}$ induced on the three-dimensional space V_3 associated with the curves (10-5.3) and given by Eq. (10-5.11), $|s$ denotes the covariant derivative formed using this metric and its inverse $-h^{rs} = g^{rs}$ and indices are raised and lowered with this metric, that is, $H_r{}^u = g_u{}^s H_{rs}$. Finally, ∇^2 is the generalized Laplacian formed $-h_{rs}$ so that

$$\nabla^2\xi \equiv (g^{rs}\xi_{,r})_{|s}.$$

If $T^{\mu\nu} = 0$ in Eq. (10-4.2) it follows by multiplication of these equations by $g^{\mu\nu}$ that $R = 0$. Consequently, in this case these equations reduce to

$$R_{\mu\nu} = 0.$$

Then, since by assumption $g_{\mu\nu}$ is everywhere regular and hence $\xi \neq 0$, it follows from Eq. (11-2.2a) that

$$\nabla^2\xi = \frac{\xi^3}{2}H^2. \tag{11-2.3}$$

[4] A. Lichnerowicz, *Théories Relativistes de la Gravitation et de L'Electromagnetisme* (Masson et Cie, Paris, 1955), Chap. VII.

We now make use of an extension of a theorem of Hopf[5] Let $d(\mathbf{x}_1, \mathbf{x}_2)$ be the lower bound of the lengths of all curves joining the points $\mathbf{x}_1$ and $\mathbf{x}_2$ in V_3. We say that a function $U(\mathbf{x})$ defined on V_3 asymptotically approaches a constant k from above if, for some interior point $\mathbf{x}_0$ of V_3, that is, a point with finite coordinates, and for every $\varepsilon > 0$ there exists a number N_ε depending only on ε such that for

$$d(\mathbf{x}_0, \mathbf{x}) \geqq N_\varepsilon,$$

$U(x)$ satisfies

$$|U(\mathbf{x}) - k| \leqq \varepsilon$$

and there exists an N such that, for $d(\mathbf{x}_0, \mathbf{x}) \leqq N$, one has $U(\mathbf{x}) \geqq k$. Then, if $U(\mathbf{x})$ is such that $\nabla^2 U(x) \geqq 0$ and $U(\mathbf{x})$ asymptotically approaches a constant k from above, $U(\mathbf{x})$ is equal to k everywhere on V_3

Referring back now to Eq. (11-2.3) we see that since both ξ and H^2 are necessarily positive (h_{rs} is positive definite), we have $\nabla^2\xi \geqq 0$. Hence, if ξ asymptotically approaches a constant from above, it follows from the above theorem that it is constant everywhere on V_3. Therefore, g_{00} is a constant on V_3 and hence a constant over the whole space-time manifold. By a suitable choice of the scale of x^0 we have, therefore, $g_{00} = 1$. Furthermore, since ξ is a nonzero constant, it follows from Eq. (11-2.3) that H^2 is zero and hence also that H_{rs} must be zero. But then we have

$$g_{0s,r} - g_{0r,s} = 0.$$

As we discussed in Section 10-5, these are the sufficient conditions that in a simply connected region of the space-time manifold there exists a mapping such that $g_{0r} = 0$. Thus in the present case where the manifold as a whole is simply connected, we can conclude that we can map so that g_{0r} vanishes everywhere.

Let us now examine Eq. (11-2.2c). When ξ is a constant and H_{rs} is zero, this equation reduces to

$${}^3R_{rs} = 0.$$

But in a three-dimensional space the vanishing of the Ricci tensor implies the vanishing of the full three-dimensional Riemann tensor. This follows from the fact that in three dimensions ${}^3R_{rsuv}$ has only six nonvanishing components as a consequence of its various symmetry properties. It therefore follows from the six equations

$${}^3R_{rs} = g^{uv}\,{}^3R_{rusv}$$

that the components of ${}^3R_{rsuv}$ are linear homogeneous functions of the components of ${}^3R_{rs}$ and hence the vanishing of the latter implies the vanishing of

[5] E. Hopf, *Preuss. Akad. Wiss. Sitz.*, **19**, 147 (1927); for a proof of the extension given here see A. Lichnerowicz, loc. cit.

the former. From our discussion of Section 3-3 we can conclude that there always exists a mapping such that $h_{rs} = \delta_{rs}$. Then, since $g_{0r} = 0$, it follows that $g_{rs} = -\delta_{rs}$ everywhere on V_3 and hence everywhere on our whole space-time manifold. Therefore, when $g_{\mu\nu}$ is stationary on the whole space-time manifold and g_{00} asymptotically approaches a constant from above, it follows that there exists a mapping such that $g_{\mu\nu} = \eta_{\mu\nu}$ everywhere.[6]

After Einstein developed the general theory of relativity he hoped that it might serve as a model for a pure field theory of elementary particles. In such a theory, which uses only fields to describe the kpt there would exist dpt that could be interpreted as particlelike solutions. For such dpt the stress-energy tensor of the theory might be significantly different from zero only in isolated regions of space-time. Einstein was motivated to look for such dpt partly because of his dislike of having to include the matter stress-energy tensor $T^{\mu\nu}$ in his field equations in order to obtain particlelike solutions. He felt that such a tensor was only a provisional element of the theory and that in a more fundamental theory it would be unnecessary to introduce such a source term for the gravitational or other fields. Because of their high degree of non-linearity he hoped that his empty-space equations might possess particlelike solutions.

The theorem of Lichnerowicz discussed above, however, makes the existence of such particlelike solutions appear unlikely. If they exist at all, one should expect that there exists a single-particle solution. But a single-particle solution would have to be stationary, to say the least, and if further we require that it be asymptotically flat or exist on a manifold that is the product of a compact V_3 and the real line, then no such solution exists. It might be argued that only many-particle solutions of the empty-space Einstein equations are everwhere regular. In general such solutions need not be stationary, since we would expect, for instance, that two particles would accelerate toward each other due to their mutual gravitational interaction. Even in such a case, however, one can argue that in some asymptotic limit when the particles are far apart from each other, the corresponding solution should reduce to the superposition of two stationary one-particle solutions.

11-3. Spherically Symmetric Exact Solutions

One of the first, and perhaps still the most important, exact solutions of the Einstein equations was obtained by Schwarzschild[7] by imposing the condition of spherical symmetry on the field $g_{\mu\nu}$ and requiring at the same time that

[6] If V_3 is compact, for example, has the topology of a sphere, torus, or so on, then it is only necessary to require that $g_{\mu\nu}$ be stationary in order that there exists a mapping such that it takes on the Minkowski values in any simply-connected domain.

[7] K. Schwarzschild, *S. B. preuss. Akad. Wiss.*, 189 (1916).

it be static. As we will see, this latter requirement is unnecessary; the requirement of spherical symmetry is sufficient to obtain this solution.

In Section 4-4 we showed that the most general spherically symmetric tensor $g_{\mu\nu}$ involved four arbitrary functions of r and x^0 and was of the form

$$g_{00}(x) = \alpha(r, x^0),$$

$$g_{0r}(x) = \beta(r, x^0)\frac{x^r}{r},$$

$$g_{rs}(x) = \gamma(r, x^0)\,\delta_{rs} + \zeta(r, x^0)\frac{x^r x^s}{r^2}\,.$$

This form will be preserved under mappings such that

$$x^r = f_1(r', x'^0)x'^r \quad \text{and} \quad x^0 = f_2(r', x'^0),$$

since in particular

$$r = f_1 r' \quad \text{and} \quad \frac{x^r}{r} = \frac{x'^r}{r'}.$$

We will now use this mapping freedom to eliminate two of the four arbitrary functions appearing in $g_{\mu\nu}$.

We can first find a mapping such that $\gamma' = -1$. To this end we consider the mapping

$$x^r = f(r', x'^0)x'^r \quad \text{and} \quad x^0 = x'^0$$

for which

$$g'_{rs} = (f_{,r}x'^u + f\,\delta_r{}^u)(f_{,s}x'^v + f\,\delta_s{}^v)\left(\gamma\,\delta_{uv} + \zeta\frac{x^u x^v}{r}\right)$$

$$= \gamma f^2\,\delta_{rs} + \left[\left(f + r'\frac{\partial}{\partial r'}f\right)^2\zeta - r'\frac{\partial}{\partial r'}\left(2f + r'\frac{\partial}{\partial r'}f\right)\gamma\right]\frac{x'^r x'^s}{r^2}.$$

We can therefore eliminate one of the arbitrary functions appearing in $g_{\mu\nu}$ by taking

$$f = -\gamma^{-1/2}.$$

A second simplification can be effected whereby g'_{0r} is made to vanish by means of a mapping of the form

$$x^r = x'^r, \qquad x^0 = f(r', x'^0).$$

Under this mapping

$$g'_{0r} = \frac{\partial f}{\partial x^0}(f_{,r}g_{00} + g_{0r}),$$

so that we can achieve our goal by taking

$$f_{,r} = \frac{x^r}{r}\frac{\partial}{\partial r'}f = -\frac{x^r}{r}\frac{\beta}{\alpha} \quad \text{or} \quad \frac{\partial}{\partial r'}f = -\frac{\beta}{\alpha}.$$

We must, of course, check that this second mapping has not undone the effect of the first mapping, that is, that γ still has the value -1. Under this mapping

$$\begin{aligned} g'_{rs} &= f_{,r}f_{,s}g_{00} + f_{,r}g_{0s} + f_{,s}g_{0r} + g_{rs} \\ &= -\delta_{rs} + \left[\alpha\left(\frac{\partial f}{\partial r'}\right)^2 + 2\left(\frac{\partial f}{\partial r'}\right)\beta + \zeta\right]\frac{x'^r x'^s}{r'^2}, \end{aligned}$$

so that the effect of the first mapping is preserved. We conclude, therefore, that it is always possible to reduce the most general spherically symmetric $g_{\mu\nu}$ to the form

$$g_{00} = \alpha(r, x^0),$$

$$g_{0r} = 0,$$

$$g_{rs} = -\delta_{rs} + \zeta(r, x^0)\frac{x^r x^s}{r^2},$$

in a finite region of space-time. We have not proved, and in general it is not true, that we can effect such a reduction everywhere. Finally it is convenient to introduce spherical coordinates by means of the mapping

$$x^1 = r \sin\phi \sin\theta,$$

$$x^2 = r \cos\phi \sin\theta,$$

$$x^3 = r \cos\theta,$$

and to introduce new functions ν and λ of r and x^0 by taking

$$e^\nu = \alpha \quad \text{and} \quad e^\lambda = (1 - \zeta),$$

so that finally

$$g_{\mu\nu} = \begin{pmatrix} e^\nu & & & \\ & -e^\lambda & & 0 \\ & & -r^2 & \\ 0 & & & -r^2\sin^2\theta \end{pmatrix} \tag{11-3.1}$$

To find the restrictions imposed on ν and λ by the Einstein field equations we first need to compute the Christoffel symbols associated with the metric (11-3.1). The nonvanishing components are

$$\begin{Bmatrix} 0 \\ 00 \end{Bmatrix} = \frac{\dot{\nu}}{2}, \qquad \begin{Bmatrix} 0 \\ 10 \end{Bmatrix} = \frac{\nu'}{2}, \qquad \begin{Bmatrix} 0 \\ 11 \end{Bmatrix} = \frac{\dot{\lambda}}{2} e^{\lambda - \nu},$$

$$\begin{Bmatrix} 1 \\ 00 \end{Bmatrix} = \frac{\nu'}{2} e^{\nu - \lambda}, \qquad \begin{Bmatrix} 1 \\ 10 \end{Bmatrix} = \frac{\dot{\lambda}}{2}, \qquad \begin{Bmatrix} 1 \\ 11 \end{Bmatrix} = \frac{\lambda'}{2},$$

$$\begin{Bmatrix} 1 \\ 22 \end{Bmatrix} = -re^{-\lambda}, \qquad \begin{Bmatrix} 1 \\ 33 \end{Bmatrix} = -r \sin^2 \theta e^{-\lambda},$$

$$\begin{Bmatrix} 2 \\ 12 \end{Bmatrix} = \frac{1}{r}, \qquad \begin{Bmatrix} 2 \\ 33 \end{Bmatrix} = -\sin\theta\cos\theta, \qquad \begin{Bmatrix} 3 \\ 23 \end{Bmatrix} = \operatorname{ctn}\theta,$$

where a dot denotes differentiation with respect to x^0 and a prime denotes differentiation with respect to r. With these Christoffel symbols we compute the following expressions for the nonvanishing components of the Einstein tensor $G_\nu{}^\mu = R_\nu{}^\mu - (1/2)R\,\delta_\nu{}^\mu$:

$$G_1{}^1 = e^{-\lambda}\left(\frac{\nu'}{r} + \frac{1}{r^2}\right) - \frac{1}{r^2}, \tag{11-3.2}$$

$$G_2{}^2 = G_3{}^3 = \frac{1}{2} e^{-\lambda}\left(\nu'' + \frac{\nu'^2}{2} + \frac{\nu' - \lambda'}{r} - \frac{\nu'\lambda'}{2}\right) - \frac{1}{2} e^{-\nu}\left(\ddot{\lambda} + \frac{\dot{\lambda}^2}{2} - \frac{\dot{\lambda}\dot{\nu}}{2}\right), \tag{11-3.3}$$

$$G_0{}^0 = e^{-\lambda}\left(\frac{1}{r^2} - \frac{\lambda'}{r}\right) - \frac{1}{r^2}, \tag{11-3.4}$$

$$G_0{}^1 = e^{-\lambda}\frac{\dot{\lambda}}{r}. \tag{11-3.5}$$

The vanishing of the Einstein tensor in the region of space-time we are working in determines λ and ν. Because of the Bianchi identities satisfied by $G_\nu{}^\mu$, the vanishing of $G_2{}^2$ is a consequence of the vanishing of the other components. Consequently, the equations that determine λ and ν are

$$e^{-\lambda}\left(\frac{\nu'}{r} + \frac{1}{r^2}\right) - \frac{1}{r^2} = 0, \tag{11-3.6}$$

$$e^{-\lambda}\left(\frac{\lambda'}{r} - \frac{1}{r^2}\right) + \frac{1}{r^2} = 0, \tag{11-3.7}$$

$$\dot{\lambda} = 0. \tag{11-3.8}$$

From Eqs. (11-3.6) and (11-3.7) it follows that $\lambda' + \nu' = 0$, so that $\lambda + \nu = h(x^0)$ If we perform a mapping such that

$$x^r = x'^r, \qquad x^0 = f(x^0),$$

then

$$g'_{00} = \dot{f}^2 g_{00}$$

so that the effect of such a mapping is to add to ν an arbitrary function of x^0 while leaving unaffected the other components of $g_{\mu\nu}$. Consequently we can always map so that $\lambda = -\nu$. Then, as a consequence of Eq. (11-3.8) we can conclude that λ and ν are functions of r alone. Finally, we can integrate Eq. (11-3.7) to obtain

$$e^{-\lambda} = e^{\nu} = 1 - \frac{2\mathfrak{m}}{r}$$

where $\mathfrak{m}$ is an integration constant. We see that for large r the solution we have found corresponds to the solution (11-1.2) of the linearized Einstein equations for a point mass located at $r = 0$, provided we choose the constant $\mathfrak{m}$ so that

$$\mathfrak{m} = \frac{Gm}{c^2} \tag{11-3.9}$$

where m is the mass of the source. The constant $2\mathfrak{m}$ is sometimes referred to as the *Schwarzschild radius* of the mass m. For the sun, $2\mathfrak{m}$ has a value 1.47 km, while for the earth it has a value 4.9 mm.

If now we gather together our results we find for the form of the Schwarzschild field, with $x^\mu = \{x^0, r, \theta, \phi\}$,

$$g_{\mu\nu} = \begin{pmatrix} 1 - \dfrac{2\mathfrak{m}}{r} & & & 0 \\ & \dfrac{-1}{1-(2\mathfrak{m}/r)} & & \\ & & -r^2 & \\ 0 & & & -r^2 \sin^2\theta \end{pmatrix}. \tag{11-3.10}$$

Mapping back to Cartesian coordinates, we obtain

$$\begin{aligned} g_{00} &= 1 - \frac{2\mathfrak{m}}{r}, \\ g_{0r} &= 0, \\ g_{rs} &= -\delta_{rs} - \frac{2\mathfrak{m}/r}{1 - 2\mathfrak{m}/r}\frac{x^r x^s}{r^2}. \end{aligned} \tag{11-3.11}$$

For some purposes it is convenient to use the so-called isotropic form of $g_{\mu\nu}$, obtained by the mapping

$$x^r = \left(1 + \frac{\mathfrak{m}}{2r'}\right)^2 x'^r, \qquad x^0 = x'^0$$

and given by

$$\begin{aligned} g_{00} &= \left(\frac{1 - \mathfrak{m}/2r}{1 + \mathfrak{m}/2r}\right)^2, \\ g_{0r} &= 0, \\ g_{rs} &= -\left(1 + \frac{\mathfrak{m}}{2r}\right)^4 \delta_{rs}. \end{aligned} \tag{11-3.12}$$

In arriving at the Schwarzschild field we have required only that $g_{\mu\nu}$ be spherically symmetric. We see that this is a very strong restriction, however, since, for a solution of the Einstein equations it leads to the conclusion that the solution is static. (Since $g_{\mu\nu}$ is independent of x^0 and in addition $g_{0r} = 0$ in all the above forms for the Schwarzschild field, it follows from the discussion of Section 10-5 that it is a static field.) We have in fact found the most general spherically symmetric solution to the Einstein equations in a region of space-time where $T^{\mu\nu} = 0$. This conclusion that if $T^{\mu\nu}$ is spherically symmetric and zero for r greater than some value a, then $g_{\mu\nu}$ is the Schwarzschild field for $r > a$, is known as *Birkhoff's theorem.*[8] Finally we point out that asymptotically $g_{\mu\nu} \sim \eta_{\mu\nu}$ as $r \to \infty$. However, it was not necessary to require this asymptotic behavior to obtain our solution.

Problem 11.3. Show that the Schwarzschild field has a Riemann-Christoffel tensor associated with it, which is type D in the Petrov classification scheme.

11-4. The Schwarzschild "Singularity"; the Topology of the Schwarzschild Field

As it stands the Schwarzschild solution to the Einstein equations is valid only for $g_{00} \geqq 0$, since it was obtained by assuming that $g_{00} = e^\nu$. Furthermore, at $r = 2\mathfrak{m}$, g_{11} diverges. For this reason the surface $r = 2\mathfrak{m}$ is sometimes called the *Schwarzschild singularity*. Nevertheless we will see that this singularity is not an intrinsic feature of the Schwarzschild solution but rather a property of the coordinate system used to express this solution. One indication is that $g = -r^4 \sin^2 \theta$ is regular at $r = 2\mathfrak{m}$. Furthermore the scalar $R_{\mu\nu\rho\sigma}R^{\mu\nu\rho\sigma}$ has the value $48m^2/r^6$ there. Since the value of a scalar, unlike

[8] G. Birkhoff, *Relativity and Modern Physics* (Harvard Univ. Press, Cambridge, Mass., 1923), p. 253.

the components of a tensor, is unaffected by a mapping, this is further evidence that the singularity at $r = 2\mathfrak{m}$ is not intrinsic.

If the Schwarzschild singularity is only a coordinate singularity, it should be possible to remove it by an appropriate mapping. Such a mapping was found by Eddington[9] and is defined by the equations

$$x^r = x'^r, \qquad x^0 = x'^0 \pm 2\mathfrak{m} \ln\left(\frac{r'}{2\mathfrak{m}} - 1\right). \tag{11-4.1}$$

The transformed components of $g_{\mu\nu}$ are given, with $x^\mu = (x^0, r, \theta, \phi)$, by

$$g'_{\mu\nu} = \begin{pmatrix} \left(1 - \frac{2\mathfrak{m}}{r}\right) & \pm\frac{2\mathfrak{m}}{r} & & 0 \\ \pm\frac{2\mathfrak{m}}{r} & -\left(1 + \frac{2\mathfrak{m}}{r}\right) & & \\ & & -r^2 & \\ & 0 & & -r^2 \sin^2\theta \end{pmatrix} \tag{11-4.2}$$

The field $g_{\mu\nu}$ is now no longer singular at $r = 2\mathfrak{m}$ and satisfies the Einstein field equations for all values of $r > 0$. Note, however, that it required a singular mapping to obtain this result. Furthermore, while the original Schwarzschild solution was time symmetric, that is, under the mapping $x'^0 = -x^0$, $x'^r = x^r$, $g'_{\mu\nu} = g_{\mu\nu}$, this is no longer true for the Eddington form (11-4.2) of the solution because g_{0r} is no longer zero.

It is instructive to examine the local light-cone structure at various points with the help of the Eddington form of the Schwarzschild field. Such a cone is defined as the locus of points $x_0{}^\mu + dx^\mu$ in the neighborhood of a given point $x_0{}^\mu$ for which $g_{\mu\nu}\, dx^\mu\, dx^\nu = 0$. We have sketched a number of these cones in Fig. 11.1 corresponding to $g_{0r} = +2\mathfrak{m}/r$ in (11-4.2). We see from Fig. 11.1 that each light cone contains a null direction for which the geodesic passing through the apex of the cone in this direction travels through the Schwarzschild surface $r = 2\mathfrak{m}$ and ultimately reaches the singular point $r = 0$. The null direction with this property is $n^\mu = -(r, x)$. The time-reflected direction $-(-r, x)$ is tangent to a geodesic that asymptotically approaches the surface $r = 2\mathfrak{m}$. The Schwarzschild surface therefore has the curious feature that it acts like a unidirectional membrane; light signals that propagate along null geodesics into the future (increasing x^0) can leave $r = 0$ but cannot reach there. If we use the minus sign in (11-4.2), the situation is reversed. Signals propagated into the future can get inside the Schwarzschild surface but cannot get out. There are thus two different Schwarzschild fields, which are the time reflections of each other.

[9] A. S. Eddington, *Nature*, **133**, 192 (1924); D. Finkelstein, *Phys. Rev.*, **110**, 965 (1958).

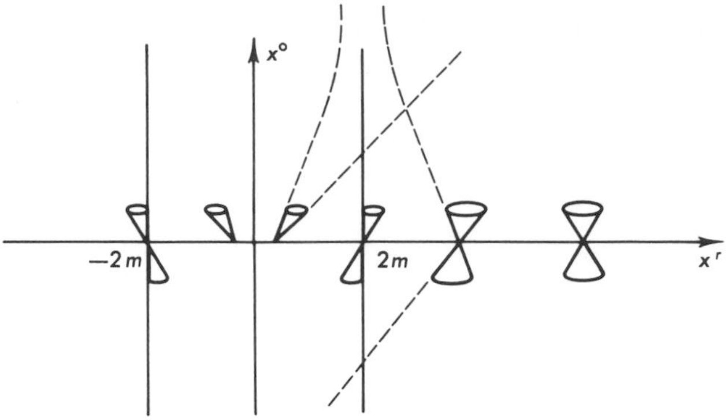

Fig. 11.1.

The Eddington form of the Schwarzschild field demonstrates that the Schwarzschild singularity is a coordinate singularity somewhat analogous to that which occurs at the antipodes of a latitude-longitude system of coordinates on the surface of a sphere. This in turn suggests that the topology of the space-time manifold of the Schwarzschild field is not equivalent to the Euclidean four-plane. A final clarification came with the work of Kruskal[10] who gave the maximal analytic extension of the Schwarzschild field. A manifold with either an affine or a metric geometry imposed on it is said to be *maximal* if every geodesic emanating from an arbitrary point of the manifold has an infinite length in both directions as measured by an affine parameter, or terminates on a physical singularity of the geometry—that is, a singularity that cannot be removed by a suitable mapping. If all geodesics emanating from a point have infinite length in both directions, the manifold is said to be *complete*. Obviously a manifold that is maximal but not complete must possess singular points. As we shall see, the Kruskal manifold is maximal but not complete.

The manifold on which the Schwarzschild field (11-3.10) is regular consists of the points for which $r > 2\mathfrak{m}$. However, this manifold is not maximal if we take $g_{\mu\nu}$ to be the metric of the space. Robertson showed that the length of a radial geodesic, starting at some finite value of $r > 2\mathfrak{m}$ and terminating at $r = 2\mathfrak{m}$, is finite even though such a geodesic only approaches $r = 2\mathfrak{m}$ asymptotically as $x^0 \to \infty$. The manifold on which the Eddington form (11-4.2) for $g_{\mu\nu}$ is nonsingular can also be shown to be incomplete in the above sense. However, if a manifold with a given geometry is not maximal, it can in general be entended to a maximal manifold in such a way that the geometry in the extension joins on smoothly and without singularities to the geometry on the original manifold.

[10] M. D. Kruskal, *Phys. Rev.*, **119**, 1743 (1960).

To find an extension of the Schwarzschild metric, Kruskal looked for a mapping such that, in the region $r > 2\mathfrak{m}$, $g_{\mu\nu}$ has the form, with $x^\mu = \{\upsilon, u, \theta, \phi\}$,

$$g_{\mu\nu} = \begin{pmatrix} f^2 & & & 0 \\ & -f^2 & & \\ & & -r^2 & \\ 0 & & & -r^2 \sin^2\theta \end{pmatrix}, \tag{11-4.3}$$

where f is a function of r and r in turn is now a transcendental function of u and υ. A mapping that transforms the Schwarzschild metric (11-3.10) into (11-4.3) is defined by

$$\upsilon = \left[\frac{r}{2\mathfrak{m}} - 1\right]^{1/2} \exp\left(\frac{r}{4\mathfrak{m}}\right) \sinh\left(\frac{x^0}{4\mathfrak{m}}\right),$$

$$u = \left[\frac{r}{2\mathfrak{m}} - 1\right]^{1/2} \exp\left(\frac{r}{4\mathfrak{m}}\right) \cosh\left(\frac{x^0}{4\mathfrak{m}}\right),$$

with the inverse mapping

$$\left[\frac{r}{2\mathfrak{m}} - 1\right] \exp\left(\frac{r}{2\mathfrak{m}}\right) = u^2 - \upsilon^2,$$

$$x^0 = 2\mathfrak{m} \tanh^{-1}\frac{\upsilon}{u}.$$

The transformed $g_{\mu\nu}$ given by (11-4.3) then agrees with the Schwarzschild $g_{\mu\nu}$ (11-2.10) in the region $r > 2\mathfrak{m}$ if we take

$$f^2 = \frac{8\mathfrak{m}}{r} e^{-r/2\mathfrak{m}}.$$

Under this mapping the entire Schwarzschild region $r > 2\mathfrak{m}$ is mapped onto the quadrant $u > |\upsilon|$.

We have depicted a number of the features of the Kruskal manifold in Fig. 11.2. In Fig. 11.2(a) the shaded area represents the region of the $x^0 r$ manifold where the Schwarzschild field (11-3.10) is regular. The shaded regions of Figs. 11.2(b) and 11.2(c) are the images of this region. We have indicated several $r = \text{const}$ curves; that is, the hyperbolas $u^2 - \upsilon^2 = \text{const}$ in Fig. 11.2(b), while in Fig. 11.2(c) we have indicated a number of $x^0 = \text{const}$ curves. For finite x^0 the "point" $r = r_0$ is mapped onto the point $u = \upsilon = 0$, while for $x^0 = \pm\infty$, it is mapped onto the lines $u = \pm\upsilon$. The region of u, υ manifold, where $g_{\mu\nu}$ given by (11-4.3) is regular, is bounded by the two hyperbolas $r = 0$; that is, $u^2 - \upsilon^2 = -1$, where the curvature scalar becomes infinite and hence represents an intrinsic singularity of the field.

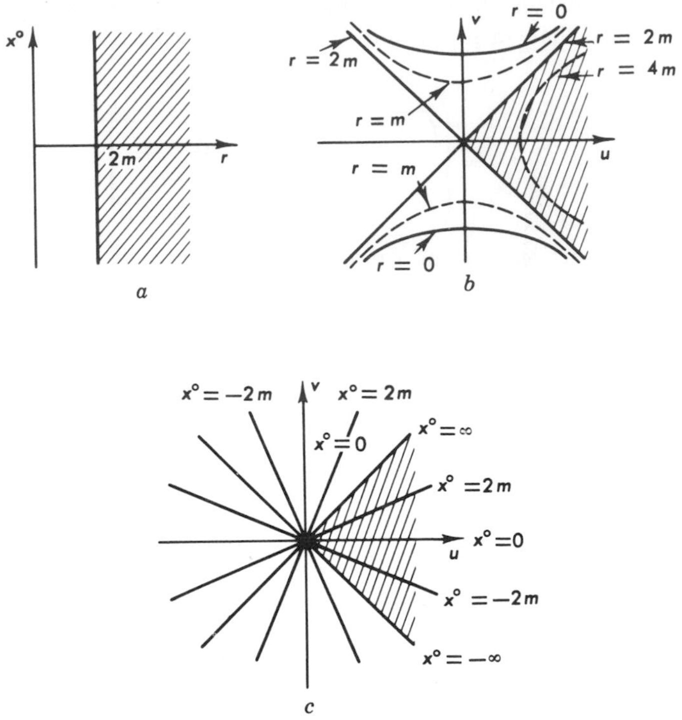

Fig. 11.2.

We must remember that the diagram of Fig. 11.2(b) represents only a part of the total space-time manifold, since we have neglected the θ, ϕ coordinates. To get an idea of the structure of the full four-dimensional manifold on which the field (11-4.3) is regular, let us consider the submanifold $v = 0$. As we move along the u axis from $+\infty$, r decreases to a minimum value $2m$ at $u = 0$ and then increases again as u goes to $-\infty$. We can draw a picture of a cross section of this manifold corresponding to $\theta = \pi/2$ by constructing a two-dimensional surface that is embedded in a flat three-dimensional space in such a way that the metric $g_{11} = f^2$, $g_{22} = r^2$, $g_{12} = g_{21} = 0$ is induced on it. Fuller and Wheeler [11] have constructed such a surface, which we reproduce in Fig. 11.3. Various $u =$ const curves traced out as ϕ runs from 0 to 2π are depicted. The $\phi =$ const. curves run from the edges of the upper surface, through the throat, and out toward the edge of the bottom surface. The same picture holds for all submanifolds defined by $x^0 =$ const that include the shaded region of Fig. 11.2(b). Thus, in many respects, the Kruskal manifold

[11] R. W. Fuller and J. A. Wheeler, *Phys. Rev.*, **128**, 919 (1962).

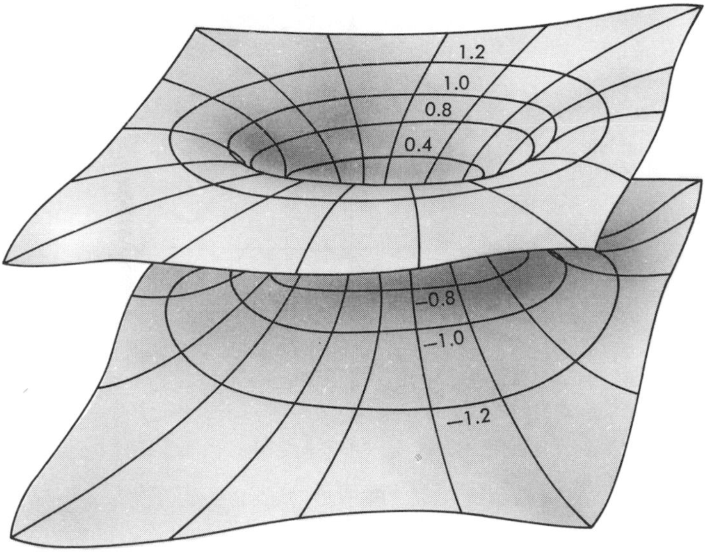

Fig. 11.3.

can be thought of as being formed by joining two Schwarzschild manifolds, an idea first suggested by Einstein and Rosen.[12] We note, however, that if we consider other $\upsilon = $ const submanifolds, the throat in Fig. 11.3 begins to pinch off as $\upsilon \to \pm 1$. Beyond this value the upper and lower surfaces completely separate. Furthermore, the regions of the Kruskal manifold for which $r < 2m$ have no counterpart in this Einstein-Rosen manifold.

Wheeler and Fuller[13] have raised the interesting question of whether an observer situated on the upper half of the surface in Fig. 11.3 can communicate with the lower half by means of light signals transmitted through the throat. Such observers reside in regions I and III, respectively, of Fig. 11.4. Because of the form (11-4.3) of $g_{\mu\nu}$, weak radial light signals that do not effect the total gravitational field, and hence travel along null-geodesics of this field, will have as representative paths the straight lines inclined at 45 deg to the υ axis in the Kruskal diagram. Consequently a signal originating in region I at point A will propagate in two directions as indicated, with one ray crossing the surface $r = 2m$ and eventually striking the intrinsic singularity at B. Thus an observer in region I cannot communicate directly with region III by means of light signals. However, the following interesting possibility exists. Suppose we require that a light signal, on reaching the point $r = 0$ be reflected there. Such a reflection represents a singular event on the path of the light ray so

[12] A. Einstein and N. Rosen, *Phys. Rev.*, **48**, 73 (1935).
[13] R. W. Fuller and J. A. Wheeler, *Phys. Rev.*, **128**, 919 (1962).

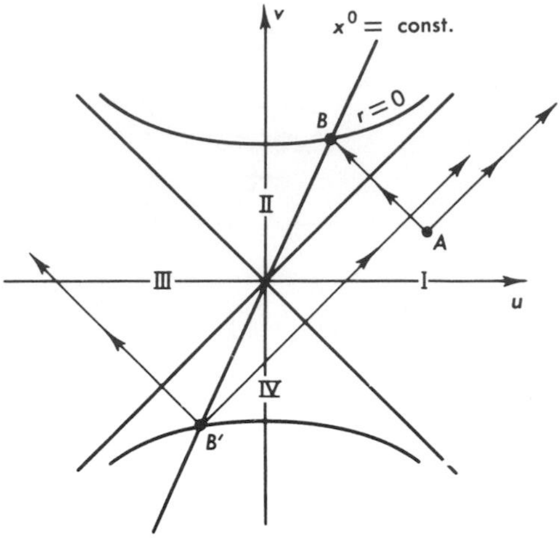

Fig. 11.4.

that this path cannot be represented by a single continuous null geodesic but rather by a combination of two such geodesics. A reflection at $r = 0$ can be represented by identifying the point B on the upper branch of the hyperbola $r = 0$ with a point on the lower branch that has the same value of x^0. This point is B' in Fig. 11.4. However, the light signal emanating from B' not only reaches region I, but also reaches region III. There are many other interesting features of the Kruskal manifold, but we cannot discuss them here.[14] We mention only one other. A free particle that travels along a radial geodesic crosses the $r = 2m$ boundary and continues to travel until it hits the singularity at $r = 0$.

11-5. Cylindrically Symmetric Fields

By definition a gravitational field is cylindrically symmetric if it can be transformed by a mapping in such a way that there exists a Killing vector of the form

$$\xi^\mu = (0, x^2, -x^1, 0).$$

If in addition the field is static, then it can be transformed to the general form

$$\begin{aligned} g_{00} &= e^{\mu}, \quad g_{0m} = 0, \quad g_{03} = 0, \\ g_{33} &= -e^{\nu-\mu}, \quad g_{3m} = 0, \\ g_{mn} &= e^{-\mu}[-\delta_{mn} + (1 - e^{-\nu})x_m x_n], \end{aligned} \tag{11-5.1}$$

[14] See J. A. Wheeler, "Geometrodynamics and the Issue of the Final State," in *Relativity, Groups and Topology* (Gordon and Breach, New York, 1964).

where $m, n = 1, 2$, $x_m = x^m/\rho$, and $\rho^2 = x^{1^2} + x^{2^2}$ and where μ and ν are arbitrary functions of ρ and $z \equiv x^3$. Solutions of the Einstein equations for fields of this form were first investigated by Weyl.[15] These solutions are interesting for both the manner in which they are obtained and the light they shed on the structure of the Einstein equations.

In those regions where $T_{\mu\nu} = 0$ the Einstein equations reduce to the following four equations:

$$\kappa_1 \equiv \frac{\partial^2 \mu}{\partial \rho^2} + \frac{1}{\rho}\frac{\partial \mu}{\partial \rho} + \frac{\partial^2 \mu}{\partial z^2} = 0, \tag{11-5.2}$$

$$\kappa_2 \equiv \frac{\partial \nu}{\partial \rho} - \frac{\rho}{2}\left[\left(\frac{\partial \mu}{\partial \rho}\right)^2 - \left(\frac{\partial \mu}{\partial z}\right)^2\right] = 0, \tag{11-5.3}$$

$$\kappa_3 \equiv \frac{\partial \nu}{\partial z} - \rho\frac{\partial \mu}{\partial \rho}\frac{\partial \mu}{\partial z} = 0, \tag{11-5.4}$$

$$\kappa_4 \equiv \frac{\partial^2 \nu}{\partial \rho^2} + \frac{\partial^2 \nu}{\partial z^2} + \frac{1}{2}\left[\left(\frac{\partial \mu}{\partial \rho}\right)^2 + \left(\frac{\partial \mu}{\partial z}\right)^2\right] = 0. \tag{11-5.5}$$

Because of the Bianchi identities these equations are not independent; the last equation is a consequence of the first three:

$$\kappa_4 \equiv \frac{\partial \kappa_2}{\partial \rho} + \frac{\partial \kappa_3}{\partial z} + \rho\frac{\partial \mu}{\partial \rho}\kappa_1 . \tag{11-5.6}$$

In addition

$$\frac{\partial \kappa_2}{\partial z} - \frac{\partial \kappa_3}{\partial \rho} \equiv \rho\frac{\partial \mu}{\partial z}\kappa_1 . \tag{11-5.7}$$

The first equation (11-5.2) is, surprisingly, nothing more than Laplace's equation in cylindrical coordinates for a function with rotational symmetry. It is a relatively simple matter to find solutions to this equation that satisfy the boundary condition $\mu = 0$ at spatial infinity, and which are finite and well behaved outside some finite spatial region where $T_{\mu\nu} \neq 0$. However, if we require that $(\partial \nu/\partial \rho)\, d\rho + (\partial \nu/\partial z)\, dz$ be an exact differential, then not all of these solutions are admissible. Using Eqs. (11-5.3) and (11-5.4) we see that μ must satisfy, in addition to Eq. (11-5.2), the equation

$$\frac{\partial}{\partial z}\left(\frac{\partial \nu}{\partial \rho}\right) - \frac{\partial}{\partial \rho}\left(\frac{\partial \nu}{\partial z}\right) = \frac{\partial}{\partial z}\left[-\frac{\rho}{2}\left\{\left(\frac{\partial \mu}{\partial \rho}\right)^2 - \left(\frac{\partial \mu}{\partial z}\right)^2\right\}\right] - \frac{\partial}{\partial \rho}\left[-\rho\frac{\partial \mu}{\partial \rho}\frac{\partial \mu}{\partial z}\right] = 0 \tag{11-5.8}$$

in regions where $T^{\mu\nu} = 0$.

[15] H. Weyl, *Ann. Physik*, **54**, 117 (1917); **59**, 185 (1919).

The solution of Eq. (11-5.2) corresponding to a point singularity located at $\rho = z = 0$ and satisfying the boundary condition $\mu = 0$ at infinity is given by

$$\mu = (\rho^2 + z^2)^{-1/2} \tag{11-5.9}$$

and is seen to satisfy Eq. (11-5.8) everywhere except at the origin. In fact one finds by direct integration that

$$\nu(\rho, z) = -\frac{\rho^2}{4(\rho^2 + z^2)^2}. \tag{11-5.10}$$

However, if there are two singularities on the z axis, that is, if

$$\mu = \frac{a_1}{r} + \frac{a_2}{r + b},$$

then in general Eq. (11-5.8) can no longer be satisfied on the line joining the two singularities.[16]

At this point one might ask why, since both μ and ν are singular only at the origin, one does not take this solution to be that of a point particle rather than the Schwarzschild solution. The answer is that the metric obtained by substituting the expressions for μ and ν from Eqs. (11-5.9) and (11-5.10) into Eq. (11-5.1) is *not* spherically symmetric. This solution must therefore correspond to a particle with a multipole structure.

Problem 11.4. Prove that $g_{\mu\nu}$ obtained from the expressions (11-5.9) and (11-5.10) is not spherically symmetric, by showing that it does not possess the necessory Killing vector structure of a spherically symmetric second-rank tensor.

Problem 11.5. Show that the axial-symmetric solution discussed here has a Riemann-Christoffel tensor that is type I in the Petrov classification scheme.

Recently Kerr[17] has found a solution of the empty-space Einstein equations that is axially symmetric but only stationary. For one form of the Kerr field, $ds^2 = g_{\mu\nu} dx^\mu \, dx^\nu$ is given by

$$ds^2 = (dx^0)^2 - (dx^1)^2 - (dx^2)^2 - dz^2 - \frac{2m\rho^3}{\rho^4 + a^2 z^2} k^2,$$

where ρ^2 is defined by

$$\rho^4 - (r^2 - a^2)\rho^2 - a^2 z^2 = 0, \qquad r^2 = (x^1)^2 + (x^2)^2 + z^2,$$

[16] For details, see P. G. Bergmann in *Handbuch der Physik*, Bd. 4 (Springer-Verlag, Heidelberg, 1962), p. 230.

[17] R. P. Kerr, *Phys. Rev. Letters*, **11**, 237 (1963).

and where k is defined by the equation

$$(\rho^2 + a^2)\rho k = \rho^2(x^1\, dx^1 + x^2\, dx^2) + a\rho(x^1\, dx^2 - x^2\, dx^1) + (\rho^2 + a^2)(z\, dz + \rho\, dx^0).$$

For $a = 0$, ds^2 reduces to

$$ds^2 = (dx^0)^2 - (dx^1)^2 - (dx^2)^2 - dz^2 - \frac{2\mathfrak{m}}{r}(dr + dx^0)^2,$$

which is the form of the Schwarzschild field given by Eddington. At present the interpretation of the Kerr field is still somewhat obscure. Kerr has suggested that it is the field of a spinning point mass with angular momentum $\mathfrak{m}a$. We mention finally that the Kerr field is type D in the Petrov classification scheme.

11-6. Null Gravitational Fields

A large class of solutions of the empty-space Einstein equations have been found for which the corresponding Riemann-Christoffel tensor is everywhere of type N in the Petrov classification scheme. The interest in these solutions lies in certain formal analogies they have with the plane-wave solutions of the empty-space Maxwell equations. But like the electromagnetic plane-wave solutions, these so-called null gravitational fields must be considered an idealization in that both types of solutions are unrelated to any source structure, as one would require of real wave solutions. The first type N fields were discovered by Brinkmann[18] in 1925, although he did not associate his solution with radiation. Later, Rosen[19] rediscovered a special N field but rejected its interpretation as a gravitational wave on the basis of physical arguments that are now considered incorrect. Finally Robinson[20] rediscovered the Brinkmann solutions and attributed to them the now accepted interpretation of "plane-fronted" waves.

The guiding idea in most discussions of wave-type solutions of the Einstein field equations is that the Riemann-Christoffel tensor $R_{\mu\nu\rho\sigma}$ plays the role of a field strength in the gravitational theory analogous to that played by the electromagnetic tensor $F_{\mu\nu}$, and that $g_{\mu\nu}$ is the "potential" of this field analogous to A_μ. The analogy is not complete, since $R_{\mu\nu\rho\sigma}$ depends on the second derivatives of $g_{\mu\nu}$ while $F_{\mu\nu}$ depends only on the first derivatives of A_μ. We must also emphasize the formal nature of this analogy. Many of the physical

[18] H. W. Brinkmann, *Math. Ann.*, **94**, 119 (1925).
[19] N. Rosen, *Phys. Z. Sowjet.*, **12**, 366 (1937).
[20] I. Robson, lecture at King's College, London, 1956.

properties of electromagnetic waves (for example, their ability to carry energy and momentum) have still to be shown to apply in the case of gravitational waves (cf. Secs. 13.1 and 13.2).

The Petrov classification of $R_{\mu\nu\rho\sigma}$ is analogous to the algebraic distinction between null and nonnull electromagnetic fields, the null fields being characterized by the conditions $F^{\mu\nu}F_{\mu\nu} = 0$ and $\varepsilon^{\mu\nu\rho\sigma}F_{\mu\nu}F_{\rho\sigma} = 0$. The plane-wave solutions discussed in Section 8-3 are seen to be null fields. Furthermore, these solutions have associated with them a lightlike, or null, vector k_μ that satisfies $F^{\mu\nu}k_\nu = 0$ and $F_{\mu[\nu}k_{\rho]} = 0$. Likewise one can prove that if $R_{\mu\nu\rho\sigma}$ is Petrov type N, then there also exists a null vector k_μ with the property that $R_{\mu\nu[\rho\sigma}k_{\lambda]} = 0$ and $R^{\mu}_{\nu\rho\sigma}k_\mu = 0$. Finally one can show[21] that in empty space, $R_{\mu\nu\rho\sigma}$ can suffer discontinuities only across a null surface. The discontinuity in $R_{\mu\nu\rho\sigma}$ has just the form that $R_{\mu\nu\rho\sigma}$ possesses in the case it is of type N.

The solutions found by Brinkmann and Robinson are all characterized by the existence of a covariantly constant vector l_μ that is colinear with the propagation vector k_μ. For reasons that we will not go into here, these solutions are called *plane-fronted* waves.[22] For such waves, $g_{\mu\nu}$ can be mapped to the form

$$g_{\mu\nu} = \begin{pmatrix} 0 & 1 & 0 & 0 \\ 1 & 2H & 0 & 0 \\ 0 & 0 & 1 & 0 \\ 0 & 0 & 0 & 1 \end{pmatrix} \tag{11-6.1}$$

where H is a function of x^1, x^2, x^3. The empty-space Einstein equations reduce in this case to the single equation

$$\frac{\partial^2 H}{\partial x^{2\,2}} + \frac{\partial^2 H}{\partial x^{3\,2}} = 0, \tag{11-6.2}$$

while $R_{\mu\nu\rho\sigma}$ has the form

$$R_{\mu\nu\rho\sigma} = \frac{1}{2}\,\delta^{0a}_{\mu\nu}\delta^{0b}_{\rho\sigma}\frac{\partial^2 H}{\partial x^a\,\partial x^b}, \qquad (a, b = 2, 3). \tag{11-6.3}$$

Therefore $g_{\mu\nu}$ is flat only if H is linear in or independent of x^2 and x^3.

Because a $g_{\mu\nu}$ of the form (11-6.1) is independent of x^0 it always has associated with it at least one Killing vector with components (1, 0, 0, 0). This vector is just the vector l_μ referred to above. Furthermore, special solutions admit motions with as many as six independent Killing vectors. The

[21] For references, see F. A. E. Pirani, "Survey of Gravitational Radiation Theory" in *Recent Developments in General Relativity* (Pergamon Press, New York, 1962), p. 95.

[22] For a discussion of the geometrical properties of plane-fronted waves, see J. Ehlers and W. Kundt, *loc. cit.*

solutions with five or six independent Killing vectors are called *plane waves*.[23] The appellation "plane wave" is used to describe these solutions because the motion group generated in the case of five Killing vectors is the same as the symmetry group of the plane waves of electrodynamics. For these plane wave solutions, H has the form

$$H(x^1, x^2, x^3) = \mathrm{Re}\{\tfrac{1}{2}A(x')e^{i\theta(x^1)}(x^2 + ix^3)^2\} \tag{11-6.4}$$

where A and θ are arbitrary functions of x^1 and can be interpreted as the amplitude and polarization angle of the wave. The manifold $-\infty < x^\mu < +\infty$ on which these plane waves are defined can be shown to be complete, thereby demonstrating the existence of complete solutions of the empty-space Einstein equations other than the flat-space solution.[24]

Problem 11.6. Show that three of the symmetries of the plane waves correspond to translations within the wave hypersurface $x^1 = \text{const}$, while the other two form a two-dimensional Abelian group of null rotations that leave the points $x^0 = \text{const}$, $x^1 = \text{const}$, $x^2 = x^3 = 0$ fixed. Compare these symmetries with those of a plane electromagnetic wave.

11-7. Solutions with Sources

So far we have considered solutions of the empty-space Einstein equations. There also exist solutions corresponding to nonvanishing $T_{\mu\nu}$, notably when $T_{\mu\nu}$ arises from an electromagnetic field or a matter fluid of some kind. In this section we discuss two such solutions, one due to the electromagnetic field and the other due to a spherically symmetric distribution of incompressible matter.

(a) Gravitational Field of a Point Electric Charge

To find the field due to a point charge we must use the Einstein equations (10-4.2) with

$$T_{\mu\nu} = \frac{1}{4\pi}\left\{\frac{1}{4}g_{\mu\nu}F_{\rho\sigma}F^{\rho\sigma} - F_{\mu\rho}F_\nu{}^\rho\right\}$$

obtained from the definition (10-4.11) and the expression (10-4.14) for the

[23] H. Bondi, F. A. E. Pirani, and I. Robinson, *Proc. Roy. Soc.* (London), **A251**, 519 (1959).

[24] J. Ehlers and W. Kundt, *loc. cit.*

action of the electromagnetic field together with the Maxwell equations (10-6.13). A solution of these equations corresponding to a point charge was first given by Reissner.[25] The gravitational field has components given by

$$g_{00} = 1 - \frac{2Gm}{r} + \frac{Ge^2}{r^2},$$

$$g_{0r} = 0, \tag{11-7.1}$$

$$g_{rs} = -\delta_{rs} + \left(1 - \frac{1}{g_{00}}\right)\frac{x^r x^s}{r^2}.$$

The last term in the expression for g_{00} arises from the electrostatic energy of the charge and is comparable in magnitude to the Schwarzschild mass term for an electron only at distances of the order of the electron radius $e^2/m \sim 10^{-13}$ cm. We note also that g_{00} has no real zeros if $e^2Gm^2 > 1$. For an electron, $e^2Gm^2 \sim 10^{40}$, so that the Reissner solution is regular all the way down to the singular point $r = 0$. The electromagnetic potential can also be obtained and is given by

$$A_\mu = \left(\frac{e^2}{r}, \mathbf{0}\right). \tag{11-7.2}$$

(b) Gravitational Field of an Incompressible Ball of Fluid

The first solution of the Einstein equations corresponding to a mass source of finite size was given by Schwarzschild.[26] He considered a sphere of incompressible fluid as this source. The stress-energy tensor $T^{\mu\nu}$ that appears in the Einstein equations can, in this case, be taken as that of a fluid given by Eq. (9-2.2), with $\eta^{\mu\nu}$ replaced by $g^{\mu\nu}$. Thus

$$T^{\mu\nu} = (e + p)u^\mu u^\nu - pg^{\mu\nu}.$$

Following Schwarzschild we look for a solution of the Einstein equations that is both static and spherically symmetric. From the discussion of Section 11-3 it follows that in this case, $g_{\mu\nu}$ can be transformed to the form given by Eq. (11-3.1), where ν and λ are now functions of r alone. Furthermore the spatial components of the four-velocity u^μ must vanish so that only u^0 is nonzero. Then, since $u_\mu u^\mu = 1$, it follows that $u^0 = u_0{}^{-1} = 1/\sqrt{g_{00}}$.

If we lower one of the indices in Eq. (10-4.2) and make use of Eqs. (11-3.2),

[25] H. Reissner, *Ann. Phys. Leipzig*, **50**, 106 (1916).

[26] K. Schwarzschild, *S. B. preuss. Akad. Wiss. Berlin* (1916), p. 424.

(11-3.3), (11-3.4), and (11-3.5) for $G_\nu{}^\mu$, we obtain the following set of equations for $g_{\mu\nu}$, e, and p:

$$G_0{}^0 = e^{-\lambda}\left(\frac{1}{r^2} - \frac{\lambda'}{r}\right) - \frac{1}{r^2} = \kappa e, \tag{11-7.3a}$$

$$G_1{}^1 = e^{-\lambda}\left(\frac{1}{r^2} + \frac{\nu'}{r}\right) - \frac{1}{r^2} = -\kappa p, \tag{11-7.3b}$$

$$G_2{}^2 = G_3{}^3 = \frac{1}{2}e^{-\lambda}\left[\nu'' + \frac{1}{2}\nu'^2 + \frac{\nu' - \lambda'}{r} - \frac{1}{2}\nu'\lambda'\right]$$
$$= -\kappa p, \tag{11-7.3c}$$

where, as before, a prime indicates differentiation with respect to r. Because of the Bianchi identities satisfied by $G_\nu{}^\mu$, it follows that $T^\mu_{\nu;\mu} = 0$. From this equation one obtains the result that

$$p' = -\tfrac{1}{2}\nu'(p + e). \tag{11-7.4}$$

We can therefore use this equation to replace one of the Eqs. (11-7.3) when it proves convenient. We see that in effect we have three equations to determine the four unknowns λ, ν, e, and p. To obtain a determinant system of equations it is therefore necessary to know the caloric equation of state for the fluid that relates e and p. Our procedure here, however, will be to assume a functional dependence of e on r, compute λ and ν from this knowledge, and then finally compute p.[27]

We can integrate Eq. (11-7.3a) directly. If we multiply this equation by r^2 we obtain

$$\kappa e r^2 = (e^{-\lambda}r)' - 1$$

so that

$$e^{-\lambda} = 1 + \frac{\kappa}{4\pi r}\mathscr{E}(r) + \text{const}\cdot\frac{1}{r}, \tag{11-7.5}$$

where

$$\mathscr{E}(r) = 4\pi\int_0^r e(r')r'^2\,dr'.$$

In order that $g_{\mu\nu}$ be regular at $r = 0$, we see that we must take the constant in Eq. (11-7.5) to be zero. Then, if $e(r) = 0$ for $r > r_0$, $e^{-\lambda}$ takes on the Schwarzschild form given by Eq. (11-3.10), with

$$\mathfrak{m} = -\frac{\kappa}{8\pi}\mathscr{E}(r_0). \tag{11-7.6}$$

[27] For a discussion of the integration of equations (11–7.3) for various equations of state that might exist in the interior of a star, see I. Iben, Jr., *Astro. J.*, **138**, 1090 (1963).

We can interpret $\mathscr{E}(r)$ as the total internal energy of the fluid contained within a sphere of radius r, so that $\mathscr{E}(r_0)$ is the total internal energy of the fluid and (divided by c^2) is equal to the mass appearing in the Schwarzschild field.

To obtain an explicit expression for v we now make the Schwarzschild assumption that e is a constant for $r \leqq r_0$. While admittedly e will not be constant throughout the interior of a large star, the general features of a more realistic equation of state will be reflected in the solutions we will obtain with this assumption. To obtain the expression for v, we first integrate Eq. (11-7.4), which we can easily do because of the assumed constancy of e. We obtain

$$e + p = \text{const } e^{-(1/2)v}. \tag{11-7.7}$$

If we then subtract Eq. (11-7.3b) from Eq. (11-7.3a) and make use of this result, we obtain the equation

$$e^{(1/2)v}e^{-\lambda}\left(\frac{\lambda'}{r} + \frac{v'}{r}\right) = \text{const.} \tag{11-7.8}$$

For $e(r) = e$, the expression (11-7.5) for $e^{-\lambda}$ can be written as

$$e^{-\lambda} = 1 - \frac{r^2}{R^2},$$

where

$$R^2 = -\frac{3}{\kappa e}.$$

If we substitute this expression for $e^{-\lambda}$ into Eq. (11-7.8) we obtain

$$e^{(1/2)v}\left(\frac{2}{R^2} + \frac{v'}{r} - \frac{rv'}{R^2}\right) = \text{const},$$

which can be integrated to give

$$e^{(1/2)v} = A - B\sqrt{1 - \frac{r^2}{R^2}} \tag{11-7.9}$$

where A and B are two constants of integration.

Let us now make use of Eq. (11-7.3b) to compute the pressure in the fluid. A short calculation leads to the result that

$$-\kappa p = \frac{1}{R^2}\left\{\frac{3B\sqrt{1 - r^2/R^2} - A}{A - B\sqrt{1 - r^2/R^2}}\right\}. \tag{11-7.10}$$

We can fix the constants A and B by requiring that $p = 0$ on the surface of the sphere and that e^{ν} join on smoothly to the Schwarzschild field on the surface. We find, then, that

$$A = \frac{3}{2}\sqrt{1 - \frac{r_0^{\,2}}{R^2}}, \qquad B = \frac{1}{2},$$

so that

$$e^{(1/2)\nu} = \frac{3}{2}\sqrt{1 - \frac{r_0^{\,2}}{R^2}} - \frac{1}{2}\sqrt{1 - \frac{r^2}{R^2}} \tag{11-7.11}$$

and

$$p = e\left\{\frac{\sqrt{1 - r^2/R^2} - \sqrt{1 - r_0^{\,2}/R^2}}{3\sqrt{1 - r_0^{\,2}/R^2} - \sqrt{1 - r^2/R^2}}\right\}. \tag{11-7.12}$$

Thus, in order for the solution to be real, we must have

$$r_0^{\,2} < R^2 = -\frac{3}{\kappa e} \tag{11-7.13}$$

This condition then ensures that the field outside the fluid will be nonsingular. If we require that the pressure inside the fluid be everywhere finite, we obtain from Eq. (11-7.12) the somewhat more restrictive condition

$$r_0^{\,2} < \tfrac{8}{9}R^2. \tag{11-7.14}$$

We see incidentally that $p > 0$ everywhere for our solution, which we would require in order that our solution be physically reasonable. The solution to the field equations given here inside the fluid is sometimes referred to as the *Schwarzschild interior solution* and the field outside is referred to as the *Schwarzschild exterior solution.*

The condition (11-7.14) tells us that for a given energy density e, there is a maximum amount of matter that can be packed into a sphere and still lead to a static solution of the Einstein equations. For an incompressible fluid Eq. (11-7.6) gives, as the total mass of the fluid,

$$m = \frac{4\pi}{3}\, e r_0^{\,2}$$

so that the critical mass for a static configuration is given by

$$m_{\text{crit}} = \left(\frac{8}{9}\right)^2 \left(\frac{4\pi}{3}\right) \left(\frac{-3}{\kappa e}\right)^{3/2}.$$

For masses greater than m_{crit} the fluid would begin to collapse as a result of the unsupported gravitational attraction between its various parts. Once begun, such a contraction would continue unabated until all fluid became concentrated in a point! Even an infinite pressure would not halt the collapse. In fact it would only accelerate it, since it would in effect correspond to an infinite interaction energy, which in turn would produce an infinite gravitational field.

While the assumption of incompressibility is unrealistic from a physical point of view, the use of a more realistic equation of state does not alter in any essential way the above conclusion. Oppenheimer and Volkoff[28] have considered the case of a cold neutron gas. Matter in this state might be imagined to exist in a large star after all thermonuclear burning had taken place and gravitational forces had overwhelmed the pressure of the electron gas. If these forces were sufficiently strong, inverse beta decay would take place and eventually all electrons would be combined with protons to form neutrons.[29] Oppenheimer and Volkoff integrated the Einstein equations when such matter acts as the source of the gravitational field and showed that no stable solution exists for a total mass exceeding 0.7 solar masses. For larger masses no static solutions of these equations exist and the star would undergo gravitational collapse.[30] What happens to matter as it is compressed into an ever decreasing volume is still an open question. Even the assumption of a hard nuclear core will not inhibit the collapse, since, at best, it would lead to the case of incompressibility, and even this extreme case is insufficient to halt the collapse once the critical mass has been exceeded. Wheeler has suggested that at sufficiently high densities, mass is completely converted into radiation. Alternatively, one can imagine that beyond a certain point it is no longer permissible to treat the gravitational field as a classical field but that it, too, must be treated as a quantized system. In any event, essentially new and at present unknown physical laws must come into play.

[28] J. R. Oppenheimer and G. Volkoff, *Phys. Rev.*, **55**, 374 (1939).

[29] For a discussion of the equation of state in large stars, see J. A. Wheeler, "The Superdense Star and the Critical Nucleon Number," in *Gravitation and Relativity*, Chiu and Hoffmann, eds. (Benjamin, New York, 1964), Chap. 10.

[30] For a discussion of the dynamics of gravitational collapse, see J. R. Oppenheimer and H. Snyder, *Phys. Rev.*, **56**, 455 (1939).

12

Experimental Tests of General Relativity

Many scientists hold that the general theory of relativity is one of the most beautiful constructs of the human mind. If so, it is all the more the pity that it has been of such limited value in our attempts to describe nature. There are, in fact, today only four pieces of experimental evidence to support the theory. Furthermore, all this evidence can be encompassed within other theories, but nowhere in as satisfactory a way as it can in general relativity. Finally, only two pieces of evidence bear directly on the form of the Einstein field equations and one of these is still poorly determined.

The most accurate test of general relativity is the observed constancy of the ratio of gravitational to inertial mass of material bodies discussed in Section 10-2. We must consider the Eötvös-type experiments to be tests of general relativity, since only in this theory do all structureless material particles move in a way that is independent of this ratio. There is, for instance, nothing in Newtonian gravitational theory that rules out the possibility of the existence of gravitationally uncharged bodies. Of course the Eötvös experiments cannot tell us anything about the structure of the Einstein field equations; they verify only the prediction of the theory concerning the coupling of material particles to the gravitational field. Likewise, observations on the so-called gravitational red shift supports the predictions of the theory concerning the coupling of the electromagnetic to the gravitational field. In both cases these predictions assumed the validity of the principle of minimal coupling.

The two observed effects that bear directly on the form of the Einstein field equations are the advance of the perihelia of the planetary orbits and the bending of light in the gravitational field of the sun. Of these, the first is the better established; to date, the bending of light is only in qualitative agreement with the theory, owing to the difficulty in observing the effect. While both effects can be accounted for by other theories, only in general relativity are the magnitudes of the effects unique. In the Whitehead theory, for instance, there is an adjustable parameter that can accommodate any amount of perihelion advance or retardation.

In this chapter we derive the predictions of general relativity that have been tested observationally to date and discuss these observations briefly. We also discuss briefly some of the recent new proposals for testing the theory.

12-1. Motion of Test Bodies in a Schwarzschild Field and the Advance of Planetary Perihelia

According to our discussion in Section 10-6 the motion of a test body in a gravitational field is governed by the geodesic equation (10-6.1). We are interested here in studying the motion of test bodies in a Schwarzschild gravitational field, we saw that for large values of the radial coordinate r, this field approximates that of a massive body in Newtonian gravitational theory. Taking the body to be the sun and the test body to be a planet, we can work out the deviations, if any, of the planetary orbits, as predicted by Newtonian theory, due to general relativistic effects.

Rather than work with the geodesic equations in the form (10-6.1) it is more convenient to use the equivalent equations

$$-\frac{d}{d\tau}(g_{\mu\nu}\dot{z}^{\mu}) + \tfrac{1}{2}g_{\rho\sigma,\nu}\dot{z}^{\rho}\dot{z}^{\sigma} = 0 \tag{12-1.1}$$

together with the supplementary condition

$$g_{\mu\nu}\dot{z}^{\mu}\dot{z}^{\nu} = 1. \tag{12-1.2}$$

For a static gravitational field $g_{\mu\nu}$ of the form (11-3.1), Eqs. (12-1.1) reduce to

$\nu = 0$

$$-\frac{d}{d\tau}(e^{\nu}\dot{z}^{0}) = 0. \tag{12-1.3a}$$

$\nu = 1$

$$\frac{d}{d\tau}(e^{\lambda}\dot{r}) + \frac{1}{2}\nu' e^{\nu}(\dot{z}^{0})^{2} - \frac{1}{2}\lambda' e^{\nu}\dot{r}^{2} - r\dot{\theta}^{2} - r\sin^{2}\theta\dot{\phi}^{2}. \tag{12-1.3b}$$

$\nu = 2$

$$\frac{d}{d\tau}(r^{2}\dot{\theta}) - r^{2}\sin\theta\cos\theta\dot{\phi}^{2} = 0. \tag{12-1.3c}$$

$\nu = 3$

$$\frac{d}{d\tau}(r^{2}\sin^{2}\theta\dot{\phi}) = 0. \tag{12-1.3d}$$

In these equations we have taken the particle coordinates to be $\{z^{0}, r, \theta, \phi\}$.

If initially $\theta = \pi/2$ and $\dot{\theta} = 0$, it follows from Eq. (12-1.3c) that $\ddot{\theta}$ and all higher derivatives vanish initially so that the orbit will be confined to the $\theta = \pi/2$ plane. Eq. (12-1.3d) can now be integrated, giving

$$r^2\dot{\phi} = h \tag{12-1.4}$$

where h is a constant. Likewise, Eq. (12-1.3a) can be integrated and yields the result

$$\dot{z}^0 = \alpha e^{-\nu} \tag{12-1.5}$$

Rather than try to integrate Eq. (12-1.3b) it is easier to use Eq. (12-1.2), which gives an integral of the equations (12-1.1) that is independent of those already found. In the present case this equation reduces to

$$e^{\nu}(\dot{z}^0)^2 - e^{\lambda}\dot{r}^2 - r^2\dot{\phi}^2 = 1. \tag{12-1.6}$$

If we now make explicit use of the Schwarzschild field (11-3.10) and the two integrals (12-1.4) and (12-1.5) we can rewrite Eq. (12-1.6) in the form

$$\left(\frac{h}{r^2}\frac{dr}{d\phi}\right)^2 + \frac{h^2}{r^2} = \alpha^2 - 1 + \frac{2\mathfrak{m}}{r}\left(1 + \frac{h^2}{r^2}\right), \tag{12-1.7}$$

or, setting $1/r = u$,

$$\left(\frac{du}{d\phi}\right)^2 + u^2 = \frac{\alpha^2 - 1}{h^2} + \frac{2\mathfrak{m}}{h^2}u + 2\mathfrak{m}u^3. \tag{12-1.8}$$

This equation is thus the equation of the orbit of the test body, that is, the equation of the projection of the trajectory of the test body onto the space defined by the timelike Killing vector of the Schwarzschild field.

We can integrate this equation directly to obtain u as a function of ϕ in terms of elliptic integrals. However, this form is difficult to work with and is not necessary for our purposes. To proceed we differentiate Eq. (12-1.8) to obtain

$$\frac{d^2u}{d\phi^2} + u = \frac{\mathfrak{m}}{h^2} + 3\mathfrak{m}u^2. \tag{12-1.9}$$

Aside from the last term on the right, this equation is the same as the corresponding Newtonian equation for an orbit in the field of a massive body, where h is given by $r^2\, d\phi/dt$. If we calculate the ratio of the last term on the right to the first, or "Newtonian," term and use the values of h and r for the planet Mercury, we find for the ratio the value $3h^2/r^2 \sim 7.7 \times 10^{-8}$. For the other planets the value of this ratio is even smaller. It is therefore reasonable to treat the term $3\mathfrak{m}u^2$ as a perturbation in solving Eq. (12-1.9).

If we neglect the term $3\mathfrak{m}u^2$, the solution to Eq. (12-1.9) is

$$u = \frac{\mathfrak{m}}{h^2}(1 + e\cos\phi), \tag{12-1.10}$$

which is the equation of an ellipse with eccentricity e. Substitutions of this solution into the term $3\mathfrak{m}u^2$ in Eq. (12-1.9) then gives us an equation for u_1, the general relativistic correction to the Newtonian orbit (12-1.10):

$$\frac{d^2u_1}{d\phi^2} + u_1 = \frac{3\mathfrak{m}^2}{h^4} + \frac{6\mathfrak{m}^3}{h^4} e \cos \phi + \frac{3\mathfrak{m}^3}{2h^4} e^2(1 + \cos 2\,\phi) \quad (12\text{-}1.11)$$

For the observed planetary eccentricities, the last term on the right side of Eq. (12-1.11) is small compared with the first two terms. Its effect is to produce small periodic variations in u. Furthermore the general solution of the equation

$$\frac{d^2u_1}{d\phi^2} + u_1 = \frac{3\mathfrak{m}^3}{h^4}$$

is of the form of the unperturbed solution (12-1.10), and so when added to this solution results in slight changes in the constants appearing therein. These new constants can then be taken to be equal to the observed values. Thus we need only a particular solution to the equation

$$\frac{d^2u_1}{d\phi^2} + u_1 = \frac{6\mathfrak{m}^3}{h^4} e \cos \phi.$$

The desired solution is

$$u_1 = \frac{6\mathfrak{m}^3}{h^2} e\, \phi \sin \phi,$$

which must be added to the unperturbed solution (12-1.11). Thus the approximate solution we seek is given by

$$u = \frac{\mathfrak{m}}{h^2}\left(1 + e \cos \phi\right) + \frac{3}{2}\frac{k^2}{h^2} e\phi \sin \phi$$

$$\simeq \frac{\mathfrak{m}}{h^2}(1 + e \cos (\phi - \delta\tilde{\omega})),$$

where

$$\delta\tilde{\omega} = 3\,\frac{\mathfrak{m}^2}{h^2}\,\phi$$

and where we have neglected terms in $(\delta\tilde{\omega})^2$. Consequently there will be an *advance* in the perihelion of the orbit per revolution (see Fig. 12.1), given by

$$\delta\tilde{\omega} = 6\pi\,\frac{\mathfrak{m}^2}{h^2}\,.$$

By using the value (11-3.9) for $\mathfrak{m}$ and the value $Gma(1 - e^2)$ of h^2 for the

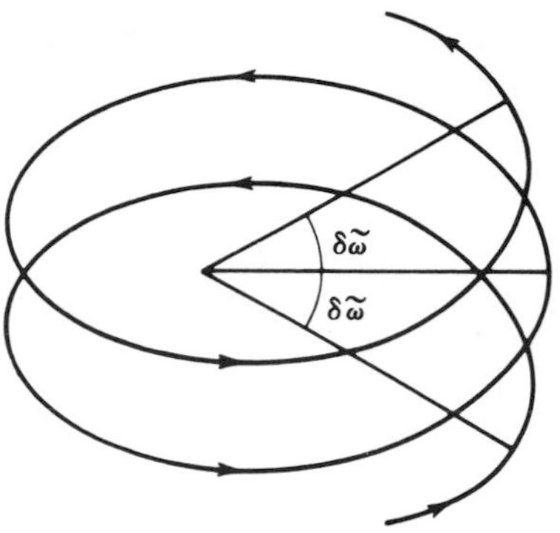

Fig. 12.1.

corresponding Newtonian orbit, where a is the semimajor axis of the orbit, we obtain for the perihelion advance[1]

$$\delta\tilde{\omega} = \frac{6\pi m G}{c^2 a(1 - e^2)} \tag{12-1.12}$$

in radians per revolution. If we make use of the relation $a^3 = GM(T/2\pi)^2$ (that is, Kepler's third law), we can also express this precession rate as

$$\delta\tilde{\omega} = \frac{24\pi^3 a^2}{c^2 T^2 (1 - e^2)}.$$

We list below the values of $\delta\tilde{\omega}$ and $e\,\delta\tilde{\omega}$ per century for the four minor planets.

Planet	$\delta\tilde{\omega}$	$e\,\delta\tilde{\omega}$
Mercury	$+43''.03$	$+8''.847$
Venus	$+8''.6$	$+0''.059$
Earth	$+3''.8$	$+0''.064$
Mars	$+1''.35$	$+0''.13$

It was already known at the time of Leverrier[2] that there was an advance in the perihelion of Mercury, which could not be accounted for on the basis of Newtonian theory by taking account of perturbations due to other planets. At the end of the past century Newcomb[3] calculated this residual advance to

[1] This result was first obtained by Einstein; see *Berlin sitz.*, 831 (1915).

[2] U. J. Leverrier, *Ann. Obs. Paris*, **5** (1859).

[3] S. Newcomb, *Astron. papers*, *Washington*, **6**, 108 (1898).

have a value of $41''.24 \pm 2''.09$. In the table below, given by Duncombe,[4] we summarize the present situation.

	Mercury	Venus	Earth
$(T)e\,\delta\tilde{\omega}$	$1142''.730 \pm 0''.040$	$34''.472 \pm 0''.006$	$103''.520 \pm 0''.004$
$(0)e\,\delta\tilde{\omega}$	$1151''.593 \pm 0''.084$	$34''.529 \pm 0''.032$	$103''.604 \pm 0''.020$
$(0-T)e\,\delta\tilde{\omega}$	$8''.863 \pm 0''.093$	$0''.057 \pm 0''.033$	$0''.084 \pm 0''.020$
$(0-T)\,\delta\tilde{\omega}$	$43''.11 \pm 0''.45$	$8''.4 \pm 4''.8$	$5''.0 \pm 1''.2$

where all values are per century. Here, T refers to the value of the relevant quantity as calculated by Newtonian theory. The errors are due to uncertainties in the known masses of the planets. Also, 0 refers to the observed values of the relevant quantities and $0 - T$ refers to the difference between the observed and calculated values. It should be borne in mind that the values quoted in this table are obtained by first subtracting off an overall precession in all observed planetary orbits of amount 5025″ per century due to the precession of the Earth's equinox. One sees that the residual precessions represent extremely small corrections to the actually observed precessions.

If we compare the two tables above, we see that the Einstein gravitational theory is able to account for the residual precessions in all cases. However, before we take this agreement as support of the Einstein theory, we must enter the following caveats. Except in the case of Mercury, the observed residual precessions represent a small fraction of the total observed precessions, which include the equinoctal precession of the Earth. Considerably more observational data must be accumulated before we can place complete confidence in the agreement between theory and observation in the case of Venus and the Earth. In this regard, it would be desirable to measure the perihelion advance for Mars, which, after Mercury, has the largest value of $e\delta\tilde{\omega}$, the actually observed quantity.

Recently Dicke[5] has suggested that at least part of the residual perihelion advances of the planets might be due to an oblateness of the sun. According to Dicke, an oblateness that lies within the present observational limits could account for 20 percent on the residual advance in the case of Mercury.

Finally it is necessary to raise the question of how one determines the reference line against which the perihelion advances, and in fact against which all planetary motions are measured. A complete discussion of this question would very quickly lead us into the realm of cosmology. Customarily the distant stars are used to fix this reference line, sometimes called the *compass of inertia*. It is known, however, that these stars, being part of the Milky Way

[4] R. L. Duncombe, *Astron. J.*, **61**, 174 (1956).

[5] R. H. Dicke, "Remarks on the Observational Basis of General Relativity," in *Gravitation and Relativity*, Chiu and Hoffmann, eds. (Benjamin, New York, 1964).

galaxy, partake of the systematic rotation of this galaxy as measured with respect to the more distant galaxies, and this fact is taken into account in determining the equinoctal precession rate. Whether the distant galaxies themselves are rotating with respect to even more distant galaxies or the universe of galaxies as a whole is rotating in some absolute sense, and how such a rotation would affect the compass of inertia are still open questions. In this connection we should mention the alternative possibility of defining the compass of inertia by means of a Foucault pendulum. Whether the two definitions agree is also not known.

A number of proposals have been made recently to test further the orbital corrections predicted by the Einstein theory.[6] For the minor planet Icarus, $\delta\tilde{\omega}$ has a theoretical value of 8″.3. However, a detailed analysis shows that the perihelion advance can be determined with an accuracy four to five times better than in the case of Mercury. It has also been suggested that an artificial planet might be put into orbit around the sun for which $\delta\tilde{\omega}$ could have a value of the order of 1000″ per century. The possibility of observing the motion of such a planet by the use of radar could also increase materially the accuracy of the relevant data. Likewise, artificial Earth satellites can have large values of $\delta\tilde{\omega}$; for the nearest satellites it can be as large as 1700″. However, in this case, our imprecise knowledge of the figure of the earth would cause difficulties in analyzing the effect.

Let us finally consider the possibility of explaining the residual perihelion advances of the planets by means other than general relativity. Eddington[7] has analyzed the motion of test bodies in radial gravitational fields other than Schwarzschild. (These fields, of course, will not in general satisfy the Einstein equations.) Taking $g_{\mu\nu}$ to have the general form (11-3.1) but with

$$e^{\nu} = \left[1 - \frac{2\mathfrak{m}}{r} + \alpha_2\left(\frac{2\mathfrak{m}}{r}\right)^2 + \cdots\right] \tag{12-1.13a}$$

and

$$e^{\lambda} = \left[1 + \beta_1\frac{2\mathfrak{m}}{r} + \beta_2\left(\frac{2\mathfrak{m}}{r}\right)^2 + \cdots\right], \tag{12-1.13b}$$

he finds that

$$\frac{\delta\tilde{\omega}}{\phi} \cong \frac{\mathfrak{m}G}{c^2 a(1-e^2)}(2 + \beta_1 - 2\alpha_2) \tag{12-1.14}$$

(The coefficient of $2\mathfrak{m}/r$ in the expression for e^{ν} has been taken as 1 in order to obtain the usual Newtonian orbits to lowest order in this quantity. This

[6] For a more detailed discussion of these proposals and references, see V. L. Ginsburg, "Experimental Verifications of the General Theory of Relativity" in *Recent Developments in General Relativity* (Pergamon Press, New York, 1962), pp. 57–71.

[7] A. S. Eddington, *The Mathematical Theory of Relativity*, 2d ed. (Cambridge Univ. Press, Cambridge, 1930), Sec. 47.

expression is seen to agree with observation if we take $\beta_1 = 1$ and $\alpha_2 = 0$, which are just the values that these constants have when $g_{\mu\nu}$ takes on its Schwarzschild values.

The Whitehead theory discussed in Section 7-21 also gives the same prediction for the perihelion advance if we take $\alpha = 2$, $\beta = -4$ in Eq. (7-21.1). However, it is able to do so by admitting an additional adjustable parameter in the theory. (The sum $\alpha + \beta$ is fixed, as we saw, by the requirement that the theory yield the usual Newtonian results in first approximation.) The Einstein theory, on the other hand, contains no such additional parameter and is therefore to be preferred on the grounds of simplicity. Nevertheless the final decision in favor of one of these two opposing theories must rest ultimately on which one better describes nature. Unfortunately differences in their predictions with regard to planetary orbits depend on terms proportional to $(2m/r)^2$, and there is little hope of detecting such effects in the foreseeable future.

Problem 12.1. Make use of the approximate solution (11-1.7) to discuss the effect of the sun's rotation on the planetary orbits. Show that the main effect is a secular shift in the perihelion given by

$$\delta\tilde{\omega}_{\text{rot}} = \frac{96}{5} \frac{\pi^3 {r_0}^2 \cos\psi}{c^2 \tau T(1 - e^2)^{3/2}} \quad \text{rad/sec}, \tag{12-1.15}$$

where r_0 is the radius of the sun, τ its period, and ψ is the angle between the plane of the orbit and the sun's equatorial plane.

12-2. Bending of Light in a Schwarzschild Field

The trajectory of a weak light ray is determined, according to the discussion of Section 10-6, by the null geodesic satisfying Eqs. (10-6.16). A ray for which initially $\theta = \pi/2$ and $\dot{\theta} = 0$ satisfies the same set of equations as those satisfied by a test body, with the single modification that the 1 on the right side of Eq. (12-1.6) is replaced by 0. An analysis similar to that used to obtain Eq. (12-1.9) leads now to the equation

$$\frac{d^2u}{d\phi^2} + u = 3mu^2 \tag{12-2.1}$$

for the orbit of the light ray.

We will solve Eq. (12-2.1) by an approximation method similar to that used in solving Eq. (12-1.9) by treating the right side as a perturbing term. Therefore, in lowest order, u satisfies

$$\frac{d^2u}{d\phi^2} + u = 0,$$

which has as a solution the straight line

$$u = \frac{\cos\phi}{R}.$$

We see that $r = 1/u$ has a minimum value R when $\phi = 0$. Substituting into the right side of Eq. (12-2.1) then gives

$$\frac{d^2u_1}{d\phi^2} + u_1 = 3\frac{\mathfrak{m}}{R^2}\cos^2\phi$$

for the perturbation of the straight-line path. A particular integral is

$$u_1 = \frac{\mathfrak{m}}{R^2}(\cos^2\phi + 2\sin^2\phi),$$

so that the desired approximate solution to Eq. (12-2.1) is

$$u = \frac{\cos\phi}{R} + \frac{\mathfrak{m}}{R^2}(\cos^2\phi + 2\sin^2\phi). \tag{12-2.2}$$

If we multiply both sides of this equation by rR and map back to Cartesian coordinates $x = r\cos\phi$, $y = r\sin\phi$, we find

$$x = R - \frac{\mathfrak{m}}{rR}\frac{x^2 + 2y^2}{\sqrt{x^2 + y^2}}. \tag{12-2.3a}$$

For large values of $|y|$, Eq. (12-2.3a) becomes

$$x \simeq R - \frac{2\mathfrak{m}}{R}|y|. \tag{12-2.3b}$$

Thus, asymptotically, the orbit is a straight line in space, as we should expect because asymptotically ($r \to \infty$), the Schwarzschild field $g_{\mu\nu} = \eta_{\mu\nu}$. The angle θ between the two asymptotes is seen to be (in radians)

$$\theta = \frac{2\mathfrak{m}}{R} = \frac{4Gm}{Rc^2} \tag{12-2.4}$$

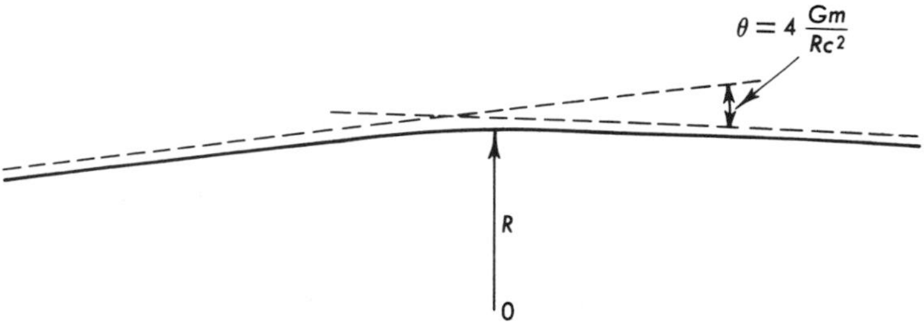

Fig. 12.2.

and represents the amount of "bending" of a light ray in passing through a Schwarzschild field (see Fig. 12.2).

On the basis of the above discussion we can expect that the light of distant stars should be deflected as it passes near a massive body such as the sun. Such a deflection should manifest itself in the relative displacement of the stars in a field of view when the sun occupied a part of this field. If we take m to be the mass of the sun and R its optical radius in Eq. (12-2.4), then θ has a value of $1''.75$. Attempts to observe such a deflection were first carried out by the British solar eclipse expeditions of 1919 and have continued down to the present time. Since it is necessary to observe a star field in the neighborhood of the sun, these observations have all been carried out during total eclipses, usually in remote parts of the earth.[8] In the following table we list the results of these observations.

Eclipse	Number of stars	R_{min}	Observer's mean value for θ in seconds of arc	Different analysis
1919	7	2	1.98 ± 0.12	2.0 to 2.2
				2.16 ± 0.14
				2.06
				1.95 ± 0.09
	5	2	1.61 ± 0.30	
1922				
	92	2.1	1.72 ± 0.11	2.2
				2.14 ± 0.18
				2.12
				2.00
				1.83 ± 0.20
	145	2.1	1.82 ± 0.15	2.1
				2.07
	14	2	1.18 to 2.35	
	18	2	1.42 to 2.16	
1929	17	1.5	2.24 ± 0.10	1.98 ± 0.20
				1.75 ± 0.13
				2.06
				1.96 ± 0.11
1936	25	2	2.71 ± 0.26	2.70 ± 0.40
1936	8	4	1.28 to 2.13	
1947	51	3.3	2.01 ± 0.27	2.01 ± 0.27
				2.20 ± 0.35
1952	10	2.1	1.70 ± 0.10	1.43 ± 0.16

[8] For a detailed analysis of these observations, see B. Bertotti, D. Brill, and R. Krotkov, *Experiments on Gravitation in Gravitation: An Introduction to Current Research*, L. Witten, ed. (Wiley, New York, 1962).

In this table R_{min} represents the minimum value of R (in solar radii) corresponding to the apparent distance of the closest observed star from the sun's center. The values of θ therefore represent an extrapolation of the data to the sun's edge. The uncertainties quoted in the fourth column and the results of different analyses of the data given in the fifth column attest to the difficulties inherent in these measurements. In all these observations the amount of bending was found to decrease as the distance of the observed star from the sun's edge increased. However, the accuracy was not sufficient to verify a $1/R$ dependence, as predicted by Eq. (12-2.4). Nevertheless the amount of bending in all cases is of the same order of magnitude as that predicted by general relativity and in no instance agrees with a prediction from Newtonian theory (see below) for θ of $0''.875$. There is, however, a clear need for further observational data. In this connection the suggestion has been made to use photoelectric methods to observe stars near the sun at times other than during an eclipse.

Let us now inquire into the possibility of explaining the observed bending of light near the sun's edge by means other than general relativity. Actually Soldner predicted in 1801, a deflection of half that of general relativity based on a corpuscular theory of light and Newtonian mechanics. By considering a gravitational field of the general form (12-1.13) and the equations of a null geodesic, Eddington found that the angle of deflection is given, to lowest order in $2m/r$, by

$$\theta = \frac{2Gm}{Rc^2}(1 + \beta_1), \tag{12-2.5}$$

which gives values of θ in agreement with observation when β_1 has the value 1, as it does for the Schwarzschild field. On the other hand, $\beta_1 = 0$ in the Newtonian limit and we obtain half the value for θ given by Eq. (12-2.4). Thus we see that the observation on the advances of the planetary perihelia and the bending of light together serve to fix the coefficients β_1 and α_2 in the expression (12-1.13) for $g_{\mu\nu}$. While the other coefficients in this expression are not fixed by present observations, they do place sufficient limits on them so that when $\alpha_2 = 0$ and $\beta_1 = 1$, the corresponding $g_{\mu\nu}$ will satisfy the Einstein field equations to a high degree of accuracy.

Recently Schiff[9] has argued that the result (12-2.4) can be derived solely from the principle of equivalence and Newtonian mechanics without recourse to the Einstein field equations. However, in the derivation it is necessary to make certain assumptions regarding the behavior of rigid rods that are being accelerated and their relation to nonaccelerated rods. Since there is still considerable disagreement on this point, we will not pursue the matter further but rather refer the interested reader to the paper by Schiff.

[9] L. I. Schiff, *Am. J. Phys.*, **28**, 340 (1960).

Before we conclude this discussion of the bending of light in a gravitational field we should mention a difficulty that arises when one tries to interpret a result such as is embodied in Eq. (12-2.4), which explicitly involves the space-time coordinates. In applying Eq. (12-2.4) we have assumed that R corresponds to the observed apparent distances of the stars from the center of the sun. Also, in computing the angle θ from the observed data, it is necessary to know the distance of the observer from the sun. In all cases these distances are determined by triangulation methods that involve the use of light rays. In computing distances by such methods one always makes the assumption that the space-time geometry is Euclidean; that is, one neglects the effect of the sun's gravitational field on light coming from it. We therefore have no *a priori* justification for taking R in Eq. (12-2.4) to be the corresponding astronomically determined distance. In principle it should be possible to re-express all such results directly in terms of observable quantities in a coordinate invariant manner.[10]

Our justification for taking R to be astronomical distance is based on the fact that the Schwarzschild field (11-3.10) reduces to the Newtonian field of a massive body for large values of the radial coordinate r, compared with the Schwarzschild radius $2\mathfrak{m}$. (For the sun, $2\mathfrak{m} = 1.47$ km and $R = 1.39 \times 10^6$ km.) Furthermore, other expressions for this field (for example, $g_{\mu\nu}$ given by Eq. (11-3.12)) that have this property all give the same result for θ to first order in $2\mathfrak{m}/r$. While this is sufficient for our purposes, it must be borne in mind that if we had to deal with strong fields, and therefore to include higher-order terms in $2\mathfrak{m}/r$ in the expression for θ, we would obtain different expressions for this quantity for different forms of the Schwarzschild field. In this case it would be necessary to analyze in detail the measurement of distance and to express θ in terms of quantities that are directly measurable, for example, angles and times of transit of light signals.

12-3. Gravitational Red Shift

We come now to the third so-called classical test of general relativity, the gravitational red shift of spectral lines. In many respects it is the most difficult of the three tests to discuss. While the first two tests devolved on the behavior of simple systems in a gravitational field, that is, free particles and light rays, any physical system that exhibits periodic motion, and hence can serve as a clock, must by its very nature be treated as a composite system, as we discussed in Section 6-13. While it is, of course, an idealization to treat a planet as a point mass as we did in Section 12-1, it was a reasonable and workable

[10] See, in this connection, J. N. Goldberg and E. Newman, *Phys. Rev*, **114**, 1391 (1959).

idealization. In the case of a clock there simply do not exist such idealizations. The closest thing to such an idealization is the light-clock discussed in Section 6-9. However, even a light clock becomes extremely difficult to discuss in general relativity. In order to discuss the problem of the gravitational red shift we will have to proceed as follows.

Suppose that we have available to us a physical system that exhibits an intrinsic periodicity in its behavior. This periodicity might, for instance, manifest itself in the repeated attainment of a maximum value of some dynamical variable of the system. Suppose further that the system is small enough that it is meaningful to assign to it an average position z^μ in space-time. Then, along its trajectory, we can mark off points corresponding to similar configurations of the system. (We do not demand, of cource, that the coordinate different Δx^0 between two such points be an invariant along the trajectory of the system.

Let us now compare the relative rates of two similar clocks moving along different trajectories in a gravitational field. In order that such a comparison be meaningful, we must decide, of course, when two clocks that are located in different regions of space-time can be considered to be similar. The mere fact that the two clocks are constructed in the same way or are synchronous when placed next to each other does not allow us to conclude that they can be considered to be similar when they move along different space-time trajectories. Before we can reach such a conclusion we must take account of two different mechanisms that can affect their behavior. In the first place, similarly constructed clocks might be affected in different ways by the gravitational fields they encounter in their respective regions of space-time. Thus their dynamics might depend on the second derivatives of $g_{\mu\nu}$. In order to make headway we must explicitly assume that this is not the case; that is, we must invoke the principle of minimal coupling for our clocks. Such an assumption is reasonable if the dimensions of the clocks are small compared with the dimensions of the space-time regions over which the gravitational field can be considered to be constant. It is furthermore necessary to make such an assumption if our previous assumption, that we can assign average positions to the clocks, is to be meaningful.

The second mechanism that can affect the internal dynamics of clocks is the forces applied to them, other than gravitational, to cause them to move along specified space-time trajectories. Thus an applied electric field can produce a Stark shift of the spectral lines of an atom. Such forces are necessary if the trajectories of our clocks are not geodesics of whatever gravitational field is present. In particular, they will be needed if the position of a clock relative to the sources of a gravitational field is to remain constant. We must therefore make the additional assumption that such forces do not affect the internal dynamics of our clocks. However, this assumption is justifiable only for sufficiently weak gravitational fields such as those one encounters in nature.

To the extent that our above assumptions are justifiable we can now make a prediction concerning the relative rates of two similar clocks in a gravitational field. Let the coordinates of two successive points on the trajectory of a clock corresponding to a repeated configuration of the clock be x^μ and $x^\mu + dx^\mu$. Then the proper time interval $d\tau^2 = g_{\mu\nu} dx^\mu\, dx^\nu$ will transform like a scalar under space-time mappings. We now take, as the mathematical statement that two clocks are similar (or that a given clock remains similar to itself along its trajectory), the equality of $d\tau$ for the clocks involved. Because of the scalar character of $d\tau$, this is an invariant characterization of the similarity of clocks under space-time mappings and is in fact the only such invariant statement that we can make. Nevertheless, it does not follow from what we have said above concerning similar clocks, and must be accepted as an assumption. In principle it should be possible to justify this assumption within the framework of the theory, but this has not been done to date.

Consider now two clocks at rest in a static gravitational field. By definition, a static field is one for which it is always possible to find a mapping such that $g_{0r} = 0$ and g_{00} and g_{rs} are independent of x^0 in a finite region of space-time. We take a clock whose trajectory is defined by $x^r = \text{const}$ to be at rest in this field. Now let one clock, located at $x = x_1$, emit a light flash each time it "ticks," that is, each time its configuration repeats. If $d\tau_0$ is the proper time between ticks and $dx_1{}^0$ the coordinate interval between ticks, then these two quantities are related by

$$\begin{aligned} d\tau_0{}^2 &= g_{\mu\nu}(1) dx_1{}^\mu\, dx_1{}^\nu \\ &= g_{00}(1)(dx_1{}^0)^2. \end{aligned} \tag{12-3.1}$$

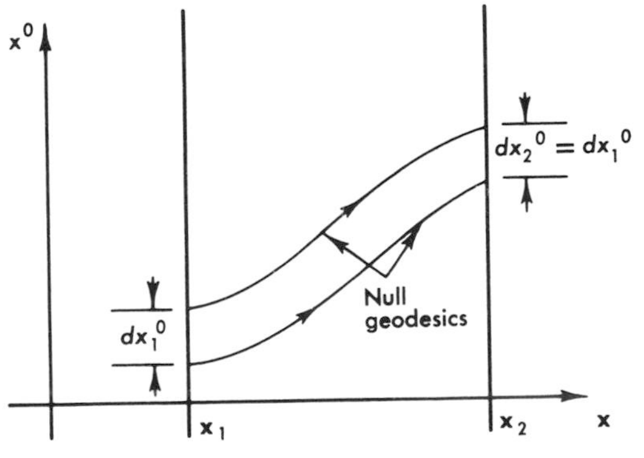

Fig. 12.3.

Since the gravitational field is static, these light flashes will be received by a second clock located at $x = x_2$ separated by a coordinate interval $dx_{20} = dx_{10}$ (see Fig. 12.3). However, during this interval the second clock will have ticked α times. As a consequence of our basic assumption that the proper time between ticks of similar clocks is a fixed constant, we have

$$(\alpha\, d\tau_0)^2 = g_{00}(2)(dx_2{}^0)^2. \tag{12-3.2}$$

It therefore follows that

$$\alpha = \sqrt{\frac{g_{00}(2)}{g_{00}(1)}} \tag{12-3.3}$$

giving the desired relation between the rates of similar clocks in a gravitational field.

The most important application of Eq. (12-3.3) is to the observed characteristic frequencies of atomic and nuclear systems. Suppose that an atomic or nuclear system is observed to have a characteristic frequency ν_0, as measured by a clock at rest with respect to it. Then the frequency ν_{12}, when it is located at x_1 and the clock is at x_2, is given by

$$\nu_{12} = \nu_0 \sqrt{\frac{g_{00}(1)}{g_{00}(2)}}. \tag{12-3.4}$$

Consequently, characteristic frequencies appear to be shifting toward the red as they pass from regions where the gravitational field is strong to regions where it is weak ($g_{00}(1) < g_{00}(2)$). This frequency shift is therefore referred to as the *gravitational red shift*.

Until quite recently the only hope of observing this gravitational red shift was to examine the spectral lines emitted by atoms on the surface of a massive body such as a star. For a gravitational field of the form given in (12-1.13) we have

$$\frac{\Delta\nu}{\nu_0} = \frac{\nu_{12} - \nu_0}{\nu_0} = -\mathfrak{m}\left(\frac{1}{r_1} - \frac{1}{r_2}\right) \tag{12-3.5}$$

to first order in $\mathfrak{m}/r$. We see from this result that to this order of accuracy, the gravitational red shift is independent of the coefficients α and β appearing in the expression (12-1.13) for $g_{\mu\nu}$. Since it is these coefficients that are determined by the particular form of the Einstein equations for $g_{\mu\nu}$, observations on the gravitational red shift, unlike those on the bending of light and the advance of planetary perihelia, cannot lead to information concerning the structure of these equations. (They must, of course, reduce to the Newtonian theory in the lowest approximation in order that Eq. (12-3.5) be valid.)

Rather, these observations must be considered to be the analogue of the Eötvös experiments for light; they can only give us information about the coupling of the electromagnetic field to the gravitational field.

If, in Eq. (12-3.5), we take r_1 to be observed radius of the sun and r_2 the radius of the Earth's orbit, and neglect the effect of the Earth's gravitational field, $\Delta\nu/\nu_0 = -2.12 \times 10^{-6}$. (For the shift in the spectral lines of light due to the Earth's gravitational field, $\Delta\nu/\nu_0 = 7 \times 10^{-10}$.) While observations of the sun's spectra near its edge seem to bear out this prediction, there are many difficulties attendant on the interpretation of the data, but we will not discuss them here.[11] Observations of the spectra of white dwarfs also tend to bear out the predictions of general relativity. The masses of these objects are of the same order as that of the sun. Their radii, however, range from 1/10 to 1/100 solar radii, so that the shift is more pronounced. Here, the great difficulty lies in the determination of the masses and radii of these objects. To date the observations on white dwarfs appear to represent the best astronomical verification of the gravitational red shift.

Recently a number of proposals have been put forward[12,13] to observe the gravitational red shift terrestrially with the help of atomic clocks. Such a possibility exists because such clocks can reach an accuracy of one part in 10^{11}. In particular, the suggestion of Singer to place such a clock in a terrestrial satellite seems promising.

Problem 12.2. Derive an expression for the apparent rate of a clock in a satellite orbiting the Earth. Include both the effect of the gravitational red shift and the transverse Doppler shift. How good a clock would be needed to observe these effects with currently obtainable satellites?

The most recent attempts to observe the gravitational red shift have made use of the Mössbauer effect. Pound and Rebka[14] (and somewhat less conclusively, Cranshaw *et al.*)[15] have used a recoilless gamma ray line in Fe^{57} with a fractional width of the order of 10^{-12} in a terrestrial gravitational red shift. The emitter and detector used by them were separated by a distance of 74 ft of vertical height for which $\Delta\nu/\nu_0 = 2.5 \times 10^{-15}$. Nevertheless they were able to observe a shift equal to 1.05 ± 0.10 of the predicted value. Thus the experiment of Pound and Rebka stands today as the best verification of the gravitational red shift.

[11] For a discussion of these observations and the observations made on white dwarf stars, see B. Bertotti, D. Brill, and R. Krotkov, *loc. cit.*

[12] C. Møller, *Nuovo Cimento Suppl.*, **6** (Ser. X), 381 (1957).

[13] S. F. Singer, *Phys. Rev.*, **104**, 11 (1956).

[14] R. V. Pound and G. A. Rebka, Jr., *Phys. Rev. Letters*, **4**, 337 (1960).

[15] T. F. Cranshaw, S. P. Schiffer, and A. B. Whitehead, *Phys. Rev. Letters*, **4**, 163 (1960).

12-4. Other Tests

Since the first tests of general relativity were proposed, a large number of technical advances have been made that have the potentiality of greatly improving the accuracy of these tests and also of allowing for the possibility of new tests. We have already discussed how the Mössbauer effect made possible an unambiguous test on the gravitational red shift and the possibility of using the greatly improved time standards we now possess in satellites to test this effect. Finally, the advent of radio telescopy has made more precise the determination of satellite and planetary orbits.

One new type of test of general relativity proposed recently is the spinning satellite experiments of Schiff.[16] He suggests that a gyroscope be put into orbit about the earth. For an observer at rest with respect to the gyroscope, the spin angular-momentum vector obeys the equation

$$\frac{d\mathbf{S}_0}{dt} = \mathbf{\Omega} \times \mathbf{S}_0 ,$$

where

$$\mathbf{\Omega} \simeq \frac{\mathbf{F} \times \mathbf{v}}{2mc^3} + \frac{3Gm}{2c^2r^3}(\mathbf{r} \times \mathbf{v}) + \frac{GI}{c^2r^3}\frac{3(\boldsymbol{\omega} \cdot \mathbf{r})\mathbf{r}}{r^2 - \boldsymbol{\omega}} .$$

In this equation m is the mass of the earth; I its moment of inertia equal to $2mr_0^2/5$, where r_0 is the radius of the earth and $\boldsymbol{\omega}$ is its angular velocity; and $\mathbf{F}$ is any nongravitational force acting on the gyroscope. This result was derived by Schiff using results obtained by Papapetrou[17] for the motion of spinning bodies in general relativity. The first term in the expression for $\mathbf{\Omega}$ is a Thomas-type precession term that is a special relativistic effect, and the last two terms are true general relativistic effects. We see that the first of these latter terms is independent of the angular velocity of the earth, while the second depends on this quantity as well as the angle of inclination of the satellite orbit to the Earth's equatorial plane. The relative magnitude of these two terms is seen to be proportional to the ratio of the Earth's angular velocity to the angular velocity of the satellite in its orbit. For satellites at moderate altitudes, Schiff calculates that the net effect of these terms is to produce a precession of the axis of the satellite of about 10^{-9} to 10^{-8} rad per revolution when the gyroscope axis is in the plane of the satellite orbit. For a gyroscope constrained to move on the Earth's surface at latitude λ, with its axis perpendicular to the earth's axis, there is a secular precession of 3.5×10^{-9} $(1 - \cos^2\lambda)$ radians per day. Because of the smallness of the precession rates

[16] L. I. Schiff, *Phys. Rev. Letters*, **4**, 215 (1960), and *Proc. Natl. Acad. Sci.*, **46**, 871 (1960).
[17] A. Papapetrou, *Proc. Roy. Soc.* (London), **A209**, 248 (1051).

in both cases it does not appear possible to observe these effects with presently available instruments.

In connection with rotational effects, one should also mention the precession of orbital perihelia due to a rotation of the source of the gravitational field, given by Eq. (12-1.15). For Mercury, $\delta\tilde{\omega}_{\mathrm{rot}} \simeq -0''.02$ per century, which is much smaller than the present observational error of about $1''$ per century. For a satellite of the Earth at moderate altitudes ($\sim 10^5$ meters), $\delta\tilde{\omega}_{\mathrm{rot}} \sim -60''$ per century, which could be conceivably measured except for the fact that the figure of the Earth is not known to sufficient accuracy (and is, in fact, being determined by satellite observations).

A rather different type of gravitational experiment that does appear to be within the range of present-day observational techniques has recently been proposed by Shapiro.[18] In this experiment one measures the time delays between the sending and returning of radar signals that are reflected by Venus or Mercury. Since, as we saw in Section 10-6, the g_{00} of a static gravitational field acts like a variable index of refraction, the velocity of light will depend on the gravitational field through which it passes. Shapiro calculates that the delay time of a radar signal that passes near the edge of the sun will be increased by about 2×10^{-4} sec over the value expected on the basis of Newtonian theory. According to him, such a difference could be measured to an accuracy of 5 to 10 percent with presently available equipment. The outcome of such observations is of interest because, unlike the recent terrestrial gravitational red-shift observations, they test the validity of the Einstein equations, although again only to the same order in $\mathfrak{m}/r$ as for the perihelion advances of the planets.

Finally we should mention the current attempts of Weber[19] to detect gravitational waves, or more specifically, the components of the Riemann-Christoffel tensor. His method is based on the fact that otherwise free particles, in moving through a gravitational field, experience relative accelerations. If η^μ is an orthogonal vector between two neighboring geodesics and λ^μ is a unit vector along one of these geodesics, one can show[20] that

$$\frac{\delta^2\eta^\mu}{\delta\tau^2} + R^\mu_{\ \nu\rho\sigma}\lambda^\nu\eta^\rho\lambda^\sigma = 0,$$

that is, the relative acceleration, or *geodesic deviation* as it is sometimes called, is proportional to the Riemann-Christoffel tensor. Weber proposes to

[18] I. I. Shapiro, *Phys. Rev. Letters*, **13**, 789 (1964).

[19] A. J. Weber, *Phys. Rev.*, **117**, 306 (1960), and "Gravitational Radiation Experiments in Relativity" in *Relativity, Groups and Topology*, C. de Witt and B. de Witt, eds. (Gordon and Breach, New York, 1964).

[20] J. L. Synge and A. Schild, *Tensor Calculus* (Univ. of Toronto, 1949), Sec. 3.3.

measure these relative strains by measuring the deformations set up in a large aluminum cylinder of mass $\sim 10^6$ grams. The ancillary observational system is estimated to be sensitive enough to detect relative displacements of the end faces "approaching root mean square values of 10^{-14} cm." Unfortunately the system is most sensitive at a frequency of 1657 cps, corresponding to the first compressional mode of the cylinder, and except in rare (?) instances the frequencies of gravitational waves produced by reasonable mechanisms (for example, double stars) correspond to periods of at least hours.

13

Further Consequences of General Relativity

In this chapter we discuss three loosely related topics in general relativity, namely, conservation laws, gravitational radiation, and direct-particle or action-at-a-distance equations of motion. Each of these topics is still the subject of active research and, as this implies, still only incompletely understood. In particular, the question of whether one can ascribe an energy density to the gravitational field as one can to fields in special relativity is still open. While a number of different expressions have been proposed for this quantity, there does not appear to be anything that favors one over the other. At the present time opinion seems to lean toward the impossibility of finding such a quantity. Likewise there is still considerable argument concerning the existence of gravitational radiation, although opinion now seems to be that the theory permits the possibility of such. Part of the difficulty in defining gravitational radiation is the lack of a definitive expression for the energy density in general relativity. While much of the research on direct-particle equations of motion has been in the direction of simplifying the lengthy calculations involved in obtaining these equations, here, too, there are significant questions that need answering, particularly concerning the convergence of the various approximations employed.

There are today a number of excellent survey articles covering recent advances in the topics discussed in this chapter. Consequently we have not felt it necessary to present here all the details of recent investigations. Rather our aim is to lay a sufficient groundwork so that the interested reader can read those articles and also the current research literature. As a consequence we have not discussed a number of the more technical advances made in these fields that may, in the future, prove to be important physical advances.

13-1. Conservation Laws in General Relativity

It was Einstein who made the first attempt to construct a stress-energy complex for the gravitational field analogous to that for fields in special relativity.[1] He started from a Lagrangian $\mathfrak{L}'$ for the gravitational field that was free of second derivatives of $g_{\mu\nu}$ and differed from $\sqrt{-g}\,R$ by a complete divergence:

$$-2\kappa\mathfrak{L}' = \sqrt{-g}\,R - \left(\sqrt{-g}\,g_{\mu\nu,\rho}\frac{\partial R}{\partial g_{\mu\nu,\rho\sigma}}\right)_{,\sigma}. \tag{13-1.1}$$

With this Lagrangian the field equations (10-4.2) take the form

$$\mathfrak{L}'^{\mu\nu} \equiv \frac{\partial\mathfrak{L}'}{\partial g_{\mu\nu}} - \left(\frac{\partial\mathfrak{L}'}{\partial g_{\mu\nu,\rho}}\right)_{,\rho} = \frac{1}{2}\sqrt{-g}\,T^{\mu\nu}. \tag{13-1.2}$$

The so-called Einstein pseudotensor ${}_E\mathrm{t}_\nu{}^\mu$ is constructed by analogy with Eq. (4-5.11) and has the form

$$\sqrt{-g}\,{}_E\mathrm{t}_\nu{}^\mu \equiv -\delta_\nu{}^\mu\mathfrak{L}' + g_{\rho\sigma,\nu}\frac{\partial\mathfrak{L}'}{\partial g_{\rho\sigma,\mu}}. \tag{13-1.3a}$$

If we take the divergence of both sides of this equation we find

$$\begin{aligned}(\sqrt{-g}\,{}_E\mathrm{t}_\nu{}^\mu)_{,\mu} &= -g_{\rho\sigma,\nu}\mathfrak{L}'^{\rho\sigma}\\ &= -\tfrac{1}{2}\sqrt{-g}\,g_{\rho\sigma,\nu}T^{\rho\sigma}.\end{aligned} \tag{13-1.3b}$$

We now make use of the fact that, as a consequence of the Bianchi identities (10-8.2), $T^{\mu\nu}$ satisfies the equation

$$T^{\mu\nu}{}_{;\mu} = 0,$$

which can be written in the form

$$\begin{aligned}(\sqrt{-g}\,T_\nu{}^\mu)_{,\mu} &= \sqrt{-g}\left\{{\rho \atop \mu\nu}\right\}T_\rho{}^\mu\\ &= \tfrac{1}{2}\sqrt{-g}\,g_{\rho\sigma,\nu}T^{\rho\sigma}.\end{aligned} \tag{13-1.4}$$

[1] A. Einstein, *Ann. d Phys.*, **49**, 769 (1916).

Consequently Eq. (13-1.4) can be rewritten finally as

$$_E\mathfrak{T}^{\mu}_{\nu,\mu} = 0, \tag{13-1.5}$$

where

$$_E\mathfrak{T}_\nu{}^\mu = \sqrt{-g}\,\{T_\nu{}^\mu + {}_E\mathfrak{t}_\nu{}^\mu\}. \tag{13-1.6}$$

While Eq. (13-1.5) is in the form of a continuity equation, $_E\mathfrak{t}_\nu{}^\mu$ is not a tensor or even a geometric object, and it is for this reason that $_E\mathfrak{t}_\nu{}^\mu$ is referred to as a pseudotensor. Since $_E\mathfrak{t}_\nu{}^\mu$ contains no higher than the first derivatives of $g_{\mu\nu}$, it can be made to vanish at a space-time point by means of a suitable mapping, which is not possible for a true tensor. Furthermore, if we form the integral $\int {}_E\mathfrak{t}_0{}^0\, d^3x$, its value is zero when $g_{\mu\nu} = \eta_{\mu\nu}$. However, if we map to spherical coordinates, the value of this integral is infinite. Likewise the value of this integral is mc^2 for a Schwarzschild field only if one maps so that asymptotically $g_{\mu\nu} \to \eta_{\mu\nu}$. For other forms of this field, for example, $g_{\mu\nu}$ given by Eq. (11-3.10), the value of the integral is infinite. Finally we note that $_E\mathfrak{T}^{\mu\nu} = g^{\mu\rho}\,{}_E\mathfrak{T}_\rho{}^\nu$ is not symmetric in μ and ν, so that we cannot form from it a conserved angular-momentum complex.

We can gain additional insight into the nature of the Einstein pseudotensor by expressing it as the divergence of a "superpotential" $_FU_\nu^{[\mu\rho]}$. P. von Freud[2] showed that $_E\mathfrak{T}_\nu{}^\mu$ could be written as

$$_E\mathfrak{T}_\nu{}^\mu = {}_FU_\nu^{[\mu\rho]}{}_{,\rho}\,, \tag{13-1.7}$$

where

$$2\kappa\sqrt{-g}\;{}_FU_\nu^{[\mu\rho]} = g_{\nu\sigma}\{g(g^{\mu\sigma}g^{\rho\lambda} - g^{\mu\lambda}g^{\rho\sigma})\}_{,\lambda} \tag{13-1.8}$$

when use is made of the field equations to replace $T_\nu{}^\mu$ in Eq. (13-1.6) by $(1/\kappa)G_\nu{}^\mu$. Since $_FU_\nu^{[\mu\rho]} = -{}_FU_\nu^{[\rho\mu]}$, the continuity equation (13-1.5) can be rewritten as

$$_FU_\nu^{[\mu\rho]}{}_{,\rho\mu} = 0. \tag{13-1.9}$$

But this equation is an identity; it holds for all $g_{\mu\nu}$, independent of whether or not they satisfy the Einstein field equations. Because of this, Klein has termed the continuity equation (13-1.5) *improper*. In general a continuity equation of the form $\mathfrak{J}_i{}^\mu{}_{,\mu} = 0$ is improper if $\mathfrak{J}_i{}^\mu$ reduces to a curl, that is, if $\mathfrak{J}_i{}^\mu = U_i^{[\mu\rho]}{}_{,\rho}$, for dynamically possible trajectories.

Before we discuss additional properties of the Einstein pseudotensor let us consider other expressions that have been proposed for the energy-momentum complex of the gravitational field. To obtain such a quantity that is symmetric

[2] P. von Freud, *Ann. Math.*, **40**, 417 (1939).

in its indices and so allows one to construct an angular-momentum complex in the usual manner, Landau and Lifshitz[3] proposed to use

$$ {}_L\mathfrak{T}^{\mu\nu} \equiv {}_L U^{\mu[\nu\sigma]}{}_{,\sigma} = (-g)(T^{\mu\nu} + {}_L\mathfrak{t}^{\mu\nu}), \tag{13-1.10}$$

where

$$ {}_L U^{\mu[\nu\sigma]}{}_{,\sigma} = \sqrt{-g}\; g^{\mu\rho}\; {}_F U_\rho^{[\nu\sigma]}. \tag{13-1.11}$$

Since ${}_L U^{\mu[\nu\sigma]} = -{}_L U^{\mu[\sigma\nu]}$, it follows that

$$ {}_L\mathfrak{T}^{\mu\nu}{}_{,\nu} = 0. \tag{13-1.12}$$

An angular-momentum complex $\mathfrak{M}^{\mu\nu\rho}$ can then be constructed as

$$ \mathfrak{M}^{\mu\nu\rho} = x^\mu\; {}_L\mathfrak{T}^{\nu\rho} - x^\nu\; {}_L\mathfrak{T}^{\mu\rho} \tag{13-1.13}$$

and is seen to satisfy

$$ \mathfrak{M}^{\mu\nu\rho}{}_{,\rho} = 0 \tag{13-1.14}$$

as a consequence of Eq. (13-1.12) and the fact that ${}_L\mathfrak{t}^{\mu\nu} = {}_L\mathfrak{t}^{\nu\mu}$. Like ${}_E\mathfrak{t}_\nu{}^\mu$, ${}_L\mathfrak{t}^{\mu\nu}$ is not a tensor, nor does ${}_L\mathfrak{t}^{\mu\nu} = g^{\nu\rho}\; {}_E\mathfrak{t}_\rho{}^\mu$; rather

$$ {}_L\mathfrak{t}^{\mu\nu} = g^{\mu\rho}\; {}_E\mathfrak{t}_\rho{}^\nu - \frac{1}{g}\,(\sqrt{-g}\; g^{\mu\rho})_{,\sigma}\; {}_F U_\rho^{[\nu\sigma]}.$$

In a like manner one can obtain numerous other conserved quantities.[4,5] We mention here only the Møller[6] stress-energy complex, given by

$$ {}_M\mathfrak{T}_\nu{}^\mu \equiv {}_M U_\nu^{[\mu\sigma]}{}_{,\sigma} = \sqrt{-g}\,(T_\nu{}^\mu + {}_M\mathfrak{t}_\nu{}^\mu), \tag{13-1.15}$$

where

$$ \begin{aligned} {}_M U_\mu^{[\nu\sigma]} &= 2{}_F U_\mu^{[\nu\sigma]} - \delta_\nu{}^\mu\; {}_F U_\rho^{[\rho\sigma]} + \delta_\mu{}^\sigma\; {}_F U_\rho^{[\rho\nu]} \\ &= \frac{\sqrt{-g}}{\kappa}\, g^{\nu\kappa} g^{\sigma\tau}(g_{\mu\kappa,\tau} - g_{\mu\tau,\kappa}). \end{aligned}$$

Møller introduced his complex in an attempt to ascribe a meaning to the local density of energy and momentum of the gravitational field. The Einstein pseudotensor certainly cannot be used for this purpose, since it can be made to vanish at a space-time point by a suitable mapping. However, the Møller complex depends on second derivatives of $g_{\mu\nu}$, so that it does not possess this

[3] L. D. Landau and E. M. Lifshitz, *The Classical Theory of Fields*, 2d ed. (Addison-Wesley, Reading, Mass., 1962), Sec. 100.

[4] J. N. Goldberg, *Phys. Rev.*, **111**, 315 (1958).

[5] P. G. Bergmann, *Phys. Rev.*, **112**, 287 (1958).

[6] C. Møller, *Ann. Physics*, **4**, 347 (1958).

property. Furthermore ${}_M\mathfrak{T}_0{}^\mu$ transforms like a vector density under mappings of the form

$$x^\mu \to x'^\mu = x^\mu + f^\mu(x^r). \tag{13-1.16}$$

As a consequence the value of the integral $\int_M \mathfrak{T}_0{}^0 \, dV$ over an $x^0 = \text{const}$ hypersurface has the same value for all mappings in that hypersurface. Furthermore, by using the Møller complex it is not necessary to map so that asymptotically $g_{\mu\nu} \to \eta_{\mu\nu}$ to obtain finite values for this integral. However, the special group of mappings only have an intrinsic significance if a timelike vector field exists on the manifold. In this case they are the mappings that leave invariant the special form $\{1, \mathbf{0}\}$ of such a vector. Unless such a vector field exists there is no reason to limit oneself to mappings of the form (13-1.16), and under a general mapping ${}_M\mathfrak{T}_0{}^\mu$ no longer transforms like a tensor density.

It is not surprising that there should be so many different expressions for quantities satisfying a continuity equation of the type (13-1.5). Given a superpotential $U^{[\rho\sigma]} = -U^{[\sigma\rho]}$ and the corresponding conserved quantity $t^\mu = U^{\rho\sigma}{}_{,\sigma}$, we can construct an infinity of other conserved quantities of the form $t'^\mu = (\xi U^{\rho\sigma})_{,\sigma}$, where ξ is some space-time function. Ultimately, the existence of so many different continuity equations is due to the symmetry group of a general relativistic system. This symmetry group is a gauge group and can be thought of as being composed of an infinity of one-parameter mappings whose infinitesimal elements have the form

$$x^\mu \to x'^\mu = x^\mu + \varepsilon\xi^\mu(x), \tag{13-1.17}$$

where $\xi^\mu(x)$ is an arbitrary space-time function. It therefore follows from Nöther's identity that there exists a continuity equation for each set of functions $\xi^\mu(x)$. Under such a mapping, $\bar{\delta} y_A$ (where the y_A include $g_{\mu\nu}$) will have the form

$$\bar{\delta} y_A = \varepsilon \eta_A, \tag{13-1.18}$$

while $\bar{\delta} t^\mu$ in Eq. (4-5.16) has the form

$$\bar{\delta} t^\mu = \varepsilon t^\mu, \tag{13-1.19}$$

where η_A and t^μ are functions of the y_A, ξ^μ, and their derivatives. For dpt, it follows from Eq. (4-5.15) that

$$t^\mu{}_{,\mu} = 0. \tag{13-1.20}$$

We will now show that the conserved quantities t^μ can be derived from a superpotential. For this purpose we make use of the identity (4-6.4):

$$(d^\mu_{A\nu}\mathfrak{L}^A)_{,\mu} - c_{A\nu}\mathfrak{L}^A \equiv 0, \tag{13-1.21}$$

corresponding to a $\bar{\delta} y_A$ of the form

$$\bar{\delta} y_A = \varepsilon\{c_{A\mu}\xi^\mu + d^\rho_{A\mu}\xi^\mu{}_{,\rho}\}. \tag{13-1.22}$$

Equation (4-5.10) then takes the form

$$t^{\mu}{}_{,\mu} - c_{A\mu}\mathfrak{L}^{A}\xi^{\mu} - d^{\rho}_{A\mu}\mathfrak{L}^{A}\xi^{\mu}{}_{,\rho} \equiv 0, \tag{13-1.23}$$

which by using Eq. (13-1.21) can be written as

$$\Theta^{\rho}{}_{,\rho} \equiv 0, \tag{13-1.24}$$

where

$$\Theta^{\rho} = t^{\rho} - d^{\rho}_{A\mu}\mathfrak{L}^{A}\xi^{\mu}. \tag{13-1.25}$$

From Eq. (13-1.24) we can infer the existence of a superpotential $U^{[\rho\sigma]} = -U^{[\sigma\rho]}$ such that

$$\Theta^{\rho} \equiv U^{[\rho\sigma]}{}_{,\sigma}, \tag{13-1.26}$$

from which it follows that

$$t^{\rho} = -d^{\rho}_{A\mu}\mathfrak{L}^{A}\xi^{\mu} + U^{[\rho\sigma]}{}_{,\sigma}. \tag{13-1.27}$$

Therefore, when the equations of motion are satisfied (that is, when $\mathfrak{L}^{A} = 0$), t^{ρ} is equal to a curl. After a somewhat lengthy calculation one finds that[7]

$$t^{\rho} \equiv \mathfrak{t}^{\rho} + \sqrt{-g}\, T_{\sigma}{}^{\rho}\xi^{\sigma} = \left\{\frac{\sqrt{-g}}{\kappa}(g^{\sigma\nu}\xi^{\rho}{}_{;\nu} - g^{\rho\nu}\xi^{\sigma}{}_{;\nu})\right\}_{,\sigma} \tag{13-1.28}$$

so that

$$U^{[\rho\sigma]} = \frac{\sqrt{-g}}{\kappa}(g^{\sigma\nu}\xi^{\rho}{}_{;\nu} - g^{\rho\nu}\xi^{\sigma}{}_{;\nu}). \tag{13-1.29}$$

The nonuniqueness of $\mathfrak{t}^{\rho}$ manifests itself in two ways. First, the ξ^{ρ} are arbitrary. By taking them to be constants, for example, one obtains the Møller superpotentials of Eq. (13-1.15). Second, by adding to both sides of Eq. (13-1.28) a curl $V^{[\rho\sigma]}{}_{,\sigma}$, where $V^{[\rho\sigma]} = -V^{[\sigma\rho]}$, one obtains a new expression for $\mathfrak{t}^{\rho}$ given by

$$\mathfrak{t}'^{\rho} = \mathfrak{t}^{\rho} + V^{[\rho\sigma]}{}_{,\sigma}$$

and a corresponding superpotential $U'^{[\rho\sigma]}$ of the form

$$U'^{[\rho\sigma]} = U^{[\rho\sigma]} + V^{[\rho\sigma]}.$$

By an appropriate choice of ξ^{ρ} and $V^{[\rho\sigma]}$ one can obtain all the various stress-energy complexes and corresponding superpotentials. But because of this nonuniqueness there is in general nothing special about any one of them that would, for example, allow us to interpret it as "the" stress-energy complex of the gravitational field. Nor do any of the various complexes transform like

[7] A. Komar, *Phys. Rev.*, **113**, 934 (1959).

geometrical objects under the MMG. Furthermore, the totality of all continuity equations is completely equivalent to the original Einstein equations. One can ultimately trace the lack of geometrically conserved quantities in general relativity to the lack of absolute objects in this theory. The great value of continuity equations in special relativity lies in the integrals of the equations of motion that follow from them. In general, no such integrals can be derived from the continuity equations associated with gauge groups of manifold mappings. However, if ξ^μ is a Killing vector associated with the gravitational field, it is possible to construct globally conserved quantities that can be given a physical interpretation.

To illustrate the above discussion, let τ^μ be a timelike Killing vector of a stationary gravitational field, and let σ be a spacelike surface and S its boundary. We now form the integral

$$P = \oint_S {}_\tau U^{[\mu\nu]} d\,\Sigma_{\mu\nu} \tag{13-1.30}$$

where ${}_\tau U^{[\mu\nu]}$ is the superpotential obtained by taking $\xi^\mu = \tau^\mu$ in Eq. (13-1.29) and $d\Sigma_{\mu\nu}$ is an element of area in S. This integral can be taken to be the total energy contained within the volume bounded by S. If we map so that $\tau^\mu = \{1, \mathbf{0}\}$ and take σ to be the surface $x^0 = \text{const}$, then

$$\begin{aligned} P &= \frac{1}{\kappa} \oint_S {}_\tau U^{[0r]} d\Sigma_r \\ &= \frac{1}{\kappa} \oint_S \sqrt{-g}\,\{g^{\sigma r}\Gamma^0_{0\sigma} - g^{\sigma 0}\Gamma^r_{0\sigma}\} d\Sigma_r \\ &= \frac{1}{\kappa} \oint_S (\sqrt{-g})^{\sigma r} g^{0\rho}(g_{0\rho,\sigma} - g_{0\sigma,\rho}) d\Sigma_r \end{aligned}$$

If in addition the field is static, one can map so that $g_{0r} = 0$, and hence

$$P = \frac{1}{\kappa} \oint_S (\sqrt{-g})^{00} g^{rs} g_{00,s} d\Sigma_r .$$

For a Schwarzschild field and S a sphere of radius a, P has the value m. independent of a.

In addition to constructing the integral appearing in Eq. (13-1.30) for a rigorous Killing vector of the gravitational field, it is also meaningful to construct such integrals when the field possesses asymptotic Killing vectors Such vectors will exist, for instance, if one can find a mapping such that, asymptotically

$$g_{\mu\nu} = \eta_{\mu\nu} + 0(r^{-1}),$$

where r is the geodesic distance from some fixed point on a spacelike surface that is asymptotically parallel to an $x^0 = \text{const}$ surface. In this case Killing's equation reduces to

$$\eta_{\mu\rho}\xi^{\rho}{}_{,\nu} + \eta_{\rho\nu}\xi^{\rho}{}_{,\mu} + 0(r^{-1}) = 0$$

Consequently a field that is asymptotically flat possesses ten asymptotic Killing vectors, the Killing vectors of a flat metric. One can then form ten integrals of the form (13-1.30) by taking the ξ^{μ} in Eq. (13-1.29) to coincide with the Killing vectors asymptotically and by taking the surface to be at spatial infinity. The integrals will then be independent of the values of ξ^{μ} inside this surface and will also be independent of x^0. They will therefore represent ten globally conserved quantities that characterize the sources of the field.

Before we conclude this discussion of conservation laws let us consider the ten conservation laws of special relativity associated with the Poincaré symmetry of that theory within the framework of general relativity. If we compare the equivalence classes of dpt of a given system, for example, the electromagnetic field in special relativity and in general relativity, there will, in general be no relation between them. However, if we construct equivalence classes of approximate dpt in general relativity by taking the gravitational coupling constant to be zero, then these approximate equivalence classes can be related to the equivalence classes in the special theory. Thus the special relativistic description of a physical system can be considered to be the correspondence limit of the general relativistic description when the coupling of the system to the gravitational field can be neglected and when the solution of the resulting Einstein equations is taken to be the flat-space gravitational field. This field, of course, possesses the ten Killing vectors associated with the Poincaré group.

If $g_{\mu\nu}$ possesses a Killing vector $\xi_{\mu}(x)$, then it follows that

$$(\sqrt{-g}T_M^{\mu\nu}\xi_{\mu})_{,\nu} = 0 \tag{13-1.31}$$

where $T_M^{\mu\nu}$ is the stress-energy tensor that stands on the right side of the Einstein equations—that is, the source of the gravitational field. The existence of such continuity equations depends only on the fact that ξ_{μ} is a Killing vector and that $T^{\mu\nu}{}_{;\nu} = 0$ and not on $g_{\mu\nu}$ being a solution of the Einstein equations. It follows, therefore, that such continuity equations will be valid when $g_{\mu\nu}$ is only an approximate solution of these equations. In particular, they will be valid in the correspondence limit discussed above. But in this case, with the ξ_{μ} the Killing vectors of the flat-space field, these continuity equations are just the ten continuity equations of special relativity associated with the Poincaré invariance of that theory as can be seen by mapping so that $g_{\mu\nu} = \eta_{\mu\nu}$. Thus we can say that these continuity equations are "broken" by the interaction of the gravitational field and the system or systems whose stress-energy tensor is $T_M^{\mu\nu}$. The fundamental reason for the breaking of these continuity equations is that the gravitational field does not, in general, possess a geometrical stress-energy complex—that is, a complex that forms the basis of a realization of the MMG. In an interaction that involves the gravitational field a system can loose energy without this energy being transmitted to the gravitational field.

13-2. Gravitational Radiation

Perhaps no other phase of gravitation theory has produced more discussion and more disagreement than that associated with the question of the existence of gravitational radiation.[8] The primary reason is obvious; to date there is no experimental evidence to guide one in defining this concept. Consequently most authors have drawn heavily upon analogies from electromagnetic theory, where the concept of radiation is reasonably well understood. Unfortunately, gravitational theory differs in several respects from electrodynamics and these differences reduce significantly the value of these analogies.

Let us consider for a moment those characteristics of the electromagnetic field that lead us to speak of electromagnetic radiation. First there exist source-free plane-wave solutions characterized by the conditions that

$$|\mathbf{E}| = |\mathbf{B}| \quad \text{and} \quad \mathbf{E} \cdot \mathbf{B} = 0,$$

or equivalently, by the condition that there exists a null vector k^μ $(k_\mu k^\mu = 0)$ such that

$$F^{\mu\nu} k_\nu = \varepsilon^{\mu\nu\rho\sigma} F_{\mu\nu} k_\rho = 0.$$

As we saw, analogous plane-wave solutions exist for the Einstein equations. More important, however, is the existence of solutions of Maxwell's equations associated with sources that asymptotically have the property of being locally plane; that is, at each asymptotic point there exists a vector k_μ satisfying the above conditions to $O(r^{-2})$, and which decrease asymptotically like r^{-1} rather than r^{-2}. Furthermore these solutions lead to a Poynting vector whose integral over a closed surface surrounding the sources correctly predicts the energy loss of these sources as measured experimentally.

If we try to carry over these ideas to gravitational theory we are confronted with a number of difficulties. First of all there are no rigorous solutions of the

[8] In recent years a number of excellent review articles containing original results have appeared. They are: F. A. E. Pirani, "Gravitational Radiation", *Gravitation: An Introduction to Current Research*, L. Witten, ed. (Wiley, New York, 1962); and "Survey of Gravitational Radiation Theory" in *Recent Developments in General Relativity.* (Pergamon Press, New York, 1962); R. K. Sachs, "Gravitational Radiations" in *Relativity, Groups and Topology* (Gordon and Breach, New York, 1964).

field equations corresponding to time-varying sources, so that one is immediately forced to make use of approximate procedures (for example, weak field solutions and multipole expansion methods). However, such methods are difficult to formulate geometrically and one is usually forced to make use of coordinate conditions. As a consequence it is difficult to decide to what extent conclusions drawn from such considerations are independent of the special coordinate conditions employed and which are not.

In addition to these difficulties, one is also faced with the problem that a clear-cut, geometrical formulation of conservation laws in general relativity is possible only when there exists one or more Killing vectors, at least asymptotically. However, conservation laws are necessary if we are to be able to calculate the amount of gravitational energy radiated by a source. Because of the difficulty in constructing such Killing vectors, most authors rely on one or another of the special conserved quantities; for example, ${}_E\mathfrak{T}_\nu{}^\mu$. However, these objects are not by themselves geometrical objects and must be supplemented by the use of coordinate conditions, whereupon all the above considerations concerning such conditions apply. An alternate procedure is to try to construct an integral that represents the mass of a system of radiating sources and to examine its temporal behavior. Under certain conditions such an integral can be shown to decrease monotonically with time as long as the system radiates.

If we have a continuity equation of the form $t^\mu{}_{,\mu} = 0$, we can integrate over a volume V in an $x^0 = \text{const}$ surface to obtain

$$\frac{d}{dx^0}\int_V t^0\, dV + \int t^r{}_{,r}\, dV = 0.$$

By using Gauss' theorem we can convert the second term to a surface integral so that

$$\frac{d}{dx^0}\int_V t^0\, dV + \oint_S t^r\, dS_r = 0, \qquad \text{(13-2.1)}$$

where S is the two-dimensional boundary of V. We can interpret this equation by saying that the rate of change of $\int_V t^0\, dV$ in V is equal to the flux of this quantity across S. Of course this interpretation is purely formal, since the surface $x^0 = \text{const}$ has no special or intrinsic significance in general. If, however, the gravitational field is sufficiently weak, it will possess as approximate Killing vectors those of the flat-space metric. In such a case an $x^0 = \text{const}$ can be characterized intrinsically, and so the conservation law (13-2.1) will have more than formal significance. In most such cases one makes use of the Einstein pseudotensor ${}_E\mathrm{t}_\nu{}^\mu$ to compute the flux of energy of a radiating system. It is reasonable to use ${}_E\mathrm{t}_\nu{}^\mu$ in this case, since it transforms like a mixed tensor under mappings generated by the approximate Killing vectors, that is, under Poincaré mappings.

For a weak gravitational field of the form (11-1.14), Eq. (13-2.1), with t^μ replaced by ${}_E t_0{}^\mu$, takes the form[9]

$$\frac{dE}{dt} = -\frac{G}{45C^5}\left(\frac{d^3Q_{rs}}{dt^3}\right)\left(\frac{d^3Q_{uv}}{dt^3}\right)\delta^{ru}\delta^{sv}, \tag{13-2.2}$$

where

$$Q_{rs} \equiv \int_V \rho\left(\mathbf{x}, t - \frac{|\mathbf{x}|}{c}\right)(3x^r x^s - \delta^{rs}x^2)\, dV$$

is the quadrupole moment of the source. For the earth-sun system, Eq. (13-2.2) yields a rate of energy loss of about 200 watts. Einstein, and later Eddington, calculated the energy radiated by a spinning rod and found it to be

$$P = \frac{32}{5}\frac{GI^2\omega^6}{c^5}$$

where I is the moment of inertia of the rod. A rod 1 meter long, spinning as fast as it can without breaking apart because of internal stresses, will radiate about 10^{-30} ergs per second. These considerations make it appear unlikely that gravitational radiation will play any role in energy transfer in physical processes except under the most extreme conditions. Dyson[10] has suggested that the components of a neutron double star spiraling toward each other as a result of gravitational radiation damping might, at the end, emit gravitational energy in excess of 10^{50} ergs in a short 200-cycle pulse.

Even if the field is not weak, one might hope to use asymptotic methods to discuss gravitational radiation. Here the problem is to formulate boundary conditions on the gravitational field so that the integrals appearing in Eq. (13-2.1) will exist but still not exclude the possibility of radiation. Trautman[11] was the first to treat the problem of radiation in this manner, making use of an outgoing radiation condition similar to the one discussed in Section 8-3. He begins by assuming that the gravitational field in question defines a scalar field $u(x)$ on the manifold whose gradient, $k_\mu = u_{,\mu}$, is null and diverging; that is, $k_\mu k^\mu = 0$ and $k^\mu{}_{;\mu} \neq 0$. The vector field k_μ can be used to construct a congruence of rays on the manifold by requiring that the tangent to the rays passing through a given point is equal to k^μ at that point. One can then define

[9] The question of gravitational radiation was first discussed in the linearized approximation by A. Einstein, *Sb. Preuss. Akad. Wiss.*, 154 (1918).

[10] F. J. Dyson, "Gravity Research Essay" (Gravity Research Foundation, New Boston, New Hampshire, 1962).

[11] A. Trautman, *Bull. Acad. Polon. Sci. Ser. Sci. Math. Astron. Phys.* **6**, 403, 407 (1958).

a "luminosity distance" r along the rays satisfying $(r^{-2}k^\mu)_{;\mu} = 0$.[12] If $g_{\mu\nu} = \eta_{\mu\nu}$ and $u = x^0 - |\mathbf{x}|$, then $k^\mu = (1, x/|\mathbf{x}|)$ and the luminosity distance is just equal to the radial distance $|\mathbf{x}|$ from the origin. In the general case the square of the luminosity distance is proportional to the area bounded by a bundle of rays as one moves out along them.

Trautman now assumes that there exists a mapping such that

$$g_{\mu\nu} = \eta_{\mu\nu} + 0(r^{-1}) \tag{13-2.3}$$

and

$$g_{\mu\nu,\rho} = i_{\mu\nu}k_\rho + 0(r^{-2}) \tag{13-2.4}$$

where $i_{\mu\nu} = 0(r^{-1})$, $k_\mu = 0(1)$, and $k_{\mu,\nu} = 0(r^{-2})$. Equation (13-2.4) implies that the analogue of the outgoing radiation condition is satisfied since $k^\rho g_{\mu\nu,\rho} = 0(r^{-2})$. In addition Trautman assumes that the asymptotic deDonder or harmonic coordinate conditions

$$(i_{\mu\nu} - \tfrac{1}{2}\eta_{\mu\nu}\eta^{\rho\sigma}i_{\rho\sigma})k^\nu = 0(r^{-2}) \tag{13-2.5}$$

are satisfied. In the asymptotic region the Einstein pseudotensor ${}_E\mathfrak{t}_\nu{}^\mu$ can then be shown to have the form

$${}_E\mathfrak{t}_\nu{}^\mu = \mathfrak{t}k^\mu k_\mu + 0(r^{-3}) \tag{13-2.6}$$

where

$$\mathfrak{t} = \frac{1}{32\pi G} i^{\mu\nu}\left(i_{\mu\nu} - \frac{1}{2}\eta_{\mu\nu}\eta^{\rho\sigma}i_{\rho\sigma}\right) \tag{13-2.7}$$

Because of Eq. (13-2.7), $\mathfrak{t}$ is nonnegative and is seen to be of order r^{-2}. Therefore a surface integral of ${}_E\mathfrak{t}_0{}^r$ of the type appearing in Eq. (13-2.1) will be finite and could represent the total energy radiated by the sources of the gravitational field. Cornish[13] has shown, in fact, that the value of this integral is insensitive to which of the various stress-energy complexes are used in its computation provided that the boundary conditions (13-2.3), (13-2.4), and (13-2.5) are satisfied.

To justify further these boundary conditions as being appropriate to fields representing gravitational radiation, Trautman computed the asymptotic form of the curvature tensor of a field satisfying them. One has

$$g_{\mu\nu,\rho\sigma} = j_{\mu\nu}k_\rho k_\sigma + 0(r^{-2}) \tag{13-2.8}$$

where $j_{\mu\nu} = 0(r^{-1})$. The Einstein field equations then require

$$(j_{\mu\nu} - \tfrac{1}{2}\eta_{\mu\nu}\eta^{\rho\sigma}j_{\rho\sigma})k^\nu = 0(r^{-2}). \tag{13-2.9}$$

[12] H. Bondi, *Nature*, **186** (1960); R. K. Sachs, *loc. cit.*

[13] F. H. J. Cornish, *Proc. Roy. Soc.* (London), **A282**, 358, 372 (1964).

With this additional information the components of the curvature tensor can be computed and has the form

$$R_{\mu\nu\rho\sigma} = \tfrac{1}{2}k_{[\mu}j_{\nu]\,[\rho}k_{\sigma]} + 0(r^{-2}). \qquad (13\text{-}2.10)$$

As a consequence it follows that $R_{\mu\nu\rho\sigma}k^{\mu} = R_{\mu\nu\,[\rho\sigma}k_{\lambda]} = 0(r^{-2})$ so that in the asymptotic limit $r \to \infty$ $R_{\mu\nu\rho\sigma}$ is type N in the Petrov classification. We saw in Section 11-6 that the plane-fronted waves were all of type N so that asymptotically gravitational fields satisfying the Trautman boundary conditions appear to have properties characteristic of plane-fronted waves.

Bondi and his coworkers[14] were the first to look for solutions of the Einstein field equations that satisfy the Trautman boundary conditions and can be considered to describe radiation from a bounded source. In doing so they considered the so-called characteristic initial value problem. Instead of giving initial data on a spacelike hypersurface to characterize a solution of the Einstein field equations, Bondi tried to characterize his radiative solutions by giving data on characteristic hypersurfaces. The possibility of characterizing solutions by giving data on characteristic hypersurfaces arises in connection with the simplest hyperbolic differential equation

$$\frac{\partial^2 \phi}{\partial t^2} - \frac{\partial^2 \phi}{\partial x^2} = 0.$$

Instead of determining ϕ by giving ϕ and $\partial\phi/\partial t$ at $t = 0$, one can equally characterize the general solution $\phi(t, x) = f(x + t) + g(x - t)$ by giving the values of f and g on the two characteristic lines $x \pm t = 0$.

To form a Bondi-type solution one first constructs a family of characteristic hypersurfaces $u =$ const and an associated ray congruence with tangent vectors $k_\mu = u_{,\mu}$ such that $k^{\mu}{}_{;\mu} \neq 0$. In addition to the null vector k_μ one also constructs another null vector m_μ such that $m^\mu k_\mu = 1$ and a complex vector $t_\mu = (1/\sqrt{2})(q_\mu + ir_\mu)$, where q_μ and r_μ are real spacelike vectors orthogonal to each other and to k_μ and m_μ. Thus

$$m^\mu k_\mu = t^\mu \bar{t}_\mu = 1 \qquad (13\text{-}2.11)$$

where a bar over a vector indicates complex conjugation, with all other products zero. One then defines a luminosity distance r as before along each ray and maps so that $x^\mu = (u, r, \theta, \phi)$ where θ and ϕ are constant along a ray. One can then divide the field equations $G_\nu{}^\mu = 0$ into three subsets:

(1) $k^\mu G_{\mu\nu} = 0, \quad t^\mu t^\nu G_{\mu\nu} = 0$ (main equations)
(2) $t^\mu \bar{t}^\nu G_{\mu\nu} = 0$ (trivial equation)
(3) $m^\mu t^\nu G_{\mu\nu} = 0, \quad m^\mu m^\nu G_{\mu\nu} = 0$ (supplementary conditions)

[14] H. Bondi, M. van der Burg and A. Metzner, *Proc. Roy. Soc.* (London), **A269**, 21 (1962).

Since a complex equation is in reality two real equations, we see that there are six main equations, one trivial equation, and three supplementary conditions. This breakup is analogous to the breakup of the field equations into the two sets $G_\mu{}^0 = 0$ and $R_s{}^r = 0$ associated with the usual initial value problem discussed in Section 10-9. One can now show that if the main equations are satisfied everywhere in a space-time region, the trivial equation is also satisfied everywhere and the supplementary conditions are satisfied everywhere if they are satisfied at one point on each ray.

In his search for a solution of the field equations that permitted the construction of the above structure, Bondi restricted himself to an axially symmetric field with a line element of the form

$$ds^2 = C\,du^2 + 2D\,du\,dr - r^2[e^\alpha(d\theta - A\,du)^2 + e^{-\alpha}\sin^2\theta\,d\phi^2] \tag{13-2.12}$$

where A, C, D, and α are independent of ϕ. The restriction to an axially symmetric field is not essential to the method which can be applied in the case of a quite general field but with considerably more difficulty. If we treat such a field as a metric the area of a wave front $u = \text{const}$, $r = \text{const}$ is equal to $4\pi r^2$ in keeping with the interpretation of r as a luminosity distance. Furthermore, such a field is the natural generalization of the form of the flat field obtained by mapping so that $x^\mu = (u = x^0 - r, r, \theta, \phi)$ with an associated line element given by

$$ds^2 = du^2 + 2du\,dr - r^2(d\theta^2 + \sin^2\theta\,d\phi^2). \tag{13-2.13}$$

By comparing the two fields Bondi was led to set

$$\begin{aligned} \alpha &= \frac{n}{r} + 0(r^{-2}) \\ A &= \frac{a}{r} + 0(r^{-2}) \\ C &= 1 - \frac{2m}{r} + 0(r^{-2}) \\ d &= 1 + \frac{d}{r} + 0(r^{-2}). \end{aligned} \tag{13-2.14}$$

where n, a, m, and d are functions of u and θ alone. One can then show, by mapping so that $g_{\mu\nu} = \eta_{\mu\nu} + 0(r^{-1})$, that this field satisfies the radiation conditions (13-2.3), (13-2.4), and (13-2.5).

With the above assumed asymptotic form of the field one now finds that given α on a $u = \text{const}$ hypersurface, the remaining quantities appearing in the expression (13-2.12) for ds^2 are determined by the main equations up to three functions of integration—that is, three functions of u and θ. These functions are then restricted by the supplementary conditions. One finds that $a = d = 0$ and m is determined from a knowledge of n, which can be a completely arbitrary function of its arguments. Bondi calls $\partial n/\partial u$ the *news function* (in the general case there are two news functions). To understand the reason for this appelation let us assume that up to some $u = \text{const}$ hypersurface a field is independent of u. Then the only way in which this field can vary beyond this hypersurface is if the change in n is taken to be nonzero beyond it. The news function can therefore be interpreted as describing the radiation due to an initially static source. On the other hand, the function m is closely related to the total energy of the system. In the static case $\partial m/\partial u = 0$ and also from the field equations it follows that $\partial m/\partial\theta = 0$. One can, in fact, show that in the static case m is just equal to the total mass of the system as calculated using one of the superpotentials. In the nonstatic case Bondi defines the mass of the system as the average of m over all angles:

$$M(u) = \frac{1}{2}\int_0^\pi m(u, \theta)\sin\theta\, d\theta. \tag{13-2.15}$$

With the help of the supplementary conditions one finds that

$$\frac{dM}{du} = -\frac{1}{2}\int_0^\pi \left(\frac{\partial n}{\partial u}\right)^2 \sin\theta\, d\theta. \tag{13-2.16}$$

Hence M decreases when there is news. The decrease in M can then be interpreted as a loss in the total energy of the system due to radiation. Sachs[15] has shown that the Riemann tensor associated with a Bondi-type field has the general form

$$R = \frac{N}{r} + \frac{III}{r^2} + \frac{D}{r^3} + \cdots \tag{13-2.17}$$

where indices have been suppressed for the sake of simplicity and where N, III, and D denote, respectively, tensors of Petrov type null, III, and degenerate. These tensors are covariantly constant along a ray and all have k_μ as their single eigenvector. We see that asymptotically a Bondi-type field leads to a null Riemann tensor, again indicating the presence of gravitational radiation. Also one finds that $N = \partial^2 n/\partial u^2$, showing once more the relation between the news function and gravitational radiation.

The expansion (13-2.17) is an example of a general theorem proven by Newman and Penrose[16] and sometimes referred to in the literature as the

[15] R. K. Sachs, *Proc. Roy. Soc.* (London), **A270**, 103 (1962).

[16] E. Newman and R. Penrose, J. Math. Phys. **3**, 566 (1962).

peeling theorem. They show that, given a set of vectors k_μ, m_μ, and t_μ satisfying Eq. (13-2.11), the field $g_{\mu\nu}$ can be written in the form

$$\tfrac{1}{2}g_{\mu\nu} = k_{(\mu}m_{\nu)} + t_{(\mu}\bar{t}_{\nu)}. \qquad (13\text{-}2.18)$$

Using this form of the field they then calculate the components of the Riemann tensor and show that (again suppressing indices)

$$R = N(k) + \mathrm{III}(k) + D(k, m) + \mathrm{III}(m) + N(m) \qquad (13\text{-}2.19)$$

where k_μ is the eigenvector associated with the tensors N and III in the first two terms, m_μ is the eigenvector of III and N in the last two terms, and k_μ and m_μ are the eigenvectors of the tensor D. Finally, Newman and Penrose show that if in empty space $N(m) = 0(r^{-5})$, then $\mathrm{III}(m) = 0(r^{-4})$, $D(k, m) = 0(r^{-3})$, $\mathrm{III}(k) = 0(r^{-2})$, and $N(k) = 0(r^{-1})$ where r is the luminosity distance along a ray in agreement with the expansion (13-2.17).

The above results and the results of the linear theory lend credence to a belief in the existence in gravitational radiation. This belief would be strengthened even further if it were possible to relate these asymptotic radiation fields with the motion of their sources, since we would then be able to correlate secular changes in this motion with gravitational radiation they produce. In the final analysis such secular changes constitute the best evidence that a system is radiating. The best that we have today in this regard is the result of Bondi expressed in Eq. (13-2.16). It is somewhat sad to think that even with all of the work that has gone into the problem the probability of detecting gravitational radiation within our lifetime is very small indeed. (I, of course, hope, with my colleagues, that I will soon have to eat these words.)

13-3. Direct-Particle Equations of Motion in General Relativity

If we compare the electromagnetic interaction between charged particles in special relativity and the gravitational interaction of massive bodies in general relativity, we find many similarities but also a number of striking dissimilarities. Let us consider the dissimilarities first.

In the electromagnetic case there are three terms in the action of the system: a free-field part, a free-particle part, and an interaction part. The first involves only the electromagnetic field variables, the second only the particle variables. The interaction term involves both sets of variables and is responsible for the electromagnetic interaction between the charged particles; without it, the particles would behave like free particles. However, there is nothing in the formalism that forces this interaction term on us, neither gauge invariance nor Poincaré invariance. In the gravitational case the situation is quite different.

The action for the system contains only *two* terms. One of these is the free-field action $-(1/2\kappa)\int\sqrt{-g}\, R\, d^4x$, and involves only the gravitational field variables $g_{\mu\nu}$. The other term, S_M, given by Eq. (10-4.13) contains both the particle variables and the $g_{\mu\nu}$. It combines in one term the free-particle action and the interaction term. Furthermore there is no way to write a free-particle interaction that leads to a theory with the MMG as a symmetry group and that does not involve $g_{\mu\nu}$. In other words, there are no free particles in general relativity. If particles exist, the principle of general invariance forces them to interact with gravitational field.

A second difference between the two cases is that it is possible to construct a direct-particle or action-at-a-distance description of the electromagnetic interaction, whereas it is not possible to do so directly in the gravitational case. Nevertheless a considerable amount of work has gone into obtaining direct-particle equations of motion from the Einstein theory. The first attempt in this direction was made by Einstein and Grommer[17] in 1927. They were able to show that, to first order, the motion of a singularity in an external gravitational field was along a geodesic of this field. The next significant advance came in 1938 when Einstein, Infeld, and Hoffmann[18] developed an approximation method that led to direct particle equations of motion for slowly moving particles. Since then, many new techniques for obtaining these equations have been developed, notably by Infeld and his school, by Fock and Papapetrou, and by Goldberg.[19]

In all the methods employed to obtain direct-particle equations of motion it is necessary to make use of coordinate conditions—that is, to destroy the invariance properties of the theory. We have already argued that generally covariant direct-particle equations cannot exist by themselves, so we should expect that something like coordinate conditions will be necessary to obtain such equations from the Einstein theory. We will now show that they are an essential part of any approximation scheme for solving the Einstein equations. This does not mean, however, that we ascribe any special significance to the coordinate conditions, since any reasonable set will do. In practice, it turns out that certain conditions—for example, the deDonder conditions—are more convenient to use than others.

The simplest way to obtain direct-particle equations of motion from a field-theory description of interacting particles is to solve the field equations for the field in terms of the particle parameters and then to substitute these solutions into the particle equations of motion. Thus, by taking $A^\mu(x)$, given

[17] A. Einstein and J. Grommer, *Sb. Preuss. Akad. Wiss.*, 2 (1927).

[18] A. Einstein, L. Infeld, and B. Hoffmann, *Ann. Math.* **39**, 65 (1938).

[19] For a comprehensive survey and references to the original literature, see J. N. Goldberg, "The Equations of Motion" in *Gravitation: An Introduction to Current Research*, L. Witten, ed. (Wiley, New York, 1962).

by Eq. (8-1.12), as the desired solution of the Maxwell equations (8-1.4) and substituting into the particle equations of motion (8-1.18), we recover the Fokker, or direct-particle, equations of electrodynamics.

Let us now try to apply this method (it is not the method used by Einstein, Infeld, and Hoffmann) to the gravitational case. The field equations are the Einstein equations

$$R^{\mu\nu} - \tfrac{1}{2}g^{\mu\nu}R = \kappa T^{\mu\nu}. \tag{13-3.1}$$

The matter tensor $T^{\mu\nu}$ and the equations of motion for the matter will vary from case to case. For pole particles we can use the matter action (10-4.13) to obtain these quantities. The equations of motion are then just the geodesic equations

$$\ddot{z}^\mu + \begin{Bmatrix} \mu \\ \rho\sigma \end{Bmatrix} \dot{z}^\rho \dot{z}^\sigma = 0, \tag{13-3.2}$$

while, using Eq. (10-4.11), we find

$$T^{\mu\nu}(x) = \sum_i m_i \int_{-\infty}^{+\infty} \frac{\delta^4(x - z_i)\dot{z}_i^{\,\mu}\dot{z}_i^{\,\nu}}{\sqrt{g_{\rho\sigma}(x)\dot{z}_i^{\,\rho}\dot{z}_i^{\,\sigma}}}\, d\lambda_i \tag{13-3.3}$$

For particles with a structure (for example, particles with dipole and higher moments), S_M and correspondingly the particle equations of motion and the expression for $T^{\mu\nu}$, will differ from these expressions.[20]

If we now try to solve Eq. (13-3.1) for $g_{\mu\nu}$ as a function of the $z_i^{\,\mu}$, we immediately encounter a difficulty. Because of the Bianchi identities satisfied by the left side of the equation, $T^{\mu\nu}$ must satisfy

$$T^{\mu\nu}{}_{;\nu} = 0. \tag{13-3.4}$$

Equation (13-3.4) represents, in fact, an integrability condition on Eq. (13-3.1). We cannot solve these equations with an arbitrary $T^{\mu\nu}$; solutions exist only for matter tensors that satisfy Eq. (13-3.4). By rewriting this equation in terms of the particle variables we see that it is completely equivalent to the particle equations of motion (13-3.2). In the more general case, Trautman[21] has shown that solutions of the particle equations of motion automatically satisfy the integrability conditions (13-3.4). Since, however, there are in general more particle equations of motion than integrability conditions, the latter restrict the particle motion but do not determine it. Only in the case of pole particles are the two sets of equations equivalent. In all cases, though,

[20] See, for instance, A. Papapetrou, *Proc. Roy. Soc.* (*London*), **A209**, 248 (1951).

[21] A. Trautman, "Conservation Laws in General Relativity" in *Gravitation: An Introduction to Current Research*, L. Witten, ed. (Wiley, New York, 1962), Sec. 5.

we seem to be presented with a dilemma: In order to solve Eq. (13-3.1) we need to know a solution to Eq. (13-3.2), which in turn requires that we need to know a solution to Eq. (13-3.1).

To obtain solutions of Eq. (13-3.1) for $g_{\mu\nu}$ as functions of the particle coordinates we need to destroy the coordinate invariance of these equations by the imposition of coordinate conditions. With their help we can modify the left side to obtain a new set of equations, which we write as

$$\tilde{G}^{\mu\nu} = \kappa T^{\mu\nu}. \tag{13-3.5}$$

Since $\tilde{G}^{\mu\nu}{}_{;\nu} \not\equiv 0$, we can now in principle solve these equations with arbitrary $T^{\mu\nu}$, that is, with arbitrary particle trajectories. This procedure of using coordinate conditions is completely equivalent to the use of the Lorentz gauge condition $A^{\mu}{}_{,\mu} = 0$ in electrodynamics to obtain the modified Maxwell equations $\Box A^{\mu} = 4\pi j^{\mu}$.

Having found a solution of Eq. (13-3.5), we must make sure that it also satisfies the coordinate conditions, just as we must make sure that a solution of $\Box A^{\mu} = 4\pi j^{\mu}$ satisfies the gauge condition $A^{\mu}{}_{,\mu} = 0$. Otherwise this solution will not satisfy the original field equations (13-3.1). Since such a solution will depend both on the space-time coordinates x^{μ} and the particle coordinates z_i^{μ}, the satisfaction of the coordinate conditions will impose restrictions on the z_i^{μ}. One must then check that these restrictions are consistent with the particle equations of motion (13-3.2).

We cannot, of course, solve the modified equations (13-3.5) rigorously but must resort to an approximation procedure for obtaining solutions. The original approximation procedure employed by Einstein, Infeld, and Hoffmann involved both a weak-field expansion of $g_{\mu\nu}$ (that is, an expansion in the gravitational coupling constant) and an expansion in $1/c$, where c is the velocity of light. This latter expansion is valid only if all quantities are slowly varying functions of the time coordinate; it is therefore referred to as the *slow-motion approximation*. However, it is not necessary to make use of this approximation in order to find solutions to Eq. (13-3.5). One need only make use of a weak-field expansion, in which case one speaks of a *fast-motion approximation.*

When using the fast-motion approximation it is most convenient to make use of the deDonder coordinate condition

$$(\sqrt{-g}\, g^{\mu\nu})_{,\nu} = 0$$

and to introduce new variables $\mathfrak{g} = \sqrt{-g}\, g^{\mu\nu}$. In terms of these new variables the deDonder conditions become just

$$\mathfrak{g}^{\mu\nu}{}_{,\nu} = 0. \tag{13-3.6}$$

Then, for fields that satisfy these conditions, the Einstein equations reduce to[22]

$$\tilde{G}^{\mu\nu} = \frac{1}{2g}\, \mathfrak{g}^{\rho\sigma}\mathfrak{g}^{\mu\nu}{}_{,\rho\sigma} + \Lambda^{\mu\nu} = \kappa T^{\mu\nu}, \qquad (13\text{-}3.7)$$

where $\Lambda^{\mu\nu}$ is a term free of second derivatives of $\mathfrak{g}^{\mu\nu}$. Let us now assume an expansion of $\mathfrak{g}^{\mu\nu}$ of the form

$$\mathfrak{g}^{\mu\nu} = \eta^{\mu\nu} + \kappa \underset{(1)}{\gamma^{\mu\nu}} + \kappa^2 \underset{(2)}{\gamma^{\mu\nu}} + \cdots. \qquad (13\text{-}3.8)$$

Thus, in lowest order, we assume that the gravitational field reduces to its special relativistic value; that is, space-time is flat. In a like manner we can expand $T^{\mu\nu}$ as

$$T^{\mu\nu} = \underset{(0)}{T^{\mu\nu}} + \kappa \underset{(1)}{T^{\mu\nu}} + \cdots. \qquad (13\text{-}3.9)$$

If we then substitute these expressions into Eq. (13-3.7) and equate to zero the coefficients of the various powers of κ, we obtain a set of equations of the form

$$\Box \underset{(i)}{\gamma^{\mu\nu}} = -2\underset{(i-1)}{T^{\mu\nu}} - 2\underset{(i)}{\Lambda^{\mu\nu}}, \qquad (13\text{-}3.10)$$

where $\underset{(i)}{T^{\mu\nu}}$ and $\underset{(i)}{\Lambda^{\mu\nu}}$ do not depend on $\underset{(i)}{\gamma^{\mu\nu}}$. Thus, at each stage of the approximation we have to solve an inhomogeneous wave equation to find $\underset{(i)}{\gamma^{\mu\nu}}$ in terms of the known quantities $\underset{(i-1)}{T^{\mu\nu}}$ and $\underset{(i)}{\Lambda^{\mu\nu}}$.

In solving an equation of the set (13-3.10) it is necessary to impose boundary conditions. In order later to obtain direct-particle equations of motion that follow from a variational principle, it will be necessary to impose boundary conditions that lead to a half-advanced, half-retarded solution:

$$\underset{(i)}{\gamma^{\mu\nu}}(x) = -\frac{1}{4\pi}\int \delta((x-x')^2)\{2\underset{(i-1)}{T^{\mu\nu}}(x') + 2\underset{(i)}{\Lambda^{\mu\nu}}(x')\}d^4x', \qquad (13\text{-}3.11)$$

where $(x-x')^2 = \eta_{\mu\nu}(x^\mu - x'^\mu)(x^\nu - x'^\nu)$. Having found this solution we must now check that it satisfies the deDonder condition (13-3.6); that is, we must check that $\underset{(i)}{\gamma^{\mu\nu}}{}_{,\nu} = 0$. Provided both $\underset{(i-1)}{T^{\mu\nu}}(x)$ and $\underset{(i)}{\Lambda^{\mu\nu}}$ decrease sufficiently rapidly for large r (distance from a fixed point on an $x^0 = \text{const}$ surface), this will be the case, provided

$$(2\underset{(i-1)}{T^{\mu\nu}} + 2\underset{(i)}{\Lambda^{\mu\nu}})_{,\nu} = 0. \qquad (13\text{-}3.12)$$

In arriving at this result we have taken the surface integrals, obtained by integrating by parts the right side of Eq. (13-3.11) after differentiating with

[22] For details of this reduction, see V. Fock, *The Theory of Space, Time, and Gravitation*, 2d ed. (Macmillan, New York, 1964), Appendix D.

respect to x^ν, to be zero. For sources that are confined to a finite region of space the contribution of $T^{\mu\nu}$ to these surface integrals will indeed give zero. The contribution from $\overset{(i-1)}{\Lambda^{\mu\nu}}$ is harder to assess in general. However, $\overset{(i)}{\Lambda^{\mu\nu}}$ is quadratic in first derivatives of the field variables. Since asymptotically $\overset{(i)}{\gamma^{\mu\nu}}$ decreases at least as fast as r^{-1}, we can expect $\overset{(i)}{\Lambda^{\mu\nu}}$ to decrease as r^{-4}. If it does, then it, too, will not contribute to the surface integrals, in which case the validity of Eq. (13-2.12) is sufficient to ensure the satisfaction of the deDonder condition to the desired order of accuracy. One can now show that Eq. (13-3.12) will be satisfied[23] if the particle trajectories satisfy Eq. (13-3.2) with $\{{}^{\mu}_{\rho\sigma}\}$ calculated using $g_{\mu\nu}$ as determined up to the ith order of approximation.

In principle, then, we have a consistent approximation procedure for solving the Einstein field equations and for obtaining direct-particle equations of motion for massive bodies up to any desired order of approximation. There are, however, still a number of essentially unsolved problems associated with this method of approximation. One class of problems centers on the role of the deDonder coordinate conditions used to obtain the set of equations (13-3.7). Since there is nothing in the rigorous theory to single out these conditions, we should in principle be able to carry through a similar approximation procedure using other coordinate conditions. It is still an unanswered questions whether the use of different coordinate conditions would lead to approximate solutions that would transform under an appropriate mapping into an approximate solution found by using the deDonder conditions. Likewise one does not know if solutions of the approximate direct-particle equations of motion obtained by using one set of coordinate conditions can be transformed into solutions obtained by using some other set of coordinate conditions.

Another question that arises in this or any other approximation procedure is that of convergence. When dealing with nonlinear equations the question of convergence is always difficult. It is especially difficult in the case of the Einstein equations because of the topological problems associated with solutions of these equations. Thus the topology of the manifold on which a solution is defined may be entirely different from that on which an approximation to this solution is defined. For example, in the case of the Schwarzschild solution we saw that the manifold of the complete solution was that described by Kruskal. However, if we expand the solution (11-3.10) in powers of m/r we see that up to any order of approximation, the corresponding manifold is the usual Euclidean manifold with the point $r = 0$ removed. Because of this

[23] J. N. Goldberg, *loc. cit.*

difficulty almost nothing is known about the convergence of this or any other approximation in general relativity. Nevertheless we can expect that as long as it is reasonable to consider the gravitational field produced by one body at the location of another as weak, the lowest orders of the approximate procedure outlined above will lead to results that correctly describe the gravitational fields present and the motions of the bodies producing these fields.

Because of the considerable computational effort involved and because it does not lead at present to experimentally verifiable predictions, we will not work out the higher approximations but content ourselves here with the first. To this order,

$$\underset{(1)}{\gamma^{\mu\nu}}(x) = -\frac{1}{2\pi}\int \delta((x - x')^2)T^{\mu\nu}(x')\, d^4x'$$

$$= -\frac{1}{2\pi}\sum_i m_i \int_{-\infty}^{+\infty} \delta((x - z_i)^2)\dot{z}_i^{\mu}\dot{z}_i^{\nu}\, d\tau_i. \qquad (13\text{-}3.13)$$

To obtain the direct-particle equations of motion to this order we can substitute this solution into Eq. (13-3.2). However, we learn more by substituting it into the action leading to these equations. For this purpose we will need to know $g_{\mu\nu}$. A short calculation shows that

$$g_{\mu\nu} = \eta_{\mu\nu} - \eta_{\mu\rho}\eta_{\nu\sigma}[\underset{(1)}{\gamma^{\rho\sigma}} - \tfrac{1}{2}\eta^{\rho\sigma}\underset{(1)}{\gamma}]$$

$$= \eta_{\mu\nu} - \kappa \sum_i m_i \int_{-\infty}^{+\infty} \delta((x - z_i)^2)(4\eta_{\mu\rho}\eta_{\nu\sigma} - 2\eta_{\mu\nu}\eta_{\rho\sigma})\dot{z}_i^{\rho}\dot{z}_i^{\sigma} \qquad (13\text{-}3.14)$$

where $\underset{(1)}{\gamma} = \eta_{\rho\sigma}\gamma^{\rho\sigma}$. If we then substitute into the action given by Eq. (10-4.13) we obtain, after subtracting off a self-interaction term, a partial action L_i that is the same as the L_i of the Whitehead theory of gravity, given by Eq. (7-21.1) with $\alpha = 2$ and $\beta = -4$. It is just these values of the parameters α and β that give the right value for the planetary perihelia advance and the bending of light as follows from the Eddington analysis discussed in Section 12-1.

14

Cosmology

The first discussions of cosmological problems based on the Einstein field equations were given by Einstein[1] and de Sitter[2] in 1917. While the results they obtained were later found not to be in agreement with observation, subsequent applications were more satisfactory in this regard. However, the Einstein equations by themselves are not sufficient to lead to unique predictions. One must, for instance, make one or another assumption concerning the stress-energy tensor appearing in these equations. One is therefore forced to choose between the various resulting cosmological models on an empirical basis, and here, again, the main problem is a lack of observational data. There are in fact only a few observable "cosmological numbers" on which to base this choice and the values assigned to them have changed over the years by as much as an order of magnitude in some cases. The chief difficulty in fixing the values of these numbers lies in the measurement of astronomical distance.

To determine the distance of a distant galaxy from us, one has to proceed in steps. One first determines the distance of nearby stars by means of parallax measurements. One then uses this information to fix the coefficient in the distance-luminosity relation of Cepheid variables, whose absolute luminosity is a known function of their period. One can then determine the distance to nearby galaxies in which Cepheid variables can be observed, so that the absolute luminosity of these galaxies can be determined. The distances to more

[1] A. Einstein, *Berlin Sitz* (1917), p. 142.

[2] W. de Sitter, *Proc. Acad. Sci. Amsterdam*, **19**, 1217 (1917); **20**, 229 (1917).

distant galaxies are then determined by observing their apparent luminosity and by making a statistical analysis of the luminosities of the galaxies in the cluster to which they presumably belong. In this way one can set up a "luminosity" distance scale in the universe for which apparent luminosity rigorously decreases with the inverse square of this luminosity distance.

By means of such a luminosity distance, Hubble[3] in 1936 gave us an important cosmological number. He observed that the spectral lines from distant galaxies were shifted toward the red and that the amount of this shift was a linear function of luminosity distance. More specifically, he found that $\Delta\lambda$, the observed shift in the wavelength of a spectral line of wavelength λ, was related to luminosity distance d_L by

$$\frac{\Delta\lambda}{\lambda} = \frac{Hd_L}{c},$$

where H is a constant, now called *Hubble's constant.* On the basis of his observations Hubble assigned a value 0.18×10^{-16} sec^{-1} to this cosmological number. One can now proceed to define a red-shift distance to the even farther away galaxies in terms of this linear relation. In this way one has been able to assign a red-shift distance to the most distant galaxies that can be observed visually. One can then use this information to determine another important cosmological number, the nebular count—that is, the number of galaxies lying at distances between L and $L + dL$ from us.

The difficulties inherent in the determination of the cosmological numbers can best be seen from the fact that the presently accepted value of the Hubble constant is almost an order of magnitude larger than the value originally determined by Hubble. In 1958, Sandage[4] pointed out that the luminosity calibration had to be revised because of the discovery of two distinct types of stellar populations. The original distance scale was fixed, using stars belonging to one population. However, the Cepheids observed in the galaxies belonged to the other population and had a different coefficient in the distance-luminosity relation. As a consequence, Sandage now recommends the value 0.24×10^{-17} sec^{-1} for H.

In order to apply the Einstein equation on a cosmological scale, one must have some knowledge of $T_\nu{}^\mu$ that enters in this equation. The basic assumption made in all cosmological applications is to ignore local irregularities (for example, clustering of galaxies) and to assume a homogeneous locally isotropic distribution of matter characterized by a mean local density ρ_0. The determination of this quantity is based on the nebular count and neglects the contribution of intergalactic dust. According to present estimates[5] this count

[3] E. Hubble, *Astrophys. J.*, **84**, 158, 270, 517 (1936).

[4] A. R. Sandage, *Astrophys. J.*, **127**, 513 (1958).

[5] *Vistas in Astronomy*, A. Beer, ed., Vol. 3 (Pergamon Press, New York, 1960), p. 320.

is 10^{-31} g/cm^3 $< \rho_0 <$ 10^{-29} g/cm^3. (For comparison purposes we note that a matter density of 10^{-29}g/cm^3 corresponds to about 10^{-5} protons/cm^3.) This estimate neglects the contribution of radiant energy. It is probably not important for the present epoch, since the ratio of the total radiant energy emitted by a star during its lifetime to its rest mass is of the order 10^{-4}. The above estimate of ρ_0 also neglects the contribution of neutrinos and antineutrinos. Present estimates based on inverse beta-decay measurements indicate that this contribution to ρ_0 may be several orders of magnitude greater than the contribution of the galaxies. However, these estimates are based on the density of local cosmic ray neutrinos and may not represent an average density in intergalactic space.

One other cosmological number that is important for evolutionary models derived from the Einstein equations is the age of the universe. One lower bound on this age is the age of the Earth and the solar system as estimated by ratios of abundances of radioactive substances to their decay products in rocks and meteorites. This age is presently put at $(4.55 \pm 0.07) \times 10^9$ years.[6] Another lower bound is the age of the oldest stars, which is estimated to be 3×10^{10} years.[7] Such a number, of course, has no relevancy for the so-called steady-state models proposed by Bondi and Gold.

In this chapter we describe a class of homogeneous, isotropic cosmological models that can be derived from the general theory of relativity. A number of these models correspond to a universe that is expanding during all or part of its lifetime and so predicts a galactic red shift. At the time of their discovery this was taken as strong support for the general theory, since it was believed that such models could arise only in this theory. However, Milne and McCrea[8] were able to show later that all these models had their counterparts in a Newtonian gravitational theory or modification thereof. Present opinion nevertheless still favors the general relativistic models, since they follow in a more natural way from this theory and do not require the special assumptions concerning the initial distributions of matter that are needed in the Newtonian models.

14-1. Homogeneous, Isotropic Cosmologies

The assertion that the gross features of the universe should be the same in all directions and at all places is known as the *cosmological principle.* It appears to be supported by present-day astronomical observations, at least

[6] C. C. Paterson, *Geochim. Cosmochim. Acta*, **10**, 280 (1960).

[7] J. G. Oke, *Astrophys. J.*, **130**, 487 (1959).

[8] E. A. Milne, *Quart. J. Math.* (*Oxford Ser.*), **5**, 64 (1934); W. H. McCrea and E. A. Milne, *Quart. J. Math.* (*Oxford Ser.*), **5**, 73 (1934).

within the rather wide limits set by these observations. It forms the basis of a large class of cosmological models and leads to essential simplifications in the Einstein equations for the gravitational field. With its help, Robertson and Walker,[9] working independently, were able to find a quite restricted form for $g_{\mu\nu}$ that made possible explicit solutions to these equations. However, we base our discussion of relativistic cosmology here on assumed properties of the smoothed-out matter distribution because, first, it leads to a somewhat larger class of cosmologies and, second, it is a somewhat more direct approach and is easier to interpret the results obtained thereby.

The simplest assumption we can make concerning the motion of the smoothed-out distribution of matter in the universe is that it is along a congruence of nonintersecting timelike world lines relative to the global gravitational field it produces. Weyl[10] was the first to introduce a form of this assumption into the study of cosmology. With its help, one can characterize solutions of the Einstein equations by the type of velocity fields associated with this motion. But more, one can, in all cases, find a mapping such that u^μ, the timelike unit vector tangent to a streamline of the motion, has the components (1, 0, 0, 0) everywhere. The corresponding coordinates are called, in the literature, *comoving*, since the matter is at rest in this coordinate system. Such a mapping always exists since, by assumption, the streamlines do not cross. It then follows that $u_\mu = g_{0\mu}$. Furthermore, the condition that $u^\mu u_\mu = 1$ leads to the result that, in comoving coordinates, $g_{00} = 1$.

In order to characterize the matter motion it is convenient to decompose the velocity field u^μ into a rotation, a shear, and a dilation as we did in our discussion of the relativistic fluid in Section 9-3. By an obvious extension of the results of that section we take

$$\omega_{\mu\nu} \equiv u_{[\mu;\nu]}, \tag{14-1.1}$$

$$\sigma_{\mu\nu} \equiv u_{(\mu;\nu)} - \tfrac{1}{3}\theta h_{\mu\nu}, \qquad h_{\mu\nu} = g_{\mu\nu} - u_\mu u_\nu, \tag{14-1.2}$$

$$\theta \equiv u^\mu{}_{;\mu}. \tag{14-1.3}$$

to be respectively the rotation tensor, the shear tensor, and the dilation scalar of the velocity field. A large class of cosmological models, called collectively the *Friedmann models*, are characterized by vanishing rotation and shear, that is, by $\omega_{\mu\nu} = \sigma_{\mu\nu} = 0$. If in addition $\theta = 0$, one obtains the Einstein model. On the other hand, the assumption of vanishing shear and dilation leads to the Gödel model. Finally, if the mean density of matter in the universe is taken to

[9] H. P. Robertson, *Astrophys. J.*, **82**, 284 (1935) and **83**, 187, 257 (1936); A. G. Walker, *Proc. London Math. Soc.*, (2), **42**, 90 (1936).
[10] H. Weyl, *Phys. ZS.*, **24**, 230 (1923).

be zero, one obtains the de Sitter models as a limiting case of the Friedmann models.

As a first restriction on the velocity field u^μ let us require that it be rotation free. If this is the case, then

$$\begin{aligned}\omega_{\mu\nu} &= u_{\mu,\nu} - u_{\nu,\mu} \\ &= g_{0\mu,\nu} - g_{0\nu,\mu} = 0. \end{aligned} \tag{14-1.4}$$

This equation is satisfied identically in comoving coordinates when $\mu = \nu = 0$. When $\mu = r$, $\nu = 0$, we see that $g_{0r,0} = 0$. Finally, when $\mu = r$, $\nu = s$, we have $g_{0r,s} - g_{0s,r} = 0$ from which it follows that g_{0r} is the gradient of a scalar:

$$g_{0r} = \phi_{,r}. \tag{14-1.5}$$

The particle trajectories are thus seen to be hypersurface orthogonal in this case, so that the coordinate x^0 serves to define a universal or "cosmic" time. The situation here is very much like that for the static gravitational fields discussed in Section 10-5. However, the gravitational fields discussed here are, of course, not static.

Because g_{0r} is a gradient we can now find another mapping such that $g_{0r} = 0$ while preserving the comoving property of the coordinates—that is, such that $u^\mu = (1, 0, 0, 0)$. Under a general mapping

$$u'^\mu = \left(\frac{\partial x'^\mu}{\partial x^\nu}\right) u^\nu = \frac{\partial x'^\mu}{\partial x^0},$$

so that a mapping of the form

$$\begin{aligned} x^0 \to x'^0 &= x^0 + f(\mathbf{x}), \\ x^r \to x'^r &= f^r(\mathbf{x}) \end{aligned}$$

will preserve the form of u^μ. If we take $f^r(\mathbf{x}) = x^r$, we have

$$g'_{0r} = \frac{\partial f}{\partial x^r} g_{00} + g_{0r}.$$

Since $g_{00} = 1$ and $g_{0r} = \phi_{,r}$, we need only take $f(\mathbf{x}) = -\phi(\mathbf{x})$ to achieve our desired result.

We next impose the condition of no shear on our velocity field:

$$\sigma_{\mu\nu} = 0.$$

Since $u_\mu = \delta_\mu{}^0$ this condition reduces to

$$-\{\mu\nu, 0\} - \frac{1}{3}(g_{\mu\nu} - u_\mu u_\nu)\frac{1}{\sqrt{-g}}(\sqrt{-g})_{,0} = 0$$

These equations are satisfied identically for $\mu = 0$. For u, $v \neq 0$ they reduce to

$$g_{rs,0} = g_{rs}(\ln \sqrt[3]{-g})_{,0}.$$

It follows that g_{rs} must be of the form

$$g_{rs}(x) = -S^2(x)\bar{\gamma}_{rs}(\mathbf{x}), \tag{14-1.6}$$

where

$$S^2(x) = \sqrt[3]{-g} \quad \text{and} \quad \det \bar{\gamma}_{rs} = 1.$$

To proceed further we must now make use of the Einstein field equations. In order to be able to include the de Sitter models in our discussion, we will include a cosmological term in these equations. Thus the gravitational field satisfies

$$R^{\mu\nu} - \tfrac{1}{2}g^{\mu\nu}R + \Lambda g^{\mu\nu} = \kappa T^{\mu\nu} \tag{14-1.7}$$

where $T^{\mu\nu}$ is the stress-energy tensor associated with the matter and radiation in the universe. If we multiply this equation by $g_{\mu\nu}$ and sum over μ and ν, we obtain the result

$$R = 4\Lambda - \kappa T$$

where $T = g_{\mu\nu}T^{\mu\nu}$. If we substitute this expression for R back into Eq. (14-1.7), we obtain an equivalent equation

$$R^{\mu\nu} = \Lambda g^{\mu\nu} + \kappa(T^{\mu\nu} - \tfrac{1}{2}g^{\mu\nu}T) \tag{14-1.8}$$

which is sometimes more convenient to use than the original equations (14-1.7).

With $g_{0\mu} = \delta_\mu{}^0$ and g_{rs} of the form (14-1.6), Eq. (14-1.8) for $\mu = 0$, $\nu = s$ reduces to

$$R_{0s} = 2(\ln S)_{,0s} = 0. \tag{14-1.9}$$

Consequently, S must be the product of a function $\mathfrak{R}^2(x^0)$ of x^0 alone times a function of $\mathbf{x}$. Therefore g_{rs} is of the form

$$g_{rs} = -\mathfrak{R}^2(x^0)\gamma_{rs}(\mathbf{x}). \tag{14-1.10}$$

Let us sum up the results obtained so far concerning the form of $g_{\mu\nu}$. This can be done most conveniently by giving an expression for the line element $ds^2 = g_{\mu\nu}\,dx^\mu\,dx^\nu$. We have, in the present case,

$$ds^2 = (dx^0)^2 - \mathfrak{R}^2(x^0)\gamma_{rs}(x)\,dx^r\,dx^s. \tag{14-1.11}$$

From this form of the line element we see that $\mathfrak{R}(x^0)$ acts like a scaling factor for all distances measured in the $x^0 = \text{const}$ hypersurfaces. Thus, while the coordinate values of matter are independent of x^0 in comoving coordinates, the actual distances between such elements increase or decrease as $\mathfrak{R}(x^0)$ changes corresponding to an expanding or contracting universe.

So far all of our results are independent of any particular form of $T^{\mu\nu}$. At this point in our discussion it becomes necessary to make an assumption concerning its form. If we neglect transport processes, the simplest assumption is that $T^{\mu\nu}$ is the stress-energy tensor of a perfect fluid. We can then use the expression (9-2.2) of special relativity for $T^{\mu\nu}$ with $\eta^{\mu\nu}$ replaced by $g^{\mu\nu}$. Thus we take

$$T^{\mu\nu} = (e + p)u^\mu u^\nu - pg^{\mu\nu}$$

where again e is the total internal energy of the matter and p is the pressure measured in the rest frame of the fluid—that is, in comoving coordinates. An equation of state relating e and p will then serve to complete the specification of $T^{\mu\nu}$. In what follows we will make the assumption that this equation of state does not depend explicitly on the spatial coordinates x^r although its form may change with time. We might, for example, start with a radiation-dominated universe in which $p \approx \frac{1}{3}e$ and evolve to a dust-filled universe in which $p \approx 0$.

We can obtain an important relation between e, p, and $\mathfrak{R}$ from the equation $T^{\mu\nu}{}_{;\nu} = 0$ which is a consequence of Eq. (14-1.7). Taking $\mu = 0$ and using the expression (14-1.10) for g_{rs} we find, after a short calculation, that

$$\frac{e_{,0}}{(e + p)} = -\frac{3}{\mathfrak{R}}\,\mathfrak{R}_{,0}\,. \tag{14-1.12}$$

Once we are given an equation of state we can use this equation to find e as a function of $\mathfrak{R}$. Thus in a radiation-dominated universe we see that

$$e \propto \mathfrak{R}^{-4}.$$

On the other hand, in a dust-filled universe we have

$$e \propto \mathfrak{R}^{-3}.$$

In all cases we see from Eq. (14-1.12) that e, and hence p, are functions of x^0 alone.

Let us now consider the remaining field equations (14-1.8). The $\mu = \nu = 0$ equation can be shown to have the form

$$R_{00} = \frac{3}{\mathfrak{R}}\,\ddot{\mathfrak{R}} = \Lambda + \frac{\kappa}{2}(e + 3p) \tag{14-1.13}$$

while the $\mu = r$, $\nu = s$ equations reduce to

$$R_{rs} = {}^3R_{rs} - (\mathfrak{R}\ddot{\mathfrak{R}} + 2\dot{\mathfrak{R}}^2)\gamma_{rs} = -\left\{\Lambda - \frac{\kappa}{2}(e - p)\right\}\mathfrak{R}^2\gamma_{rs} \tag{14-1.14}$$

where ${}^3R_{rs}$ is the three-dimensional Ricci tensor formed from γ_{rs}. Since ${}^3R_{rs}$ and γ_{rs} are functions of the x^r while the coefficients of γ_{rs} in this latter equation are functions of x^0, it follows that the sum of these coefficients must be a constant. By an appropriate scaling of the x^r we see, from the expression (14-1.10) for g_{rs}, that we can take this constant to be -2ε, where $\varepsilon = 1, 0$ or -1. Therefore we obtain from Eq. (14-1.14) the two equations

$$-\left\{\Lambda - \frac{\kappa}{2}(e - p)\right\}\mathfrak{R}^2 + \mathfrak{R}\ddot{\mathfrak{R}} + 2\dot{\mathfrak{R}}^2 = -2\varepsilon \tag{14-1.15}$$

and

$$^3R_{rs} = -2\varepsilon\gamma_{rs}. \tag{14-1.16}$$

If we multiply Eq. (14-1.16) by γ^{rs} and sum, we find that

$$^3R = -6\varepsilon \tag{14-1.17}$$

where ^3R is the curvature scalar formed from γ_{rs}. If γ_{rs} is taken to be the metric of a three-dimensional space it follows that this space has a constant curvature. (The four-dimensional curvature R is, of course, not constant, having the value $4\Lambda - \kappa(\varepsilon - 3p)$ which will, in general, depend on x^0.) For such spaces γ_{rs} is uniquely determined, modulo coordinate mappings, once we decide on the value to assign to ε. In the next section we will give expressions for γ_{rs} and show that in all cases it admits a six-parameter group of motions that correspond to local translations and rotations. Such fields are therefore said to be homogeneous and isotropic.

We can now make use of either Eq. (14-1.13) or (14-1.15) together with Eq. (14-1.12) and an equation of state to determine $\mathfrak{R}(x^0)$. As we saw above, the latter two equations serve to determine e as a function of $\mathfrak{R}$. If then we substitute for $\ddot{\mathfrak{R}}$ in Eq. (14-1.14) the expression for this quantity gotten from Eq. (14-1.13), we obtain the equation

$$\frac{3}{\mathfrak{R}^2}[\varepsilon + \dot{\mathfrak{R}}^2] = \Lambda - \kappa e(\mathfrak{R}). \tag{14-1.18}$$

This equation can then be integrated to give

$$x^0 = \int \frac{d\mathfrak{R}}{\sqrt{(\mathfrak{R}^2/3)[\Lambda - \kappa e(\mathfrak{R})] - \varepsilon}} + \text{const.} \tag{14-1.19}$$

For a large class of cosmological models one makes the assumption that the matter present in the universe can be considered to be a cold dust. In this case the thermal energy of the matter is negligible compared to its rest energy so that $e \approx \rho_0$ where ρ_0 is the rest density of the matter. Also in this case $p \ll \rho_0$. It then follows from Eq. (14-1.12) that

$$\frac{4\pi}{3}\mathfrak{R}^3\rho_0 = \mathfrak{M} \tag{14-1.20}$$

where $\mathfrak{M}$ is a positive constant. For such matter Eq. (14-1.18) takes the form

$$\dot{\mathfrak{R}}^2 = -\frac{\kappa}{4\pi}\frac{\mathfrak{M}}{\mathfrak{R}} + \frac{\Lambda}{3}\mathfrak{R}^2 - \varepsilon. \tag{14-1.21}$$

This is the famous Friedmann differential equation[11] for $\mathfrak{R}$. Depending on the values assigned to $\mathfrak{M}$, Λ, and ε we get the various so-called Friedmann model universes. Before we examine the properties of these solutions it will be instructive to discuss the geometrical and topological properties of the three-dimensional space with a metric γ_{rs}.

14-2. Structure of Homogeneous, Isotropic Spaces

For a space of constant curvature such as we have in the case of the homogeneous, isotropic cosmologies, one can show[12] that the metric form $d\sigma^2$ can be transformed to isotropic form:

$$d\sigma^2 = \gamma_{rs}\,dx^r\,dx^s = \left(1 + \frac{\varepsilon}{4}r^2\right)^{-2}((dx^1)^2 + (dx^2)^2 + (dx^3)^2), \tag{14-2.1}$$

where ε is defined by Eqs. (14-1.15).

We can gain an intuitive picture of the geometry of such a space by embedding it in a flat four-dimensional Euclidean space whose points have coordinates $\{\bar{x}^1, \bar{x}^2, \bar{x}^3, \bar{x}^4\}$. (This space is, of course, a fictitious space and is not related to physical space.) The line element of this space can be written as

$$d\bar{s}^2 = \sum_{i=1}^{4}(d\bar{x}^i)^2. \tag{14-2.2}$$

Consider now a hypersurface in this space defined by

$$\sum_{i=1}^{4}(\bar{x}^i)^2 = \varepsilon^{-1}. \tag{14-2.3}$$

On this surface

$$\sum_{i=1}^{4}\bar{x}^i\,d\bar{x}^i = 0.$$

[11] A. Friedmann, *Z. Physik*, **10**, 377 (1922); **21**, 326 (1924).

[12] L. P. Eisenhart, *Riemannian Geometry* (Princeton Univ. Press, Princeton, 1949), Sec. 27.

We can solve this equation for $d\bar{x}^4$ and substitute the result into Eq. (14-2.2) to obtain

$$d\bar{s}^2 = \sum_{r=1}^{3} (d\bar{x}^r)^2 + \frac{\sum_{r=1}^{3} \bar{x}^r\, d\bar{x}^r}{\varepsilon^{-1} - \sum_{r=1}^{3} (\bar{x}^r)^2}. \tag{14-2.4}$$

By introducing "spherical" coordinates $\bar{r}$, θ, ϕ where

$$\bar{x}^1 = \bar{r}\cos\phi\sin\theta, \qquad \bar{x}^2 = \bar{r}\sin\phi\sin\theta, \qquad \bar{x}^3 = \bar{r}\cos\theta,$$

we can bring $d\bar{s}^2$ into the form

$$d\bar{s}^2 = \frac{d\bar{r}^2}{1 - \varepsilon\bar{r}^2} + \bar{r}^2\, d\Omega^2. \tag{14-2.5}$$

Finally we introduce a new coordinate r by

$$\bar{r} = \frac{r}{1 + (\varepsilon/4)r^2}$$

and "Cartesian" coordinates by

$$x^1 = r\cos\phi\sin\theta, \qquad x^2 = r\sin\phi\sin\theta, \qquad x^3 = r\cos\theta$$

to obtain

$$d\bar{s}^2 = \left(1 + \frac{\varepsilon}{4}r^2\right)^{-2}((dx^1)^2 + (dx^2)^2 + (dx^3)^2), \tag{14-2.6}$$

which is seen to coincide with the expression (14-2.1) for $d\sigma^2$. Consequently the case $\varepsilon = +1$ can be described as being the geometry of a three-dimensional sphere of radius $+1$. The space with $\varepsilon = -1$ corresponds to a three-dimensional pseudosphere, while $\varepsilon = 0$ corresponds to a three-dimensional hyperplane.

An especially convenient expression for $d\sigma^2$ can be obtained in terms of four-dimensional spherical coordinates:

$$\bar{x}^1 = \frac{1}{\sqrt{\varepsilon}}\sin\sqrt{\varepsilon}\,\chi\sin\theta\cos\phi, \qquad \bar{x}^2 = \frac{1}{\sqrt{\varepsilon}}\sin\sqrt{\varepsilon}\,\chi\sin\theta\sin\phi,$$

$$\bar{x}^3 = \frac{1}{\sqrt{\varepsilon}}\sin\sqrt{\varepsilon}\,\chi\cos\theta, \qquad \bar{x}^4 = \frac{1}{\sqrt{\varepsilon}}\cos\sqrt{\varepsilon}\,\chi.$$

If we substitute these expressions into Eq. (14-2.2), we obtain

$$d\sigma^2 = d\chi^2 + S^2(\chi)\, d\Omega^2, \tag{14-2.7}$$

where

$$S(x) = \frac{1}{\sqrt{\varepsilon}} \sin \sqrt{\varepsilon}\, \chi = \begin{cases} \sin \chi & \text{for } \varepsilon = +1 \\ \chi & \text{for } \varepsilon = 0, \\ \sinh \chi & \text{for } \varepsilon = -1 \end{cases}$$

and where the range of χ is

$$0 \leqq \chi \leqq \pi, \qquad \varepsilon = +1,$$
$$0 \leqq \chi < \infty, \qquad \varepsilon = 0, -1.$$

The geometry of the x^0 = const surfaces can be further explored by inquiring after their symmetries. A theorem of differential geometry states that an n-dimensional space of constant curvature admits an $n(n+1)/2$ parameter group of motions. Thus, with $n = 3$, we have a six-parameter group of motions. These motions are most easily found by noting that any rotation in the flat four-dimensional space leaves invariant the sphere defined by Eq. (14-2.3). An infinitesimal rotation is defined by

$$\delta \bar{x}^i = \varepsilon^{ij} \bar{x}^j,$$

where $\varepsilon^{ij} = -\varepsilon^{ji}$. Since ε^{ij} is antisymmetric 4×4 metric, it has six arbitrary elements. If $\varepsilon^{4j} = 0$ we have a rotation about the "origin" of three-space. Alternatively, if $\varepsilon^{rs} = 0$, we have "translations" at the origin, since in this case

$$\delta \bar{x}^r = \varepsilon^{r4} \bar{x}^4 = \varepsilon^{r4}(\varepsilon^{-1} - \bar{r}^2)^{1/2}.$$

Such translations carry any point on the hypersurface over onto any other point. Hence the space is homogeneous and isotropic.

While the above analysis illustrates some of the geometrical properties of spaces of constant curvature, it does not give us anything like a complete picture of the topological or global properties of these spaces.[13] Thus, while a space with $\varepsilon = +1$ can be embedded as a sphere in a higher dimensional space, it can also be embedded in it as an ellipse. There are in fact infinitely many topologically different space forms of a three-dimensional space of constant positive curvature, all of which are closed. For the flat spaces with $\varepsilon = 0$, there are just 18 such forms, not all of which are open. Thus, by identifying opposite edges of a cube, one obtains a closed, flat hypertorus. Finally, for

[13] For a detailed discussion of these topological questions, see W. Rinow, *Die innere Geometrie der metrischen Räume* (Berlin, 1960).

spaces of negative constant curvature, there are again an infinity of such forms and again some of them are closed. It is therefore incorrect to say that a space of zero or negative constant curvature is a priori open, as is sometimes done.

14-3. Observational Consequences

In the introduction to this chapter we discussed the difficulties that attend all cosmological observations. Only a few numbers can be obtained from such observations of distant galaxies, of which the apparent luminosity and red shift are the most important. For the cosmological models discussed here it is possible to derive a relation between these two quantities. We shall follow here the derivation of this relation as given by Heckmann and Schücking.[14]

The electromagnetic radiation emitted by a galaxy or star can be characterized by a stress-energy tensor $t^{\mu\nu}_{em}$ with the following properties: It is symmetric, traceless, and has a vanishing covariant divergence; that is,

$$t^{\mu\nu}_{em} = t^{\nu\mu}_{em}, \qquad t^{\mu}_{em\,\mu} = 0, \qquad \text{and} \quad t^{\mu\nu}_{em;\nu} = 0.$$

Furthermore it is assumed that the radiation is sufficiently weak so that it does not distort the cosmological gravitational field. The energy and momentum density P^μ of the radiation field as measured by a comoving observer (that is, an observer at rest relative to the smoothed-out matter distribution in the universe) is given by

$$P^\mu = t^{\mu\nu}_{em} u_\nu ,$$

where $u_\mu = \delta_\mu{}^0$ is the four-velocity of the observer. For the homogeneous, isotropic cosmologies P^μ can be shown to satisfy the following continuity equation:

$$(\mathfrak{R}P^\mu)_{;\mu} = \frac{1}{\sqrt{-g}}(\sqrt{-g}\,\mathfrak{R}P^\mu)_{,\mu} = 0. \tag{14-3.1}$$

It follows therefore that $\mathfrak{R}$ times the density of electromagnetic energy is conserved.

Let us now define the absolute luminosity $\mathfrak{L}$ of a source as the energy radiated by it per second per unit solid angle. Such a source will then emit an amount

$$dQ = 4\pi\mathfrak{R}(t_1)\mathfrak{L}\, dt_1 \tag{14-3.2}$$

[14] O. Heckmann and E. Schücking, "Relativistic Cosmology" in *Gravitation: An Introduction to Current Research*, L. Witten, ed. (Wiley, New York, 1962).

of the conserved quantity in an interval dt_1 at the time t_1, assuming that the radiation is emitted isotropically. We next define the apparent luminosity $\mathfrak{l}$ of the source as the energy received per second per unit area normal to the direction of propagation. If we use the expression (14-2.7) for $d\sigma^2$ and take the source to be located at the origin of coordinates $\chi = 0$, the energy emitted by the source at t_1 will reach a sphere of radius $\Re(t_2)S(\chi)$ at a later time t_2. The conserved quantity received during the time interval dt_2 will therefore be

$$dQ = 4\pi[\Re(t_2)S(\chi)]^2\Re(t_2)\mathfrak{l}\, dt_2. \tag{14-3.3}$$

Since Q is conserved, it follows from Eqs. (14-3.2, 3) that

$$\mathfrak{l} = \frac{\mathfrak{L}}{[\Re(t_2)S(\chi)]^2} \frac{\Re(t_1)\, dt_1}{\Re(t_2)\, dt_2}. \tag{14-3.4}$$

The propagation of light from the source can be considered to be along radial null geodesics. For such geodesics it follows from Eq. (14-1.11), with $ds^2 = 0$, that

$$\frac{d\sigma}{dt} = \Re(t)^{-1}.$$

Consequently, if a signal emitted at time t_1 is received at the point χ at time t_2, then t_1 and t_2 are related by

$$\int_0^\chi d\sigma = \int_{t_1}^{t_2} \frac{dt}{\Re(t)}. \tag{14-3.5}$$

Likewise, a signal emitted at $t_1 = dt_1$ will be received at x at a time $t_2 = dt_2$, where

$$\int_0^\chi d\sigma = \int_{t_1+dt_1}^{t_2+dt_2} \frac{dt}{\Re(t)}.$$

Therefore, energy radiated by a source during an interval dt_1 will be received at χ during an interval dt_2, where

$$\frac{dt_2}{\Re(t_2)} = \frac{dt_1}{\Re(t_1)}$$

or

$$\frac{dt_2}{dt_1} = \frac{\Re(t_1)}{\Re(t_2)} \equiv 1 + z.$$

If two successive minima of a particular spectral line are emitted at t_1 and $t_1 + \Delta t_1$, they will be received at χ at t_2 and $t_2 + \Delta t_2$. The wavelength of the emitted light, $\lambda_1 = c\,\Delta t$, and the received light, $\lambda_2 = c\,\Delta t_2$, are thus related by

$$\frac{\lambda_2}{\lambda_1} = \frac{\Re(t_2)}{\Re(t_1)} = 1 + z. \tag{14-3.6}$$

This is the red-shift formula for a homogeneous, isotropic cosmological model.

Once $\mathfrak{R}(t)$ corresponding to a particular model is given, one can make use of Eqs. (14-3.4), (14-3.5), and (14-3.6) to derive a relation between $\mathfrak{l}$ and z and so obtain an observationally testable consequence of the model. If

$$\alpha \equiv (t_2 - t_1)/\mathfrak{R}(t_2)$$

is small we can obtain an explicit expression for this relation. To this end we expand $\mathfrak{R}^{-1}(t)$ in powers of α:

$$\frac{1}{\mathfrak{R}(t)} = \frac{1}{\mathfrak{R}_2} - \frac{\dot{\mathfrak{R}}_2}{\mathfrak{R}_2}\alpha + \left\{\frac{\dot{\mathfrak{R}}_2{}^2}{\mathfrak{R}_2} - \frac{\ddot{\mathfrak{R}}_2}{2}\right\}\alpha^2 + O(\alpha^3),$$

where $\mathfrak{R}_2 \equiv \mathfrak{R}(t_2)$. With this expansion, from Eq. (14-3.5) we obtain

$$\chi = \alpha + \tfrac{1}{2}\dot{\mathfrak{R}}_2\alpha^2 + O(\alpha^3),$$

while from Eq. (14-3.6) we obtain

$$z = \dot{\mathfrak{R}}_2\alpha + \mathfrak{R}_2\left\{\frac{\dot{\mathfrak{R}}_2{}^2}{\mathfrak{R}_2} - \frac{\ddot{\mathfrak{R}}_2}{2}\right\}\alpha^2 + O(\alpha^3).$$

We can thus express χ as a function of z, or vice versa, by eliminating α from these two equations. This gives

$$z = \dot{\mathfrak{R}}_2\chi + \tfrac{1}{2}(\dot{\mathfrak{R}}_2{}^2 - \ddot{\mathfrak{R}}_2\mathfrak{R}_2)\chi^2 + O(x^3) \tag{14-3.7}$$

or

$$\chi = \frac{z}{\dot{\mathfrak{R}}_2}\left\{1 + \frac{1}{2}\left(\frac{\ddot{\mathfrak{R}}_2\mathfrak{R}_2}{\dot{\mathfrak{R}}_2{}^2} - 1\right)z + \cdots\right\} + O(z^3). \tag{14-3.8}$$

In using the above results in conjunction with Eq. (14-3.4) it is convenient to use magnitudes rather than luminosities, defined by

$$m = -2.5 \log \mathfrak{l} + k$$

$$M = -2.5 \log \mathfrak{L} + k$$

where all logs are to the base 10 and k is a constant that fixes the zero of the magnitude scale. Using these equations and Eq. (14-3.6) we can rewrite Eq. (14-3.4) as

$$m = M + 5\log(1 + z) + 5\log[S(\chi)\mathfrak{R}_2]. \tag{14-3.9}$$

Then, since in all cases

$$S(\chi) = \chi + O(\chi^3),$$

we obtain, with the help of Eq. (14-3.8),

$$m = M + 5\log(1+z) + 5\log\frac{\mathfrak{R}_2}{\dot{\mathfrak{R}}_2} + 5\log\left\{1 + \frac{1}{2}\left(\frac{\ddot{\mathfrak{R}}_2\mathfrak{R}_2}{\dot{\mathfrak{R}}_2{}^2} - 1\right)z + \cdots\right\}$$

$$= M + 5\log\frac{\mathfrak{R}_2}{\dot{\mathfrak{R}}_2} + 5\log z + (1.086)\left(\frac{\ddot{\mathfrak{R}}_2\mathfrak{R}_2}{\dot{\mathfrak{R}}_2{}^2} + 1\right)z + \cdots. \tag{14-3.10}$$

If we define a luminosity distance d_L by

$$\mathfrak{l} = \frac{\mathfrak{L}}{d_L{}^2},$$

we see that, to lowest order in z, Eq. (14-3.10) reduces to

$$5\log d_L = 5\log\frac{\mathfrak{R}_2}{\dot{\mathfrak{R}}_2}z \quad \text{or} \quad z = \frac{\dot{\mathfrak{R}}_2}{\mathfrak{R}_2}d_L. \tag{14-3.11}$$

Then, since $z = (\lambda_2/\lambda_1) - 1 = \Delta\lambda/\lambda_1$, we obtain the linear luminosity-distance red-shift relation mentioned in the introduction, with the Hubble constant H given by

$$H = \frac{\dot{\mathfrak{R}}_2}{\mathfrak{R}_2}. \tag{14-3.12}$$

One can use Eq. (14-3.10) to determine $\ddot{\mathfrak{R}}_2/\mathfrak{R}_2$ as well as $\dot{\mathfrak{R}}_2/\mathfrak{R}_2$, although the value of the former is subject to even more uncertainty than the latter. Most determinations indicate that it is approximately equal to $-(\dot{\mathfrak{R}}_2/\mathfrak{R}_2)^2$. Finally, from a knowledge of these two quantities plus a knowledge of ρ_0, one can determine values of Λ, $\mathfrak{R}_2$, and ε from Eqs. (14-1.13) and (14-1.15). In particular we see that a knowledge of the Hubble constant and ρ_0 would allow us to determine whether ε is positive, zero, or negative. Unfortunately the present observational data, and especially those leading to a value for ρ_0, are so imprecise as to leave this interesting question unanswered.

14-4. The Static Einstein Model

One of the first cosmological solutions of the field equations was the static Einstein space. This space is characterized by the requirement that the velocity field, in addition to having a vanishing shear and rotation, has a vanishing dilation:

$$\theta = u^\mu{}_{;\mu} = 0. \tag{14-4.1}$$

Since the conditions $\omega_{\mu\nu} = \sigma_{\mu\nu} = 0$ allowed us to find a mapping such that $g_{0\mu} = \delta_{0\mu}$, and g_{rs} had the form given by Eq. (14-1.10), for these forms of $g_{\mu\nu}$ Eq. (14-4.1) reduces to

$$\begin{Bmatrix} \mu \\ 0\mu \end{Bmatrix} = \frac{1}{2} g^{rs} g_{rs,0} = 0,$$

so that

$$\dot{\mathfrak{R}} = 0.$$

From Eq. (14-1.13) it follows that in a dust-filled universe

$$\Lambda = -\frac{\kappa \rho_0}{2} = \frac{4\pi G^2}{c^4} \rho_0 > 0, \tag{14-4.2}$$

so that the cosmological constant is determined by the mean density of matter in the universe. Finally, from Eq. (14-1.21) it follows that

$$\mathfrak{R}^2 = \frac{2\varepsilon}{\kappa \rho_0}, \tag{14-4.3}$$

so that we must have $\varepsilon = +1$ in order that $\text{sig}(g_{\mu\nu}) = -2$. The Einstein space is thus a static space of constant positive curvature.

If we take $\rho_0 \approx 10^{-29}$ g/cm^3, we have $\mathfrak{R} \approx 10^{10}$ light-years (l.y.) and $\Lambda \approx 10^{-20}$ (l.y.)$^{-2}$. The inclusion of a cosmological term of this order of magnitude in the gravitational field equations would be unobservable on a noncosmological scale. Such a term would result in a modification of the Schwarzschild field (11-3.10) in which

$$g_{00} = g_{11}^{-1} = \left(1 - \frac{2m}{r} - \frac{\Lambda}{3} r^2\right)$$

and is thus effective if at all, only for large r. At the orbit of Neptune, $(\Lambda r^2/3)/(2m/r) \simeq 10^{-19}$. However, the Einstein model is unsatisfactory for several reasons. Lemaître[15] showed that it was unstable against fluctuations in the matter density. Even more important is the fact that it does not predict a galactic red shift.

14-5. The de Sitter Models and the Steady-State Cosmology

The de Sitter models can be thought of as the limiting case of the homogeneous, isotropic models, with $\Lambda > 0$, in which $\rho_0 = 0$. There are three

[15] G. Lemaître, *Proc. Akad. Wetensch. Amsterdam*, **19**, 1217 (1917).

possibilities, depending on the value of ε. Solutions of Eq. (14-1.21) with $\mathfrak{M} = 0$ then lead to the following forms for ds^2:

$\varepsilon = +1$

$$ds^2 = dt^2 - \frac{1}{\alpha^2}\cosh^2 \alpha t[d\chi^2 + \sin^2 \chi\, d\Omega^2], \qquad 0 \leqq \chi \leqq \pi; \tag{14-5.1a}$$

$\varepsilon = 0$

$$ds^2 = dt^2 - e^{2\alpha t}[(dx^1)^2 + (dx^2)^2 + (dx^3)^2]; \tag{14-5.1b}$$

$\varepsilon = -1$

$$ds^2 = dt^2 - \frac{1}{\alpha^2}\sinh^2 \alpha t[d\chi^2 + \sinh^2 \chi\, d\Omega^2]; \qquad 0 \leqq \chi < \infty, \tag{14-5.1c}$$

where

$$\alpha^2 = \frac{\Lambda}{3}, \qquad 0 \leqq \theta \leqq \pi, \quad 0 \leqq \phi < 2\pi.$$

Of the three possibilities, only the second has been considered seriously as a cosmological model. Lemaître,[16] and later Robertson,[17] showed that it could be obtained from the original de Sitter model of 1917 by a suitable mapping.

If we are to take seriously the de Sitter model we must imagine that the actual global gravitational field differs from it only in the neighborhood of massive bodies, for example, the galaxies. In this respect the de Sitter space must be considered to be an absolute space analogous to that of Newtonian theory or special relativity. Furthermore, like these latter spaces, it possesses a ten-parameter group of motions. This can be seen by embedding the de Sitter space in a flat five-dimensional space, with coordinates z^A, $A = 1, \ldots, 5$, with a metric form

$$ds^2 = \sum_{A=1}^{5} (dz^A)^2.$$

If we take

$$z^1 = x^1 e^{\alpha t}, \qquad z^2 = x^2 e^{\alpha t}, \qquad z^3 = x^3 e^{\alpha t},$$

$$z^4 = \frac{1}{\alpha}\left(\cosh \alpha t - \frac{1}{2}\alpha^2 r^2 e^{\alpha t}\right),$$

$$z^5 = \frac{i}{\alpha}\left(\sinh \alpha t + \frac{1}{2}\alpha^2 r^2 e^{\alpha t}\right),$$

[16] G. Lemaître, *J. Math. Phys.* (*MIT*), **4**, 188 (1925).
[17] H. P. Robertson, *Phil. Mag.*, **5**, 835 (1925).

we obtain the de Sitter metric form when

$$\sum_{A=1}^{5} (z^A)^2 = \alpha^{-2}.$$

Thus the de Sitter space can be pictured as the surface of a four-dimensional sphere of radius α^{-1} in a flat five-dimensional space. The symmetries of the de Sitter space are therefore the rotations in five-dimensions, which form a ten-parameter group. In the limit $\alpha \to 0$, this group goes over into the Poincaré group and for this reason has been considered of late in connection with elementary particle theory.[18]

Unlike the Einstein model, the de Sitter model allows for the possibility of a galactic red shift. To obtain such a shift we assume that in the mean, the galaxies are at rest in the de Sitter space; that is, they move along the trajectories $\chi =$ const. It follows from our discussion in Section 14-1, or by direct calculation, that such trajectories are geodesics of the de Sitter metric so that the galaxies can be considered to be "freely falling" in the de Sitter model as in the Friedmann models. We can therefore use Eq. (14-3.12) to determine the amount of the red shift. Accordingly we have, with $\mathfrak{R}(t) = e^{\alpha t}$,

$$H = \alpha.$$

Then, since $\alpha^2 = \Lambda/3$, we obtain a relation between the Hubble constant and the cosmological constant. The value of Λ thus obtained is again $\approx 10^{-20}$ $(\text{l.y.})^{-2}$.

We see that the value of the Hubble constant is independent of time in the de Sitter model, which is not the case for the Friedmann models in general. For this reason Bondi and Gold[19] used the de Sitter model as the basis of their steady-state cosmology. They based this cosmology on what they called the *perfect cosmological principle*: Not only should the universe appear the same when viewed from anywhere in space—that is, should be homogeneous and locally isotropic—but it should retain the same appearance throughout the course of time. This means that the density of matter should also remain constant in time. However, this requirement is inconsistent with the existence of a galactic red shift unless one is willing to give up the law of conservation of energy, which Bondi and Gold did. The reason for this inconsistency is that the galactic red shift can be thought of as a Doppler shift resulting from a recession of the galaxies from each other. This recession in turn is due to the expanding distance scale in the de Sitter space. Thus, in a

[18] See in this connection, F. Gursey, "Introduction to Group Theory" in *Relativity, Groups and Topology* (Gordon and Breach, New York, 1964).

[19] H. Bondi and T. Gold, *Monthly Notices Roy. Astron. Soc.*, **108**, 252 (1948).

de Sitter space, a galaxy with spatial coordinates (r, θ, ϕ) can be considered to have a velocity of recession given by

$$v = Hr.$$

Consequently the amount of matter in a sphere of radius r will decrease at a rate $4\pi r^2 v \rho_0$. To compensate for this decrease, Bondi and Gold postulated a spontaneous creation of matter density Q, determined by

$$4\pi r^2 v \rho_0 = \frac{4\pi}{3} r^2 Q.$$

Thus Q has a value $Q = 3\rho_0 H$, which is equivalent to $10^{-(15\pm2)}$ hydrogen atoms per cm^3-year, making such a creation rate completely undetectable. The steady-state cosmology has the epistemological advantage that it does not suppose a beginning or end to the universe, as do the Friedmann models. However, this advantage is offset by the requirement of continuous creation. Furthermore, in its original form, the steady-state theory made no attempt to relate the global gravitational field with the matter in the universe, which is a defect if one takes Mach's principle seriously. A later version of the theory, due to Hoyle,[20] was able to remedy this situation somewhat by introducing an additional field into the theory that was responsible for the creation of matter and by subtracting from the matter tensor $T^{\mu\nu}$ in the Einstein equations a tensor $C^{\mu\nu}$ due to this additional field. Recently[21] Hoyle has abandoned the steady-state cosmology in favor of the evolutionary Friedmann models because of new observational data.

14-6. The Friedmann Models

The Friedmann models all assume a metric form of the Robertson-Walker type, given by Eq. (14-1.11), with $\mathfrak{R}$ determined by the Friedmann differential equation (14-1.21). The general solution of this equation was first given by Lemaître[22] in terms of elliptic functions. For the case $\Lambda = 0$ the solutions can be written in the form

$\varepsilon = 0$

$$\mathfrak{R} = \frac{G\mathfrak{M}}{2}\tau \qquad x^0 = \frac{G\mathfrak{M}}{6}\tau^3 \tag{14-6.1a}$$

$\varepsilon = -1$

$$\mathfrak{R} = G\mathfrak{M}(\cosh\tau - 1), \qquad x^0 = \pm G\mathfrak{M}(\sinh\tau - \tau); \tag{14-6.1b}$$

$\varepsilon = +1$

$$\mathfrak{R} = G\mathfrak{M}(1 - \cos\tau), \qquad x^0 = \pm G\mathfrak{M}(\tau - \sin\tau); \tag{14-6.1c}$$

[20] F. Hoyle, *Monthly Notices Roy. Astron. Soc.*, **108**, 372 (1949).
[21] F. Hoyle, *Nature*, **208**, 111 (1965).
[22] G. Lemaître, *Ann. Soc. Sci. Bruxelles* A **47**, 49 (1927).

where

$$\tau = \int \frac{dx^0}{\Re(x^0)}.$$

The solution (14-6.1a) leads to the original nonstatic model proposed by Friedmann in 1922. The Hubble constant, given by $H \simeq \dot{\Re}/\Re = 2/(3x^0)$, is a linearly decreasing function of time in this model. It follows from Eq. (14-1.21) that the Hubble constant and the density of matter in this model are related by

$$8\pi G\rho_0 = 3H^2$$

so that, with $H = 0.24 \times 10^{-17}$ sec^{-1}, ρ_0 has a value 1.1×10^{-29} g/cm^2, which is within the present range of values for this quantity. The fact that a solution to the gravitational field equations without a cosmological term exists and corresponds to a nonzero mass density led Einstein to abandon the cosmological term completely, as we discussed in Section 10-4. We see that for $t = 0$, $\Re = 0$ in this model, implying that there was a "birth" of the universe at that time, followed by a continuous expansion. For this reason cosmologies based on this model are called "big-bang" theories. However, one should not take this model seriously in the neighborhood $t = 0$, since for such times $\rho_0 \to \infty$. Consequently it is inappropriate to treat the matter in the universe as being composed of noninteracting particles for these times. One would certainly have to include pressure effects in the stress-energy tensor as well as contributions from the radiation and neutrino fields, both of which would then be appreciable.[23]

The model based on the solution (14-6.1a) is somewhat unsatisfactory from the epistemological point of view, since it postulates an act of creation of the whole universe in the finite past. We see in fact that the present "age" of the universe is numerically equal to the inverse of the Hubble constant, which amounts to 1.3×10^{10} years. This age is somewhat too small to accommodate the oldest stars if we accept the estimate of 3×10^{10} years for their age. However, because of the uncertainty in the value of the Hubble constant, we cannot rule out this model at the present time.

The model based on the solution (14-6.1b) has essentially the same features as one based on the solution (14-6.1a); that is, it starts at $t = 0$ with $\Re = 0$ and then expands indefinitely. An essentially different model results from the solution (14-6.1c), since $\Re(t)$ is a cycloid. Here $\Re(t)$ starts from zero, reaches a maximum, and then drops back down to zero, where the whole process is repeated. Such a model leads, therefore, to an oscillating universe. It was strongly favored by Einstein, who felt it did not require an act of spontaneous

[23] For a discussion of the possible physical phenomena that come into play at those times, see J. A. Wheeler, "Geometrodynamics and the Issue of the Final State", in *Relativity Groups and Topology* (Gordon and Breach, New York, 1964).

creation as do the other models discussed above. The same considerations concerning the stress-energy tensor of course apply to this model when $\mathfrak{R} \approx 0$ as apply to these other models at their beginning.

The three models discussed above all have an interesting property that is worth pointing out here; namely, they all possess *event horizons* during part or all of their evolution. An observer situated anywhere in such a universe will receive signals from distant matter that left this matter at times earlier than when it was received due to the finite velocity of light, and the more distant this matter, the earlier will be this time. Furthermore, this light will progressively redden due to the galactic red shift. Finally, a point will be reached when the light that reaches the observer will be emitted by matter at its birth and the red shift will be infinite corresponding to $\mathfrak{R}(t_1) = 0$ in Eq. (14-3.6). Since no light will reach the observer from beyond such points, we call the locus of these points an *event horizon*. This does not mean that there is nothing beyond this horizon, only that an observer cannot see it.

If we make a change of variable in the expression (14-1.11) for ds^2 from x^0 to $\tau = \int dx^0/\mathfrak{R}(x^0)$ and use the expression (14-2.7) for $d\sigma^2$, we see that the equation for a radial light ray is of the form

$$\tau \pm \chi = \text{const.}$$

Therefore, the event horizon for an observer in a Friedmann universe with coordinates $\tau = \tau_0$ and $\chi = 0$ is the locus of points for which $\chi = \tau_0$. In the cases $\varepsilon = -1$ and 0 the event horizon increases monotonically with time. In the case $\varepsilon = +1$ the situation is even more interesting. The event horizon will also increase for a while but when $\tau_0 = \pi$, corresponding to the time of maximum expansion, the observer will see the whole universe. Beyond this time the observer will see two (!) images of every galaxy in opposite directions to each other and at the moment of final collapse when $\tau_0 = 2\pi$ he will be able to see the back of his own head as it looked at the moment of initial expansion.

Solutions of the Friedmann equation for $\Lambda \neq 0$ have much the same qualitative behavior as the solutions (14-6.1), differing mainly in their quantitative predictions. Thus the solutions corresponding to $\varepsilon = 0$ and -1 are oscillatory for $\Lambda < 0$ and monotonically increasing from $\mathfrak{R} = 0$ for $\Lambda \geqq 0$. The case $\varepsilon = +1$ is somewhat more complicated. For $\Lambda < 0$, $\mathfrak{R}(t)$ is oscillatory. For $0 < \Lambda \leqq \Lambda_c$, where $\Lambda_c = (3G\mathfrak{G})^{-2}$, there are two solutions. One again is oscillatory. The other solution passes through a minimum at a finite time and has an asymptotic solution

$$\mathfrak{R}(t) \sim e^{\alpha t},$$

so that it approaches a de Sitter model in the infinite past and future.

For $\Lambda = \Lambda_c$ there are three solutions, one of which corresponds to the

Einstein static model. A second solution starts out from $\mathfrak{R} = 0$ and asymptotically approaches the Einstein solution, while the third solution starts from the Einstein model in the infinite past and continues to expand into the infinite futute, asymptotically approaching the de Sitter model. This particular model was studied by Eddington and Lemaître.[24] For $\Lambda > \Lambda_c$ we have the Lemaître model, which starts out with $\mathfrak{R} = 0$ at $t = 0$, increasing as $t^{2/3}$ at first. The expansion slows down for a while but ultimately increases and the model asymptotically approaches the de Sitter model.[25]

14-7. The Gödel Model

From time to time other homogeneous, as well as inhomogeneous, models have been put forward.[26] However, the observational data are at present insufficient to justify one or another of these models. Furthermore, what data are available can be encompassed by the Friedmann models. The interest in these models lies in the fact that they include those for which the rotation tensor $\omega_{\mu\nu}$ is nonzero. Gödel[27] was the first to construct such a model. In this model, matter is again taken to be composed of noninteracting particles so that $T^{\mu\nu} = \rho_0 u^\mu u^\nu$. Its shear and dilation are taken to be zero but its rotation does not vanish. For this model the metric form is given by

$$ds^2 = a^2[(dx^0 + e^{x^1}\, dx^2)^2 - (dx^1)^2 - \tfrac{1}{2}e^{2x^1}(dx^2)^2 - (dx^3)^2],$$

where a is a nonzero constant and

$$\frac{1}{a^2} = -\kappa\rho, \qquad \Lambda = \frac{1}{2a^2}.$$

The angular velocity vector

$$W^\mu \equiv \frac{1}{\sqrt{-g}}\, \varepsilon^{\mu\nu\rho\sigma} u_\nu u_{\rho,\sigma}$$

has a nonvanishing component

$$W^3 = \frac{\sqrt{2}}{a^2}.$$

[24] A. S. Eddington, *Monthly Notes Roy. Astron. Soc.*, **90**, 668 (1930); G. Lemaître, *Monthly Notices Roy. Astron. Soc.*, **91**, 490 (1931).

[25] For a detailed analysis of these various models, see R. C. Tolman, "Relativity" in *Thermodynamics and Relativity* (Oxford Univ. Press, London and New York, 1934).

[26] For a brief discussion of these models and references to the literature, see O. Heckmann and E. Schücking, *loc. cit.*

[27] K. Gödel, *Rev. Mod. Phys.*, **21**, 447 (1949).

The Gödel model has several interesting features. Since the rotation vector does not vanish, the particle trajectories are not hypersurface orthogonal as they are in the Friedmann models. (The vanishing of the rotation vector is a necessary condition for the trajectories to be hypersurface orthogonal.) Hence one cannot introduce a cosmological "time" in the Gödel model as one can in the Friedmann models. Furthermore there exist closed, everywhere timelike curves in this model, which allows for the possibility of a person to travel into his own past. But perhaps the most interesting feature of this model is the fact that the matter in it can be said to be in a state of absolute rotation. This rotation manifests itself in the motion of test bodies in the Gödel universe. Consider a particle that starts out initially from the origin of coordinates with a finite velocity in the x^1 direction. In a nonrotating universe, for example, a Friedmann universe, such a particle would continue to move along the x^1 axis. However, in the Gödel universe the particle can be shown to spiral outward relative to the matter in the universe. A Foucault pendulum would thus appear to rotate relative to the distant galaxies. Such a situation is clearly in contradiction to Mach's principle, which would require an external force to produce such a motion, and shows further that this principle is not an automatic consequence of the Einstein gravitational equations.

14-8. Further Developments

Recently two problems have come to the fore in the discussion of cosmological models. One of these has to do with the formation of galaxies in an expanding universe, the other with the existence of physical singularities in cosmological models in general. We will indicate here the kinds of considerations that arise in connection with these two problems.

A number of authors[28] have discussed the possibility of galaxies forming in an expanding universe due to instabilities in such a universe. The technique employed in these discussions is the standard one of examining the behavior of small perturbations off a solution of a differential equation—in this case the Einstein field equations. Starting with one or another of the Friedmann models one finds that density perturbations grow only as $t^{2/3}$ if they grow at all.

It is argued that the only spontaneous density perturbations are those due to statistical fluctuations. Since a typical large galaxy will contain 10^{11} sun masses or about 10^{68} particles, we can expect fluctuations in this number of particles of the order of 10^{34} particles. Therefore an initial density fluctuation in

[28] E. M. Lifshitz, *J. Phys. U.S.S.R.*, **10**, 116 (1946); W. B. Bonner, *Mon. Not. Roy. Astro. Soc.*, **117**, 104 (1957); E. M. Lifshitz and I. N. Khalatnikov, *Ad. in Phys.*, **12**, 185 (1963); W. Irvine, *Ann. Phys.*, **32**, 322 (1965).

this amount of matter will correspond to a value of $(\delta\rho/\rho)_{\text{initial}}$ of 10^{-34} where ρ is the density of the background material. If the density fluctuations grow as $t^{2/3}$, we have

$$\left(\frac{\delta\rho}{\rho}\right) = \left(\frac{\delta\rho}{\rho}\right)_{\text{initial}} \left(\frac{t}{t_0}\right)^{2/3}$$

where t_0 is the time from the singular point $\mathfrak{R} = 0$ at which the fluctuation initially occurred. Since the linear analysis breaks down when $\delta\rho/\rho \approx 1$ we can ask how long it would take for an initial density fluctuation to reach this value. (At the present epoch $\delta\rho/\rho \approx 10^6$ for galactic matter.) We see that even if we take t to be the present age of the universe, namely, 4×10^{17} sec, t_0 would have to be of the order of 10^{-34} sec. Most authors feel that this is too early in the evolution for such fluctuations to form. Thus in a Friedmann universe with $\Lambda = \varepsilon = 0$ and a present value for ρ of 10^{-31} gm/cm^3, ρ would have a value of 10^{71} gm/cm^3 at this time. However, long before such densities were reached entirely new, and at present unknown, laws governing the behavior of matter would be needed. On the other hand, if the fluctuations occurred when matter had reached the density of nuclear matter ($\approx 10^{15}$ gm/cm^3) and we could begin to believe that a Friedmann model would correctly describe the subsequent evolution of the universe, t_0 would have a value of 10^5 sec and $(\delta\rho/\rho)_{\text{present}}$ would have a value of only about 10^{-26}. Put another way, $(\delta\rho/\rho)_{\text{initial}}$ would have to have a value of 10^{-8} in order for $(\delta\rho/\rho)_{\text{present}}$ to have a value of unity, and there does not appear to be any known mechanism available that would account for such large fluctuations. (It has been suggested that nuclear matter would break up into parts containing a sufficient number of particles to form a galaxy.) The problem of how galaxies formed is at this date still an open question.

The second problem of interest in current cosmological speculations is that of the existence of singularities in the solutions of the Einstein equations. We saw that all of the Friedmann models with $\Lambda = 0$ possessed real physical singularities corresponding to $\mathfrak{R} = 0$. Likewise, if the motion of matter is rotation free[29] or if the universe is homogeneous but not isotropic,[30] there will be singularities present. While there are nonsingular models when $\Lambda \neq 0$, the introduction of a cosmological constant is a rather ad hoc way to solve the singularity problem. Recently Lifshitz and Khalatnikov[31] claimed that the special symmetries of the Friedmann and other models were responsible for the existence of singularities. They argued that if such models were perturbed slightly, the perturbation would grow as one went back in time and would prevent the occurrence of a physical singularity. However, Hawking[32] has

[29] A. Raychaudhuri, *Phys. Rev.*, **98**, 1123 (1955).
[30] S. W. Hawking and G. F. R. Ellis, *Phys. Letters*, **17**, 247 (1965).
[31] E. M. Lifshitz and I. N. Khalatnikov, *op. cit.*
[32] S. W. Hawking, *Phys. Rev. Letters*, **15**, 689 (1965).

given a quite general proof that in the Friedmann models with $\varepsilon = -1$ or 0 such perturbations would not prevent the growth of singularities in the finite past thereby disproving the Lifshitz-Khalatnikov claim. This is, of course, distressing because it means that somewhere in the early history of our universe the known laws of physics would cease to apply and so, unless we can discover by some means what laws would apply, we shall have to give up the interesting task of constructing a complete description of the evolution of the universe.

Appendix

The Dirac Delta Function

In the text we have made use of a number of properties of the Dirac delta function without proof. We will supply these proofs in this appendix. The properties of this "function" follow from its primary definition that for any function $g(x)$ that is well defined at the point $x = x'$,

$$\int_{-\infty}^{+\infty} g(x)\delta(x - x')\,dx = g(x').$$

Formally, therefore, $\delta(x)$ is a function with values given by

$$\delta(x - x') = \begin{matrix} 0 & \text{if } x \neq x' \\ \infty & \text{if } x = x' \end{matrix}$$

such that

$$\int_{-\infty}^{+\infty} \delta(x - x')\,dx = 1.$$

From its definition follow a number of properties of the delta function which we list below:

$$\delta(x) = \delta(-x) \tag{A1}$$

$$\delta(ax) = \frac{1}{|a|}\,\delta(x) \tag{A2}$$

$$f(x)\delta(x - a) = f(a)\delta(x - a) \tag{A3}$$

$$\int \delta(x - y)\delta(y - a)\,dy = \delta(x - a) \tag{A4}$$

$$\delta(x) = \frac{1}{2\pi} \int_{-\infty}^{+\infty} e^{ikx}\, dk \tag{A5}$$

$$\delta(g(x)) = \sum_n \frac{1}{|g'(x_n)|}\, \delta(x - x_n) \tag{A6}$$

where, in Eq. (A6) the sum is extended over all of the roots x_n of $g(x)$ and $g'(x_n) \neq 0$. These equalities are all to be understood to hold in the sense that the one side can be replaced by the other when it is multiplied by a regular function and integrated over x.

A rigorous proof of the above relations can be given with the help of distribution theory.[1] They can also be proved formally (but not rigorously) by showing that they lead to an equality when multiplied by any regular function and integrated over x. Thus, to prove Eq. (A6) one subdivides the interval $(-\infty, +\infty)$ into subintervals such that each one contains only one root of $g(x)$ and makes a change of variable from x to $g(x)$ in each such interval. Likewise, one can prove that the derivative, $\delta'(x)$, satisfies

$$\delta'(x) = -\delta'(-x) \tag{A7}$$

and

$$x\delta'(x) = -\delta(x). \tag{A8}$$

As a special case of Eq. (A6) let $g(x) = x^2 - a^2$. It then follows that

$$\delta(x^2 - a^2) = \frac{1}{2|a|}\{\delta(x - a) + \delta(x + a)\} \tag{A9}$$

If we take x to be the timelike coordinate x^0 and $a = |\mathbf{x}|$, we have the result stated in Eq. (9-9.6), namely,

$$\delta(x^{0^2} - |\mathbf{x}|^2) = \frac{1}{2|\mathbf{x}|}\{\delta(x^0 + |\mathbf{x}|) + \delta(x^0 - |\mathbf{x}|)\}.$$

Let us now take the Dalembertian of both sides of this equation. We have

$$\Box\delta(x^{0^2} - |\mathbf{x}|^2) = -\nabla^2(\tfrac{1}{2}|\mathbf{x}|)\{\delta(x^0 + |\mathbf{x}|) + \delta(x^0 - |\mathbf{x}|)\}$$
$$+ (\tfrac{1}{2}|\mathbf{x}|)\Box\{\delta(x^0 + |\mathbf{x}|) + \delta(x^0 - |\mathbf{x}|\}.$$

By making use of Eqs. (A3) and (A8) one finds that the second term on the right side vanishes. Furthermore, since $\nabla^2(1/|\mathbf{x}|) = -4\pi\delta^3(x)$ we have finally

$$\Box\delta(x^{0^2} - |\mathbf{x}|^2) = 4\pi\delta^3(\mathbf{x})\{\delta(x^0 + |\mathbf{x}|) + \delta(x^0 - |\mathbf{x}|)$$
$$= 4\pi\delta^3(\mathbf{x})\delta(x^0)$$

[1] Cf. L. Schwartz, *Théorie des distributions* (Paris, 1950); also I. Halperin, *Introduction to the Theory of Distributions* (Univ. of Toronto, 1952).

where we have again made use of Eq. (A3) to obtain this last result. It also follows from this derivation that

$$\Box\delta^{\pm}(x) = 4\pi\delta^4(x)$$

where

$$\delta^{\pm}(x) = \frac{1}{|\mathbf{x}|}\,\delta(x^0 \pm |\mathbf{x}|).$$

Author Index

Subject Index